Springer Proceedings in Mathematics & Statistics

Volume 347

This book series features volumes composed of selected contributions from workshops and conferences in all areas of current research in mathematics and statistics, including operation research and optimization. In addition to an overall evaluation of the interest, scientific quality, and timeliness of each proposal at the hands of the publisher, individual contributions are all refereed to the high quality standards of leading journals in the field. Thus, this series provides the research community with well-edited, authoritative reports on developments in the most exciting areas of mathematical and statistical research today.

More information about this series at http://www.springer.com/series/10533

Melvyn B. Nathanson
Editor

Combinatorial and Additive Number Theory IV

CANT, New York, USA, 2019 and 2020

Springer

Editor
Melvyn B. Nathanson
Department of Mathematics
Lehman College (CUNY)
Bronx, NY, USA

ISSN 2194-1009 ISSN 2194-1017 (electronic)
Springer Proceedings in Mathematics & Statistics
ISBN 978-3-030-67995-8 ISBN 978-3-030-67996-5 (eBook)
https://doi.org/10.1007/978-3-030-67996-5

Mathematics Subject Classification: 11B75, 11B13, 11B24, 43A46, 47A68, 11P99, 11B34, 11N25, 05C55, 05D10, 11P70, 13A05, 20M13, 11P32, 11M26, 11M41, 05B10, 05E15, 20M14, 11P81, 11B30, 11L03

This Springer imprint is published by the registered company Springer Nature Switzerland AG
The registered company address is: Gewerbestrasse 11, 6330 Cham, Switzerland

Preface: Math in the Time of Cholera

Workshops on Combinatorial and Additive Number Theory (CANT) have been organized at the CUNY Graduate Center in New York every year since 2003. The CANT conferences have usually met for four days in May, in the week immediately preceding or immediately following Memorial Day. They have become a fixed point in the number theory calendar.

CANT 2019 met at the Graduate Center from Tuesday, May 21, to Friday, May 24, 2019. In addition to the regular program of lectures in combinatorial and additive number theory and related topics, there were two special sessions, organized with Alex Iosevich, in memory of Jean Bourgain and devoted to his work in number theory.

CANT 2020 was different. Because of the coronavirus pandemic, CANT 2020 was held entirely on Zoom. It was a five-day meeting that ran from Monday, June 1, to Friday, June 5, 2020. There were 73 lectures by speakers in North America, Europe, and Asia. I thank Alex Iosevich, Steve Miller, and Kevin O'Bryant for their help with the technical complexities of Zoom conferencing.

Has the social distancing caused by the pandemic helped or hurt mathematics? Certainly, Zoom technology has vastly enlarged the range of mathematics now available everywhere on the planet to everyone with Internet access. Zoom created the opportunity for CANT 2020 to have more lectures and more participants than any previous CANT conference. Across the world, specialized seminars that would normally meet in small rooms on campus now routinely take place on Zoom, and anyone interested can see and hear the talks. The CANT workshops are organized by the New York Number Theory Seminar. The seminar began in 1981, and for 40 years had met every Thursday afternoon at the CUNY Graduate Center. Now it is an international Zoom seminar.

Today, established researchers continue their work, but, because of COVID, there are many more opportunities for isolated or less connected researchers to learn new methods and results and become part of the international mathematics community. When the covid pandemic finally recedes, the Zoom-enabled expansion of science will persist, and, like the human–machine interface in computer science, the human–Zoom interface will continue to benefit mathematics.

This volume contains papers from the CANT 2019 and CANT 2020 workshops. There are 25 papers on important topics in number theory and related parts of mathematics. These topics include sumsets, partitions, convex polytopes and discrete geometry, Ramsey theory, and analytic number theory.

I thank the National Science Foundation, Springer, and the Journal of Number Theory (Elsevier) for their support of CANT.

I am grateful to Springer and mathematics editors Dahlia Fisch and Robinson dos Santos for making possible the publication of the proceedings of the CANT 2019 and CANT 2020 workshops.

Previous volumes in this series are [1–3].

Short Hills, NJ, USA
December 2020

Melvyn B. Nathanson

References

1. M. B. Nathanson, editor, *Combinatorial and Additive Number Theory–CANT 2011 and 2012*, Springer Proc. Math. Stat., vol. 101, Springer, New York, 2014.
2. M. B. Nathanson, editor, *Combinatorial and Additive Number Theory II–CANT 2015 and 2016*, Springer Proc. Math. Stat., vol. 220, Springer, New York, 2017.
3. M. B. Nathanson, editor, *Combinatorial and Additive Number Theory III–CANT 2017 and 2018*, Springer Proc. Math. Stat., vol. 297, Springer, New York, 2020.

Contents

Extremal Sequences for Some Weighted Zero-Sum Constants for Cyclic Groups

S. D. Adhikari, Md Ibrahim Molla, and Shameek Paul

Dedicated to Prof. Melvyn B. Nathanson on the occasion of his 75th birthday.

Abstract A particular weighted generalization of some classical zero-sum constants became popular and some applications of this weighted generalization have also been found. After some introductory remarks, we here take up some questions regarding inverse problems related to the values of a weighted zero-sum constant for some particular weights for a finite cyclic group.

Keywords Extremal sequences · Zero-sum constants

1 Introduction

Given a sequence $S = (g_1, \ldots, g_k)$ of elements of an abelian group G (written additively), for any $1 \leq l \leq k$, by an *l-sum* one means a sum $g_{i_1} + \cdots + g_{i_l}$ of a subsequence of length l of S, and S is called a *zero-sum sequence* if $g_1 + \cdots + g_k = 0$, where 0 is the identity element of the group.

(Formerly at Harish-Chandra Research Institute)

S. D. Adhikari (✉) · M. I. Molla · S. Paul
Department of Mathematics, Ramakrishna Mission Vivekananda Educational and Research Institute, Belur 711202, India
e-mail: adhikari@hri.res.in; adhikarisukumar@gmail.com

M. I. Molla
e-mail: mdibimya@gmail.com

S. Paul
e-mail: shameek.rkmvu@gmail.com

M. B. Nathanson (ed.), *Combinatorial and Additive Number Theory IV*,
Springer Proceedings in Mathematics & Statistics 347,
https://doi.org/10.1007/978-3-030-67996-5_1

For a finite abelian group G, the *Davenport constant* $D(G)$ is defined to be the smallest natural number k such that any sequence S of k elements in G has a non-empty zero-sum subsequence.

The study of this constant was initiated by K. Rogers [18] and H. Davenport (see [17], for instance) in connection with factorization in an algebraic number field K. Later it became useful in other problems including the proof of the infinitude of Carmichael numbers by Alford et al. [5].

For a finite abelian group G of exponent n, and a positive integer m, the arithmetical invariant $\mathsf{s}_{mn}(G)$ is defined (see [11], for instance) to be the least integer k such that any sequence S with length k of elements in G has a zero-sum subsequence of length mn.

When $m = 1$, it is called *the Erdős-Ginzburg-Ziv constant or the EGZ constant* and is denoted by $\mathsf{s}(G)$. When $mn = |G|$, that is, when one requires a zero-sum subsequence of length $|G|$, it is denoted by $E(G)$ and is called *Gao constant* by some authors. One observes that for the cyclic group $\mathbb{Z}_n$, $\mathsf{s}(\mathbb{Z}_n) = E(\mathbb{Z}_n)$.

For known results and conjectures related to these classical zero-sum constants one may look into the expository articles [4, 7, 8, 11], for instance.

Here we are going to talk about some problems related to a particular weighted generalization of the above zero-sum constants $D(G)$ and $E(G)$.

If G is a finite abelian group of exponent m and $A \subset [1, m-1]$, then the *Davenport constant of G with weight A*, denoted by $D_A(G)$, is defined to be the least positive integer k such that any sequence $(x_1, x_2, \ldots, x_k)$ over G of length k, has a non-empty *A-weighted zero-sum subsequence*; that is, there exists a non-empty subsequence $(x_{j_1}, \ldots, x_{j_l})$ and $a_1, \ldots a_l \in A$ such that $\sum_{i=1}^{l} a_i x_{j_i} = 0$.

Similarly, for any such weight set A, for a finite abelian group G of order n, *the constant* $E_A(G)$ is defined to be the least $t \in \mathbb{N}$ such that for any sequence $(x_1, \ldots, x_t)$ of t elements with $x_i \in G$, there exists an A-weighted zero-sum subsequence of length n. The case $A = \{1\}$ corresponds to $D(G)$ and $E(G)$.

We are going to take up some questions regarding some inverse problems related to the values of $E_A(G)$ for some particular weights when G is a finite cyclic group.

In the classical case from the EGZ theorem, we have $E(\mathbb{Z}_n) = 2n - 1$ (see [9] and [16], for instance). Regarding inverse problems, here it is known (see [6], for instance) that for any positive integer $n \geq 2$, given a sequence $a_1, a_2, \ldots, a_{2n-2}$, of elements in $\mathbb{Z}_n$, such that there are no n indices $1 \leq i_1 < i_2 < \cdots < i_n \leq 2n-2$ with $a_{i_1} + a_{i_2} + \cdots + a_{i_n} = 0 \in \mathbb{Z}_n$, then there are 2 elements of $\mathbb{Z}_n$ such that $n-1$ of the a_i's are equal to one of them and the remaining $n-1$ of the a_i's are equal to the other element. It is easy to see that if the 2 elements of $\mathbb{Z}_n$ appearing in an extremal sequence are a and b, then $\gcd((a-b), n) = 1$. We shall look for similar inverse results related to $E_A(\mathbb{Z}_n)$ for some 'weight-sets' A; the methods involved will include the Cauchy-Davenport Theorem and combinatorial and elementary number theoretic arguments.

Before taking up the inverse results, in the next section we shall see some necessary preliminaries.

2 Some Preliminaries

In the very first paper due to Adhikari et al. [2], introducing the weighted zero-sum constant $E_A(G)$, the main result was related to the weight $A = \{1, -1\}$. By the pigeonhole principle (see [2]), $D_A(\mathbb{Z}_n) \le [\log_2 n] + 1$; and considering the sequence $(1, 2, \ldots, 2^r)$, where r is defined by $2^{r+1} \le n < 2^{r+2}$, it follows that $D_A(\mathbb{Z}_n) \ge [\log_2 n] + 1$.

In this case, it was shown in [2] that $E_A(\mathbb{Z}_n) = n + [\log_2 n]$.

This, and results with some more weight-sets lead people (Adhikari and Rath [3], Adhikari and Chen [1], Thangadurai [19] , Griffiths [13]) to guess and conjecture that for a finite abelian group G of order n, the relation $E(G) = D(G) + n - 1$ of Gao [10] (see also [12], Proposition 5.7.9) also holds for the weighted generalizations as defined above; that is,

$$E_A(G) = D_A(G) + n - 1. \tag{1}$$

It was established for the group $\mathbb{Z}_p$ by Adhikari and Rath [3] (a conditional general result of Adhikari and Chen [1] also implies the result for the group $\mathbb{Z}_p$), for general cyclic groups by Yuan and Zeng [20] and for general finite abelian groups by Grynkiewicz, Marchan and Ordaz [14].

For a finite abelian group G of order n and a weight-set A, for an extremal sequence of length $D_A(G) + n - 2$ over G, not having an A-weighted zero-sum subsequence of length n, a natural candidate is the following:

$$(\underbrace{0, 0, \ldots, 0}_{(n-1)\text{ times}}, a_1, a_2, \ldots a_{D_A(G)-1}), \tag{2}$$

where $a_1, a_2, \ldots a_{D_A(G)-1}$ is a sequence over G, not having any non-empty A-weighted zero-sum subsequence (which exists by the definition of $D_A(G)$).

For a particular weight-set A, the inverse problem here is to characterize the sequences of length $D_A(G) + n - 2$ over G, not having an A-weighted zero-sum subsequence of length n. A special case of characterizing such sequences of the form (2) boils down to the characterization of extremal sequences for $D_A(G)$, that is, sequences of length $D_A(G) - 1$ not having any non-empty A-weighted zero-sum subsequence. As was mentioned in the previous section, here we consider the problem for the cyclic group $\mathbb{Z}_n$ and some particular weight-sets. We do not distinguish between sequences differing by a permutation.

If the weight-set A is a group under multiplication and if a sequence $S : (x_1, \ldots, x_k)$ over $\mathbb{Z}_n$ does not have any non-empty A-weighted zero-sum subsequence, then for any $a_1, \ldots, a_k \in A$ and for any $c \in \mathbb{Z}_n^*$, the sequence $c(a_1x_1, \ldots, a_kx_k)$ too does not have any non-empty A-weighted zero-sum subsequence.

This motivates the following definition.

If A is a group under multiplication, two sequences $(x_1, \ldots, x_k)$ and $(y_1, \ldots, y_k)$ over $\mathbb{Z}_n$ are said to be *equivalent with respect to* A if there are $a_1, \ldots, a_k \in A$ and $c \in \mathbb{Z}_n^*$ and a permutation $\sigma \in S_k$ such that $y_{\sigma(i)} = ca_ix_i$ for $i = 1, 2, \ldots k$.

3 The Case $A = \mathbb{Z}_n \setminus \{0\}$

In the very first paper [2] introducing the weighted zero-sum constant $E_A(G)$, the main result was obtaining the value of $E_{\{\pm 1\}}(\mathbb{Z}_n)$. Towards the end of the paper, there was the rather easy observation that taking $A = \mathbb{Z}_n \setminus \{0\}$, $E_A(\mathbb{Z}_n) = n + 1$.

Considering the weight-set $A = \mathbb{Z}_n \setminus \{0\}$, here we intend to look into the inverse problem related to the result $E_A(\mathbb{Z}_n) = n + 1$, that is, we want to characterize the sequences of length n over the group $\mathbb{Z}_n$ which are not A-weighted zero-sum sequences. Clearly, the sequence

$$(\underbrace{0, 0, \ldots, 0}_{(n-1)\text{ times}}, 1)$$

is not an A-weighted zero-sum sequence. For $n = 2$, the only possibility is that $A = \{1\}$ so that we are in the classical case and it is easy to observe that the only sequence of length 2 over $\mathbb{Z}_2$ which is not a zero-sum sequence is $(0, 1)$. So, we proceed to determine all sequences of length n over the group $\mathbb{Z}_n$, with $n \geq 3$, with this property. We start with the following observations.

Observation 1. If a non-zero element a in $\mathbb{Z}_n$ happens to be a non-unit, then there is a divisor $d > 1$ of n such that $d|a$. Now, $b = \frac{n}{d}$ is in A and $ba = 0$.

Observation 2. If $a, b \in \mathbb{Z}_n$, with $1 \leq a,\ b \leq n - 1$, then $ab + (n - b)a = 0$ so that the sequence (a, b) is an A-weighted zero-sum sequence.

Observation 3. If $n \geq 3$ and $u, v, w \in \mathbb{Z}_n$ are units, then $uu^{-1} + vv^{-1} - 2ww^{-1} = 0$.

Now, given a sequence $S : a_1, a_2, \ldots, a_n$, over $\mathbb{Z}_n$, if there are an even number of non-zero elements in S, then by Observation 2, S is an A-weighted zero-sum sequence.

If the number of non-zero elements in S is odd and there is a non-zero non-unit in S, then by Observations 1 and 2, S is an A-weighted zero-sum sequence.

If the number t of non-zero elements in S is odd and all these elements are units, then if $t \geq 3$, by Observations 2 and 3, S is an A-weighted zero-sum sequence.

Hence we have the following.

Theorem 1 *Taking $A = \mathbb{Z}_n \setminus \{0\}$, a sequence $S : a_1, a_2, \ldots, a_n$, over $\mathbb{Z}_n$, is not an A-weighted zero-sum sequence if and only if exactly one of the a_i's is non-zero and that non-zero element is a unit.*

4 The Case Where A is the Set of Quadratic Residues Modulo p

In this section we consider the case where A is the set of quadratic residues modulo p, for an odd prime p. That is, A consists of all the squares in $\mathbb{Z}_p^*$. In this case, it was proved in [3] that $D_A(\mathbb{Z}_p) = 3$, and $E_A(\mathbb{Z}_p) = p + 2$.

As given in [3], an extremal sequence of length $p+1$, not having an A-weighted zero-sum subsequence of length p is

$$(\underbrace{0, 0, \ldots, 0}_{(p-1)\text{ times}}, v_1, -v_2),$$

where v_1 is a quadratic residue and v_2 a quadratic non-residue.

The following theorem shows that for $p \geq 5$, essentially these are the only such extremal sequences. For the case $p = 3$, $A = \{1\}$, so that we are in the classical case and from the known results as mentioned towards the end of the introduction (can also be seen by an elementary argument), the only sequences of length 4 over $\mathbb{Z}_3$ not having any A-weighted zero-sum subsequence of length 3 are $(0, 0, 1, 1)$, $(0, 0, 2, 2)$ and $(1, 1, 2, 2)$.

Theorem 2 *Let $p \geq 5$ be an odd prime and A consists of all the squares in $(\mathbb{Z}_p)^*$. A sequence of length $p+1$ over $\mathbb{Z}_p$ does not have any A-weighted zero-sum subsequence of length p if and only if, it is equivalent with respect to A to*

$$(\underbrace{0, 0, \ldots, 0}_{(p-1)\text{ times}}, v_1, -v_2)$$

where v_1 is a quadratic residue and v_2 a quadratic non-residue.

Proof If v_1 is a quadratic residue and v_2 a quadratic non-residue, then for $a_1, a_2 \in A$, $a_1 v_1 + a_2(-v_2) = 0$ implies $a_1 v_1 = a_2 v_2$, which is not possible as $a_1 v_1$ is a quadratic residue and $a_2 v_2$ a non-residue.

Hence a sequence as in the statement of the theorem cannot have any A-weighted zero-sum subsequence of length p.

Now, consider any sequence $(x_1, \ldots, x_{p+1})$ over $\mathbb{Z}_p$ which does not have any A-weighted zero-sum subsequence of length p.

Obviously, there must be at least two non-zero elements in the sequence. We claim that the given sequence cannot have more than two non-zero elements in it, so that it has exactly two non-zero elements. Assume the contrary and without loss of generality let $x_i \neq 0$, for $i = 1, 2, 3$.

Now $|x_i A| = \frac{p-1}{2}$, for $i = 1, 2, 3$. Since by the Cauchy-Davenport Theorem (see [16], for instance) if $A_1, A_2, \ldots, A_h$ are non-empty subsets of $\mathbb{Z}_p$, then

$$|A_1 + A_2 + \cdots + A_h| \geq \min\left\{p, \sum_{i=1}^{h} |A_i| - h + 1\right\},$$

for $p \geq 7$ we have

$$\begin{aligned}|x_1A + x_2A + x_3A| &\geq \min\left\{p, \sum_{i=1}^{3} |x_iA| - 2\right\} \\ &\geq \min\left\{p, \frac{3(p-1)}{2} - 2\right\} \\ &= p.\end{aligned}$$

□

Therefore, when $p \geq 7$, there are a_1, a_2, a_3 in A such that

$$a_1x_1 + a_2x_2 + a_3x_3 = -\sum_{i=4}^{p} x_i,$$

giving us an A-weighted zero-sum subsequence of length p—a contradiction to our assumption. So the sequence is of the form

$$(\underbrace{0, 0, \ldots, 0}_{(p-1)\text{ times}}, y, z),$$

where y and z are non-zero. This sequence will have an A-weighted zero-sum subsequence of length p, if and only if $a_1y + a_2z = 0$, for a_1, a_2 in A. Therefore, in order not to have any A-weighted zero-sum subsequence of length p, if y is a square, then $Ay = A$ will imply that $-a_2z$ has to be a non-square and we are through.

When $p = 5$, by the argument using the Cauchy-Davenport Theorem, the given sequence cannot have more than three non-zero elements in it. However, if there are three non-zero elements, then at least one of the sets $\{1, -1\}$ or $\{2, -2\}$ will have two of them and the set of squares being $\{1, -1\}$, those two elements along with three zeros will give an A-weighted zero-sum subsequence of length 5. Hence, in this case too there will be exactly two non-zero elements in the given sequence and we are through.

5 The Case $A = \mathbb{Z}_n^*$

For the case $A = \mathbb{Z}_n^* = \{a \in \mathbb{Z}_n : (a, n) = 1\}$, settling a conjecture from [2], it was proved by Luca [15] that $E_A(\mathbb{Z}_n) = n + \Omega(n)$ (so that $D_A(\mathbb{Z}_n) = \Omega(n) + 1$), where $\Omega(n)$ denotes the number of prime factors of n, multiplicity included. Soon after that, this result was also established by Griffiths in [13]. Writing $n = p_1 \cdots p_s$ as a product of $s = \Omega(n)$ not necessarily distinct primes, the following sequence of the form (2) of length $n + \Omega(n) - 1$ does not have any A-weighted zero-sum subsequence of length n:

$$(\underbrace{0, 0, \ldots, 0}_{(n-1)\text{ times}}, 1, p_1, p_1p_2, \ldots, p_1p_2 \cdots p_{s-1}).$$

First, we observe that when $n = p^s$ is an odd prime power, then up to multiplication by units, the above one is essentially the only such extremal sequence of length $n + \Omega(n) - 1$.

We need the following ([13], Lemma 2.1 (ii)):

Lemma 1 *Let p be an odd prime. If a sequence $(y_1, \ldots, y_\ell)$ over $\mathbb{Z}_{p^s}$ has at least two terms $\not\equiv 0 \pmod{p}$, then there exist units $\alpha_1, \ldots, \alpha_\ell$ of $\mathbb{Z}_{p^s}$ such that*

$$\alpha_1 y_1 + \cdots + \alpha_\ell y_\ell = 0.$$

Theorem 3 *Let p be an odd prime and $n = p^s$ and $A = \mathbb{Z}_n^*$. A sequence of length $n + \Omega(n) - 1 = n + s - 1$ over $\mathbb{Z}_{p^s}$ does not have any A-weighted zero-sum subsequence of length n if and only if it is equivalent with respect to A to*

$$(\underbrace{0, 0, \ldots, 0}_{(n-1) \text{ times}}, 1, p, p^2, \ldots, p^{s-1}).$$

Proof It is clear that a sequence of the form as given in the statement of the theorem cannot have any A-weighted zero-sum subsequence of length n.

We proceed to prove the converse. Writing $t = n + \Omega(n) - 1 = n + s - 1$, given a sequence $(x_1, \ldots, x_t)$ over $\mathbb{Z}_{p^s}$, any $x_i \neq 0$ can be written as $x_i = p^j \alpha$ where $0 \leq j < s$ and α is a unit of $\mathbb{Z}_{p^s}$. If the number of elements x_i, corresponding to a particular $j, 0 \leq j < s$, is more than one, then by Lemma 1, those elements form an A-weighted zero-sum subsequence. In fact, if there are d such j with $1 \leq d < s$, then the given sequence has A-weighted zero-sum subsequences for any length k with $2d \leq k \leq t - f$ where f is the number of those powers of p appearing only once in the sequence. So, if $f < s$, then $t - f > t - s = n - 1$ and hence there is an A-weighted zero-sum subsequence of length n. Therefore, the non-zero elements in the given sequence are precisely $\alpha_0, p\alpha_1, \ldots, p^{s-1}\alpha_{s-1}$, where α_i are units. That is, if unit multiple of each power p^j appears exactly once, then only a sequence of length $n + s - 1$ does not have an A-weighted zero-sum subsequence for length n.

□

If p is an odd prime and $n = p^s$, and the weight-set is $A = \mathbb{Z}_n^*$, in the above theorem we have determined the sequences of length $n + \Omega(n) - 1 = n + s - 1$ over $\mathbb{Z}_{p^s}$ not having any A-weighted zero-sum subsequence of length n. For integers n, not of the above type, we now take up characterization of sequences of the form (2) not having any zero-sum subsequence of length n; this, as mentioned in Section 2, leads to characterization of extremal sequences for $D_A(G)$.

To deal with the case $n = 2^s$, we need the following ([13], Lemma 2.3).

Lemma 2 *If x, y are odd (i.e. are units of $\mathbb{Z}_{2^s}$), then given any even $t \in \mathbb{Z}_{2^s}$, there exist units α_1, α_2 such that $\alpha_1 x + \alpha_2 y = t$.*

Theorem 4 *Let $n = 2^s$ and $A = \mathbb{Z}_n^*$. Then a sequence of length $\Omega(n)$ over $\mathbb{Z}_{2^s}$ does not have any non-empty A-weighted zero-sum subsequence if and only if it is equivalent with respect to A to $(1, 2, 2^2, \ldots, 2^{s-1})$.*

Proof Here $\Omega(n) = s$. Clearly a sequence of the form as given in the statement of the theorem cannot have any non-empty A-weighted zero-sum subsequence.

Now, let $S : (x_1, \ldots, x_s)$ be a sequence of length s over $\mathbb{Z}_{2^s}$, such that S does not have any non-empty A-weighted zero-sum subsequence. Any $x_i \neq 0$ can be written as $x_i = 2^j\alpha$ where $0 \leq j < s$ and α is odd. If for some j, there are $i_1 \neq i_2$ such that $x_{i_1} = 2^j\alpha_1$ and $x_{i_2} = 2^j\alpha_2$ where α_1 and α_2 are odd, then by Lemma 2, x_{i_1}, x_{i_2} is an A-weighted zero-sum subsequence. Hence for any j with $0 \leq j < s$, number of elements x_i of the form $2^j\alpha$ with an odd α can be at most one. Since S is a sequence of length s, it cannot therefore happen that there is no x_i corresponding to a particular $j,\ 0 \leq j < s$. This proves the theorem.

□

Writing $n = p_1 \cdots p_s$ as product of $s = \Omega(n)$ not necessarily distinct primes, the following is an example of a sequence of length $\Omega(n)$ that does not have any non-empty A-weighted zero-sum subsequence:

$$(1, p_1, p_1p_2, \ldots, p_1p_2 \ldots p_{s-1}).$$

If $n = p_1 \ldots p_s$ is square-free, writing $\hat{p}_i = p_1p_2 \ldots p_s/p_i$, the following is another such example:

$$(\hat{p}_1, \hat{p}_2, \ldots, \hat{p}_s).$$

We will give the general form of such extremal sequences, the above two examples being among them.

We proceed to address the general case for an odd integer n having the prime factorization $n = p_1^{s_1} \ldots p_k^{s_k}$. As $\mathbb{Z}_n$ is isomorphic to the product $\mathbb{Z}_{p_1}^{s_1} \times \cdots \times \mathbb{Z}_{p_k}^{s_k}$, one may write $x \in \mathbb{Z}_n$ as the tuple $(x^{(1)}, \ldots, x^{(k)})$ where $x^{(j)} \equiv x \pmod{p_j^{s_j}}$, for $j,\ 1 \leq j \leq k$.

Observation I. As was pointed out in Observation 2.2 in [13], given a sequence $S : (x_1, \ldots, x_m)$ over $\mathbb{Z}_n$, there exist units $\alpha_1, \ldots \alpha_m$ of $\mathbb{Z}_n$ with $\alpha_1x_1 + \cdots + \alpha_mx_m = 0$ in $\mathbb{Z}_n$ if and only if, for each $j = 1, \ldots, k$,

$$\alpha_1^{(j)}x_1^{(j)} + \cdots + \alpha_m^{(j)}x_m^{(j)} = 0 \in \mathbb{Z}_{p_j}^{s_j},$$

$\alpha_i^{(j)}$ being units in $\mathbb{Z}_{p_j}^{s_j}$.

For simplicity of notations, first we state and prove the result in the square-free case.

Theorem 5 *Let $n = p_1 \ldots p_k$ be the prime factorization of an odd integer n, where $p_1, p_2, \ldots, p_k$ are distinct primes and $A = \mathbb{Z}_n^*$. A sequence $S : (x_1, \ldots, x_k)$ over $\mathbb{Z}_n$, of length $k = \Omega(n)$ does not have any non-empty A-weighted zero-sum subsequence if and only if it is of the following form:*

$$a_1b_1,\ p_{i_1}a_2b_2,\ p_{i_1}p_{i_2}a_3b_3, \ldots,\ p_{i_1}p_{i_2} \cdots p_{i_{k-1}}a_kb_k,$$

where a_js are units in $\mathbb{Z}_n$ and b_js are co-prime to p_{i_j} for $j = 1, 2, \ldots, k$ where $p_{i_1}, \ldots, p_{i_k}$ is a permutation of the primes $p_1, p_2, \ldots, p_k$.

Proof With the weight-set $A = \mathbb{Z}_n^*$, one has $D_A(\mathbb{Z}_n) = \Omega(n) + 1$. The theorem characterizes the extremal sequences of length $k = \Omega(n)$.

It is easy to observe that a sequence of the given form does not have any non-empty A-weighted zero-sum subsequence.

Now, let $S : (x_1, \ldots, x_k)$ be a sequence over $\mathbb{Z}_n$, such that it does not have any non-empty A-weighted zero-sum subsequence. By Observation I and Lemma 1, for at least one p_{i_1} with $i_1 \in \{1, \ldots, k\}$, not more than one x_i is co-prime to p_{i_1}. So, either each x_i is divisible by p_{i_1} or without loss of generality, let x_1 be the only element co-prime to p_{i_1}. In the second case, $x_1 = a_1 b_1$, where a_1 is a unit in $\mathbb{Z}_n$ and b_1 is such that the set of primes dividing b_1 is a subset of $\{p_1, p_2, \ldots, p_k\} \setminus \{p_{i_1}\}$.

In the first case, let $x_i = p_{i_1} y_i$, $i = 1, \ldots, k$. Since the length of $(y_1, \ldots, y_k)$ is $k = \Omega(n) = \Omega(n/p_{i_1}) + 1$, writing $n_1 = n/p_{i_1}$, there is a non-empty subsequence of $(y_1, \ldots, y_k)$ summing to 0 modulo n_1 with weights in $Z_{n_1}^*$. By the Chinese Remainder Theorem, this will lead to a non-empty subsequence of $(y_1, \ldots, y_k)$ summing to 0 modulo n_1 with weights in Z_n^*. This will imply that there is a non-empty subsequence of $(x_1, \ldots, x_k)$ summing to 0 modulo n with weights in Z_n^*, contradicting the assumption. So the first case does not arise.

Now, the sequence $S_1 : (x_2, \ldots, x_k)$ does not have any non-empty A-weighted zero-sum subsequence. Writing $x_i = p_{i_1} y_i$, $i = 2, \ldots, k$ and $n_1 = n/p_{i_1}$, by the above argument $(y_2, \ldots, y_k)$ cannot have a non-empty subsequence summing to 0 modulo n_1 with weights in $Z_{n_1}^*$ and hence there is at least one prime $p_{i_2} \in \{p_1, p_2, \ldots, p_k\} \setminus \{p_{i_1}\}$ such that there is exactly one y_j, $j = 2, \ldots, k$, which is co-prime to p_{i_2}. Without loss of generality, let y_2 be co-prime to p_{i_2}. Clearly, $y_2 = a_2 b_2$, where a_2 is a unit in $\mathbb{Z}_n$ and b_2 is such that the set of primes dividing b_2 is a subset of $\{p_1, \ldots, p_k\} \setminus \{p_{i_2}\}$.

Proceeding in this way we obtain our assertion. □

We shall now state the general result.

Theorem 6 *Let an odd integer $n = p_1 \ldots p_k$ be a product of $k = \Omega(n)$ not necessarily distinct primes and let $A = \mathbb{Z}_n^*$. A sequence $S : (x_1, \ldots, x_k)$ over $\mathbb{Z}_n$ of length $k = \Omega(n)$ does not have any non-empty A-weighted zero-sum subsequence if and only if it is equivalent with respect to A to a sequence of the following form:*

$$b_1, q_1 b_2, q_1 q_2 b_3, \ldots, q_1 q_2 \ldots q_{k-1} b_k$$

where $q_1, \ldots, q_k$ is a permutation of $p_1, \ldots, p_k$ and $\forall\, i,\ b_i$ is coprime to q_i.

Proof The proof would be along similar lines as in the previous theorem. □

The corresponding result for an even integer and the problem of determining extremal sequences related to $E_A(\mathbb{Z}_n)$ for the units and some other important weight-sets A, will be interesting to be taken up.

Acknowledgements The first author would like to acknowledge a fellowship (under MATRICS) from SERB, Department of Science & Technology, Govt. of India, and the second author would like to acknowledge CSIR, Govt. of India, for a research fellowship.

References

1. S. D. Adhikari and Y. G. Chen, *Davenport constant with weights and some related questions - II*, J. Combinatorial Theory, Ser. A **115**, No. 1, (2008) 178–184.
2. S. D. Adhikari, Y. G. Chen, J. B. Friedlander, S. V. Konyagin and F. Pappalardi, Contributions to zero-sum problems, *Discrete Math.*, **306**, (2006) 1–10.
3. Sukumar Das Adhikari and Purusottam Rath, *Davenport constant with weights and some related questions*, Integers, **6** (2006), A30.
4. S. D. Adhikari and P. Rath, *Zero-sum problems in combinatorial number theory*, Ramanujan Math. Soc. Lect. Notes Ser., 2, Ramanujan Math. Soc., Mysore (2006), 1–14.
5. W. R. Alford, A. Granville and C. Pomerance, *There are infinitely many Carmichael numbers*, Annals of Math., **139 (2)** no. 3 (1994), 703–722.
6. A. Bialostocki and P. Dierker, *On the Erdős-Ginzburg-Ziv theorem and the Ramsey numbers for stars and matchings*, Discrete Math., **110**, no. 1-3 (1992), 1–8.
7. Y. Caro, *Zero-sum problems, a survey*, Discrete Math., **152** (1996), 93–113.
8. Y. Edel, C. Elsholtz, A. Geroldinger, S. Kubertin, L. Rackham, *Zero-sum problems in finite abelian groups and affine caps*, Quart. J. Math., **58** (2007), 159–186.
9. P. Erdős, A. Ginzburg and A. Ziv, *Theorem in the additive number theory*, Bull. Research Council Israel, **10F**, (1961), 41–43.
10. W. D. Gao, *A combinatorial problem on finite abelian groups*, J. Number Theory, **58**, (1996), 100–103.
11. W. D. Gao and A. Geroldinger, *Zero-sum problems in finite abelian groups: a survey*, Expo. Math. **24** (2006), 337–369.
12. A. Geroldinger and F. Halter-Koch, *Non-Unique Factorizations*, Chapman & Hall, CRC (2006).
13. Simon Griffiths, *The Erdős-Ginzburg-Ziv theorem with units*, Discrete Math., **308**, No. 23, (2008), 5473–5484.
14. D. J. Grynkiewicz, L. E. Marchan and O. Ordaz, A weighted generalization of two theorems of Gao, *Ramanujan J.*, **28** (2012), no. 3, 323–340.
15. Florian Luca, *A generalization of a classical zero-sum problem*, Discrete Math. 307, No. 13, (2007), 1672–1678.
16. Melvyn B. Nathanson, *Additive Number Theory: Inverse Problems and the Geometry of Sumsets*, Springer, 1996.
17. J. E. Olson, *On a combinatorial problem of finite Abelian groups, I*, J. Number Theory, **1** (1969), 8–10.
18. K. Rogers, *A Combinatorial problem in Abelian groups,* Proc. Cambridge Phil. Soc., **59** (1963), 559–562.
19. R. Thangadurai, *A variant of Davenport's constant*, Proc. Indian Acad. Sci. (Math. Sci.) **117**, No. 2, (2007), 147 –158.
20. P. Yuan, X. Zeng, Davenport constant with weights, *European Journal of Combinatorics*, **31** (2010), 677–680.

On a Zero-Sum Problem Arising From Factorization Theory

Aqsa Bashir, Alfred Geroldinger, and Qinghai Zhong

Abstract We study a zero-sum problem dealing with minimal zero-sum sequences of maximal length over finite abelian groups. A positive answer to this problem yields a structural description of sets of lengths with maximal elasticity in transfer Krull monoids over finite abelian groups.

Keywords Zero-sum sequences · Sets of lengths · Elasticity · Transfer Krull monoids

2020 Mathematics Subject Classification 11B75 · 11P70 · 13A05 · 20M13

1 Introduction

Let G be an additively written, finite abelian group and $G_0 \subset G$ be a subset. By a sequence $S = g_1 \cdot \ldots \cdot g_\ell$ over G_0, we mean a finite sequence of terms from G_0, where the order is disregarded and repetition is allowed. We say that S has sum zero if $g_1 + \ldots + g_\ell = 0$ and that S is a minimal zero-sum sequence if no proper subsum equals zero (i.e., $\sum_{i \in I} g_i \neq 0$ for all $\emptyset \neq I \subsetneq [1, \ell]$). The set of all zero-sum sequences is a (multiplicative) monoid with concatenation of sequences as operation. The empty sequence is the identity element of this monoid and the minimal zero-sum sequences are the irreducible elements. The Davenport constant $\mathsf{D}(G)$ of G is the

This work was supported by the Austrian Science Fund FWF, Project Numbers W1230 and P33499-N.

A. Bashir · A. Geroldinger (✉) · Q. Zhong
Institut für Mathematik und Wissenschaftliches Rechnen, Karl-Franzens-Universität Graz, NAWI Graz, Heinrichstraße 36, Graz 8010, Austria
e-mail: alfred.geroldinger@uni-graz.at

A. Bashir
e-mail: aqsa.bashir@uni-graz.at

Q. Zhong
e-mail: qinghai.zhong@uni-graz.at

M. B. Nathanson (ed.), *Combinatorial and Additive Number Theory IV*,
Springer Proceedings in Mathematics & Statistics 347,
https://doi.org/10.1007/978-3-030-67996-5_2

maximal length of a minimal zero-sum sequence over G (equivalently, $\mathsf{D}(G)$ is the smallest integer $\ell \in \mathbb{N}$ such that every sequence over G of length at least ℓ has a non-empty zero-sum subsequence).

In this note we study a conjecture stemming from factorization theory. We first formulate it in basic terms. Its background and significance will be discussed in Sect. 2, when we have more terminology at our disposal (Theorem 2 and Corollary 1).

Conjecture 1 Let G be a finite abelian group, which is neither cyclic nor an elementary 2-group. Then, for every minimal zero-sum sequence $U = g_1 \cdot \ldots \cdot g_\ell$ of length $|U| = \ell = \mathsf{D}(G)$, there are $k \in \mathbb{N}$ and minimal zero-sum sequences $U_1, \ldots, U_k$, $V_1, \ldots, V_{k+1}$ with terms from $\{g_1, \ldots, g_\ell, -g_1, \ldots, -g_\ell\}$ such that $U_1 \cdot \ldots \cdot U_k = V_1 \cdot \ldots \cdot V_{k+1}$.

Let G be a cyclic group of order $|G| = n \geq 3$. Then $\mathsf{D}(G) = n$ and every minimal zero-sum sequence over G of length n consists of an element g of order n repeated n times. Thus all distances $s - r$, occurring in equations $U_1 \ldots U_r = V_1 \ldots V_s$ over minimal zero-sum sequences with terms from $\{-g, g\}$, is a multiple of $n - 2$. Similarly, if G is an elementary 2-group of rank $r \geq 2$, then $\mathsf{D}(G) = r + 1$ and all distances $s - r$ are multiples of $r - 1$. Thus, the above conjecture neither holds for cyclic groups nor for elementary 2-groups with Davenport constant greater than or equal to four.

To describe the challenge of the above conjecture, suppose that $G \cong C_{n_1} \oplus \ldots \oplus C_{n_r}$, where $r = \mathsf{r}(G) = \max\{\mathsf{r}_p(G) \colon p \in \mathbb{P}\}$ is the rank of G, $\mathsf{r}_p(G)$ is the p-rank of G for every prime p, and $1 < n_1 \mid \ldots \mid n_r$ are positive integers. Then

$$\mathsf{D}^*(G) := 1 + \sum_{i=1}^{r} (n_i - 1) \leq \mathsf{D}(G) \,. \tag{1}$$

It is known since the 1960s that equality holds for p-groups and for groups of rank $\mathsf{r}(G) \leq 2$. There are further sparse series of groups where equality holds and groups where equality does not hold (see [1, 9, 10, 12] for recent progress). Even less is known for the associated inverse question asking for the structure of minimal zero-sum sequences of length $\mathsf{D}(G)$. However, the full structural description of minimal zero-sum sequences of length $\mathsf{D}(G)$ is not always needed in order to settle the Conjecture 1. We summarize what is known so far.

Conjecture 1 is settled for groups of rank two and for groups isomorphic to $C_2 \oplus C_2 \oplus C_{2n}$ with $n \geq 2$. These proofs heavily depend on the complete structural description of minimal zero-sum sequences of length $\mathsf{D}(G)$. Furthermore, the conjecture is proved for groups isomorphic to $C_{p^k}^r$, where p is a prime, $r \geq 2$, and $k \in \mathbb{N}$ such that $p^k > 2$, although for these groups there is not even a conjecture concerning the structure of minimal zero-sum sequences of maximal length (for all these results see [7]). We formulate a main result of the present paper.

Theorem 1 *Conjecture 1 holds true for the following non-cyclic finite abelian groups G.*

(a) *G is a p-group such that* $\gcd(\exp(G)-2, \mathsf{D}(G)-2)=1$.
(b) $G \cong C_{p^{s_1}}^{r_1} \oplus C_{p^{s_2}}^{r_2}$, *where p is a prime and* $r_1, r_2, s_1, s_2 \in \mathbb{N}$ *such that* s_1 *divides* s_2.
(c) *G is a group with exponent* $\exp(G)=pq$, *where p, q are distinct primes satisfying one of the four properties.*
   (i) $\gcd(pq-2, \mathsf{D}(G)-2)=1$.
   (ii) $\gcd(pq-2, p+q-3)=1$.
   (iii) $q=2$ *and* $p-1$ *is a power of* 2.
   (iv) $q=2$ *and* $\mathsf{r}_p(G)=1$.
(d) *G is a group with exponent* $\exp(G) \in [3, 11] \setminus \{8\}$.

Since Conjecture 1 does not hold for groups G with $\exp(G)=2$, groups that are sums of two elementary p-groups (as listed in (c)) and groups with small exponents, as listed in (d), are extremal cases for the validity of the conjecture. Statement (a) has a simple proof. However, since for p-groups we have $\mathsf{D}(G)=\mathsf{D}^*(G)$, it yields a variety of groups satisfying the conjecture. The precise value of the Davenport constant is not known in general for groups with exponent $\exp(G)=pq$, where p and q are distinct primes. To mention a few examples of what is known so far, let $G \cong C_2^r \oplus C_6$ with $r \in \mathbb{N}$. Then $\mathsf{D}(G)=\mathsf{D}^*(G)$ (i.e., equality holds in (1)) if and only if $r \in [1, 3]$ (see [6, Corollary 2] and [2]). Moreover, if a group G with $\exp(G)=6$ has a subgroup isomorphic to $C_2^i \oplus C_6^{5-i}$ for some $i \in [1, 4]$, then $\mathsf{D}(G) > \mathsf{D}^*(G)$ by [4, Theorem 3.1].

We proceed as follows. In Sect. 2, we present the background from factorization theory which motivates the above conjecture. We formulate a conjecture and a theorem in terms of factorization theory (Conjecture 2 and Theorem 2), associated to the ones given in the Introduction. In Corollary 1, we establish the significance of the two conjectures for the structure of sets of lengths having maximal elasticity. In Sect. 3, we prove Theorems 1 and 2.

2 Background on Sets of Lengths

For integers $a, b \in \mathbb{Z}$, we denote by $[a, b]=\{x \in \mathbb{Z}\colon a \le x \le b\}$ the discrete interval between a and b. Let $L=\{m_1, \ldots, m_k\} \subset \mathbb{Z}$ be a finite nonempty subset with $k \in \mathbb{N}$ and $m_1 < \ldots < m_k$. Then $\Delta(L)=\{m_i - m_{i-1}\colon i \in [2, k]\} \subset \mathbb{N}$ denotes the set of distances of L. If $L' \subset \mathbb{Z}$ is a finite subset, then $L+L'=\{a+a'\colon a \in L, a' \in L'\}$ is the sumset of L and L'. If $L \subset \mathbb{N}$ consists of positive integers, then $\rho(L)=\max L/\min L$ denotes the elasticity of L and for convenience we set $\rho(\{0\})=1$. Let G be an additively written finite abelian group. If $G_0 \subset G$ is a subset, then $\langle G_0 \rangle$ is the subgroup generated by G_0. Let $r \in \mathbb{N}$ and $(e_1, \ldots, e_r)$ be an r-tuple of elements of G. Then $(e_1, \ldots, e_r)$ is said to be independent if $e_i \neq 0$

for all $i \in [1, r]$ and if for all $m_1, \ldots, m_r \in \mathbb{Z}^r$ an equation $m_1 e_1 + \ldots + m_r e_r = 0$ implies that $m_i e_i = 0$ for all $i \in [1, r]$. Furthermore, $(e_1, \ldots, e_r)$ is a basis of G if it is independent and $G = \langle e_1 \rangle \oplus \ldots \oplus \langle e_r \rangle$. A subset $G_0 \subset G$ is independent if the tuple $(g)_{g \in G_0}$ is independent. We recall some basics of the arithmetic of monoids and of zero-sum sequences. Our notation and terminology are consistent with [5, 11].

Arithmetic of Monoids. By a monoid, we mean a commutative cancellative semigroup with identity element. Let H be a multiplicatively written monoid. We denote by $\mathcal{A}(H)$ the set of atoms (irreducible elements) of H and say that H is *atomic* if every non-invertible element can be written as a finite product of atoms. If $a = u_1 \cdot \ldots \cdot u_k$, where $k \in \mathbb{N}$ and $u_1, \ldots, u_k \in \mathcal{A}(H)$, then k is a factorization length of a, and

$$\mathsf{L}_H(a) = \mathsf{L}(a) = \{k \colon k \text{ is a factorization length of } a\} \subset \mathbb{N}$$

denotes the *set of lengths* of a. It is usual to set $\mathsf{L}(a) = \{0\}$ if $a \in H$ is invertible. The family

$$\mathcal{L}(H) = \{\mathsf{L}(a) \colon a \in H\}$$

is called the *system of sets of lengths* of H and

$$\rho(H) = \sup\{\rho(L) \colon L \in \mathcal{L}(H)\} \in \mathbb{R}_{\geq 1} \cup \{\infty\}$$

denotes the *elasticity* of H. Furthermore,

$$\Delta(H) = \bigcup_{L \in \mathcal{L}(H)} \Delta(L) \subset \mathbb{N}$$

is the *set of distances* of H. By definition, $\rho(H) = 1$ if and only if $\Delta(H) = \emptyset$, and otherwise we have $\min \Delta(H) = \gcd \Delta(H)$.

Zero-sum Sequences. Let G be an additively written finite abelian group and $G_0 \subset G$ be a subset. We denote by $\mathcal{F}(G_0)$ the (multiplicatively written) free abelian monoid with basis G_0, called the *monoid of sequences* over G_0. Let

$$S = g_1 \cdot \ldots \cdot g_\ell = \prod_{g \in G_0} g^{\mathsf{v}_g(S)} \in \mathcal{F}(G_0)$$

be a sequence over G_0. Then, for every $g \in G_0$, $\mathsf{v}_g(S) \in \mathbb{N}_0$ is the multiplicity of g in S, $\operatorname{supp}(S) = \{g_1, \ldots, g_\ell\} \subset G_0$ is the support of S, $|S| = \ell = \sum_{g \in G_0} \mathsf{v}_g(S) \in \mathbb{N}_0$ is the length of S, $\Sigma(S) = \{\sum_{i \in I} g_i \colon \emptyset \neq I \subset [1, \ell]\}$ is the set of subsequence sums of S, and $\sigma(S) = g_1 + \ldots + g_\ell \in G$ is the sum of S. We say that S is zero-sum free if $0 \notin \Sigma(S)$. The set

$$\mathcal{B}(G_0) = \{S \in \mathcal{F}(G_0) \colon \sigma(S) = 0\} \subset \mathcal{F}(G_0)$$

is a submonoid of $\mathcal{F}(G_0)$, called the *monoid of zero-sum sequences* over G_0. We set

$$\mathcal{L}(G_0) := \mathcal{L}(\mathcal{B}(G_0)),\ \Delta(G_0) := \Delta(\mathcal{B}(G_0)),\ \rho(G_0) := \rho(\mathcal{B}(G_0)),$$

and so on.
Transfer Krull monoids. A monoid H (resp. a domain D) is said to be a *transfer Krull monoid* (resp. a transfer Krull domain) over a finite abelian group G if there exists a transfer homomorphism $\theta\colon H \to \mathcal{B}(G)$ (resp. $\theta\colon D \setminus \{0\} \to \mathcal{B}(G)$). The classical example of a transfer Krull domain is the ring of integers $\mathcal{O}_K$ of an algebraic number field K, and in this case G is the ideal class group of $\mathcal{O}_K$. We refer to the survey [8] for formal definitions and further examples. The crucial property of a transfer homomorphism $\theta\colon H \to \mathcal{B}(G)$ is that it preserves the system of sets of lengths. We have $\mathcal{L}(H) = \mathcal{L}(G)$, whence all invariants describing the structure of sets of length coincide. In particular, we have

$$\Delta(H) = \Delta(G) \subset [1, \mathsf{D}(G) - 2] \quad \text{and} \quad \rho(H) = \rho(G) = \mathsf{D}(G)/2. \tag{2}$$

We refer to the survey [14] for what is known on the system $\mathcal{L}(G)$ and on associated invariants. Let H be a transfer Krull monoid over G. It is classical that $|L| = 1$ for all $L \in \mathcal{L}(H)$ if and only if $|G| \le 2$. Suppose that $|G| \ge 3$. Then there is $a \in H$ such that $|\mathsf{L}(a)| > 1$. For every $n \in \mathbb{N}$, the n-fold sumset

$$\mathsf{L}(a) + \ldots + \mathsf{L}(a) \subset \mathsf{L}(a^n),$$

whence $|\mathsf{L}(a^n)| > n$. Thus, sets of lengths in $\mathcal{L}(H)$ can be arbitrarily large.
On $\Delta_\rho(H)$. Now we define the crucial invariant of the present paper (see [7, Definition 2.1]). Let $\Delta_\rho(H)$ denote the set of all $d \in \mathbb{N}$ with the following property: for every $k \in \mathbb{N}$, there is some $L_k \in \mathcal{L}(H)$ with $\rho(L_k) = \rho(H)$ and which has the form

$$L_k = y + (L' \cup \{0, d, \ldots, \ell d\} \cup L'') \subset y + d\mathbb{Z} \tag{3}$$

where $y \in \mathbb{Z}$, $\ell \ge k$, $\max L' < 0$, and $\min L'' > \ell D$. If H is a transfer Krull monoid over a finite abelian group G, then $\Delta_\rho(H) = \Delta_\rho(G)$ and there is a constant $M \in \mathbb{N}_0$ such that $L' \subset [-M, -1]$, and $L'' \subset \ell d + [1, M]$ ([7, Lemma 2.3]). The following conjecture was first formulated in [7, Conjecture 3.20].

Conjecture 2 Let H be a transfer Krull monoid over a finite abelian group G with $|G| > 4$. Then $\Delta_\rho(H) = \{1\}$ if and only if G is neither cyclic nor an elementary 2-group.

In the present note we study $\Delta_\rho(G)$ and obtain the following result.

Theorem 2 *Let H be a transfer Krull monoid over a finite abelian non-cyclic group G. Then $\Delta_\rho(H) = \{1\}$ for the following groups.*

(a) G is a p-group such that $\gcd(\exp(G) - 2, \mathsf{D}(G) - 2) = 1$.
(b) $G \cong C_{p^{s_1}}^{r_1} \oplus C_{p^{s_2}}^{r_2}$, where p is a prime and $r_1, r_2, s_1, s_2 \in \mathbb{N}$ such that s_1 divides s_2.

(c) G is a group with exponent $\exp(G) = pq$, *where* p, q *are distinct primes satisfying one of the four properties.*

(i) $\gcd(pq - 2, \mathsf{D}(G) - 2) = 1$.
(ii) $\gcd(pq - 2, p + q - 3) = 1$.
(iii) $q = 2$ *and* $p - 1$ *is a power of* 2.
(iv) $q = 2$ *and* $\mathsf{r}_p(G) = 1$.

(d) G is a group with exponent $\exp(G) \in [3, 11] \setminus \{8\}$.

The proof of Theorem 2 will be given in Sect. 3. We derive a corollary, which demonstrates the significance of the Conjecture 2 and of Theorem 2. It states that, if $\Delta_\rho(H) = \{1\}$, then all sets of lengths L with maximal elasticity $\rho(L) = \rho(H)$ are intervals, apart from their globally bounded initial and end parts.

Corollary 1 *Let H be a transfer Krull monoid over a finite abelian group G and suppose that* $\Delta_\rho(H) = \{1\}$. *Then there exists a constant* $M^* \in \mathbb{N}_0$ *such that every* $L \in \mathcal{L}(H)$ *with* $\rho(L) = \rho(H)$ *has the form*

$$L = y + (L' \cup [0, \ell] \cup L'') ,$$

where $y \in \mathbb{Z}$, $\ell \in \mathbb{N}_0$, $L' \subset [-M^*, -1]$, *and* $L'' \subset \ell + [1, M^*]$.

Proof Since $\mathcal{L}(H) = \mathcal{L}(G)$, it is sufficient to prove the claim for the monoid $\mathcal{B}(G)$ of zero-sum sequences over G. If $\mathsf{D}(G) \le 3$, then $\Delta(G) \subset \{1\}$, whence all $L \in \mathcal{L}(G)$ are intervals and the claim holds with $M^* = 0$. Suppose that $\mathsf{D}(G) \ge 4$ and recall that $\Delta(G) \subset [1, \mathsf{D}(G) - 2]$ (see (2)). We proceed in four steps.
1. By [3, Sect. 4.7], there is a constant $M_1 \in \mathbb{N}_0$ such that every $L \in \mathcal{L}(G)$ has the form

$$L = y + (L' \cup L^* \cup L'') \subset y + (\mathcal{D} + d\mathbb{Z}) , \tag{4}$$

where $y \in \mathbb{Z}$ is a shift parameter,

- $d \in \Delta(G) \subset [1, \mathsf{D}(G) - 2]$ and $\{0, d\} \subset \mathcal{D} \subset [0, d]$,
- L^* is finite nonempty with $\min L^* = 0$ and $L^* = (\mathcal{D} + d\mathbb{Z}) \cap [0, \max L^*]$,
- $L' \subset [-M_1, -1]$, and $L'' \subset \max L^* + [1, M_1]$.

As a side remark, we recall that the above description is best possible, as it was shown by a realization result of Schmid ([13]).
2. Let $G_0 \subset G$ be a subset with $\Delta(G_0) \ne \emptyset$. By [3, Theorem 4.3.6] (applied to the monoid $\mathcal{B}(G_0)$), there are constants $\psi(G_0)$ and $M_2(G_0) \in \mathbb{N}_0$ such that for every $A \in \mathcal{B}(G_0)$ with $\mathsf{v}_g(A) \ge \psi(G_0)$ for all $g \in G_0$,

$$\mathsf{L}(A) = y_A + (L'_A \cup \{0, d_A, 2d_A, \ldots, s_A d_A\} \cup L''_A) \subset y_A + d_A\mathbb{Z} , \tag{5}$$

where $y_A \in \mathbb{Z}$, $d_A = \min \Delta(G_0)$, $s_A \ge M_1 + \mathsf{D}(G)$, $L'_A \subset [-M_2(G_0), -1]$, and $L''_2 \subset s_A d_A + [1, M_2(G_0)]$. Since G has only finitely many subsets G_0 with $\Delta(G_0) \ne \emptyset$, we let ψ be the maximum over all $\psi(G_0)$ and let M_2 be the maximum over all

$M_2(G_0)$. Then the structural statement (5) holds with constants ψ and M_2 for all subsets $G_0 \subset G$ with $\Delta(G_0) \neq \emptyset$.
3. Clearly, it is sufficient to prove the claim of the corollary for all $A \in \mathcal{B}(G)$ with $\rho(\mathsf{L}(A)) = \mathsf{D}(G)/2$, for which $\max \mathsf{L}(A) - \min \mathsf{L}(A)$ is sufficiently large. Indeed, suppose that there are constants $M_3, M_4 \in \mathbb{N}_0$ such that the claim holds for all A with $\mathsf{L}(A) \not\subset \min \mathsf{L}(A) + [0, M_3]$ and with bound M_4 for the initial and end parts of $\mathsf{L}(A)$. Then the claim holds for all A with bound $\max\{M_3, M_4\}$ for the initial and end parts of $\mathsf{L}(A)$.
4. Now let $A \in \mathcal{B}(G)$ with $\rho(\mathsf{L}(A)) = \mathsf{D}(G)/2$. By [7, Lemma 3.2.(a)], there are $k, \ell \in \mathbb{N}$ and $U_1, \ldots, U_k, V_1, \ldots, V_\ell \in \mathcal{A}(G)$ with $|U_1| = \ldots = |U_k| = \mathsf{D}(G)$ and $|V_1| = \ldots = |V_\ell| = 2$ such that $A = U_1 \cdot \ldots \cdot U_k = V_1 \cdot \ldots \cdot V_\ell$. Then $k = \min \mathsf{L}(A)$ and $\ell = \max \mathsf{L}(A) = k\mathsf{D}(G)/2$. By **3.**, we may suppose that $k \geq |\mathcal{A}(G)|\psi$. Then there is $i \in [1, k]$, say $U_i = U$, such that U^ψ divides A. This implies that $(-U)^\psi U^\psi$ divides A, say $A = (-U)^\psi U^\psi B_\psi$ for some $B_\psi \in \mathcal{B}(G)$. By **2.** (applied to the subset $\operatorname{supp}\big((-U)U\big)$),

$$\mathsf{L}((-U)^\psi U^\psi) = y_U + (L'_U \cup \{0, d_U, 2d_U, \ldots, s_U d_U\} \cup L''_U) \subset y_U + d_U\mathbb{Z}, \quad (6)$$

where $y_U \in \mathbb{Z}, s_U \in \mathbb{N}$ with $s_U \geq M_1 + \mathsf{D}(G), d_U = \min \Delta(G_U), L'_U \subset [-M_2, -1]$, and $L''_U \subset s_U d_U + [1, M_2]$. Since $\Delta_\rho(G) = \{1\}$, [7, Corollary 3.3] implies that $d_U = 1$. Since

$$\mathsf{L}(B_\psi) + \mathsf{L}((-U)^\psi U^\psi) \subset \mathsf{L}(A),$$

$\mathsf{L}(A)$ contains an interval $[t, t + s_U]$ for some $t \in \mathbb{N}_0$. By **3.**, we may assume that $\mathsf{L}(A)$ is not contained in $\min \mathsf{L}(A) + [0, 2M_1 + \mathsf{D}(G)]$. Thus, by comparing the two representations (4) and (6), we infer that the period $\mathcal{D}$ in (4) is an interval. Thus, L^* is an interval, whence $\mathsf{L}(A)$ has the required form. □

Remark 1 Let G be a finite abelian group. If $\Delta(G) = \{1\}$ (which, for example, holds if $G \cong C_3 \oplus C_3$), then all sets of lengths are intervals. In particular, Corollary 1 holds with $M^* = 0$. Suppose that $G = C_p^r$ is an elementary p-group with $p \geq 5$ and $r \geq 2$.

1. Let $(e_1, \ldots, e_r)$ be a basis of G and $e_0 = e_1 + \ldots + e_r$. Then $U = e_1^{p-1} \cdot \ldots \cdot e_r^{p-1} e_0 \in \mathcal{A}(G)$ with $|U| = \mathsf{D}(G)$. For every $k \in \mathbb{N}$, we set $A_k = (-U)^k U^k$. Then $\rho(\mathsf{L}(A_k)) = \mathsf{D}(G)/2$ and $\min \mathsf{L}(A_k) = 2k$. It is easy to see that $2k + 1 \notin \mathsf{L}(A_k)$, whence the constant M^*, occurring in Corollary 1, cannot be zero but is strictly positive.

2. Every nonzero element $g \in G$ can be extended to a basis. Thus, every nonzero element of G occurs in the support of a minimal zero-sum sequence of length $\mathsf{D}(G)$. Therefore, for every $k \in \mathbb{N}$, there is $B_k \in \mathcal{B}(G)$ with $\rho(\mathsf{L}(B_k)) = \mathsf{D}(G)/2$, $\operatorname{supp}(B_k) = G \setminus \{0\}$, and $\min \mathsf{L}(B_k) \geq k$. Since $\mathsf{L}(B)$ is an interval for all $B \in \mathcal{B}(G)$ with $\operatorname{supp}(B) = G \setminus \{0\}$ ([3, Theorem 7.6.9]), all sets $\mathsf{L}(B_k)$ are intervals with elasticity $\mathsf{D}(G)/2$.

3 Proof of Theorems 1 and 2

In this section, we prove Theorems 1 and 2. We start with two lemmas.

Lemma 1 *Let G be a finite abelian group with rank $\mathsf{r}(G) \geq 2$ and $\exp(G) \geq 3$, and let $U \in \mathcal{A}(G)$ with $|U| = \mathsf{D}(G)$. If there exist an independent tuple $(e_1, \ldots, e_t) \in G^t$ with $t \geq 2$ and an element g such that $\{e_1, \ldots, e_t, g\} \subset \operatorname{supp}(U)$ and $ag = k_1e_1 + \ldots + k_te_t$ for some $a \in [1, \operatorname{ord}(g) - 1] \setminus \{\frac{\operatorname{ord}(g)}{2}\}$ and with $k_i \in [1, \operatorname{ord}(e_i) - 1]$ for all $i \in [1, t]$, then $\min \Delta\big(\operatorname{supp}\big((-U)U\big)\big) = 1$. In particular, if $\operatorname{supp}(U)$ contains a basis of G, then $\min \Delta\big(\operatorname{supp}\big((-U)U\big)\big) = 1$.*

Proof See [7, Lemma 3.10]. □

Lemma 2 *Let G be a finite abelian group such that $G \cong C_{p^{s_1}}^{r_1} \oplus C_{p^{s_2}}^{r_2}$, where p is a prime, $r_1, r_2, s_1, s_2 \in \mathbb{N}$ with $s_1 < s_2$, and let $G_0 \subset G$ be a subset with $\langle G_0 \rangle = G$. Then there is a subset $G_0' \subset G_0$ such that $\langle G_0' \rangle \cong C_{p^{s_2}}^{r_2}$ and G_0' is a basis of $\langle G_0' \rangle$.*

Proof Let G_1 and G_2 be subgroups of G such that $G = G_1 \oplus G_2$, $G_1 \cong C_{p^{s_1}}^{r_1}$, and $G_2 \cong C_{p^{s_2}}^{r_2}$. Then every element $g \in G_0$ can be written uniquely as $g = u_g + v_g$, where $u_g \in G_1$ and $v_g \in G_2$. Hence $\langle v_g : g \in G_0 \rangle = G_2$ and $\{v_g : g \in G_0\}$ contains a basis of G_2 by [3, Lemma A.7.3]. We choose elements $g_1, \ldots, g_{r_2} \in G_0$ such that $(v_1 = v_{g_1}, \ldots, v_{r_2} = v_{g_{r_2}})$ is a basis of G_2. Note that $\operatorname{ord}(v_i) = \operatorname{ord}(g_i) = p^{s_2}$ for every $i \in [1, r_2]$. If $k_1, \ldots, k_{r_2} \in [0, p^{s_2} - 1]$ are such that $k_1g_1 + \ldots + k_{r_2}g_{r_2} = 0$, then $k_1v_1 + \ldots + k_{r_2}v_{r_2} = 0$, whence the independence of $(v_1, \ldots, v_{r_2})$ implies that $k_1 = \ldots = k_{r_2} = 0$. It follows that $(g_1, \ldots, g_{r_2})$ is independent and hence $\langle g_1, \ldots, g_{r_2} \rangle \cong C_{p^{s_2}}^{r_2}$. □

Proof of Theorems 1 and 2 Let H be a monoid, G be a finite abelian non-cyclic group, and let $\theta \colon H \to \mathcal{B}(G)$ be a transfer homomorphism. Then $\Delta_\rho(H) = \Delta_\rho(G)$. In order to show that $\Delta_\rho(G) = \{1\}$, it is sufficient to show that

$$\min \Delta\big(\operatorname{supp}\big((-U)U\big)\big) = 1 \tag{7}$$

for every minimal zero-sum sequence U over G with $|U| = \mathsf{D}(G)$ (see [7, Corollary 3.3.2]). Note that (7) is precisely the statement of Conjecture 1. Let U be a minimal zero-sum sequence over G with $|U| = \mathsf{D}(G)$. We set $d = \min \Delta\big(\operatorname{supp}\big((-U)U\big)\big)$ and have to show that $d = 1$. Since $|U| = \mathsf{D}(G)$, we have $G = \langle \operatorname{supp}(U) \rangle$ by [3, Proposition 5.1.4].

Let $A \subset \operatorname{supp}(U)$ be a minimal subset such that for every element $g \in \operatorname{supp}(U) \setminus A$, there exists $h \in A$ such that $g \in \langle h \rangle$. Thus, for any two elements $g_1, g_2 \in A$, we have $g_1 \notin \langle g_2 \rangle$ and $\langle A \rangle = \langle \operatorname{supp}(U) \rangle = G$ is not cyclic, whence $|A| \geq 2$. Assume to the contrary that A is independent. We set $A = \{g_1, \ldots, g_m\}$ and $W_i = \prod_{g \in \langle g_i \rangle} g^{\mathsf{v}_g(U)}$ for every $i \in [1, m]$, where $m = |A| \geq 2$. Then $U = \prod_{i \in [1,m]} W_i$ and $\sigma(W_i) \in \langle g_i \rangle$ for every $i \in [1, m]$. Since A is independent and U is a zero-sum sequence, we obtain that W_i are zero-sum sequences for all $i \in [1, m]$, a contradiction to the minimality of U. Thus A is not independent.

We start with two simple observations. Let V be a minimal zero-sum sequence over $\operatorname{supp}\big(U(-U)\big)$. Since $(-V)V$ has a factorization of length $|V|$, it follows that

$$d \text{ divides } |V|-2\,. \quad \text{In particular, } d \text{ divides } \mathsf{D}(G)-2\,. \tag{8}$$

If $g \in \operatorname{supp}\big((-U)U\big)$ with $\operatorname{ord}(g) = n$, then $V = g^n$ is a minimal zero-sum sequence over $\operatorname{supp}\big(U(-U)\big)$, whence (8) implies that $d \mid (n-2)$. Thus we obtain that

$$d \text{ divides } \operatorname{ord}(g)-2 \text{ for all } g \in \operatorname{supp}\big((-U)U\big)\,. \tag{9}$$

We distinguish four cases. Whenever it is convenient, an elementary p-group will be considered as a vector space over the field with p elements.

CASE 1: G is a p-group such that $\gcd(\exp(G)-2, \mathsf{D}(G)-2) = 1$.

By [3, Corollary 5.1.13], $\operatorname{supp}(U)$ contains an element of order $\exp(G)$. Thus $d = 1$ by (9) and (8).

CASE 2: $G \cong C_{p^{s_1}}^{r_1} \oplus C_{p^{s_2}}^{r_2}$, where p is a prime and $r_1, r_2, s_1, s_2 \in \mathbb{N}$ such that s_1 divides s_2.

If $s_1 = s_2$, then the assertion follows from [7, Theorem 3.11]. Suppose that $s_1 < s_2$. Then $\exp(G) = p^{s_2} \ge 4$. By Lemma 2, there is a subset $A_2 \subset A$ such that $\langle A_2\rangle \cong C_{p^{s_2}}^{r_2}$ and A_2 is a basis of $\langle A_2\rangle$, say $A_2 = \{g_1, \ldots, g_{r_2}\}$. Since $\langle A_2\rangle$ is a direct summand of G, there is a subgroup G_1 of G with $G = G_1 \oplus \langle A_2\rangle$, whence $G_1 \cong C_{p^{s_1}}^{r_1}$. Every element g of A can be written uniquely as $g = u_g + v_g$, where $u_g \in G_1$ and $v_g \in \langle A_2\rangle$. Hence $\langle u_g \colon g \in A\rangle = G_1$ and $\{u_g \colon g \in A\}$ contains a basis of G_1 by [3, Lemma A.7.3]. We choose $h_1, \ldots, h_{r_1} \in A$ such that $(u_1 = u_{h_1}, \ldots, u_{r_1} = u_{h_{r_1}})$ is a basis of G_1. We distinguish two cases.

Suppose $\operatorname{ord}(h_i) = p^{s_1}$ for every $i \in [1, r_1]$. Then the tuple $(h_1, \ldots, h_{r_1})$ is independent, whence the tuple $(h_1, \ldots, h_{r_1}, g_1, \ldots, g_{r_2})$ forms a basis of G. Then the assertion follows by Lemma 1.

Suppose there exists $i \in [1, r_1]$ such that $\operatorname{ord}(h_i) \ne p^{s_1}$. Then $0 \ne p^{s_1}h_i$ and p^{s_1} is the minimal integer such that $p^{s_1}h_i \in \langle g_1, \ldots, g_{r_2}\rangle$. Set $h = h_i$. There exist $\emptyset \ne I \subset [1, r_2]$ and $k_i \in [1, p^{s_2}-1]$ for every $i \in I$ such that $p^{s_1}h = \sum_{i\in I} k_i g_i$. After renumbering if necessary, we may assume that $I = [1, t]$ for some $t \in [1, r_2]$.

If $t = 1$, then $(-h)^{p^{s_1}} g_1^{k_1}$ and $h^{p^{s_1}} g_1^{p^{s_2}-k_1}$ are both minimal zero-sum sequences. It follows by $\big(h^{p^{s_1}} g_1^{p^{s_2}-k_1}\big)\big((-h)^{p^{s_1}} g_1^{k_1}\big) = \big(h(-h)\big)^{p^{s_1}} g_1^{p^{s_2}}$ that d divides $p^{s_1}-1$. Since d divides $p^{s_2}-2$ by (9), it follows by the fact that s_1 divides s_2 that $d = 1$.

If $t \ge 2$ and $p^{s_1} \ne \frac{\operatorname{ord}(h)}{2}$, then $d = 1$ by Lemma 1.

If $t \ge 2$ and $\operatorname{ord}(h) = 2p^{s_1}$, then $p = 2$. Since $h^{2^{s_1}} g_1^{k_1} \cdot \ldots \cdot g_t^{k_t}$ is a minimal zero-sum sequence, we obtain that $g_1^{2k_1} \cdot \ldots \cdot g_t^{2k_t}$ is a zero-sum sequence, whence the independence of $(g_1, \ldots, g_t)$ implies that $k_i = 2^{s_2-1}$ for all $i \in [1, t]$. Since $(-h)^{2^{s_1}} g_1^{k_1} \cdot \ldots \cdot g_t^{k_t}$ is a minimal zero-sum sequence,

$$\big(h^{2^{s_1}} g_1^{k_1} \cdot \ldots \cdot g_t^{k_t}\big)^2 = h^{2^{s_1+1}} g_1^{2^{s_2}} \cdot \ldots \cdot g_t^{2^{s_2}}\,,$$

$$\text{and } \big(h^{2^{s_1}} g_1^{k_1} \cdot \ldots \cdot g_t^{k_t}\big)\big((-h)^{2^{s_1}} g_1^{k_1} \cdot \ldots \cdot g_t^{k_t}\big) = \big(h(-h)\big)^{2^{s_1}} g_1^{2^{s_2}} \cdot \ldots \cdot g_t^{2^{s_2}}\,,$$

we obtain that d divides $(t+1-2)-(2^{s_1}+t-2)=1-2^{s_1}$. Since d divides $2^{s_2}-2$ by (9), it follows by the fact that s_1 divides s_2 that $d=1$.
CASE 3: G is the sum of two elementary p-groups, say $G=C_p^r\oplus C_q^s$, where p,q are distinct primes and $r,s\in\mathbb{N}$.

We set $U=U_pU_qU_{pq}$, where $U_p\in\mathcal{F}(G)$ consists of elements of order p, $U_q\in\mathcal{F}(G)$ consists of elements of order q, and $U_{pq}\in\mathcal{F}(G)$ consists of elements of order pq. Since U is a minimal zero-sum sequence and $0=\sigma(U)=\sigma(U_p)+\sigma(U_q)+\sigma(U_{pq})$, it follows that U_{pq} cannot be the empty sequence. Thus, supp(U) contains an element of order $pq=\exp(G)$ and hence by (9), we have

$$d \text{ divides } pq-2\,. \tag{10}$$

CASE 3.(i): $\gcd(pq-2,\mathsf{D}(G)-2)=1$.

Then $d=1$ by (8) and (10).

CASE 3.(ii): $\gcd(pq-2,p+q-3)=1$.

We first note that $\gcd(pq-2,p+q-3)=1$ implies that p,q are both odd, $\gcd(pq-2,p-2)=1$, $\gcd(pq-2,q-2)=1$, and

$$\gcd(pq-2,p-1)=1\,. \tag{11}$$

If there exists an element $h\in\text{supp}(U)$ such that $\text{ord}(h)\neq pq$, then d divides $p-2$ or $q-2$ by (9). In each case we infer that $d=1$ by (10). Now, we assume that every element of supp(U) has order pq.

Let G_1 and G_2 be subgroups of G such that $G=G_1\oplus G_2$, $G_1\cong C_p^r$, and $G_2\cong C_q^s$. Since G is not cyclic, we have $r\geq 2$ or $s\geq 2$. By symmetry, we may assume that $r\geq s$. Every element g of A can be written uniquely as $g=u_g+v_g$, where $u_g\in G_1$ and $v_g\in G_2$. Hence $\langle u_g\colon g\in A\rangle=G_1$ and $\{u_g\colon g\in A\}$ contains a basis of G_1. We choose $g_1,\ldots,g_r\in A$ such that $(u_1=u_{g_1},\ldots,u_r=u_{g_r})$ is a basis of G_1. We distinguish two cases.

First, suppose that $(v_1=v_{g_1},\ldots,v_r=v_{g_r})\in G_2^r$ is independent. Since $r\geq s=\mathsf{r}(G_2)$, we infer that $r=s$. Therefore, $G\cong C_{pq}^r$ and $(g_1,\ldots,g_r)$ is a basis of G. The assertion follows by Lemma 1.

Now, suppose that $(v_1=v_{g_1},\ldots,v_r=v_{g_r})$ is not independent. Since $\langle v_1,\ldots,v_r\rangle$ is an elementary q-group, there exists $I\subset[1,r]$ such that $(v_i)_{i\in I}$ is a basis of $\langle v_1,\ldots,v_r\rangle$. After renumbering if necessary, we may assume that $I=[1,y]$, where $y\in[1,r-1]$. Then $(g_1,\ldots,g_y)$ is independent and p is the minimal integer such that $pg_r\in\langle g_1,\ldots,g_y\rangle$. If there exists $i\in[1,y]$ such that $pg_r\in\langle g_i\rangle$, then there exists $k\in[1,pq-1]$ such that $g_r^pg_i^k$ and $(-g_r)^pg_i^{pq-k}$ are atoms, whence it follows by $(g_r^pg_i^k)\,((-g_r)^pg_i^{pq-k})=(g_r(-g_r))^p\,g_i^{pq}$ that d divides $p-1$. The assertion follows by (11). Otherwise there exist $J\subset[1,y]$ with $|J|\geq 2$ and $k_j\in[1,pq-1]$ for every $j\in J$ such that $pg_r=\sum_{j\in J}k_jg_j$. Now the assertion follows by Lemma 1.
CASE 3. (iii): $q=2$ and $p-1$ is a power of 2.

If there is an element $g\in\text{supp}(U)$ such that $\text{ord}(g)=p$, then d divides $\gcd(p-2,2p-2)=1$ by (9) and (10), whence $d=1$. Now, we suppose that supp(U)

contains no element of order p. Since d divides $2(p-1)$ and $p-1$ is a power of 2, it suffices to prove that d divides an odd number.

Let G_1 and G_2 be subgroups of G such that $G = G_1 \oplus G_2$, $G_1 \cong C_p^r$, and $G_2 \cong C_2^s$. Then every element g of $\text{supp}(U)$ can be written uniquely as $g = u_g + v_g$, where $u_g \in G_1$ and $v_g \in G_2$. Hence $\langle u_g : g \in \text{supp}(U) \rangle = G_1$ and $\{u_g : g \in \text{supp}(U)\}$ contains a basis of G_1. We choose $g_1, \ldots, g_r \in \text{supp}(U)$ such that $(u_1 = u_{g_1}, \ldots, u_r = u_{g_r})$ is a basis of G_1. Since $\text{supp}(U)$ has no element of order p, it follows that $\text{ord}(g_i) = 2p$ for all $i \in [1, r]$.

We set

$$T_0 = \prod_{g \in \text{supp}(U) \text{ with } \text{ord}(g)=2} g$$

$$\text{and} \quad T_i = \prod_{g \in \langle g_i \rangle \text{ with } \text{ord}(g)=2p} g^{\mathsf{v}_g(U)} \quad \text{for all } \ i \in [1, r] \,.$$

Assume to the contrary that $U = T_0 T_1 \cdot \ldots \cdot T_r$. Then $\sigma(T_i)$ has order 2 for every $i \in [1, r]$. If $|T_i| \ge p + 1 = \mathsf{D}(C_p) + 1$, then there exists a subsequence T_i' of T_i such that $1 \le |T_i'| \le p$ and $\sigma(T_i')$ has order 2, which implies that T_i' or $T_i T_i'^{-1}$ is a nonempty zero-sum sequence, a contradiction to the minimality of U. Thus $|T_i| \le p$ for every $i \in [1, r]$. Since $T_0 \sigma(T_1) \cdot \ldots \cdot \sigma(T_r)$ is a minimal zero-sum sequence consisting of elements of order 2, we infer that $|T_0| + r \le \mathsf{D}(G_2) = s + 1$. Hence $|U| \le (s - r + 1) + pr = (p-1)r + s + 1 < \mathsf{D}^*(G) \le \mathsf{D}(G)$, a contradiction.

Therefore, $U \ne T_0 T_1 \cdot \ldots \cdot T_r$, whence there is an element $h \in \text{supp}(U) \setminus \{g_1, \ldots, g_r\}$ such that $\text{ord}(h) = 2p$ and $h \notin \langle g_i \rangle$ for any $i \in [1, r]$. Let $I \subset [1, r]$ be a minimal subset such that $u_h \in \langle u_i : i \in I \rangle$. After renumbering if necessary, we may assume that $I = [1, x]$, where $x \in [1, r]$.

Suppose $x = 1$. Note that $h \notin \langle g_1 \rangle$ and $2h \in \langle g_1 \rangle$. Then there exists $k \in [1, 2p-1]$ such that $h^2 g_1^k$ is a minimal zero-sum sequence, and then the same is true for $(-h)^2 g_1^{2p-k}$. Since $(h^2 g_1^k)((-h)^2 g_1^{2p-k}) = (h(-h))^2 g_1^{2p}$, we obtain $d = 1$.

Suppose $x \ge 2$ and $v_h \in \langle v_1, \ldots, v_x \rangle$, where $v_i = v_{g_i}$ for all $i \in [1, x]$. For all $i \in [1, x]$, let $k_i \in [1, p-1]$ be such that $u_h = k_1 u_1 + \ldots + k_x u_x$. The elements $v_h, v_1, \ldots, v_x$ have order 2. After renumbering if necessary, we may assume that $v_h v_1 \cdot \ldots \cdot v_y$ is a minimal zero-sum sequence over G_2 for some $y \in [1, x]$. Therefore, the tuple $(v_1, \ldots, v_y)$ is independent, whence $(g_1, \ldots, g_y)$ is independent. If $y = x$, then it follows by $u_h \in \langle u_1, \ldots, u_x \rangle$ and $v_h \in \langle v_1, \ldots, v_x \rangle$ that $h \in \langle g_1, \ldots, g_x \rangle$, whence the assertion follows by Lemma 1. Suppose $y < x$. Then $h \notin \langle g_1, \ldots, g_y \rangle$ and p is the minimal integer such that $ph \in \langle g_1, \ldots, g_y \rangle$. If y is even, then $h^p g_1^p \cdot \ldots \cdot g_y^p$ is a minimal zero-sum sequence of odd length, whence d divides an odd number by (8). Suppose y is odd. We replace g_i by $-g_i$, u_i by $-u_i$, and k_i by $p - k_i$, if necessary, in order to make k_i to be odd for all $i \in [1, y]$ and k_j to be even for all $j \in [y+1, x]$. Then

$$W = (-h) g_1^{k_1} \cdots g_x^{k_x} \quad \text{and} \quad V = h^{p-1} (-g_1)^{p-k_1} \cdots (-g_y)^{p-k_y} g_{y+1}^{k_{y+1}} \cdot \ldots \cdot g_x^{k_x}$$

are minimal zero-sum sequences over $\mathrm{supp}(U(-U))$, $T = g_{y+1}^{pk_{y+1}} \cdot \ldots \cdot g_x^{pk_x}$ is a zero-sum sequence,

$$W^p = T\ ((-h)g_1 \ldots g_y)^p \prod_{i=1}^{y} (g_i^{2p})^{\frac{k_i-1}{2}}$$

$$\text{and} \quad V^p = T\ ((-h)^{2p})^{\frac{p-1}{2}} \prod_{i=1}^{y} ((-g_i)^{2p})^{\frac{p-k_i}{2}} .$$

If $\ell_0 \in \mathsf{L}(T)$, then d divides

$$\left(\ell_0 + \frac{p-1}{2} + \sum_{i\in[1,y]} \frac{p-k_i}{2} - p\right) - \left(\ell_0 + 1 + \sum_{i\in[1,y]} \frac{k_i-1}{2} - p\right)$$
$$= \frac{p-1}{2} + \frac{p+1}{2} y - \sum_{i=1}^{y} k_i - 1.$$

Since y is odd and k_i are odd for all $i \in [1, y]$, it follows that $\frac{p-1}{2} + \frac{p+1}{2} y - \sum_{i=1}^{y} k_i - 1 \equiv 1 \pmod 2$, whence d divides an odd number.

Suppose $x \geq 2$ and $v_h \notin \langle v_1, \ldots, v_x \rangle$. Then $h \notin \langle g_1, \ldots, g_x \rangle$ and $2h \in \langle g_1, \ldots, g_x \rangle$. Let $2u_h = k_1 u_1 + \ldots + k_x u_x$, where $k_i \in [1, p-1]$. We replace g_i by $-g_i$, u_i by $-u_i$, and k_i by $p - k_i$, if necessary, in order to make k_i to be even for all $i \in [1, x]$. If $(g_1, \ldots, g_x)$ is independent, then the assertion follows by Lemma 1. Otherwise the tuple $(v_1 = v_{g_1}, \ldots, v_x = v_{g_x})$ is not independent. After renumbering if necessary, we may assume that $v_1 \cdot \ldots \cdot v_y$ is a minimal zero-sum sequence over G_2, where $y \in [2, x]$, whence $g_1^p \cdot \ldots \cdot g_y^p$ is a minimal zero-sum sequence. If y is odd, then d divides an odd number by (8). Suppose y is even. Then $W = (-h)^2 g_1^{k_1} \cdot \ldots \cdot g_x^{k_x}$ and $V = (-h)^2 (-g_1)^{p-k_1} \cdot \ldots \cdot (-g_y)^{p-k_y} g_{y+1}^{k_{y+1}} \cdot \ldots \cdot g_x^{k_x}$ are minimal zero-sum sequences, $T = (-h)^{2p} g_{y+1}^{pk_{y+1}} \cdot \ldots \cdot g_x^{pk_x}$ is a zero-sum sequence,

$$W^p = T \prod_{i=1}^{y} (g_i^{2p})^{\frac{k_i}{2}} \quad \text{and} \quad V^p = T\ ((-g_1) \cdot \ldots \cdot (-g_y))^p \prod_{i=1}^{y} ((-g_i)^{2p})^{\frac{p-k_i-1}{2}} .$$

If $\ell_0 \in \mathsf{L}(T)$, then d divides

$$\left(\ell_0 + 1 + \sum_{i\in[1,y]} \frac{p-k_i-1}{2} - p\right) - \left(\ell_0 + \sum_{i\in[1,y]} \frac{k_i}{2} - p\right) = 1 + \frac{p-1}{2} y - \sum_{i=1}^{y} k_i .$$

Since y is even and k_i are even for all $i \in [1, y]$, it follows that $1 + \frac{p-1}{2} y - \sum_{i=1}^{y} k_i \equiv 1 \pmod 2$, whence d divides an odd number.
CASE 3. (iv): $q = 2$ and $r = 1$.

Let $h \in \text{supp}(U)$ such that $\text{ord}(h) = 2p$. Since $\langle h \rangle$ is a direct summand of G, there is a subgroup G_1 of G with $G \cong G_1 \oplus \langle h \rangle$, whence $G_1 \cong C_2^{s-1}$. Every element g of $\text{supp}(U)$ can be written uniquely as $g = u_g + v_g$, where $u_g \in G_1$ and $v_g \in \langle h \rangle$. Hence $\langle u_g : g \in \text{supp}(U) \setminus \{h\} \rangle = G_1$ and $\{u_g : g \in \text{supp}(U) \setminus \{h\}\}$ contains a basis of G_1. We choose $g_1, \ldots, g_{s-1} \in \text{supp}(U) \setminus \{h\}$ such that $(u_1 = u_{g_1}, \ldots, u_{s-1} = u_{g_{s-1}})$ is a basis of G_1. We distinguish two cases.

Suppose $\text{ord}(g_i) = 2$ for every $i \in [1, s-1]$. Then the tuple $(g_1, \ldots, g_{s-1})$ is independent, whence the tuple $(g_1, \ldots, g_{s-1}, h)$ forms a basis of G. Then the assertion follows by Lemma 1.

Suppose there exists $i \in [1, s-1]$ such that $\text{ord}(g_i) = 2p$. Then 2 is the minimal integer such that $2g_i \in \langle h \rangle$, whence there exists $k \in [1, 2p-1]$ such that both $g_i^2 h^k$ and $(-g_i)^2 h^{2p-k}$ are minimal zero-sum sequences. Then $d = 1$ because

$$\left(g_i^2 h^k\right)\left((-g_i)^2 h^{2p-k}\right) = \left(g_i(-g_i)\right)^2 h^{2p} .$$

CASE 4: G is a group with $\exp(G) \in [3, 11] \setminus \{8\}$.

If $\exp(G)$ is prime, then the claim follows from CASE 2 (with $s_1 = s_2 = 1$). The case, when $\exp(G) \in \{4, 9\}$, is also handled in CASE 2, and the case $\exp(G) \in \{6, 10\}$ is handled in CASE 3.(iii). □

Acknowledgements We thank the reviewers for their careful reading.

References

1. G. Bhowmik and J.-C. Schlage-Puchta, *Davenport's constant for groups of the form $\mathbb{Z}_3 \oplus \mathbb{Z}_3 \oplus \mathbb{Z}_{3d}$*, Additive Combinatorics (A. Granville, M.B. Nathanson, and J. Solymosi, eds.), CRM Proceedings and Lecture Notes, vol. 43, American Mathematical Society, 2007, pp. 307 – 326.
2. F. Chen and S. Savchev, *Long minimal zero-sum sequences in the groups $C_2^{r-1} \oplus C_{2k}$*, Integers **14** (2014), Paper A23.
3. A. Geroldinger and F. Halter-Koch, *Non-Unique Factorizations. Algebraic, Combinatorial and Analytic Theory*, Pure and Applied Mathematics, vol. 278, Chapman & Hall/CRC, 2006.
4. A. Geroldinger, M. Liebmann, and A. Philipp, *On the Davenport constant and on the structure of extremal sequences*, Period. Math. Hung. **64** (2012), 213–225.
5. A. Geroldinger and I. Ruzsa, *Combinatorial Number Theory and Additive Group Theory*, Advanced Courses in Mathematics - CRM Barcelona, Birkhäuser, 2009.
6. A. Geroldinger and R. Schneider, *On Davenport's constant*, J. Comb. Theory, Ser. A **61** (1992), 147 – 152.
7. A. Geroldinger and Q. Zhong, *Long sets of lengths with maximal elasticity*, Can. J. Math. **70** (2018), 1284–1318.
8. A. Geroldinger and Q. Zhong, *Factorization theory in commutative monoids*, Semigroup Forum **100** (2020), 22–51.
9. B. Girard, *An asymptotically tight bound for the Davenport constant*, J. Ec. Polytech. Math. **5** (2018), 605–611.
10. B. Girard and W.A. Schmid, *Direct zero-sum problems for certain groups of rank three*, J. Number Theory **197** (2019), 297–316.

11. D.J. Grynkiewicz, *Structural Additive Theory*, Developments in Mathematics 30, Springer, Cham, 2013.
12. Chao Liu, *On the lower bounds of Davenport constant*, J. Comb. Theory, Ser. A **171** (2020), 105162, 15pp.
13. W.A. Schmid, *A realization theorem for sets of lengths*, J. Number Theory **129** (2009), 990 – 999.
14. W.A. Schmid, *Some recent results and open problems on sets of lengths of Krull monoids with finite class group*, in Multiplicative Ideal Theory and Factorization Theory, Springer, 2016, pp. 323–352.

Conditional Bounds on Siegel Zeros

Gautami Bhowmik and Karin Halupczok

Dedicated to Melvyn Nathanson.

Abstract We present an overview of bounds on zeros of L-functions and obtain some improvements under weak conjectures related to the Goldbach problem.

Keywords Siegel zero · Goldbach problem · Congruences · Dirichlet L-function · Generalised Riemann hypothesis

2010 Mathematics Subject Classification 11P32 · 11M26 · 11M41

1 Introduction

The existence of non-trivial real zeros of a Dirichlet L-function would contradict the Generalised Riemann Hypothesis. One possible counter-example, called the Landau–Siegel zero, is real and simple and the region in which it could eventually exist is important to determine. In 1936 Siegel gave a quantitative estimate on the distance of an exceptional zero from the line $\Re s = 1$. The splitting into cases depending on whether such an exceptional zero exists or not happens to be an important technique

G. Bhowmik (✉)
Laboratoire Paul Painlevé, UMR 8524, Université de Lille,
59655 Villeneuve d'Ascq Cedex, France
e-mail: gautami.bhowmik@univ-lille.fr

K. Halupczok
Mathematisches Institut der Heinrich-Heine-Universität Düsseldorf,
Universitätsstr. 1, 40225 Düsseldorf, Germany
e-mail: karin.halupczok@uni-duesseldorf.de

M. B. Nathanson (ed.), *Combinatorial and Additive Number Theory IV*,
Springer Proceedings in Mathematics & Statistics 347,
https://doi.org/10.1007/978-3-030-67996-5_3

often used in analytic number theory, for example in the theorem of Linnik. In the first section we discuss properties of the Siegel zero and results assuming classical and more recent hypotheses. This part of the paper is expository.

In the second section we present a conditional bound. In 2016 Fei improved Siegel's bound for certain moduli under a weakened Hardy–Littlewood conjecture on the Goldbach problem of representing an even number as the sum of two primes. In Theorem 11 and Corollary 1 we further weaken this conjecture and enlarge the set of moduli to include more Dirichlet characters.

2 Background

Consider a completely multiplicative, periodic arithmetic function $\chi : \mathbb{Z} \to \mathbb{C}$ where for $q \geq 1$ there exists a group homomorphism $\tilde{\chi} : (\mathbb{Z}/q\mathbb{Z})^\times \to \mathbb{C}^\times$ such that $\chi(n) = \tilde{\chi}(n (\text{mod } q))$ for n coprime to q and $\chi(n) = 0$ if not. We call χ a Dirichlet character (mod q). In fact, if $(n, q) = 1$, $\chi(n)$ is a $\phi(q)^{th}$ complex root of unity. We denote the *principal* character mod q, whose value $\chi(n)$ is always 1 for n coprime to q, by χ_0 (mod q). The *order* of χ is the least positive integer n such that $\chi^n = \chi_0$, both characters having the same modulus. A non-principal character is called *quadratic* if $\chi^2 = \chi_0$. In the case where χ always takes a real value, the possibilities being only 0 or ± 1, it is called a *real* character, otherwise it is called *complex*. A character modulo q is termed *primitive* and q its *conductor* if it cannot be factored as $\chi = \chi' \chi_0$, where χ_0 is a principal character and χ' a character of modulus strictly less than q. For a given χ (mod q), there is a unique primitive character $\tilde{\chi}(\text{mod } \tilde{q})$ with least possible $\tilde{q}$, where $\tilde{q} \mid q$, that *induces* χ, such that χ and $\tilde{\chi}$ have the same value at all n coprime to q.

The L series were introduced in 1837 by Dirichlet who used them to prove an analytic formula for the class number and the infinitude of primes in any arithmetic progression. For $s = \sigma + it$, $\sigma > 1$, and a Dirichlet character χ, we consider the Dirichlet L-function

$$L(s, \chi) = \sum_{n=1}^{\infty} \frac{\chi(n)}{n^s} = \prod_{p \text{ prime}} \left(1 - \frac{\chi(p)}{p^s}\right)^{-1}.$$

Note that since $L(s, \chi) = L(s, \tilde{\chi}) \prod_{p|q} (1 - \tilde{\chi}(p) p^{-s})$ there is no loss in considering only primitive characters for obtaining analytic properties.

Let $\rho_\chi = \beta_\chi + i\gamma_\chi$ be the non-trivial zeros of the L-function. These are known to be contained in the strip $0 < \Re(s) < 1$ though according to the Generalised Riemann Hypothesis (GRH), the only possible value of β_χ is $1/2$. While the GRH remains out of reach, much work has been directed towards finding zero-free regions for L-functions.

Around hundred years ago it was proved that real zeros close to $\Re s = 1$ are indeed rare. More precisely,

Theorem 1 (Landau–Page) *There is an absolute constant $c > 0$ such that for any $Q, T \geq 2$, the product $\prod_{q \leq Q} \prod^{\star}_{\chi \bmod q} L(s, \chi)$ has at most one zero of an L-function in the region*

$$|t| \leq T,\ 1 - \sigma \leq \frac{c}{\log(QT)};$$

where * runs over all primitive real characters of modulus q. If such a zero exists, then it is real and associated to a unique, quadratic $\chi \bmod q$.

This eventual 'bad' zero contradicting the GRH is called the exceptional or Siegel or *Landau–Siegel zero* and the corresponding character is called the *exceptional* character. We denote the Landau–Siegel zero by β_χ or simply β.

The work of Landau and Siegel provide bounds on the proximity of such a zero on the real axis from $s = 1$. Quantitatively,

Theorem 2 (Siegel) *For an exceptional zero β associated to a primitive character χ of conductor q and any $\epsilon > 0$ there is a constant $c(\epsilon) > 0$ such that*

$$1 - \beta \geq \frac{c(\epsilon)}{q^\epsilon}. \tag{1}$$

Unfortunately, the constant $c(\epsilon)$ cannot be computed effectively for any $\epsilon < 1/2$, which is a serious difficulty for many applications. In 1951, Tatuzawa [17] did improve on Siegel's theorem to give an effective version for almost all cases by proving that for any positive ϵ there does exist an effectively computable positive constant $c(\epsilon)$ such that for all quadratic characters χ, with at most one exception, $L(s, \chi)$ has no zeros in the interval $[1 - c(\epsilon)/q^\epsilon, 1]$.

2.1 Repulsion Property

The possible exceptional zero would force all other zeros, real or otherwise, of all L-functions of the same modulus away from the real axis. We state a quantitative version of the Deuring–Heilbronn result of 1933–1934.

Theorem 3 *There exist effective constants $c, c' > 0$ such that for any $T \geq 2$ and any $q \geq 1$, if for some quadratic $\chi(\text{mod } q)$, $L(s, \chi)$ has an exceptional zero $\beta \in [1 - c/\log(qT), 1]$, then $\prod_\chi L(s, \chi)$, the product over all characters of modulus q including the exceptional one, has no other zero in the domain*

$$\sigma \geq 1 - \frac{c' |\log((1 - \beta) \log qT)|}{\log qT},\ t \leq T.$$

Here is a reformulated version of the repulsion phenomenon also due to Linnik in 1944 which he used to find the size of the least prime in an arithmetic progression.

Theorem 4 *If there exists an exceptional zero β with $1-\beta=\frac{\varepsilon}{\log q}$ for ε sufficiently small, then all other zeroes $\sigma+it$ of L-functions of modulus q are such that*

$$1-\sigma \geq c\frac{\log\frac{1}{\varepsilon}}{\log(q(2+|t|))}$$

for an absolute positive constant c.

Compared to the classical estimate $\sigma \geq 1-\frac{c}{\log(q(2+|t|))}$ with some absolute positive constant c, for a region where $L(s,\chi)$, for any χ mod q, contains no zeros except at most one eventual exception, we now have a zero-free region wider by a factor of $\log\frac{1}{\varepsilon}$.

Theorem 4 above was strengthened by Bombieri [4] to

Theorem 5 *Let $T \geq 2$ and β be an exceptional zero with respect to the otherwise zero-free region $\sigma \geq \frac{c}{\log T}$, $|t| \leq T$, then there exist constants c_1, c_2 such that if $(1-\beta)\log T \leq c_2/e$, then for any zero $\sigma+it \neq \beta$ of $L(s,\chi)$, we have*

$$1-\sigma \geq c_1\frac{\log\frac{c_2}{(1-\beta)\log T}}{\log T},$$

where $|t| \leq T$ for every primitive χ of modulus $q \leq T$.

This can be written in terms of a density estimate. Let $N(\alpha, q, T)$ denote the number of zeroes, counted with multiplicity, of any L function of modulus q with $\alpha \leq \sigma \leq 1$ and $0 \leq t \leq T$ and let N' denote the case when β is omitted. Then there is an improvement

$$N'(\alpha, q, T) \ll (1-\beta)(\log q)\Big(1+\frac{\log T}{\log q}\Big)(qT)^{O(1-\alpha)}$$

with effective implied constants over Linnik's density estimate

$$N(\alpha, q, T) \ll (1-\beta)(\log qT)(qT)^{O(1-\alpha)}.$$

In [7] Friedlander and Iwaniec state the 'ultimate Deuring–Heilbronn property' as

Theorem 6 *Let $\chi(\text{mod } q)$ be a real primitive character of conductor q with the largest real zero β and let $\eta=\frac{1}{(1-\beta)\log q} \geq 3$. Then $L(s,\chi)$ has no zeros other than β in the region $\sigma > 1-\frac{c\log\eta}{\log q(|t|+1)}$ where c is an absolute positive constant.*

2.2 *Bounds for* $L(1, \chi)$

We know at least since Hecke and Landau that zeros of $L(s, \chi)$ and its value at $s = 1$ are closely related. If $L(1, \chi)$ is sufficiently small relative to the conductor, then there is a Siegel zero and conversely. More precisely, if $L(1, \chi) \le \frac{c}{\log q}$ for a small constant $c > 0$, then $1 - \beta \le \frac{1}{\log q}$. Using the Deuring–Heilbronn repulsion property, Friedlander and Iwaniec recently proved [7] that

$$\{1 - \beta \ll (\log q)^{-3} \log\log q\} \implies \{L(1, \chi) \ll (\log q)^{-1}\}. \tag{2}$$

Goldfeld [8] provided an asymptotic result for the location of the Siegel zero. In fact, when $1 - \beta < \frac{c}{\log q}$, he obtained a precise asymptotic formula

$$1 - \beta \sim \frac{6}{\pi^2} L(1, \chi) \Big(\sum a^{-1}\Big)^{-1} \tag{3}$$

where the summation is over all reduced quadratic forms (a, b, c) of discriminant $-q$.

Dirichlet expressed the value of $L(1, \chi)$ where $\chi(n) = (\frac{-q}{n})$ is the real primitive character of conductor q in terms of the number $h(q)$ of equivalence classes of binary quadratic forms of discriminant q, which can equivalently be formulated in terms of the number of ideal classes of an imaginary quadratic number field $\mathbb{K} = \mathbb{Q}(\sqrt{-q})$. From the class number formula $L(1, \chi) = \frac{\pi h(q)}{\sqrt{-q}}$ (here $q < -4$) one obtains the non-vanishing of $L(1, \chi)$ which is not that obvious when χ is real, and leads to the prime number theorem in arithmetic progression. Another obvious consequence of the above formula is the elementary lower bound

$$L(1, \chi) \gg \frac{1}{q^{1/2}}.$$

Bounding $L(1, \chi)$ is equivalent to estimating the size of the class number of the imaginary quadratic field $\mathbb{K} = \mathbb{Q}(\sqrt{-q})$, another important question in number theory. The corresponding formula is $L(1, \chi) = \frac{2\pi h_{\mathbb{K}}}{w_K \sqrt{d_{\mathbb{K}/\mathbb{Q}}}}$ for the class number $h_{\mathbb{K}}$ and discriminant $d_{\mathbb{K}/\mathbb{Q}}$ of $\mathbb{K}$, w_K the order of the group of units with regulator being 1 and the LHS being the residue of the Dedekind zeta function at $s = 1$.

Good effective lower bounds are more difficult to obtain. Goldfeld [9] in 1976 using known cases of the Birch and Swinnerton-Dyer conjecture for elliptic curves showed that

$$L(1, \chi) \gg \frac{\log q}{\sqrt{q}(\log\log q)}$$

for $q \ge 3$, the implied constant being effective. This together with Gross–Zagier's work of 1983 is a major step in the Gauß class number problem.

Theorem 7 (Goldfeld–Gross–Zagier) *For every $\epsilon > 0$ there exists an effectively computable positive constant c such that $h(-q) > (c \log q)^{1-\epsilon}$.*

This corresponds to a zero-free region of $L(s, \chi)$ of size $[1 - c_0 \frac{\log^{c_1}(q)}{\sqrt{q}}, 1]$ for some effective positive constants c_0, c_1 and for all real primitive characters.

Oesterlé's calculation of the involved constant in 1985 makes it possible to state this bound for $q > 0$ as

$$L(1, \chi) > \frac{\pi}{55\sqrt{q}} \log q \prod_{p|q} \Big(1 - \frac{2\sqrt{p}}{p+1}\Big).$$

Rather recently Bennett et al. [2] proved that if χ is a primitive quadratic character with conductor $q > 6677$, then $L(1, \chi) > \frac{12}{\sqrt{q}}$.

We are still far from the plausible lower bound $L(1, \chi) \gg (\log q)^{-1}$ which holds in many cases, for example for complex characters with an effective constant.

Aisleitner et al. [1] in 2019 showed the existence, for q sufficiently large, of an extremal non-principal character which satisfies, for constants C and γ, the inequality $|L(1, \chi)| \geq e^{\gamma}(\log\log q + \log\log\log q - C)$ using the method of resonance for detecting large values of the Riemann zeta function. Up to the constant, this corresponds to the predicted order of the extremal values.

A simple unconditional upper bound is $L(1, \chi) \ll \log q$. The implied constants have been worked on by a variety of methods. For example, for complex characters Granville and Soundararajan [10] determine the constant c_k for primitive characters of order k for which the bounds $|L(1, \chi)| \leq (c_k + o(1)) \log q$ hold true. For real primitive characters, the constant $c_2 = \frac{1}{4}(2 - \frac{2}{\sqrt{e}} + o(1)) \log q$ was obtained by Stephens for prime characters [16] and Pintz [14] extended this to non-prime characters.

2.2.1 Conditional Bounds

The optimal bounds of $L(1, \chi)$ under the condition of GRH are

$$(\log\log q)^{-1} \ll L(1, \chi) \ll \log\log q$$

where the implied constants are effective. Precisely speaking,

Theorem 8 (Littlewood 1928) *If the Generalised Riemann Hypothesis is true then*

$$\Big(\frac{1}{2} + o(1)\Big) \frac{\pi^2}{6e^{\gamma} \log\log q} \leq L(1, \chi) \leq (2 + o(1)) e^{\gamma} \log\log q$$

where γ is Euler's constant.

Only the implied constants in the above can be improved because there actually exist infinitely many q for which the special value of the corresponding character at $s = 1$

correspond to the above magnitude of orders. The classical unconditional Ω results, that $(1+o(1))\frac{\pi^2}{6e^\gamma \log\log q} \geq L(1,\chi)$ and $L(1,\chi) \geq (1+o(1))e^\gamma \log\log q$ hold for infinitely many q [5] show that there is a factor of 2 that remains undetermined for the extreme values.

We cite one example of a recent refined upper and lower bound established by Lamzouri et al. [12] assuming the GRH for characters of large conductor and studying certain character sums. For $q \geq 10^{10}$, the bounds obtained therein can be written in a simplified manner as

$$\frac{\pi^2}{12e^\gamma \log\log q} < |L(1,\chi)| < 2e^\gamma \log\log q.$$

The lower bound can be improved a lot by admitting the existence of Landau–Siegel zeros, and thus weakening the GRH. One such assumption, sometimes called the Modified Generalised Riemann Hypothesis (MGRH), is that all the zeros of $L(s,\chi)$ lie either on the critical line or on the real axis. Sarnak and Zaharescu [15] showed that if all Dirichlet $L(s,\chi)$ with χ real satisfy the MGRH then

$$L(1,\chi) \geq \frac{c^\epsilon}{(\log|q|)^\epsilon}$$

for any positive ϵ. The above constant is ineffective but the bounds can be made effective under certain additional conditions. These bounds use the explicit formula with an appropriately constructed kernel function.

Assuming that the GRH holds except for one possible exception, Friedlander and Iwaniec obtained an improved version of (2). They proved that:

Theorem 9 ([7]) *Let the GRH be true except for only one* $\beta > 3/4$. *Now if* $1-\beta \ll (\log\log q)^{-1}$, *then*

$$1-\beta \ll L(1,\chi) \ll (1-\beta)(\log\log q)^2.$$

3 Better Siegel Zero Bounds from Weak Goldbach Conjectures

Connections between Siegel zeros and the Goldbach problem were studied, for example in [3, 6]. Among the classical conjectures of the Goldbach problem is one due to Hardy and Littlewood in 1923 that predicts an equivalence between the number of representations of an even number as a sum of two primes and a singular series, $g(n) = \sum_{n=p_1+p_2} 1 \sim \mathcal{S}(n)$, where

$$\mathcal{S}(n) := \frac{n}{\varphi(n)} \prod_{p \nmid n} \Big(1 - \frac{1}{(p-1)^2}\Big) \cdot \frac{n}{\log^2 n} = 2C_2 \Big(\prod_{\substack{p \mid n \\ p>2}} \frac{p-1}{p-2}\Big) \cdot \frac{n}{\log^2 n}$$

with the twin prime constant

$$C_2 = \prod_{p>2} \Big(1 - \frac{1}{(p-1)^2}\Big)$$

which is approximately 0.66.

Fei [6] obtained an upper bound for β under a weakened form of the Hardy–Littlewood conjecture (WHL), namely

Conjecture 1 (WHL) There exists a positive contant δ such that $g(n) \geq \frac{\delta n}{\log^2 n}$ for every even integer $n > 2$.

Considering the size of $\mathcal{S}(n)$ we could even expect a $\delta > 1.32$, but in this weakened form of the Hardy–Littlewood conjecture, we assume only the existence of some small positive δ.

We now state Fei's theorem.

Theorem 10 ([6]) *If the WHL-conjecture is true and if there is an exceptional zero* β *for a character* χ *with a prime modulus* $q \equiv 3 \mod 4$, *then there exists a positive constant* c *such that* $1 - \beta \geq \frac{c}{\log^2 q}$.

Here, the corresponding region for the exceptional zero β is meant to be that of Theorem 1 with $T = q$. We will keep to this convention for the rest of this section.

One would like to know if it is possible to include other moduli in the above result or to relax the assumed WHL-conjecture.

Here we generalise Fei's result in these two aspects and obtain a conditional improvement of Siegel's bound for certain exceptional characters which includes Fei's modulus condition. Our result still assumes the weak Hardy–Littlewood conjecture but allows certain exceptions (WHLE) making it weaker than the WHL. Our proof is similar to that of Fei's but exploits, apart from the use of the WHLE, the generalisation to suitable composite moduli q.

Conjecture 2 (WHLE) Suppose that x is sufficiently large, and $q \leq x/4$. Then we have, with at most $x/8q$ exceptions,

$$g(n) \gg \frac{n}{\log^2 n}$$

for the multiples n of q in the interval $x/2 < n \leq x$.

Theorem 11 *Assume the WHLE Conjecture to be true. Let* q *be a sufficiently large integer and* χ *be a primitive character mod* q *with* $\chi(-1) = -1$ *such that there is an exceptional zero* β *of* $L(s, \chi)$. *Then there exists an effective constant* $c > 0$ *such that* $1 - \beta \geq \frac{c\varphi(q)}{q \log^2(q)}$.

Proof of Theorem 11. Step 1. We prove the following lower bound for the sum

$$S = \sum_{k=1}^{q} \Big(\sum_{2<p\le x} \mathrm{e}\Big(\frac{kp}{q}\Big)\Big)^2 \ge \frac{\delta x^2}{8\log^2 x} \tag{4}$$

for any sufficiently large real $x > 2$ and some small $\delta > 0$.

For this, we first note note that

$$\begin{aligned} S &= \sum_{k=1}^{q} \sum_{2<p_1,p_2\le x} \mathrm{e}\Big(\frac{k(p_1+p_2)}{q}\Big) \\ &= \sum_{n\le 2x} \sum_{k=1}^{q} \mathrm{e}\Big(\frac{kn}{q}\Big) \sum_{\substack{2<p_1,p_2\le x\\ p_1+p_2=n}} 1 = \sum_{\substack{n\le 2x\\ n\equiv 0(q)}} q \sum_{\substack{2<p_1,p_2\le x\\ p_1+p_2=n}} 1. \end{aligned}$$

Let x be large enough with

$$q \le x/4. \tag{5}$$

Hence under the assumption of Conjecture 2, for all even n in the interval $x/2 < n \le x$ that are divisible by q, we have

$$\sum_{\substack{2<p_1,p_2\le x\\ p_1+p_2=n}} 1 \ge \delta \frac{x}{\log^2 x}$$

for some constant $\delta > 0$, with the possible exception of at most $x/8q$ such n. Let $\mathcal{E}$ be the set of these exceptions.

Keeping this in mind, we obtain the lower bound

$$\begin{aligned} S &\ge q \sum_{\substack{x/2<n\le x\\ 2|n\\ n\equiv 0(q)}} \sum_{p_1+p_2=n} 1 \ge q \sum_{\substack{x/2<n\le x\\ 2|n\\ n\equiv 0(q)\\ n\notin\mathcal{E}}} \delta \frac{x}{\log^2 x} \\ &\ge \frac{q\delta x}{\log^2 x}\Big(\frac{x}{4q} - \frac{x}{8q}\Big) = \frac{\delta x^2}{8\log^2 x} \end{aligned} \tag{6}$$

as was to be shown.

Step 2. We now prove a more explicit expression for the sum S in the first step. In (4), the sum over p is subdivided into parts depending on whether or not q is divisible by p, i.e.

$$S_1 = \sum_{2<p\le x} \mathrm{e}\Big(\frac{kp}{q}\Big) = \sum_{\substack{2<p\le x\\ p\nmid q}} \mathrm{e}\Big(\frac{kp}{q}\Big) + \sum_{\substack{2<p\le x\\ p\mid q}} \mathrm{e}\Big(\frac{kp}{q}\Big) = \sum_{\substack{2<p\le x\\ (p,q)=1}} \mathrm{e}\Big(\frac{kp}{q}\Big) + \mathcal{O}(\omega(q))$$

$$= \sum_{\substack{1\le a\le q\\ (a,q)=1}} \mathrm{e}\Big(\frac{ka}{q}\Big) \sum_{\substack{2<p\le x\\ p\equiv a(q)}} 1 + \mathcal{O}(\omega(q)) = c_q(k) \sum_{\substack{2<p\le x\\ p\equiv a(q)}} 1 + \mathcal{O}(\omega(q)),$$

where $c_q(k)$ denotes the above Ramanujan sum

$$\sum_{\substack{1\le a\le q\\ (a,q)=1}} e(\tfrac{ka}{q})$$

while $\omega(q) = \#\{p \mid q\}$. Using the prime number theorem in arithmetic progressions [13], the last sum over p is written out as

$$\sum_{\substack{2<p\le x\\ p\equiv a(q)}} 1 = \frac{\mathrm{li}(x)}{\varphi(q)} - \frac{\chi(a)}{\varphi(q)} \int_2^x \frac{u^{\beta-1}}{\log u} du + \mathcal{O}(x \exp(-\tilde{c}\sqrt{\log x})) \tag{7}$$

for some constant $\tilde{c} > 0$, which holds uniformly in

$$q \le \exp(C\sqrt{\log x}) \tag{8}$$

for any $C > 0$ (this range is consistent with (5), though we could have chosen a larger x). Hence, this gives

$$S_1 = M + c_q(k)\frac{\mathrm{li}(x)}{\varphi(q)} + \mathcal{O}\Big(qx \exp(-\tilde{c}\sqrt{\log x})\Big) \tag{9}$$

with the term M being

$$M = \frac{-1}{\varphi(q)} \sum_{a=1}^{q} \mathrm{e}\Big(\frac{ak}{q}\Big)\chi(a) \int_2^x \frac{u^{\beta-1}}{\log u} du = \frac{-\tau_k(\chi)}{\varphi(q)} \int_2^x \frac{u^{\beta-1}}{\log u} du,$$

since $\chi(a) = 0$ if $(a, q) > 1$, with the Gauß sum

$$\tau_k(\chi) = \sum_{a=1}^{q} \mathrm{e}\Big(\frac{ak}{q}\Big)\chi(a).$$

Inserting the expansion $\int_2^x \frac{u^{\beta-1}}{\log u} du = \frac{x^\beta}{\beta \log x} + \mathcal{O}(\frac{x^\beta}{\log^2 x})$ yields

$$M = \frac{-\tau_k(\chi)}{\varphi(q)} \cdot \frac{x^\beta}{\beta \log x} + \mathcal{O}\Big(\frac{q^{1/2}}{\varphi(q)} \cdot \frac{x^\beta}{\log^2 x}\Big)$$

from the estimate $\tau_k(\chi) \ll q^{1/2}$.

We substitute this expression in (9) and the resulting approximation for S_1 into S to get

$$\begin{aligned} S = \sum_{k=1}^{q} S_1^2 = \sum_{k=1}^{q} \Big(c_q(k)\frac{\mathrm{li}(x)}{\varphi(q)} - \frac{\tau_k(\chi)}{\varphi(q)} \cdot \frac{x^\beta}{\beta \log x} \\ + \mathcal{O}\Big(qx \exp(-\tilde{c}\sqrt{\log x}) + \frac{q^{1/2}}{\varphi(q)} \cdot \frac{x^\beta}{\log^2 x}\Big)\Big)^2 \\ = \sum_{k=1}^{q} \Big(c_q^2(k)\frac{\mathrm{li}^2(x)}{\varphi^2(q)} + \frac{\tau_k^2(\chi)}{\varphi^2(q)} \cdot \frac{x^{2\beta}}{\beta^2 \log^2 x}\Big) \\ + \mathcal{O}\Big(\frac{q^{1/2}x^\beta}{\varphi(q)\log x} q^2 x \exp(-\tilde{c}\sqrt{\log x}) + \frac{q^2 x^{2\beta}}{\beta \varphi^2(q) \log^3 x} + q^3 x^2 \exp(-2\tilde{c}\sqrt{\log x})\Big), \end{aligned} \tag{10}$$

where all the mixed terms containing the Ramanujan sum have now disappeared since

$$\begin{aligned} \sum_{k=1}^{q} c_q(k)\tau_k(\chi) = \sum_{k=1}^{q} \sum_{\substack{1\le a\le q \\ (a,q)=1}} \sum_{\substack{1\le b\le q \\ (b,q)=1}} \mathrm{e}\Big(\frac{ak}{q}\Big)\mathrm{e}\Big(\frac{bk}{q}\Big)\chi(b) \\ = \sum_{\substack{1\le a\le q \\ (a,q)=1}} \sum_{\substack{1\le b\le q \\ (b,q)=1 \\ b\equiv -a(q)}} \chi(b) q = q \sum_{\substack{1\le a\le q \\ (a,q)=1}} \chi(-a) = 0 \end{aligned}$$

and $\sum_{k=1}^{q} c_q(k) = 0$. The $\mathcal{O}$-term in (10) simplifies to

$$E_{expl} = \mathcal{O}\Big(\frac{x^2}{\log^3 x}\frac{q^2}{\varphi(q)^2} + q^3 x^2 \exp(-\tilde{c}\sqrt{\log x})\Big). \tag{11}$$

For the main term in (10), we use properties of Gauß sums ([13, p.287]),

$$\tau_k(\chi) = \begin{cases} \bar{\chi}(k)\tau_1(\chi), & (k, q) = 1 \\ 0, & \text{else}, \end{cases}$$

so that the sum over $\tau_k^2(\chi)$ in (10) becomes

$$\frac{1}{\varphi^2(q)} \sum_{\substack{k=1 \\ (k,q)=1}}^{q-1} \tau_1^2(\chi)\bar{\chi}^2(k) = \frac{q}{\varphi(q)}\chi(-1)$$

since $\tau_1^2(\chi) = \chi(-1)q$ and $\chi^2 = \chi_0$. Similarly, we have

$$\sum_{k=1}^{q} c_q^2(k) = q\varphi(q),$$

see [13, p. 113]. Hence

$$S = \frac{q}{\varphi(q)}\,\mathrm{li}^2(x) + \frac{q}{\varphi(q)}\chi(-1)\frac{x^{2\beta}}{\beta^2 \log^2 x} + E_{expl}. \tag{12}$$

Comment. We note that an alternative approach via the identity

$$\frac{q}{\varphi(q)}\sum_{\chi(q)} \chi(-1)|\psi(x,\chi)|^2 = \sum_{k=1}^{q}\Big(\sum_{p\le x}\Lambda(p)e\Big(\frac{kp}{q}\Big)\Big)^2 + o(\log^3 x)$$

and the use of an explicit formula for $\psi(x,\chi)$ might avoid the use of Gauß sums in Step 2 of the proof.

Step 3. Now we compare the lower bound from Step 1 with the explicit evaluation from Step 2. With the assumption $\chi(-1) = -1$ we get the inequality

$$\frac{x^{2\beta}}{\beta^2 \log^2 x} \le \Big(1 - \frac{\delta}{8}\cdot\frac{\varphi(q)}{q}\Big)\frac{x^2}{\log^2 x} + E_{\mathrm{expl}},$$

where E_{expl} is the error term of (12) above. This yields

$$x^{2\beta-2} \le \Big(1 - \frac{\delta}{8}\cdot\frac{\varphi(q)}{q}\Big) + \mathcal{O}\Big(\frac{q}{\varphi(q)\log x} + q^3\exp(-\tilde{c}\sqrt{\log x})\Big).$$

We may now choose x such that $(\frac{4\log q}{\tilde{c}})^2 \le \log x \le c_3 \log^2 q$ for some $c_3 > (4/\tilde{c})^2$, so that x is not too large compared to q, but still such that the choice is admissible with our previous assumptions (5) and (8). Hence with $1/\log x \le \tilde{c}^2/16\log^2 q$, we obtain

$$x^{2\beta-2} \le \Big(1 - \frac{\delta}{8}\cdot\frac{\varphi(q)}{q}\Big) + \frac{c_1}{\log^2 q}\frac{q}{\varphi(q)}$$

for some positive constant c_1, since the expression $q/\varphi(q)\log x$ dominates the error term due to

$$q^3\exp(-\tilde{c}\sqrt{\log x}) \le q^3\exp\Big(-\tilde{c}\frac{4}{\tilde{c}}\log q\Big) = q^3\exp(-4\log q) = q^{-1}.$$

Assume now that q is large enough so that $\frac{16c_1}{\log^2 q} \le \delta\frac{\varphi^2(q)}{q^2}$. This means that

$$x^{2\beta-2} \leq \left(1 - \frac{\delta}{8} \cdot \frac{\varphi(q)}{q}\right) + \frac{\delta}{16}\frac{\varphi(q)}{q} \leq 1 - \frac{\delta}{16}\frac{\varphi(q)}{q}$$

for $\delta < 8$. This gives the inequality of Theorem 11, since

$$\beta - 1 \leq \frac{\log(1 - \frac{\delta}{16} \cdot \frac{\varphi(q)}{q})}{2\log x} \leq \frac{-c\varphi(q)}{q\log^2 q} \text{ with } c = \frac{\delta}{32c_3} > 0,$$

where we use the upper bound for $\log x$, which is $\log x \leq c_3 \log^2 q$. □

We emphasize that all constants in the above proof are *effectively computable* since the constant $\tilde{c}$ in (7) coming from the prime number theorem in progressions is itself so and all other constants in the proof can be chosen effectively depending on $\tilde{c}$.

The next corollary gives a criterion for a composite modulus q to satisfy Theorem 11.

Corollary 1 *Assume the WHLE Conjecture and that q is a sufficiently large integer with $\#\{t \mid q\} \cap (\{4\} \cup \{p \equiv 3 (\mathrm{mod}\ 4);\ p \text{ prime}\}) = 1$. If there is an exceptional zero β for a character mod q, then $1 - \beta \geq \frac{c\varphi(q)}{q\log^2 q}$ for some (effective) constant $c > 0$.*

Proof For the moduli q in question there is a single real primitive character such that $\chi(-1) = -1$. This is certainly true if $q \in \mathcal{S} := \{4\} \cup \{p \equiv 3 (\mathrm{mod}\ 4)\}$, and $q = tm$ with $t \in \mathcal{S} \cup \{2\}$ and with m having only prime divisors $p \equiv 1 (\mathrm{mod}\ 4)$. Then there is only a single real χ with $\chi(-1) = -1$, namely the one that is induced by that mod t. For the others, $\chi(-1) = 1$, since this equation holds for every prime modulus $p \equiv 1 (\mathrm{mod}\ 4)$.

Hence, if we assume that β exists for an exceptional character χ, we know that χ is real and primitive, and necessarily $\chi(-1) = -1$. Then Theorem 11 applies to give the assertion. □

We recover Fei's Theorem from Corollary 1 when $q = p \equiv 3 (\mathrm{mod}\ 4)$.

Under GRH except for a possible $\beta > 3/4$ and the WHLE Conjecture we can deduce the conditional bound

$$L(1, \chi) \gg 1 - \beta \gg \frac{\varphi(q)}{q\log^2 q}.$$

using Theorem 9 of Friedlander and Iwaniec, and using Theorem 11, supposing $\chi(-1) = -1$ for the exceptional character χ mod q. In fact, with Goldfeld's asymptotic formula (3) we can relax the assumption of GRH with one exception to obtain.

Corollary 2 *Let the WHLE Conjecture be true. Assuming $L(1, \chi) = o(\log^{-1} q)$, we have $L(1, \chi) \gg R_q\varphi(q)q^{-1}\log^{-2} q$ for an exceptional character χ mod q with $\chi(-1) = -1$. Here $R_q = \sum_{(a,b,c)} a^{-1}$ with the sum running over all reduced quadratic forms (a, b, c) of discriminant $-q$.*

A reduced quadratic form of discriminant $-q$ is an integer tuple (a, b, c) with $b^2 - 4ac = -q$ and $-a < b \leq a < \frac{1}{4}\sqrt{q}$, see [9, p. 624].

Proof By (3) from [9, p. 624] and Theorem 11, we have

$$L(1,\chi) \sim \frac{\pi^2}{6}\Big(\sum_{(a,b,c)} a^{-1}\Big)(1-\beta) \gg \frac{R_q \varphi(q)}{q\log^2 q}.$$

□

Note that for a prime modulus $q = p \equiv 3 (\text{mod } 4)$, we have $R_q = \sum_{(a,b,c)} a^{-1} \geq 1$ since then there is a reduced quadratic form $(1, 1, c)$ with $a = 1$. Then our bound states $L(1, \chi) \gg \log^{-2} q$ under the assumptions of Corollary 2.

3.1 Questions

We would like to know if the case $\chi(-1) = 1$ could be handled as well. But we have not been able to combine the results of [3] together with Fei's approach which seems to be a natural way to proceed.

One could also ask if the WHLE Conjecture itself can be obtained from existing results. We were not able to find anything appropriate. Even when averaging the assertion over moduli $q \leq Q$, we only reach a special case of Conjecture 1 from [11], which seems to be out of reach.

Acknowledgements We thank Andrew Granville and Lasse Grimmelt for helpful comments and the referee for an improved presentation.
Note added: After the paper was accepted we learnt of a preprint of Chaohua Jia (arXiv:2010.14161) where a mistake in an earlier version of our paper is corrected.

References

1. C. Aistleitner, K. Mahatab, M. Munsch, A. Peyrot, On large values of $L(\sigma, \chi)$. *Q. J. Math.* **70** (2019), 831–848.
2. M.A. Bennett; G. Martin, K. O'Bryant, A. Rechnitzer, Explicit bounds for primes in arithmetic progressions *Illinois J. Math.* **62** (2018), 427–532. arXiv:1809.06920.
3. G. Bhowmik, K. Halupczok, K. Matsumoto and Y. Suzuki, Goldbach Representations in Arithmetic Progressions and zeros of Dirichlet L-functions, *Mathematika* **65** (2019), 57–97.
4. E. Bombieri, Le grand crible dans la théorie analytique des nombres, *Astérisque,* **18**, Soc. Math. France, Paris, 1987/1974.
5. S. Chowla, Improvement of a theorem of Linnik and Walfisz, *Proc. London Math. Soc. (2)* **50** (1949), 423–429.
6. J. H. Fei, An application of the Hardy–Littlewood conjecture, *J. Number Theory* **168** (2016), 39–44.
7. J.B. Friedlander, H. Iwaniec, A note on Dirichlet L-functions, *Expo. Math.* **36** (2018), 343–350.

8. D. Goldfeld, An asymptotic formula relating the Siegel zero and the class number of quadratic fields. *Ann. Scuola Norm. Sup. Pisa Cl. Sci.* (4) 2 (1975), no. 4, 611–615.
9. D. Goldfeld, The class number of quadratic fields and the conjectures of Birch and Swinnerton-Dyer, *Ann. Scuola Norm. Sup. Pisa* **3, 4** (1976), 623–663.
10. A. Granville, K. Soundararajan, Upper bounds for $|L(1,\chi)|$. *Q. J. Math.* **53** (2002) 265–284.
11. K. Halupczok, Goldbach's problem with primes in arithmetic progressions and in short intervals, *J. Théor. Nombres Bordeaux*, **25** no. 2 (2013), 331–351.
12. Y. Lamzouri, X. Li and K. Soundararajan, Conditional bounds for the least quadratic non-residue and related problems, *Math. Comp* **84** (2015), 2391–2412.
13. H. L. Montgomery, R.C. Vaughan, Multiplicative Number Theory I, Classical Theory, Cambridge, 2007.
14. J. Pintz, Elementary methods in the theory of L-functions. VIII. Real zeros of real L-functions. *Acta Arith.* **33** (1977), 89–98.
15. P. Sarnak, A. Zaharescu, Some remarks on Landau–Siegel zeros, *Duke Math. J.* **111** (2002), 495–507.
16. P. J. Stephens, Optimizing the size of $L(1,\chi)$. *Proc. London Math. Soc.* **24** (1972), 1–14.
17. T. Tatuzawa, On a theorem of Siegel, *Jap. J. Math.* **21** (1951), 163–178.

Infinite Co-minimal Pairs in the Integers and Integral Lattices

Arindam Biswas and Jyoti Prakash Saha

Abstract Given two nonempty subsets A, B of a group G, they are said to form a co-minimal pair if $A \cdot B = G$, and $A' \cdot B \subsetneq G$ for any $\emptyset \neq A' \subsetneq A$ and $A \cdot B' \subsetneq G$ for any $\emptyset \neq B' \subsetneq B$. The existence of co-minimal pairs is a stronger criterion than the existence of minimal complements. In this work, we show several new results about them. The existence and the construction of co-minimal pairs in the integers, with both the subsets A and B (A is not a translate of B) of infinite cardinality was unknown. We show that such pairs exist and give the first explicit construction of these pairs. The constructions also satisfy a number of algebraic properties. Further, we prove that for any $d \geq 1$, the group $\mathbb{Z}^{2d}$ admits infinitely many automorphisms such that for each such automorphism σ, there exists a subset A of $\mathbb{Z}^{2d}$ such that A and $\sigma(A)$ form a co-minimal pair.

Keywords Additive complements · Minimal complements · Sumsets · Representation of integers · Additive number theory

Subject Classifications: 11B13 · 05B10 · 11P70 · 05E15

1 Introduction

Let $(G, +)$ be an abelian group and $W \subseteq G$ be a nonempty subset. A nonempty set $W' \subseteq G$ is said to be an *additive complement* to W if $W + W' = G$. Additive complements have been studied since a long time in the context of representations of the integers, e.g., they appear in the works of Erdős, Hanani, Lorentz and others.

A. Biswas (✉)
Department of Mathematics, Technion - Israel Institute of Technology, 32000 Haifa, Israel
e-mail: biswas@campus.technion.ac.il; arin.math@gmail.com

J. P. Saha
Department of Mathematics, Indian Institute of Science Education and Research Bhopal, Bhopal Bypass Road, Bhauri, Bhopal 462066, Madhya Pradesh, India
e-mail: jpsaha@iiserb.ac.in

M. B. Nathanson (ed.), *Combinatorial and Additive Number Theory IV*,
Springer Proceedings in Mathematics & Statistics 347,
https://doi.org/10.1007/978-3-030-67996-5_4

See [6, 7, 9], etc. In [10], Nathanson introduced the notion of *minimal additive complements* for nonempty subsets of groups. An additive complement W' to W is said to be minimal if no proper subset of W' is an additive complement to W, i.e.,

$$W + W' = G \text{ and } W + (W' \setminus \{w'\}) \subsetneq G \ \forall w' \in W'.$$

Nathanson was interested in a number of questions from metric geometry arising in the context of discrete groups. As an example, the existence of minimal nets in groups is strongly related to the existence of minimal additive complements of generating sets, see [10, §1] (see also [10, Problem 1]) and [1, 2, 5]. A notion stronger to minimal additive complements is that of *additive co-minimal pairs.*[1] Given two nonempty subsets A, B of a group G, they are said to form a *co-minimal pair* if $A \cdot B = G$, and $A' \cdot B \subsetneq G$ for any $\emptyset \neq A' \subsetneq A$ and $A \cdot B' \subsetneq G$ for any $\emptyset \neq B' \subsetneq B$. Thus, they are pairs $(A \subseteq G, B \subseteq G)$ such that each element in a pair is a minimal additive complement to the other. The notion of a co-minimal pair was first considered in a prior work of the authors, see [4, Definition 1.2]. Henceforth, by a complement we shall mean an additive complement. If we mean set-theoretic complement, we shall explicitly state it.

It is a challenging task to classify the co-minimal pairs in a given group. Even in the context of the group of integers $\mathbb{Z}$, they are not completely understood. In [4], it was shown that non-empty finite subsets in free abelian groups (not necessarily of finite rank) belong to co-minimal pairs. However, the existence and the construction of infinite co-minimal pairs in $\mathbb{Z}$, i.e., co-minimal pairs (A, B) where both A and B are infinite (and A is not a translate of B) has been unknown. One of our motivations in this article is to show the existence and give explicit constructions of these pairs. Further, we also study co-minimal pairs in free abelian groups of higher rank. Our constructions satisfy certain nice combinatorial and group theoretic properties. They are mainly motivated by the following questions:

Question 1 Does there exist infinite subsets A, B of $\mathbb{Z}$ which form a co-minimal pair and one of them is bounded below and the other is bounded above?

Question 2 Does there exist infinite subsets A, B of $\mathbb{Z}$ which form a co-minimal pair and at least one of them is a symmetric subset of $\mathbb{Z}$?

In [8], [4], it was established that (A, A) is a co-minimal pair in an abelian group G, if and only if $A + A = G$ and A avoids 3-term arithmetic progressions. This motivates the following two questions when we note that the trivial automorphism $\sigma(g) = g, \forall g \in G$ fixes any subset $A \subseteq G$.

Question 3 Given an automorphism σ of a group G, does there exist subsets A in G such that $(A, \sigma(A))$ is a co-minimal pair?

---

[1] See the discussion about minimal pairs in [4, Sect. 1] and [4, Lemma 2.1] for the exact sense in which it is stronger.

Question 4 Given an automorphism σ of $\mathbb{Z}^d$, does there exist subsets A in G such that A is contained in a quadrant[2] and $(A, \sigma(A))$ is a co-minimal pair?

Note that Questions 3, 4 are related to Questions 2, 1. Indeed, having affirmative answers to Questions 2, 1 allows us to answer Questions 3, 4 for certain automorphisms of free abelian groups. For instance, if $\mathcal{U}, \mathcal{V}$ are infinite subsets of $\mathbb{Z}$ forming a co-minimal pair and $\mathcal{V}$ is symmetric, then taking $A = \mathcal{U} \times \mathcal{V}, \mathcal{V} \times \mathcal{U}$, we obtain an affirmative answer to Question 3 when $G = \mathbb{Z}^2$ and $\sigma = \left(\begin{smallmatrix} 0 & 1 \\ -1 & 0 \end{smallmatrix}\right), \left(\begin{smallmatrix} 0 & -1 \\ 1 & 0 \end{smallmatrix}\right)$. Moreover, if $\mathcal{S}, \mathcal{T}$ are infinite subsets of $\mathbb{Z}$ forming a co-minimal pair and $\mathcal{S}$ (resp. $\mathcal{T}$) is bounded above (resp. below) by 0, then taking $A = (-\mathcal{S}) \times \mathcal{T}$, we obtain an affirmative answer to Question 4 when $d = 2$ and $\sigma = \left(\begin{smallmatrix} 0 & -1 \\ -1 & 0 \end{smallmatrix}\right)$. In the following, for $? \in \{<, \leq, >, \geq\}$ and $x \in \mathbb{Z}$, the set $\{n \in \mathbb{Z} \mid n?x\}$ is denoted by $\mathbb{Z}_{?x}$. Let X be a subset of $\mathbb{Z}$ of the form $\{n \in \mathbb{Z} \mid a \leq n \leq b\}$ for some $a, b \in \mathbb{Z}$. If X contains an even number of elements (i.e., if $b - a + 1$ is even), then the subset $\{x \in X \mid x \leq \frac{b+a-1}{2}\}$ (resp. $\{x \in X \mid x \geq \frac{b+a+1}{2}\}$) of X is called the left half (resp. right half) of X. If $b - a + 1$ is divisible by 4, then the subsets $\{x \in X \mid x \leq a - 1 + \frac{b-a+1}{4}\}$, $\{x \in X \mid a - 1 + \frac{b-a+1}{4} < x \leq a - 1 + \frac{b-a+1}{2}\}$, $\{x \in X \mid a - 1 + \frac{b-a+1}{2} < x \leq a - 1 + \frac{3(b-a+1)}{4}\}$, $\{x \in X \mid a - 1 + \frac{3(b-a+1)}{4} < x\}$ of X are called the first quarter, the second quarter, the third quarter, and the fourth quarter of X respectively. The first quarter (resp. the fourth quarter) of X is also called the left quarter (resp. the right quarter) of X.

1.1 Statement of Results

In Theorems 1, 2, we prove that Questions 1, 2 admit answers in the affirmative. Theorems 1, 2 follow from Theorems 5, 7.

Theorem 1 *Let $\mathcal{T}$ denote the subset $\{1, 2, 2^2, \ldots\}$ of $\mathbb{Z}$. Then there exists an infinite subset $\mathcal{S}$ of $\mathbb{Z}_{\leq -1}$ such that $(\mathcal{S}, \mathcal{T})$ is a co-minimal pair.*

Theorem 2 *Let $\mathcal{V}$ denote the subset of $\mathbb{Z}$ defined as*

$$\mathcal{V} := \{1, 2, 2^2, 2^3, \ldots\} \cup \{-1, -2, -2^2, -2^3, \ldots\}.$$

Then there exists an infinite subset $\mathcal{U}$ of $\mathbb{Z}_{\leq -1}$ such that $(\mathcal{U}, \mathcal{V})$ is a co-minimal pair.

Using these two results, we establish Theorems 8, 9, which prove that Questions 3, 4 admit answers in the affirmative for an infinite class of automorphisms of free abelian groups. An immediate consequence of Theorem 8 is stated below.

Theorem 3 *If σ is an automorphism of $\mathbb{Z}^2$, i.e., an element of $\mathrm{GL}_2(\mathbb{Z})$ having exactly two nonzero entries, then there exists a subset A of $\mathbb{Z}^2$ such that $(A, \sigma(A))$ forms a*

[2] A subset X of $\mathbb{Z}^d$ is said to be *contained in a quadrant* if for any given $1 \leq i \leq d$, the i-th coordinate of all the points of X is either positive or negative.

co-minimal pair in $\mathbb{Z}^2$. *For any* $d \geq 1$, *the group* $\mathbb{Z}^{2d}$ *admits infinitely many automorphisms such that for each such automorphism* σ, *there exists a subset* A *of* $\mathbb{Z}^{2d}$ *such that* A *and* $\sigma(A)$ *form a co-minimal pair.*

2 A Co-minimal Pair Involving a Bounded Below Subset

In this section, we establish Theorem 1, which follows from Theorem 5. Consider the subsets $\{J_n\}_{n\geq 0}, \{K_n\}_{n\geq 0}, \{I_n\}_{n\geq 0}, \{\mathcal{I}_n\}_{n\geq 0}, S$ of $\mathbb{Z}$ defined by

$$J_n = \begin{cases} \{1\} & \text{if } n = 0, \\ \{1\} & \text{if } n = 1, \\ \{1, 2, 3, \ldots, 2^{n-2}\} \cup (2^{n-1} + J_{n-1}) & \text{if } n \geq 2, \end{cases}$$

$$K_n = \begin{cases} J_0 & \text{if } n = 0, \\ J_1 & \text{if } n = 1, \\ J_2 \setminus \{1\} & \text{if } n = 2, \\ \{2^{n-3}+1, 2^{n-3}+2, 2^{n-3}+3, \ldots, 2^{n-2}\} & \\ \quad \cup (2^{n-1} + 2^{n-2} + J_{n-2}) & \text{if } n \geq 3, \end{cases}$$

$$I_n = K_n - (1 + 2^{n+1}) \quad \text{if } n \geq 0,$$

$$\mathcal{I}_n = \begin{cases} \{-2, -1\} & \text{if } n = 0, \\ \{1, 2, 3, \ldots, 2^n\} - (1 + 2^{n+1}) & \text{if } n \geq 1, \end{cases}$$

$$S = \cup_{n\geq 0} I_n.$$

Proposition 1 *The set* $\mathcal{T}$ *is an additive complement of* S *in* $\mathbb{Z}$.

Proof For any $m \geq 3$,

$$\cup_{k=3}^{m} (I_k + 2^{k+1}) \supseteq \cup_{k=3}^{m} \{2^{k-3}, \ldots, 2^{k-2} - 1\} = \{1, 2, \ldots, 2^{m-2} - 1\}$$

holds, which implies that $S + \mathcal{T}$ contains $\mathbb{Z}_{\geq 1}$. Note that I_n is contained in $\mathcal{I}_n$ for all $n \geq 0$, and the union of the sets $\{\mathcal{I}_n\}_{n\geq 0}$ is equal to $\mathbb{Z}_{\leq -1}$ and these sets lie next to each other in the sense that $\min \mathcal{I}_n = 1 + \max \mathcal{I}_{n+1}$ for all $n \geq 0$. Thus, to prove that $S + \mathcal{T}$ contains $\mathbb{Z}$, it suffices to show that $S + \mathcal{T}$ contains $\{0\}$ and $\mathcal{I}_n$ for all $n \geq 0$, which we establish by proving the following statements.

1. $S + \mathcal{T}$ contains the left half of $\mathcal{I}_n$ for all $n \geq 1$,
2. $S + \mathcal{T}$ contains the left quarter of the right half of $\mathcal{I}_n$ for all $n \geq 3$,
3. $S + \mathcal{T}$ contains the second quarter of the right half of $\mathcal{I}_n$ for all $n \geq 3$,
4. $S + \mathcal{T}$ contains the right quarter of $\mathcal{I}_n$ for all $n \geq 4$,
5. $S + \mathcal{T}$ contains $\{0\}, \mathcal{I}_0, \mathcal{I}_1$, the right half of $\mathcal{I}_2$ and the right quarter of $\mathcal{I}_3$.

Note that the sets $I_0 + \{1, 2\}$, $I_1 + \{1, 2\}$, $I_2 + \{1, 2\}$, $I_3 + 2^2$ contain $\{-1, 0\}$, $\{-3, -2\}$, $\{-5, -4\}$, $\{-6\}$ respectively. So, $S + \mathcal{T}$ contains $\mathcal{I}_1 \cup \mathcal{I}_0 \cup \{0\}$ and the right half of $\mathcal{I}_2$. Since $I_3 + 1$ contains $\{-9\}$ and $I_4 + 2^3$ contains $\{-10\}$, the set $S + \mathcal{T}$ contains the right quarter of $\mathcal{I}_3$. This establishes the fifth statement.

Note that for any $n \geq 1$, the left half of $\mathcal{I}_n$ is contained in $I_{n+3} + 2^{n+3}$, i.e., the inclusions

$$\begin{aligned} I_{n+3} + 2^{n+3} &= K_{n+3} - (1 + 2^{n+4}) + 2^{n+3} \\ &= K_{n+3} - (1 + 2^{n+3}) \\ &\supseteq (2^{n+2} + 2^{n+1} + J_{n+1}) - (1 + 2^{n+3}) \\ &= J_{n+1} - (1 + 2^{n+1}) \\ &\supseteq \{1, 2, 3, \ldots, 2^{n-1}\} - (1 + 2^{n+1}) \end{aligned}$$

hold, thus establishing the first statement.

Note that for any $n \geq 3$, the left quarter of the right half of $\mathcal{I}_n$ is contained in $I_{n+1} + 2^n$, i.e., the inclusions

$$\begin{aligned} I_{n+1} + 2^n &= K_{n+1} - (1 + 2^{n+2}) + 2^n \\ &= K_{n+1} - (1 + 2^{n+2}) + 2^n \\ &\supseteq (2^n + 2^{n-1} + J_{n-1}) - (1 + 2^{n+2}) + 2^n \\ &= (2^{n-1} + J_{n-1}) - (1 + 2^{n+2}) + 2^{n+1} \\ &= (2^{n-1} + J_{n-1}) - (1 + 2^{n+1}) \\ &\supseteq \{2^{n-1} + 1, 2^{n-1} + 2, 2^{n-1} + 3, \ldots, 2^{n-1} + 2^{n-3}\} - (1 + 2^{n+1}) \end{aligned}$$

hold, the second quarter of the right half of $\mathcal{I}_n$ is contained in $I_n + 2^{n-1}$, i.e., the inclusions

$$\begin{aligned} I_n + 2^{n-1} &= K_n - (1 + 2^{n+1}) + 2^{n-1} \\ &\supseteq \{2^{n-3} + 1, 2^{n-3} + 2, 2^{n-3} + 3, \ldots, 2^{n-2}\} - (1 + 2^{n+1}) + 2^{n-1} \\ &= \{2^{n-1} + 2^{n-3} + 1, 2^{n-1} + 2^{n-3} + 2, 2^{n-1} + 2^{n-3} + 3, \\ &\quad \ldots, 2^{n-1} + 2^{n-2}\} - (1 + 2^{n+1}) \end{aligned}$$

hold. This establishes the second and the third statement.

For $n \geq 3$, the points in the right quarter of $\mathcal{I}_n$ that lie I_n are contained in $I_{n+1} + 2^n$, i.e., the inclusions

$$\begin{aligned} I_{n+1} + 2^n &= (2^{n-1} + J_{n-1}) - (1 + 2^{n+1}) \\ &\supseteq (2^{n-1} + 2^{n-2} + J_{n-2}) - (1 + 2^{n+1}) \end{aligned}$$

hold. Note that for any $m \geq 2$,

$$J_m + \{0, 1, 2, 2^2, \dots, 2^{m-2}\} \supseteq \{1, 2, 3, \dots, 2^m\}$$

holds. Indeed, it holds for $m = 2$, and assuming that

$$J_k + \{0, 1, 2, 2^2, \dots, 2^{k-2}\} \supseteq \{1, 2, 3, \dots, 2^k\}$$

holds for some integer $k \geq 2$, it follows that the inclusions

$$\begin{aligned}
&J_{k+1} + \{0, 1, 2, 2^2, \dots, 2^{k-2}, 2^{k-1}\} \\
&\supseteq (\{1, 2, \dots, 2^{k-1}\} \cup (2^k + J_k)) + \{0, 1, 2, 2^2, \dots, 2^{k-2}, 2^{k-1}\} \\
&\supseteq (\{1, 2, \dots, 2^{k-1}\} + \{0, 2^{k-1}\}) \cup ((2^k + J_k) + \{0, 1, 2, 2^2, \dots, 2^{k-2}\}) \\
&= \{1, 2, \dots, 2^k\} \cup (2^k + (J_k + \{0, 1, 2, 2^2, \dots, 2^{k-2}\})) \\
&\supseteq \{1, 2, \dots, 2^k\} \cup (2^k + \{1, 2, 3, \dots, 2^k\}) \\
&= \{1, 2, \dots, 2^k\} \cup \{2^k + 1, 2^k + 2, 2^k + 3, \dots, 2^{k+1}\} \\
&= \{1, 2, \dots, 2^{k+1}\}
\end{aligned}$$

hold. Consequently, for any $n \geq 4$, the points in the right quarter of $\mathcal{I}_n$ are contained in $(I_{n+1} + 2^n) \cup (I_n + \{1, 2, 2^2, \dots, 2^{n-4}\})$, i.e., the inclusions

$$\begin{aligned}
&(I_{n+1} + 2^n) \cup (I_n + \{1, 2, 2^2, \dots, 2^{n-4}\}) \\
&\supseteq ((2^{n-1} + 2^{n-2} + J_{n-2}) - (1 + 2^{n+1})) \\
&\quad \cup (((2^{n-1} + 2^{n-2} + J_{n-2}) - (1 + 2^{n+1})) + \{1, 2, 2^2, \dots, 2^{n-4}\}) \\
&= ((2^{n-1} + 2^{n-2} + J_{n-2}) - (1 + 2^{n+1})) + \{0, 1, 2, 2^2, \dots, 2^{n-4}\} \\
&= (2^{n-1} + 2^{n-2} + (J_{n-2} + \{0, 1, 2, 2^2, \dots, 2^{n-4}\})) - (1 + 2^{n+1}) \\
&\supseteq (2^{n-1} + 2^{n-2} + \{1, 2, 3, \dots, 2^{n-2}\}) - (1 + 2^{n+1}) \\
&= \{2^{n-1} + 2^{n-2} + 1, 2^{n-1} + 2^{n-2} + 2, 2^{n-1} + 2^{n-2} + 3, \dots, 2^n\} - (1 + 2^{n+1})
\end{aligned}$$

hold. This proves the fourth statement. So, the set $S + \mathcal{T}$ contains $\mathbb{Z}$. □

In the above proof, we established that for any $n \geq 3$, the left quarter of the right half of $\mathcal{I}_n$ is contained in $I_{n+1} + 2^n$, and the second quarter of the right half of $\mathcal{I}_n$ is contained in $I_n + 2^{n-1}$. In fact, the points of I_{n+1} (which are precisely the points in the left quarter of the right quarter of $\mathcal{I}_{n+1}$) that yield the left quarter of the right half of $\mathcal{I}_n$ and the points of I_n (which are precisely the points in the second quarter of the left half of $\mathcal{I}_n$) that yield the second quarter of the right half of $\mathcal{I}_n$ cannot be removed from S in a sense made precise in the following result.

Theorem 4 *Let $\mathscr{S}, \mathscr{T}$ be nonempty subsets of $S, \mathcal{T}$ respectively such that $\mathscr{S} + \mathscr{T} = \mathbb{Z}$. Then for $n \geq 3$, the set $\mathscr{T}$ contains 2^{n-1}, $\mathscr{S}$ contains the points in the second quarter of the left half of $\mathcal{I}_n$, i.e.,*

$$\mathscr{S} \supseteq \{2^{n-3}+1, 2^{n-3}+2, 2^{n-3}+3, \ldots, 2^{n-2}\} - (1+2^{n+1}) \quad \textit{for } n \geq 3, \tag{1}$$

and the points in the left quarter of the right quarter of $\mathcal{I}_{n+1}$, i.e.,

$$\mathscr{S} \supseteq \{2^n + 2^{n-1} + 1, 2^n + 2^{n-1} + 2, 2^n + 2^{n-1} + 3, \ldots, 2^n + 2^{n-1} + 2^{n-3}\} - (1 + 2^{n+2}) \quad \textit{for } n \geq 3. \tag{2}$$

Moreover, the set $\mathscr{T}$ contains $1, 2$, and the set $\mathscr{S}$ contains $-2, -4$. Consequently, $\mathcal{T}$ is a minimal complement of S.

Proof We claim that for any $n \geq 2$, no point in the right half of $\mathcal{I}_n$ lie in $\cup_{m \neq n, n+1} I_m + \mathcal{T}$, i.e.,

$$\left(\{2^{n-1}+1, 2^{n-1}+2, 2^{n-1}+3, \ldots, 2^n\} - (1+2^{n+1})\right) \cap \left(\cup_{m\neq n,n+1} I_m + \mathcal{T}\right) = \emptyset \quad \text{for } n \geq 2. \tag{3}$$

Since the inclusions

$$\begin{aligned}(\cup_{0\leq m<n} I_m) + \mathcal{T} &\subseteq (\cup_{0\leq m<n} \mathcal{I}_m) + \mathcal{T} \\ &\subseteq \{-2^n, -2^n+1, -2^n+2, -2^n+3, \ldots, -1\} + \mathcal{T}\end{aligned}$$

hold and the elements of $\mathcal{T}$ are positive, it follows that the smallest element of $(\cup_{0\leq m<n} I_m) + \mathcal{T}$ is greater than -2^n, which is greater than the largest element of $\{2^{n-1}+1, 2^{n-1}+2, 2^{n-1}+3, \ldots, 2^n\} - (1+2^{n+1})$, and hence

$$\left(\{2^{n-1}+1, 2^{n-1}+2, 2^{n-1}+3, \ldots, 2^n\} - (1+2^{n+1})\right) \cap \left(\cup_{0\leq m<n} I_m + \mathcal{T}\right) = \emptyset \quad \text{for } n \geq 1. \tag{4}$$

For any $n \geq 2$, the inclusion $I_n \subseteq \mathcal{I}_n = \{1, 2, 3, \ldots, 2^n\} - (1+2^{n+1})$ yields that $I_n \cap \mathbb{Z}_{\geq 1} = \emptyset$, $I_n \cap \mathbb{Z}_{\leq -2^{n+1}-1} = \emptyset$, and for any $m \geq n+2$, the inclusions

$$\begin{aligned}I_m + 2^k &\subseteq \mathcal{I}_m + 2^k \\ &\subseteq (\{1, 2, 3, \ldots, 2^m\} - (1+2^{m+1})) + 2^k \\ &= \{1, 2, 3, \ldots, 2^m\} + 2^k - 2^{m+1} - 1 \\ &\subseteq \mathbb{Z}_{\geq 1}\end{aligned}$$

hold for any $k \geq m+1$, and the inclusions

$$\begin{aligned}
I_m + 2^k &\subseteq \mathbb{Z}_{\le 2^m-(1+2^{m+1})} + 2^k \\
&= \mathbb{Z}_{\le 2^m-(1+2^{m+1})+2^k} \\
&= \mathbb{Z}_{\le -1-2^m+2^k} \\
&\subseteq \mathbb{Z}_{\le -1-2^m+2^{m-1}} \\
&\subseteq \mathbb{Z}_{\le -1-2^{m-1}} \\
&\subseteq \mathbb{Z}_{\le -2^{n+1}-1}
\end{aligned}$$

hold for $0 \le k < m$, and hence

$$\begin{aligned}
&\left(\{2^{n-1}+1, 2^{n-1}+2, 2^{n-1}+3, \dots, 2^n\} - (1+2^{n+1})\right) \\
&\qquad\qquad \cap \left(\cup_{m \ge n+2}(I_m + (\mathcal{T} \setminus \{2^m\}))\right) = \emptyset \quad \text{for } n \ge 2. \qquad (5)
\end{aligned}$$

For any $m \ge 3$, the inclusions

$$\begin{aligned}
I_m + 2^m &\subseteq (\{2^{m-3}+1, 2^{m-3}+2, 2^{m-3}+3, \dots, 2^{m-2}\} \\
&\qquad \cup (2^{m-1} + 2^{m-2} + J_{m-2})) - (1+2^{m+1}) + 2^m \\
&\subseteq (J_m \cup (2^{m-1} + 2^{m-2} + J_{m-2})) - (1+2^m) \\
&\subseteq (J_m \cup (2^{m-1} + J_{m-1})) - (1+2^m) \\
&= J_m - (1+2^m) \\
&= (\{1, 2, \dots, 2^{m-2}\} - (1+2^m)) \cup \dots \cup (\{1, 2, 3, 2^2\} - (1+2^4)) \\
&\qquad \cup (\{1, 2\} - (1+2^3)) \cup (\{1\} - (1+2^2)) \cup (\{1\} - (1+2)) \\
&= \left(\cup_{k=0}^{m-2}(\{1, 2, 3, \dots, 2^k\} - (1+2^{k+2}))\right) \cup (\{1\} - (1+2))
\end{aligned}$$

hold, which implies that for $n \ge 2$ and $m \ge n+1$, the inclusions

$$\begin{aligned}
&(I_m + 2^m) \cap \mathcal{I}_n \\
&= (I_m + 2^m) \cap (\{1, 2, 3, \dots, 2^n\} - (1+2^{n+1})) \\
&\subseteq \left(\left(\cup_{k=0}^{m-2}(\{1, 2, 3, \dots, 2^k\} - (1+2^{k+2}))\right) \cup (\{1\} - (1+2))\right) \cap \mathcal{I}_n \\
&= \left(\left(\cup_{k=1}^{m-1}(\{1, 2, 3, \dots, 2^{k-1}\} - (1+2^{k+1}))\right) \cup (\{1\} - (1+2))\right) \cap \mathcal{I}_n \\
&= \{1, 2, 3, \dots, 2^{n-1}\} - (1+2^{n+1}),
\end{aligned}$$

hold, which yields

$$\begin{aligned}
&\left(\{2^{n-1}+1, 2^{n-1}+2, 2^{n-1}+3, \dots, 2^n\} - (1+2^{n+1})\right) \\
&\qquad\qquad \cap \left(\cup_{m \ge n+1}(I_m + 2^m)\right) = \emptyset \quad \text{for } n \ge 2. \qquad (6)
\end{aligned}$$

Consequently, Eq. (3) follows from Eqs. (4)–(6).

We claim that for any $n \ge 3$, no point in the second quarter of the right half of $\mathcal{I}_n$ lie in $I_{n+1} + \mathcal{T}$, i.e.,

$$\big(\{2^{n-1}+2^{n-3}+1, 2^{n-1}+2^{n-3}+2, 2^{n-1}+2^{n-3}+3, \\ \cdots, 2^{n-1}+2^{n-2}\} - (1+2^{n+1})\big) \cap (I_{n+1}+\mathcal{T}) = \emptyset. \quad (7)$$

This claim follows since for $n \geq 3$, the inclusions

$$\begin{aligned} I_{n+1}+2^k &\subseteq \mathcal{I}_{n+1}+2^k \\ &\subseteq \mathbb{Z}_{\leq 2^{n+1}-(1+2^{n+2})+2^k} \\ &\subseteq \mathbb{Z}_{\leq 2^k-(1+2^{n+1})} \\ &\subseteq \mathbb{Z}_{\leq 2^{n-1}-(1+2^{n+1})} \end{aligned}$$

hold for $0 \leq k \leq n-1$, the inclusions

$$\begin{aligned} & I_{n+1}+2^n \\ &= K_{n+1}-(1+2^{n+2})+2^n \\ &\subseteq \big(\{2^{n-2}+1, 2^{n-2}+2, \ldots, 2^{n-1}\} \cup (2^n+2^{n-1}+J_{n-1})\big) - (1+2^{n+2})+2^n \\ &= \big(\{2^{n-2}+1, 2^{n-2}+2, \ldots, 2^{n-1}\} - (1+2^{n+2})+2^n\big) \\ &\quad \cup \big((2^n+2^{n-1}+J_{n-1}) - (1+2^{n+2})+2^n\big) \\ &\subseteq \mathbb{Z}_{\leq 2^{n-1}-(1+2^{n+2})+2^n} \cup \big((2^n+2^{n-1}+J_{n-1}) - (1+2^{n+2})+2^n\big) \\ &\subseteq \mathbb{Z}_{\leq 2^n-(1+2^{n+2})+2^n} \\ &\quad \cup \big((2^n+2^{n-1}+(\{1,2,\ldots,2^{n-3}\} \cup (2^{n-2}+J_{n-2}))) - (1+2^{n+2})+2^n\big) \\ &\subseteq \mathbb{Z}_{\leq -(1+2^{n+1})} \cup \mathbb{Z}_{\leq 2^n+2^{n-1}+2^{n-3}-(1+2^{n+2})+2^n} \\ &\quad \cup \mathbb{Z}_{\geq 2^n+2^{n-1}+2^{n-2}+1-(1+2^{n+2})+2^n} \\ &\subseteq \mathbb{Z}_{\leq -(1+2^{n+1})} \cup \mathbb{Z}_{\leq 2^{n-1}+2^{n-3}-(1+2^{n+1})} \cup \mathbb{Z}_{\geq 2^{n-1}+2^{n-2}+1-(1+2^{n+1})} \end{aligned}$$

hold, and the inclusions

$$\begin{aligned} I_{n+1}+2^k &\subseteq \mathcal{I}_{n+1}+2^k \\ &\subseteq \mathbb{Z}_{\geq 1-(1+2^{n+2})+2^k} \\ &= \mathbb{Z}_{\geq 2^k-2^{n+2}} \\ &\subseteq \mathbb{Z}_{\geq 0} \end{aligned}$$

hold for $k \geq n+2$.

We claim that for any $n \geq 3$, no point in the second quarter of the right half of $\mathcal{I}_n$ lie in $I_n + (\mathcal{T} \setminus \{2^{n-1}\})$, i.e.,

$$\big(\{2^{n-1}+2^{n-3}+1, 2^{n-1}+2^{n-3}+2, \ldots, 2^{n-1}+2^{n-2}\} - (1+2^{n+1})\big) \\ \cap \big(I_n + (\mathcal{T} \setminus \{2^{n-1}\})\big) = \emptyset. \quad (8)$$

This claim follows since for $n \geq 3$,

$$I_n = (\{2^{n-3}+1, 2^{n-3}+2, 2^{n-3}+3, \dots, 2^{n-2}\} - (1+2^{n+1})) \\ \cup ((2^{n-1}+2^{n-2}+J_{n-2}) - (1+2^{n+1})),$$

the inclusions

$$((2^{n-1}+2^{n-2}+J_{n-2}) - (1+2^{n+1})) + \mathscr{T} \subseteq \mathbb{Z}_{\geq 2^{n-1}+2^{n-2}+1-(1+2^{n+1})} + \mathbb{Z}_{\geq 1} \\ \subseteq \mathbb{Z}_{\geq 2^{n-1}+2^{n-2}+1-(1+2^{n+1})}$$

hold, the inclusions

$$(\{2^{n-3}+1, 2^{n-3}+2, 2^{n-3}+3, \dots, 2^{n-2}\} - (1+2^{n+1})) + 2^k \\ \subseteq \mathbb{Z}_{\geq 2^{n-3}+1-(1+2^{n+1})+2^k} \\ \subseteq \mathbb{Z}_{\geq 2^{n-3}+1-(1+2^{n+1})+2^n} \\ = \mathbb{Z}_{\geq 2^{n-3}+1-(1+2^{n+1})+2^{n-1}+2^{n-2}+2^{n-2}} \\ \subseteq \mathbb{Z}_{\geq 2^{n-1}+2^{n-2}+1-(1+2^{n+1})}$$

hold for $k \geq n$, and the inclusions

$$(\{2^{n-3}+1, 2^{n-3}+2, 2^{n-3}+3, \dots, 2^{n-2}\} - (1+2^{n+1})) + 2^k \\ \subseteq \mathbb{Z}_{\leq 2^{n-2}-(1+2^{n+1})+2^k} \\ \subseteq \mathbb{Z}_{\leq 2^{n-2}-(1+2^{n+1})+2^{n-2}} \\ = \mathbb{Z}_{\leq 2^{n-1}-(1+2^{n+1})} \\ \subseteq \mathbb{Z}_{\leq 2^{n-1}+2^{n-3}-(1+2^{n+1})}$$

hold for $k \leq n-2$.

We now show that Eq. (1) holds. From Eqs. (3), (7), it follows that

$$(\mathscr{S} \cap I_n) + \mathscr{T} \supseteq \{2^{n-1}+2^{n-3}+1, 2^{n-1}+2^{n-3}+2, \\ 2^{n-1}+2^{n-3}+3, \dots, 2^{n-1}+2^{n-2}\} - (1+2^{n+1}).$$

Using Eq. (8), it follows that $2^{n-1} \in \mathscr{T}$ and

$$(\mathscr{S} \cap I_n) + 2^{n-1} \supseteq \{2^{n-1}+2^{n-3}+1, 2^{n-1}+2^{n-3}+2, \\ 2^{n-1}+2^{n-3}+3, \dots, 2^{n-1}+2^{n-2}\} - (1+2^{n+1}).$$

Thus Eq. (1) holds.

We claim that for any $n \geq 3$, no point in the left quarter of the right half of $\mathcal{I}_n$ lie in $I_n + \mathcal{T}$, i.e.,

$$\big(\{2^{n-1}+1, 2^{n-2}+2, \dots, 2^{n-1}+2^{n-3}\} - (1+2^{n+1})\big) \\ \cap (I_n + \mathcal{T}) = \emptyset \quad \text{for } n \geq 3. \qquad (9)$$

This claim follows since for $n \geq 3$, the inclusions

$$\begin{aligned}
& I_n + 2^k \\
& \subseteq K_n - (1 + 2^{n+1}) + 2^k \\
& \subseteq \left(\{2^{n-3} + 1, 2^{n-3} + 2, \ldots, 2^{n-2}\} \cup (2^{n-1} + 2^{n-2} + J_{n-2})\right) - (1 + 2^{n+1}) + 2^k \\
& \subseteq \left(\mathbb{Z}_{\leq 2^{n-2}} \cup \mathbb{Z}_{\geq 2^{n-1}+2^{n-2}+1}\right) - (1 + 2^{n+1}) + 2^k \\
& = \mathbb{Z}_{\leq 2^{n-2}-(1+2^{n+1})+2^k} \cup \mathbb{Z}_{\geq 2^{n-1}+2^{n-2}+1-(1+2^{n+1})+2^k} \\
& \subseteq \mathbb{Z}_{\leq 2^{n-2}-(1+2^{n+1})+2^{n-2}} \cup \mathbb{Z}_{\geq 2^{n-1}+2^{n-2}+1-(1+2^{n+1})} \\
& \subseteq \mathbb{Z}_{\leq 2^{n-1}-(1+2^{n+1})} \cup \mathbb{Z}_{\geq 2^{n-1}+2^{n-2}+1-(1+2^{n+1})}
\end{aligned}$$

hold for $k \leq n - 2$, and the inclusions

$$\begin{aligned}
& I_n + 2^k \\
& = K_n - (1 + 2^{n+1}) + 2^k \\
& \subseteq \left(\{2^{n-3} + 1, 2^{n-3} + 2, \ldots, 2^{n-2}\} \cup (2^{n-1} + 2^{n-2} + J_{n-2})\right) - (1 + 2^{n+1}) + 2^k \\
& \subseteq \left(\mathbb{Z}_{\geq 2^{n-3}+1} \cup \mathbb{Z}_{\geq 2^{n-1}+2^{n-2}+1}\right) - (1 + 2^{n+1}) + 2^k \\
& \subseteq \mathbb{Z}_{\geq 2^{n-3}+1} - (1 + 2^{n+1}) + 2^k \\
& = \mathbb{Z}_{\geq 2^{n-3}+1-(1+2^{n+1})+2^k} \\
& \subseteq \mathbb{Z}_{\geq 2^{n-3}+1-(1+2^{n+1})+2^{n-1}} \\
& = \mathbb{Z}_{\geq 2^{n-1}+2^{n-3}+1-(1+2^{n+1})}
\end{aligned}$$

hold for $k \geq n - 1$.

We claim that for any $n \geq 3$, no point in the left quarter of the right half of $\mathcal{I}_n$ lie in $I_{n+1} + (\mathcal{T} \setminus \{2^n\})$, i.e.,

$$\begin{aligned}
\left(\{2^{n-1} + 1, 2^{n-1} + 2, \ldots, 2^{n-1} + 2^{n-3}\} - (1 + 2^{n+1})\right)& \\
\cap \left(I_{n+1} + (\mathcal{T} \setminus \{2^n\})\right) = \emptyset. & \qquad (10)
\end{aligned}$$

This claim follows since for $n \geq 3$, the inclusions

$$\begin{aligned}
I_{n+1} + 2^k & \subseteq \mathcal{I}_{n+1} + 2^k \\
& = (\{1, 2, 3, \ldots, 2^{n+1}\} - (1 + 2^{n+2})) + 2^k \\
& \subseteq \mathbb{Z}_{\geq 1-(1+2^{n+2})+2^k} \\
& = \mathbb{Z}_{\geq 2^k - 2^{n+2}} \\
& \subseteq \mathbb{Z}_{\geq 0}
\end{aligned}$$

hold for $k \geq n + 2$, the inclusions

$$\begin{aligned}
&I_{n+1} + 2^{n+1} \\
&= K_{n+1} - (1 + 2^{n+2}) + 2^{n+1} \\
&= \left(\{2^{n-2}+1, 2^{n-2}+2, \ldots, 2^{n-1}\} \cup (2^n + 2^{n-1} + J_{n-1})\right) - (1 + 2^{n+2}) + 2^{n+1} \\
&= \left(\{2^{n-2}+1, 2^{n-2}+2, \ldots, 2^{n-1}\} \cup (2^n + 2^{n-1} + J_{n-1})\right) - (1 + 2^{n+1}) \\
&\subseteq \left(\mathbb{Z}_{\leq 2^{n-1}} \cup \mathbb{Z}_{\geq 2^n + 2^{n-1} + 1}\right) - (1 + 2^{n+1}) \\
&\subseteq \mathbb{Z}_{\leq 2^{n-1} - (1+2^{n+1})} \cup \mathbb{Z}_{\geq 2^n + 1 - (1+2^{n+1})}
\end{aligned}$$

hold, and the inclusions

$$\begin{aligned}
I_{n+1} + 2^k &\subseteq \mathcal{I}_{n+1} + 2^k \\
&= (\{1, 2, 3, \ldots, 2^{n+1}\} - (1 + 2^{n+2})) + 2^k \\
&\subseteq \mathbb{Z}_{\leq 2^{n+1} - (1+2^{n+2}) + 2^k} \\
&= \mathbb{Z}_{\leq 2^k - (1+2^{n+1})} \\
&\subseteq \mathbb{Z}_{\leq 2^{n-1} - (1+2^{n+1})}
\end{aligned}$$

hold for $k \leq n - 1$.

We now show that Eq. (2) holds. From Eqs. (3), (9), it follows that

$$(\mathscr{S} \cap I_{n+1}) + \mathscr{T} \supseteq \{2^{n-1}+1, 2^{n-1}+2, 2^{n-1}+3, \ldots, 2^{n-1}+2^{n-3}\} - (1 + 2^{n+1}).$$

Using Eq. (10), it follows that $2^n \in \mathscr{T}$ and

$$(\mathscr{S} \cap I_{n+1}) + 2^n \supseteq \{2^{n-1}+1, 2^{n-1}+2, 2^{n-1}+3, \ldots, 2^{n-1}+2^{n-3}\} - (1 + 2^{n+1}).$$

So the inclusions

$$\begin{aligned}
\mathscr{S} &\supseteq \{2^{n-1}+1, 2^{n-1}+2, 2^{n-1}+3, \ldots, 2^{n-1}+2^{n-3}\} - (1 + 2^{n+1}) - 2^n \\
&= \{2^n + 2^{n-1} + 1, 2^n + 2^{n-1} + 2, \\
&\qquad 2^n + 2^{n-1} + 3, \ldots, 2^n + 2^{n-1} + 2^{n-3}\} - (1 + 2^{n+1}) - 2^n - 2^n \\
&= \{2^n + 2^{n-1} + 1, 2^n + 2^{n-1} + 2, \\
&\qquad 2^n + 2^{n-1} + 3, \ldots, 2^n + 2^{n-1} + 2^{n-3}\} - (1 + 2^{n+2}).
\end{aligned}$$

hold, which establishes Eq. (2).

Note that $S + (\mathcal{T} \setminus \{1\})$ does not contain -1. It follows that $a = 0$. So $1 \in \mathscr{T}$ and $-2 \in \mathscr{S}$. The integer $-5 - (1 + 2^5)$ does not lie in $S + (\mathcal{T} \setminus \{2\})$. Hence $\mathscr{T}$ contains 2. Note that $(S \setminus \{-4\}) + \mathcal{T}$ does not contain -3. It follows that $-4 \in \mathscr{S}$. Consequently, $\mathcal{T}$ is a minimal complement of S. □

By Theorem 4, it follows that $\mathcal{T}$ is a minimal complement of S. However, it turns out that S is not a minimal complement of $\mathcal{T}$. In fact,

$$S \setminus \left\{ 2^{2n} + 2^{2n-1} + 2^{2n-2} + \cdots + 2^4 + 2^3 + \frac{2^2}{2} - (1 + 2^{2n+2}) \mid n \geq 3 \right\}$$

is also an additive complement of $\mathcal{T}$. In the following result, we prove that S contains a subset $\mathcal{S}$ such that $\mathcal{S}$ is a minimal complement of $\mathcal{T}$.

Theorem 5 *Let $S = \{-x_1, -x_2, -x_3, \ldots\}$ where $x_1 < x_2 < x_3 < \cdots$. Define $S_1 = S$ and for each positive integer $i \geq 1$, define*

$$S_{i+1} := \begin{cases} S_i \setminus \{-x_i\} & \text{if } S_i \setminus \{-x_i\} \text{ is a complement to } \mathcal{T}, \\ S_i & \text{otherwise.} \end{cases}$$

Let $\mathcal{S}$ denote the subset $\cap_{i \geq 1} S_i$ of $\mathbb{Z}$. The subsets $\mathcal{S}$ and $\mathcal{T}$ of $\mathbb{Z}$ form a co-minimal pair.

Proof By Theorem 4, for $i \geq 1$, S_i contains the points in the second quarter of the left half of $\mathcal{I}_n$, i.e.,

$$S_i \supseteq \{2^{n-3} + 1, 2^{n-3} + 2, 2^{n-3} + 3, \ldots, 2^{n-2}\} - (1 + 2^{n-1}) \quad \text{for } n \geq 3, \tag{11}$$

thus $\mathcal{S}$ contains these points and hence $\mathbb{Z}_{\geq 1}$ is contained in $\mathcal{S} + \mathcal{T}$. By Theorem 4, for any $i \geq 1$, S_i contains $-2, -4$. So $\{-3, -2, -1, 0\} \in \mathcal{S} + \mathcal{T}$. By Theorem 4, for any $i \geq 1$, S_i contains the points in the left quarter of the right quarter of $\mathcal{I}_n$, i.e.,

$$\begin{aligned} S_i \supseteq \{&2^{n-1} + 2^{n-2} + 1, 2^{n-1} + 2^{n-2} + 2, 2^{n-1} + 2^{n-2} + 3, \\ &\ldots, 2^{n-1} + 2^{n-2} + 2^{n-4}\} - (1 + 2^{n+1}) \quad \text{for } n \geq 4, \end{aligned} \tag{12}$$

thus $\mathcal{S}$ contains these points and hence the left half of $\mathcal{I}_n$ is contained in $\mathcal{S} + \mathcal{T}$ for any $n \geq 1$. It remains to show that the second half of $\mathcal{I}_n$ is also contained in $\mathcal{S} + \mathcal{T}$ for any $n \geq 2$.

For any element $y \in \mathbb{Z}_{\leq -1}$ that lie in the right half of some $\mathcal{I}_n$ for some $n \geq 2$ and for any $i \geq 1$, there exist elements $s_{y,i} \in S_i$, $t_{y,i} \in \mathcal{T}$ such that $y = s_{y,i} + t_{y,i}$. By Eq. (3), it follows that for some element $s_y \in S$, the equality $s_y = s_{y,i}$ holds for infinitely many i and for such integers i, we have $t_y = t_{y,i}$ where $t_y := y - s_y \in \mathcal{T}$. Thus $y - t_y = s_{y,i}$ holds for infinitely many i. Hence, for each integer $i \geq 1$, there exists an integer $m_i \geq i$ such that $y - t_y = s_{y,m_i}$, which yields $y \in t_y + S_{m_i} \subseteq t_y + S_i$. Thus y lies in $t_y + \mathcal{S}$. This proves that the second half of $\mathcal{I}_n$ is also contained in $\mathcal{S} + \mathcal{T}$ for any $n \geq 2$. Hence $\mathcal{S}$ is an additive complement of $\mathcal{T}$. It follows that $\mathcal{S}$ is a minimal complement to $\mathcal{T}$.

By Theorem 4, it follows that $\mathcal{T}$ is a minimal complement to $\mathcal{S}$. This proves that $(\mathcal{S}, \mathcal{T})$ is a co-minimal pair. □

3 A Co-minimal Pair Involving an Infinite Symmetric Subset

In this section, we establish Theorem 2, which follows from Theorem 7. Consider the subsets $\{\mathcal{I}_n\}_{n\geq 0}$, $\{U_n\}_{n\geq 0}$ of $\mathbb{Z}$ defined by

$$\mathcal{I}_n = \begin{cases} \{-2, -1\} & \text{if } n = 0, \\ \{1, 2, 3, \ldots, 2^n\} - (1 + 2^{n+1}) & \text{for } n \geq 1, \end{cases}$$

$$U_n = \begin{cases} \{-2, -1\} & \text{if } n = 0, \\ \emptyset & \text{if } n = 1, 2, \\ \{6\} - (1 + 2^4) & \text{if } n = 3, \\ \{3, 4, 14\} - (1 + 2^5) & \text{if } n = 4, \\ ((2^{n-3} + \{1, 2, 3, \ldots, 2^{n-3}\}) & \\ \quad \cup (2^{n-1} + 2^{n-2} + \{1, 2, 3, \ldots, 2^{n-4}\})) - (1 + 2^{n+1}) & \text{if } n \geq 5. \end{cases}$$

Denote the union $\cup_{n\geq 0} U_n$ by U.

Proposition 2 *The set $\mathcal{V}$ is an additive complement of U in $\mathbb{Z}$.*

Proof Since the inclusion

$$U_0 + \{2\} \supseteq \{0, 1\}$$

holds and the inclusions

$$\begin{aligned} \cup_{k=4}^{m}(U_k + 2^{k+1}) &\supseteq \cup_{k=4}^{m}(2^{k-3} + \{0, 1, 2, \ldots, 2^{k-3} - 1\}) \\ &\supseteq \{2, 3, 4, \ldots, 2^{m-2} - 1\} \end{aligned}$$

hold for any $m \geq 4$, it follows that $U + \mathcal{V}$ contains $\mathbb{Z}_{\geq 0}$.

Note that

$$U_n \subseteq \mathcal{I}_n \quad \text{for all } n \geq 0,$$

and

$$\bigcup_{n\geq 0} \mathcal{I}_n = \mathbb{Z}_{\leq -1},$$

and the sets $\mathcal{I}_n$ lie next to each other in the sense that

$$\min \mathcal{I}_n = 1 + \max \mathcal{I}_{n+1} \quad \text{for all } n \geq 0.$$

Thus, to prove that $U + \mathcal{V}$ is equal to $\mathbb{Z}$, it remains to show that $U + \mathcal{V}$ contains $\mathcal{I}_n$ for all $n \geq 0$.

The left half of $\mathcal{I}_n$ is contained in $U_{n+3} + 2^{n+3}$ for any $n \geq 1$, i.e., the inclusions

$$\begin{aligned} U_{n+3}+2^{n+3} &\supseteq (2^{n+2}+2^{n+1}+\{1,2,3,\ldots,2^{n-1}\})-(1+2^{n+4})+2^{n+3} \\ &= (\{1,2,3,\ldots,2^{n-1}\}-(1+2^{n+1}))+(1+2^{n+1}) \\ &\quad +2^{n+2}+2^{n+1}-(1+2^{n+4})+2^{n+3} \\ &= \{1,2,3,\ldots,2^{n-1}\}-(1+2^{n+1}) \end{aligned}$$

hold for any $n \geq 1$. The third quarter of $\mathcal{I}_n$ is contained in the set $(U_{n+1}+2^n) \cup (U_n+2^{n-1})$ for $n \geq 4$, i.e., the inclusions

$$\begin{aligned} &(U_{n+1}+2^n) \cup (U_n+2^{n-1}) \\ &\supseteq \big((2^n+2^{n-1}+\{1,2,3,\ldots,2^{n-3}\})-(1+2^{n+2})+2^n\big) \\ &\quad \cup \big((2^{n-3}+\{1,2,3,\ldots,2^{n-3}\})-(1+2^{n+1})+2^{n-1}\big) \\ &\supseteq ((2^{n-1}+\{1,2,3,\ldots,2^{n-3}\})-(1+2^{n+1})) \\ &\quad \cup ((2^{n-1}+2^{n-3}+\{1,2,3,\ldots,2^{n-3}\})-(1+2^{n+1})) \\ &\supseteq (2^{n-1}+\{1,2,3,\ldots,2^{n-2}\})-(1+2^{n+1}) \end{aligned}$$

hold for $n \geq 4$. The left half of the right quarter of $\mathcal{I}_4$ is contained in the set $(U_3+(-2^3)) \cup (U_4+(-1))$, and the left half of the right quarter of $\mathcal{I}_n$ is contained in the set $(U_n+(-2^{n-5})) \cup (U_n+2^{n-5}) \cup (U_n+2^{n-4})$ for $n \geq 5$, i.e., the inclusions

$$\begin{aligned} &(U_n+(-2^{n-5})) \cup (U_n+2^{n-5}) \cup (U_n+2^{n-4}) \\ &\supseteq ((2^{n-1}+2^{n-2}+\{1,2,3,\ldots,2^{n-4}\})-(1+2^{n+1}))+\{-2^{n-5},2^{n-5},2^{n-4}\} \\ &\supseteq \big(((2^{n-1}+2^{n-2}+\{2^{n-5}+1,2^{n-5}+2, \\ &\quad 2^{n-5}+3,\ldots,2^{n-4}\})-(1+2^{n+1}))-2^{n-5}\big) \\ &\quad \cup \big(((2^{n-1}+2^{n-2}+\{1,2,3,\ldots,2^{n-5}\})-(1+2^{n+1}))+2^{n-5}\big) \\ &\quad \cup \big(((2^{n-1}+2^{n-2}+\{1,2,3,\ldots,2^{n-4}\})-(1+2^{n+1}))+2^{n-4}\big) \\ &\supseteq ((2^{n-1}+2^{n-2}+\{1,2,3,\ldots,2^{n-4}\})-(1+2^{n+1})) \\ &\quad \cup ((2^{n-1}+2^{n-2}+2^{n-4}+\{1,2,3,\ldots,2^{n-4}\})-(1+2^{n+1})) \\ &\supseteq (2^{n-1}+2^{n-2}+\{1,2,3,\ldots,2^{n-3}\})-(1+2^{n+1}), \end{aligned}$$

hold for $n \geq 5$. The right half of the right quarter of $\mathcal{I}_4$ is contained in the set $U_0+(-2^4)$, and the right half of the right quarter of $\mathcal{I}_n$ is contained in the set $(U_n+2^{n-3}) \cup (U_{n-1}-2^{n-3})$ for $n \geq 5$, i.e., the inclusions

$$\begin{aligned} &(U_n+2^{n-3}) \cup (U_{n-1}-2^{n-3}) \\ &\supseteq \big(((2^{n-1}+2^{n-2}+\{1,2,3,\ldots,2^{n-4}\})-(1+2^{n+1}))+2^{n-3}\big) \\ &\quad \cup \big(((2^{n-4}+\{1,2,3,\ldots,2^{n-4}\})-(1+2^n))-2^{n-3}\big) \\ &= \big(((2^{n-1}+2^{n-2}+2^{n-3}+\{1,2,3,\ldots,2^{n-4}\})-(1-2^{n+1})\big) \end{aligned}$$

$$
\begin{aligned}
&\cup\big(((2^{n-1}+2^{n-2}+2^{n-3}+2^{n-4}+\{1,2,3,\ldots,2^{n-4}\})-(1+2^n))\\
&\quad -2^{n-3}-2^{n-1}-2^{n-2}-2^{n-3}\big)\\
&=\big(((2^{n-1}+2^{n-2}+2^{n-3}+\{1,2,3,\ldots,2^{n-4}\})-(1+2^{n+1})\big)\\
&\quad\cup\big((2^{n-1}+2^{n-2}+2^{n-3}+2^{n-4}+\{1,2,3,\ldots,2^{n-4}\})-(1+2^{n+1})\big)\\
&=(2^{n-1}+2^{n-2}+2^{n-3}+\{1,2,3,\ldots,2^{n-3}\})-(1+2^{n+1})
\end{aligned}
$$

hold for $n \geq 5$. This proves that $\mathcal{I}_n$ is contained in $U + \mathcal{V}$ for $n \geq 4$, and hence $U + \mathcal{V}$ contains $\mathbb{Z}_{\leq 2^4-(1+2^5)} = \mathbb{Z}_{\leq -1-2^4}$. Also note that the left half of $\mathcal{I}_1, \mathcal{I}_2, \mathcal{I}_3$ is contained in $U + \mathcal{V}$. From the inclusions

$$
\begin{aligned}
&(U_4 \cup U_3 \cup U_0) + \{-4,-2,-1,1,2,4,8\}\\
&\supseteq ((U_4+\{8\}) \cup (U_3+\{-1,1,2\}))\\
&\qquad \cup (U_0+\{-4\}) \cup (U_0+\{-2\}) \cup (U_0+\{-1,1\})\\
&\supseteq (\{5,6,7,8\}-(1+2^4)) \cup (\{3,4\}-(1+2^3)) \cup (\{1,2\}-(1+2^2)) \cup \{-2,-1\},
\end{aligned}
$$

it follows that the subsets $\mathcal{I}_0, \mathcal{I}_1$ and the right half of $\mathcal{I}_2, \mathcal{I}_3$ is contained in $U + \mathcal{V}$. Hence $\mathcal{V}$ is an additive complement of U. □

Theorem 6 *The set $\mathcal{V}$ is a minimal complement of U in $\mathbb{Z}$.*

Proof By Proposition 2, $\mathcal{V}$ is an additive complement of U. To prove that $\mathcal{V}$ is a minimal complement of U in $\mathbb{Z}$, it suffices to prove that if $\mathscr{V}$ is nonempty subset of $\mathcal{V}$ satisfying $U + \mathscr{V} = \mathbb{Z}$, then $\mathscr{V} = \mathcal{V}$, which follows from the nine claims below. Indeed, the third claim below implies that $2^n \in \mathscr{V}$ for all $n \geq 2$, the fourth (resp. fifth, sixth, seventh, eighth) claim implies that $\mathscr{V}$ contains 2 (resp. $1, -1, -2, -4$), and the ninth claim implies that $\mathscr{V}$ contains -2^n for all $n \geq 3$. Thus, from the following claims, it follows that $\mathscr{V} = \mathcal{V}$ and hence $\mathcal{V}$ is a minimal complement to U.

We claim the following.

Claim 1 *For any $n \geq 2$, no point in the right half of $\mathcal{I}_n$ lie in $\cup_{m\geq n+2} U_m + \mathcal{V}$, i.e.,*

$$
\begin{aligned}
&\big(\{2^{n-1}+1, 2^{n-1}+2, 2^{n-1}+3, \ldots, 2^n\} - (1+2^{n+1})\big)\\
&\qquad\qquad \cap \big(\cup_{m\geq n+2} U_m + \mathcal{V}\big) = \emptyset \quad \textit{for } n \geq 2. \qquad (13)
\end{aligned}
$$

Claim 2 *For any $n \geq 3$, no point in the right half of $\mathcal{I}_n$ lie in $U_{n+1} + (\mathcal{V} \setminus \{2^n\})$, i.e.,*

$$
\begin{aligned}
&\big(\{2^{n-1}+1, 2^{n-1}+2, 2^{n-1}+3, \ldots, 2^n\} - (1+2^{n+1})\big)\\
&\qquad\qquad \cap \big(U_{n+1} + (\mathcal{V} \setminus \{2^n\})\big) = \emptyset \quad \textit{for } n \geq 3. \qquad (14)
\end{aligned}
$$

Claim 3 *For any $n \geq 5$, the largest element of the third quarter of the right quarter of $\mathcal{I}_n$ does not lie in $U + (\mathcal{V} \setminus \{2^{n-3}\})$, more precisely,*

$$\begin{aligned} &2^{n-1} + 2^{n-2} + 2^{n-3} + 2^{n-4} - (1 + 2^{n+1}) \\ &\notin (U_n + (\mathcal{V} \setminus \{2^{n-3}\})) \cup (\cup_{m \geq 0, m \neq n} U_m + \mathcal{V}) \quad \textit{for } n \geq 5. \end{aligned} \tag{15}$$

Claim 4

$$1 \notin ((U \setminus \{-1\}) + \mathcal{V}) \cup (-1 + (\mathcal{V} \setminus \{2\})). \tag{16}$$

Claim 5

$$26 - (1 + 2^6) \notin ((U \setminus \{25 - (1 + 2^6)\}) + \mathcal{V}) \cup (25 - (1 + 2^6) + (\mathcal{V} \setminus \{1\})). \tag{17}$$

Claim 6

$$25 - (1 + 2^6) \notin ((U \setminus \{26 - (1 + 2^6)\}) + \mathcal{V}) \cup (26 - (1 + 2^6) + (\mathcal{V} \setminus \{-1\})). \tag{18}$$

Claim 7

$$-4 \notin ((U \setminus \{-2\}) + \mathcal{V}) \cup (-2 + (\mathcal{V} \setminus \{-2\})). \tag{19}$$

Claim 8

$$-6 \notin ((U \setminus \{-2\}) + \mathcal{V}) \cup (-2 + (\mathcal{V} \setminus \{-4\})). \tag{20}$$

Claim 9 *For any $n \geq 6$, the third largest element of $\mathcal{I}_n$, i.e., the integer $2^n - 2 - (1 + 2^{n+1})$ does not belong to $U + (\mathcal{V} \setminus \{-2^{n-3}\})$, i.e.,*

$$2^n - 1 - (1 + 2^{n+1}) \notin U + (\mathcal{V} \setminus \{-2^{n-3}\}). \tag{21}$$

First, we establish Eq. (13). For any $n \geq 2$, the inclusion $U_n \subseteq \mathcal{I}_n = \{1, 2, 3, \ldots, 2^n\} - (1 + 2^{n+1})$ yields that

$$U_n \cap \mathbb{Z}_{\geq 0} = \emptyset, \, U_n \cap \mathbb{Z}_{\leq -2^{n+1}-1} = \emptyset,$$

and for any $m \geq n + 2$ with $n \geq 2$, the inclusions

$$\begin{aligned} U_m + v &\subseteq \mathcal{I}_m + v \\ &\subseteq (\{1, 2, 3, \ldots, 2^m\} - (1 + 2^{m+1})) + v \\ &= \{1, 2, 3, \ldots, 2^m\} + v - 2^{m+1} - 1 \\ &\subseteq \mathbb{Z}_{\geq 0} \end{aligned}$$

hold for any $v \in \mathcal{V}$ with $v \geq 2^{m+1}$, and the inclusions

$$\begin{aligned} U_m + v &\subseteq \mathbb{Z}_{\leq 2^m-(1+2^{m+1})} + v \\ &= \mathbb{Z}_{\leq 2^m-(1+2^{m+1})+v} \\ &= \mathbb{Z}_{\leq -1-2^m+v} \\ &\subseteq \mathbb{Z}_{\leq -1-2^m+2^{m-1}} \\ &\subseteq \mathbb{Z}_{\leq -1-2^{m-1}} \\ &\subseteq \mathbb{Z}_{\leq -(1+2^{n+1})} \end{aligned}$$

hold for any $v \in \mathcal{V}$ with $v \leq 2^{m-1}$, and hence

$$\begin{aligned} &\big(\{2^{n-1}+1, 2^{n-1}+2, 2^{n-1}+3, \ldots, 2^n\} - (1+2^{n+1})\big) \\ &\qquad \cap \big(\cup_{m\geq n+2}(U_m + (\mathcal{V} \setminus \{2^m\}))\big) = \emptyset \quad \text{for } n \geq 2. \qquad (22) \end{aligned}$$

For any $m \geq 4$, the inclusions

$$\begin{aligned} U_m + 2^m &\subseteq ((2^{m-3} + \{1, 2, 3, \ldots, 2^{m-3}\}) \\ &\quad \cup (2^{m-1} + 2^{m-2} + \{1, 2, 3, \ldots, 2^{m-4}\})) - (1 + 2^{m+1}) + 2^m \\ &= ((2^{m-3} + \{1, 2, 3, \ldots, 2^{m-3}\}) \\ &\quad \cup (2^{m-1} + 2^{m-2} + \{1, 2, 3, \ldots, 2^{m-4}\})) - (1 + 2^m) \\ &\subseteq (\{1, 2, 3, \ldots, 2^{m-2}\} - (1 + 2^m)) \\ &\quad \cup (2^{m-1} + 2^{m-2} + \{1, 2, 3, \ldots, 2^{m-4}\} - (1 + 2^m)) \\ &= (\{1, 2, 3, \ldots, 2^{m-2}\} - (1 + 2^m)) \cup (\{1, 2, 3, \ldots, 2^{m-4}\} - (1 + 2^{m-2})) \end{aligned}$$

hold, which implies that for $n \geq 3$ and $m \geq n + 1$, the set $U_m + 2^m$ is contained in the union of the left half of $\mathcal{I}_{m-1}$ and the left half of $\mathcal{I}_{m-3}$, and hence

$$\begin{aligned} &\big(\{2^{n-1}+1, 2^{n-1}+2, 2^{n-1}+3, \ldots, 2^n\} - (1+2^{n+1})\big) \\ &\qquad \cap \big(\cup_{m\geq n+1}(U_m + 2^m)\big) = \emptyset \quad \text{for } n \geq 3. \qquad (23) \end{aligned}$$

From Eqs. (22), (23), Eq. (13) follows.

Now, we prove that Eq. (14) holds. For any $m \geq 4$, the inclusions

$$\begin{aligned} U_m + v &\subseteq \mathbb{Z}_{\geq 1-(1+2^{m+1})} + v \\ &= \mathbb{Z}_{\geq 1-(1+2^{m+1})+v} \\ &\subseteq \mathbb{Z}_{\geq 1-(1+2^{m+1})+2^{m+1}} \\ &= \mathbb{Z}_{\geq 0} \end{aligned}$$

hold for $v \in \mathcal{V}$ with $v \geq 2^{m+1}$, and the inclusions

$$\begin{aligned} U_m + v &\subseteq \mathbb{Z}_{\leq 2^m-(1+2^{m+1})} + v \\ &= \mathbb{Z}_{\leq 2^m-(1+2^{m+1})+v} \\ &\subseteq \mathbb{Z}_{\leq 2^m-(1+2^{m+1})+2^{m-2}} \\ &= \mathbb{Z}_{\leq 2^{m-2}-(1+2^m)} \end{aligned}$$

hold for $v \in \mathcal{V}$ with $v \leq 2^{m-2}$, and thus Eq. (23) yields Eq. (14).

Now, we show that Eq. (15) holds. Note that the inclusions

$$\begin{aligned} \mathcal{I}_n \cap (\cup_{0\leq m<n} U_m + v) &\subseteq \mathcal{I}_n \cap (\cup_{0\leq m<n} \mathcal{I}_m + v) \\ &\subseteq \mathcal{I}_n \cap (\mathbb{Z}_{\leq -1} + v) \\ &\subseteq \mathbb{Z}_{\geq 1-(1+2^{n+1})} \cap (\mathbb{Z}_{\leq -(1+2^{n+1})}) \\ &\subseteq \emptyset \end{aligned}$$

hold for any $n \geq 5$ and for any $v \in \mathcal{V}$ with $v \leq -2^{n+1}$. The inclusions

$$\begin{aligned} &(\{2^{n-1} + 2^{n-2} + 2^{n-3} + k \mid 1 \leq k \leq 2^{n-4}\} - (1 + 2^{n+1})) \cap (\cup_{0\leq m<n} U_m - 2^n) \\ &\subseteq (\{2^{n-1} + 2^{n-2} + 2^{n-3} + k \mid 1 \leq k \leq 2^{n-4}\} - (1 + 2^{n+1})) \\ &\quad \cap (((U_{n-1} \cup U_{n-2} \cup U_{n-3}) \cup U_{n-4} \cup (\cup_{0\leq m\leq n-5} U_m)) - 2^n) \\ &\subseteq (\{2^{n-1} + 2^{n-2} + 2^{n-3} + k \mid 1 \leq k \leq 2^{n-4}\} - (1 + 2^{n+1})) \\ &\quad \cap (((\mathcal{I}_{n-1} \cup \mathcal{I}_{n-2} \cup \mathcal{I}_{n-3}) \cup U_{n-4} \cup (\cup_{0\leq m\leq n-5} \mathcal{I}_m)) - 2^n) \\ &\subseteq (\{2^{n-1} + 2^{n-2} + 2^{n-3} + k \mid 1 \leq k \leq 2^{n-4}\} - (1 + 2^{n+1})) \\ &\quad \cap (((\mathbb{Z}_{\leq 2^{n-4}-1-(1+2^{n-3})}) \cup (\mathbb{Z}_{\geq 1-(1+2^{n-4})})) - 2^n) \\ &\subseteq (\{2^{n-1} + 2^{n-2} + 2^{n-3} + k \mid 1 \leq k \leq 2^{n-4}\} - (1 + 2^{n+1})) \\ &\quad \cap ((\mathbb{Z}_{\leq 2^{n-4}-1-(1+2^{n-3})} - 2^n) \cup ((\mathbb{Z}_{\geq 1-(1+2^{n-4})}) - 2^n)) \\ &\subseteq (\{2^{n-1} + 2^{n-2} + 2^{n-3} + k \mid 1 \leq k \leq 2^{n-4}\} - (1 + 2^{n+1})) \\ &\quad \cap (\mathbb{Z}_{\leq 2^{n-4}-1-(1+2^{n-3})-2^n} \cup \mathbb{Z}_{\geq 1-(1+2^{n-4})-2^n}) \\ &\subseteq (\{2^{n-1} + 2^{n-2} + 2^{n-3} + k \mid 1 \leq k \leq 2^{n-4}\} - (1 + 2^{n+1})) \\ &\quad \cap (\mathbb{Z}_{\leq 2^{n-1}+2^{n-2}+2^{n-3}+2^{n-4}-1-(1+2^{n+1})} \cup \mathbb{Z}_{\geq 2^{n-1}+2^{n-2}+2^{n-3}+2^{n-4}+1-(1+2^{n+1})}) \\ &\subseteq (\{2^{n-1} + 2^{n-2} + 2^{n-3} + k \mid 1 \leq k \leq 2^{n-4}\} - (1 + 2^{n+1})) \\ &\quad \cap \mathbb{Z}_{\leq 2^{n-1}+2^{n-2}+2^{n-3}+2^{n-4}-1-(1+2^{n+1})} \\ &\subseteq \mathbb{Z}_{\leq 2^{n-1}+2^{n-2}+2^{n-3}+2^{n-4}-1-(1+2^{n+1})} \end{aligned}$$

hold for $n \geq 5$. The inclusions

$$
\begin{aligned}
&\mathcal{I}_n \cap (\cup_{0\le m<n} U_m - 2^{n-1}) \\
&\subseteq (U_{n-1} - 2^{n-1}) \cup (\mathcal{I}_n \cap (\cup_{0\le m\le n-2} U_m - 2^{n-1})) \\
&\subseteq \Big((((2^{n-4} + \{1, 2, 3, \ldots, 2^{n-4}\}) \\
&\quad \cup (2^{n-2} + 2^{n-3} + \{1, 2, 3, \ldots, 2^{n-5}\})) - (1 + 2^n)) - 2^{n-1} \Big) \\
&\quad \cup (\mathcal{I}_n \cap (\cup_{0\le m\le n-2} \mathcal{I}_m - 2^{n-1})) \\
&\subseteq \Big(((2^{n-4} + \{1, 2, 3, \ldots, 2^{n-4}\}) - (1 + 2^n + 2^{n-1})) \\
&\quad \cup ((2^{n-2} + 2^{n-3} + \{1, 2, 3, \ldots, 2^{n-5}\}) - (1 + 2^n + 2^{n-1}))) \Big) \\
&\quad \cup (\mathcal{I}_n \cap (\mathbb{Z}_{\ge 1-(1+2^{n-1})} - 2^{n-1})) \\
&= \Big(((2^{n-1} + 2^{n-4} + \{1, 2, 3, \ldots, 2^{n-4}\}) - (1 + 2^{n+1})) \\
&\quad \cup ((2^{n-1} + 2^{n-2} + 2^{n-3} + \{1, 2, 3, \ldots, 2^{n-5}\}) - (1 + 2^{n+1}))) \Big) \\
&\quad \cup (\mathcal{I}_n \cap (\mathbb{Z}_{\ge -2^n})) \\
&= \Big(\mathbb{Z}_{\le 2^{n-1}+2^{n-4}+2^{n-4}-(1+2^{n+1})} \cup \mathbb{Z}_{\le 2^{n-1}+2^{n-2}+2^{n-3}+2^{n-5}-(1+2^{n+1})} \Big) \\
&\quad \cup (\mathcal{I}_n \cap (\mathbb{Z}_{\ge 2^n+1-(1+2^{n+1})})) \\
&= \mathbb{Z}_{\le 2^{n-1}+2^{n-3}-(1+2^{n+1})} \cup \mathbb{Z}_{\le 2^{n-1}+2^{n-2}+2^{n-3}+2^{n-5}-(1+2^{n+1})}
\end{aligned}
$$

hold for any $n \ge 5$. The inclusions

$$
\begin{aligned}
&\mathcal{I}_n \cap (\cup_{0\le m<n} U_m - 2^{n-2}) \\
&\subseteq (U_{n-1} - 2^{n-2}) \cup (\mathcal{I}_n \cap (\cup_{0\le m\le n-2} U_m - 2^{n-2})) \\
&\subseteq \Big((((2^{n-4} + \{1, 2, 3, \ldots, 2^{n-4}\}) \\
&\quad \cup (2^{n-2} + 2^{n-3} + \{1, 2, 3, \ldots, 2^{n-5} + 1\})) - (1 + 2^n)) - 2^{n-2} \Big) \\
&\quad \cup (\mathcal{I}_n \cap (\cup_{0\le m\le n-2} \mathcal{I}_m - 2^{n-2})) \\
&\subseteq \Big(((2^{n-4} + \{1, 2, 3, \ldots, 2^{n-4}\}) - (1 + 2^n + 2^{n-2})) \\
&\quad \cup ((2^{n-2} + 2^{n-3} + \{1, 2, 3, \ldots, 2^{n-5} + 1\}) - (1 + 2^n + 2^{n-2}))) \Big) \\
&\quad \cup (\mathcal{I}_n \cap (\mathbb{Z}_{\ge 1-(1+2^{n-1})} - 2^{n-2})) \\
&= \Big(((2^{n-1} + 2^{n-2} + 2^{n-4} + \{1, 2, 3, \ldots, 2^{n-4}\}) - (1 + 2^{n+1})) \\
&\quad \cup ((2^{n-3} + \{1, 2, 3, \ldots, 2^{n-5} + 1\}) - (1 + 2^n))) \Big) \cup (\mathcal{I}_n \cap (\mathbb{Z}_{\ge -2^{n-1}-2^{n-2}})) \\
&= \Big(((2^{n-1} + 2^{n-2} + 2^{n-4} + \{1, 2, 3, \ldots, 2^{n-4}\}) - (1 + 2^{n+1})) \\
&\quad \cup ((2^{n-1} + 2^{n-2} + 2^{n-3} + 2^{n-2} + \{1, 2, 3, \ldots, 2^{n-5} + 1\}) - (1 + 2^{n+1}))) \Big) \\
&\quad \cup (\mathcal{I}_n \cap (\mathbb{Z}_{\ge 2^n+(1+2^{n-2})-(1+2^{n+1})})) \\
&= \Big(((2^{n-1} + 2^{n-2} + 2^{n-4} + \{1, 2, 3, \ldots, 2^{n-4}\}) - (1 + 2^{n+1})) \\
&\quad \cup ((2^{n-1} + 2^{n-2} + 2^{n-3} + 2^{n-2} + \{1, 2, 3, \ldots, 2^{n-5} + 1\}) - (1 + 2^{n+1}))) \Big)
\end{aligned}
$$

hold for any $n \geq 5$. The inclusions

$$\begin{aligned}\cup_{0\leq m<n} U_m + v &\subseteq \mathbb{Z}_{\geq 2^{n-4}+1-(1+2^n)} + v \\ &\subseteq \mathbb{Z}_{\geq 2^{n-4}+1-(1+2^n)-2^{n-3}} \\ &\subseteq \mathbb{Z}_{\geq 2^{n-1}+2^{n-2}+2^{n-3}+2^{n-4}+1-(1+2^{n+1})}\end{aligned}$$

hold for any $n \geq 5$ and for any $v \in \mathcal{V}$ with $v \geq -2^{n-3}$. These inclusions prove that for $n \geq 5$, the largest element of the third quarter of the right quarter of $\mathcal{I}_n$ does not belong to $\cup_{0\leq m<n} U_m + \mathcal{V}$, i.e.,

$$2^{n-1} + 2^{n-2} + 2^{n-3} + 2^{n-4} - (1 + 2^{n+1}) \notin \cup_{0\leq m<n} U_m + \mathcal{V}.$$

Combining the above with Eqs. (13), (14), we obtain

$$2^{n-1} + 2^{n-2} + 2^{n-3} + 2^{n-4} - (1 + 2^{n+1}) \notin \left(\cup_{m\geq 0, m\neq n, n+1} U_m + \mathcal{V}\right) \cup (U_{n+1} + (\mathcal{V} \setminus \{2^n\}))$$

for $n \geq 5$. Since the inclusions

$$\begin{aligned}U_{n+1} + 2^n &\subseteq \mathbb{Z}_{\leq 2^n+2^{n-1}+2^{n-3}-(1+2^{n+2})} + 2^n \\ &\subseteq \mathbb{Z}_{\leq 2^n+2^{n-1}+2^{n-3}-(1+2^{n+2})+2^n} \\ &\subseteq \mathbb{Z}_{\leq 2^{n-1}+2^{n-3}-(1+2^{n+1})}\end{aligned}$$

hold for $n \geq 5$, it follows that

$$2^{n-1} + 2^{n-2} + 2^{n-3} + 2^{n-4} - (1 + 2^{n+1}) \notin \left(\cup_{m\geq 0, m\neq n} U_m + \mathcal{V}\right) \quad \text{for } n \geq 5.$$

Note that the inclusions

$$\begin{aligned}U_n + v &\subseteq \mathbb{Z}_{\geq 1-(1+2^{n+1})} + v \\ &\subseteq \mathbb{Z}_{\geq 1-(1+2^{n+1})+v} \\ &\subseteq \mathbb{Z}_{\geq 1-(1+2^{n+1})+2^n} \\ &\subseteq \mathbb{Z}_{\geq 1-(1+2^n)}\end{aligned}$$

hold for any $v \in \mathcal{V}$ with $v \geq 2^n$, the inclusions

$$\begin{aligned}U_n + 2^{n-1} &\subseteq (\mathbb{Z}_{\leq 2^{n-3}+2^{n-3}-(1+2^{n+1})} \cup \mathbb{Z}_{\geq 2^{n-1}+2^{n-2}+1-(1+2^{n+1})}) + 2^{n-1} \\ &\subseteq (\mathbb{Z}_{\leq 2^{n-2}-(1+2^{n+1})} \cup \mathbb{Z}_{\geq 2^{n-1}+2^{n-2}+1-(1+2^{n+1})}) + 2^{n-1} \\ &\subseteq \mathbb{Z}_{\leq 2^{n-2}-(1+2^{n+1})+2^{n-1}} \cup \mathbb{Z}_{\geq 2^{n-1}+2^{n-2}+1-(1+2^{n+1})+2^{n-1}} \\ &\subseteq \mathbb{Z}_{\leq 2^{n-1}+2^{n-2}-(1+2^{n+1})} \cup \mathbb{Z}_{\geq 1+2^{n-2}-(1+2^n)}\end{aligned}$$

hold, the inclusions

$$\begin{aligned}U_n + 2^{n-2} &\subseteq (\mathbb{Z}_{\leq 2^{n-3}+2^{n-3}-(1+2^{n+1})} \cup \mathbb{Z}_{\geq 2^{n-1}+2^{n-2}+1-(1+2^{n+1})}) + 2^{n-2}\\ &\subseteq (\mathbb{Z}_{\leq 2^{n-2}-(1+2^{n+1})} \cup \mathbb{Z}_{\geq 2^{n-1}+2^{n-2}+1-(1+2^{n+1})}) + 2^{n-2}\\ &\subseteq \mathbb{Z}_{\leq 2^{n-2}-(1+2^{n+1})+2^{n-2}} \cup \mathbb{Z}_{\geq 2^{n-1}+2^{n-2}+1-(1+2^{n+1})+2^{n-2}}\\ &\subseteq \mathbb{Z}_{\leq 2^{n-1}-(1+2^{n+1})} \cup \mathbb{Z}_{\geq 1-(1+2^{n})}\end{aligned}$$

hold, and the inclusions

$$\begin{aligned}U_n + v &\subseteq \mathbb{Z}_{\leq 2^{n-1}+2^{n-2}+2^{n-4}-(1+2^{n+1})} + v\\ &\subseteq \mathbb{Z}_{\leq 2^{n-1}+2^{n-2}+2^{n-4}-(1+2^{n+1})+v}\\ &\subseteq \mathbb{Z}_{\leq 2^{n-1}+2^{n-2}+2^{n-4}-(1+2^{n+1})+2^{n-4}}\\ &\subseteq \mathbb{Z}_{\leq 2^{n-1}+2^{n-2}+2^{n-3}-(1+2^{n+1})}\end{aligned}$$

hold for any $v \in \mathcal{V}$ with $v \leq 2^{n-4}$. This yields Eq. (15).

Now, we show that Eq. (16) holds. The inclusions

$$\begin{aligned}U_n + v &\subseteq \mathbb{Z}_{\leq 2^{n}-(1+2^{n+1})} + v\\ &\subseteq \mathbb{Z}_{\leq 2^{n}-(1+2^{n+1})+v}\\ &\subseteq \mathbb{Z}_{\leq 2^{n}-(1+2^{n+1})+2^{n}}\\ &\subseteq \mathbb{Z}_{\leq -1}\end{aligned}$$

hold for any $n \geq 1$ and $v \in \mathcal{V}$ with $v \leq 2^n$. The inclusions

$$\begin{aligned}U_n + 2^{n+1} &\subseteq \mathbb{Z}_{\geq 3-(1+2^{n+1})} + 2^{n+1}\\ &\subseteq \mathbb{Z}_{\geq 3-(1+2^{n+1})+2^{n+1}}\\ &\subseteq \mathbb{Z}_{\geq 2}\end{aligned}$$

hold for any $n \geq 1$. The inclusions

$$\begin{aligned}U_n + v &\subseteq \mathbb{Z}_{\geq 1-(1+2^{n+1})} + v\\ &\subseteq \mathbb{Z}_{\geq 1-(1+2^{n+1})+v}\\ &\subseteq \mathbb{Z}_{\geq 1-(1+2^{n+1})+2^{n+2}}\\ &\subseteq \mathbb{Z}_{\geq 2^{n+1}}\end{aligned}$$

hold for any $n \geq 1$ and $v \in \mathcal{V}$ with $v \geq 2^{n+2}$. This proves that

$$1 \notin (U \setminus U_0) + \mathcal{V} = (U \setminus \{-2, -1\}) + \mathcal{V}.$$

Note that

$$\begin{aligned}-2+\mathcal{V} &\subseteq (-2+\{1\}) \cup (-2+(\mathcal{V}\setminus\{1\}))\\ &\subseteq \{-1\} \cup (-2+(\mathcal{V}\setminus\{1\}))\end{aligned}$$

and the elements of $-2+(\mathcal{V}\setminus\{1\})$ are even. So 1 does not belong to $-2+\mathcal{V}$. So $1 \notin (U\setminus\{-1\})+\mathcal{V}$, and hence Eq. (16) holds.

Now, we prove that Eqs. (17), (18) hold. By Eqs. (13), (14), we obtain

$$(\{25,26\}-(1+2^6)) \cap \big((U_6+(\mathcal{V}\setminus\{2^5\})) \cup (\cup_{m\geq 7} U_m+\mathcal{V})\big) = \emptyset.$$

It follows that

$$(\{25,26\}-(1+2^6)) \cap (U_6+2^5) = \emptyset,$$

$$(\{25,26\}-(1+2^6)) \cap ((U_0 \cup \cdots \cup U_4)+\mathcal{V}) = \emptyset,$$

and hence

$$(\{25,26\}-(1+2^6)) \cap (\cup_{m\neq 5} U_m+\mathcal{V}) = \emptyset.$$

It also follows that

$$(\{26\}-(1+2^6)) \cap ((U_5\setminus\{25-(1+2^6)\})+\mathcal{V}) = \emptyset,$$

$$(\{26\}-(1+2^6)) \cap (25-(1+2^6)+(\mathcal{V}\setminus\{1\})) = \emptyset,$$

$$(\{25\}-(1+2^6)) \cap ((U_5\setminus\{26-(1+2^6)\})+\mathcal{V}) = \emptyset,$$

$$(\{25\}-(1+2^6)) \cap (26-(1+2^6)+(\mathcal{V}\setminus\{-1\})) = \emptyset.$$

Consequently, Eqs. (17), (18) hold.

By considering the representation of the integer -4 as a sum of an element of U and an element of $\mathcal{V}$, Eq. (19) follows.

Using Eqs. (13), (14), we obtain

$$\{-6\} \cap \big((U_4+(\mathcal{V}\setminus\{2^3\}) \cup (\cup_{m\geq 5} U_m+\mathcal{V})\big) = \emptyset.$$

Then Eq. (20) follows from considering the representation of -6.

Now, we prove that Eq. (21) holds. Note that the inclusions

$$\begin{aligned}\cup_{0\leq m\leq n-2} U_m+v &\subseteq \mathbb{Z}_{\geq 1-(1+2^{n-1})}+v\\ &\subseteq \mathbb{Z}_{\geq 1-(1+2^{n-1})+v}\\ &\subseteq \mathbb{Z}_{\geq 1-(1+2^{n-1})-2^{n-1}}\\ &\subseteq \mathbb{Z}_{\geq 1-(1+2^{n})}\end{aligned}$$

hold for $n \geq 6$ and for any $v \in \mathcal{V}$ with $v \geq -2^{n-1}$. The inclusions

$$\begin{aligned}\cup_{0\leq m\leq n-2} U_m + v &\subseteq \mathbb{Z}_{\leq -1} + v \\ &\subseteq \mathbb{Z}_{\leq -1+v} \\ &\subseteq \mathbb{Z}_{\leq -1-2^{n+1}}\end{aligned}$$

hold for $n \geq 6$ and for any $v \in \mathcal{V}$ with $v \leq -2^{n+1}$. The inclusions

$$\begin{aligned}\cup_{0\leq m\leq n-2} U_m - 2^n &\subseteq (\cup_{3\leq m\leq n-2} \mathcal{I}_m - 2^n) \cup (\{-2,-1\} - 2^n) \\ &\subseteq (\mathbb{Z}_{\leq -9} - 2^n) \cup (\{-1,0\} - (1+2^n)) \\ &\subseteq (\mathbb{Z}_{\leq -8-(1+2^n)}) \cup (\{-1,0\} - (1+2^n)) \\ &= (\mathbb{Z}_{\leq 2^n-8-(1+2^{n+1})}) \cup (\{2^n-1, 2^n\} - (1+2^{n+1}))\end{aligned}$$

hold for $n \geq 6$. These inclusions prove that for $n \geq 6$, the third largest element of $\mathcal{I}_n$ does not belong to $\cup_{0\leq m\leq n-2} U_m + \mathcal{V}$, i.e.,

$$2^n - 2 - (1+2^{n+1}) \notin \cup_{0\leq m\leq n-2} U_m + \mathcal{V}.$$

Combining the above with Eqs. (13), (14), we obtain

$$2^n - 2 - (1+2^{n+1}) \notin \big(\cup_{m\geq 0, m\neq n-1,n,n+1} U_m + \mathcal{V}\big) \cup (U_{n+1} + (\mathcal{V} \setminus \{2^n\})) \quad \text{for } n \geq 6.$$

Since the inclusions

$$\begin{aligned}U_{n+1} + 2^n &\subseteq \mathbb{Z}_{\leq 2^n+2^{n-1}+2^{n-3}-(1+2^{n+2})} + 2^n \\ &\subseteq \mathbb{Z}_{\leq 2^n+2^{n-1}+2^{n-3}-(1+2^{n+2})+2^n} \\ &\subseteq \mathbb{Z}_{\leq 2^{n-1}+2^{n-3}-(1+2^{n+1})} \\ &\subseteq \mathbb{Z}_{\leq 2^{n-1}+2^{n-1}-3-(1+2^{n+1})} \\ &\subseteq \mathbb{Z}_{\leq 2^n-3-(1+2^{n+1})}\end{aligned}$$

hold for $n \geq 6$, it follows that

$$2^n - 2 - (1+2^{n+1}) \notin \big(\cup_{m\geq 0, m\neq n-1,n} U_m + \mathcal{V}\big) \quad \text{for } n \geq 6.$$

The inclusions

$$\begin{aligned}U_n + v &\subseteq \mathbb{Z}_{\geq 1-(1+2^{n+1})} + v \\ &\subseteq \mathbb{Z}_{\geq 1-(1+2^{n+1})+v} \\ &\subseteq \mathbb{Z}_{\geq 1-(1+2^{n+1})+2^n} \\ &= \mathbb{Z}_{\geq 2^n+1-(1+2^{n+1})}\end{aligned}$$

hold for any $n \geq 5$ and for any $v \in \mathcal{V}$ with $v \geq 2^n$. The inclusions

$$\begin{aligned}
U_n + 2^{n-1} &\subseteq \left(\mathbb{Z}_{\leq 2^{n-3}+2^{n-3}-(1+2^{n+1})} \cup \mathbb{Z}_{\geq 2^{n-1}+2^{n-2}+1-(1+2^{n+1})}\right) + 2^{n-1} \\
&= \mathbb{Z}_{\leq 2^{n-3}+2^{n-3}+2^{n-1}-(1+2^{n+1})} \cup \mathbb{Z}_{\geq 2^{n-1}+2^{n-2}+1+2^{n-1}-(1+2^{n+1})} \\
&= \mathbb{Z}_{\leq 2^{n-2}+2^{n-1}-(1+2^{n+1})} \cup \mathbb{Z}_{\geq 2^{n}+2^{n-2}+1-(1+2^{n+1})} \\
&= \mathbb{Z}_{\leq 2^{n}-2^{n-2}-(1+2^{n+1})} \cup \mathbb{Z}_{\geq 2^{n}+2^{n-2}+1-(1+2^{n+1})} \\
&\subseteq \mathbb{Z}_{\leq 2^{n}-3-(1+2^{n+1})} \cup \mathbb{Z}_{\geq 2^{n}+2^{n-2}+1-(1+2^{n+1})}
\end{aligned}$$

hold for any $n \geq 6$. The inclusions

$$\begin{aligned}
U_n + 2^{n-2} &\subseteq \left(\mathbb{Z}_{\leq 2^{n-3}+2^{n-3}-(1+2^{n+1})} \cup \mathbb{Z}_{\geq 2^{n-1}+2^{n-2}+1-(1+2^{n+1})}\right) + 2^{n-2} \\
&= \mathbb{Z}_{\leq 2^{n-3}+2^{n-3}+2^{n-2}-(1+2^{n+1})} \cup \mathbb{Z}_{\geq 2^{n-1}+2^{n-2}+1+2^{n-2}-(1+2^{n+1})} \\
&= \mathbb{Z}_{\leq 2^{n-1}-(1+2^{n+1})} \cup \mathbb{Z}_{\geq 2^{n}+1-(1+2^{n+1})}
\end{aligned}$$

hold for any $n \geq 6$. The inclusions

$$\begin{aligned}
U_n + v &\subseteq \mathbb{Z}_{\leq 2^{n-1}+2^{n-2}+2^{n-4}-(1+2^{n+1})} + v \\
&\subseteq \mathbb{Z}_{\leq 2^{n-1}+2^{n-2}+2^{n-4}-(1+2^{n+1})+v} \\
&\subseteq \mathbb{Z}_{\leq 2^{n-1}+2^{n-2}+2^{n-4}-(1+2^{n+1})+2^{n-3}} \\
&= \mathbb{Z}_{\leq 2^{n-1}+2^{n-2}+2^{n-3}+2^{n-4}-(1+2^{n+1})} \\
&= \mathbb{Z}_{\leq 2^{n}-2^{n-4}-(1+2^{n+1})} \\
&\subseteq \mathbb{Z}_{\leq 2^{n}-3-(1+2^{n+1})}
\end{aligned}$$

hold for any $n \geq 6$ and for any $v \in \mathcal{V}$ with $v \leq 2^{n-3}$. Consequently, we obtain

$$2^n - 2 - (1 + 2^{n+1}) \notin \left(\cup_{m \geq 0, m \neq n-1} U_m + \mathcal{V}\right) \quad \text{for } n \geq 6.$$

The inclusions

$$\begin{aligned}
U_{n-1} + v &\subseteq \mathbb{Z}_{\geq 2^{n-4}+1-(1+2^{n})} + v \\
&\subseteq \mathbb{Z}_{\geq 2^{n-4}+1-(1+2^{n})-2^{n-4}} \\
&= \mathbb{Z}_{\geq 1-(1+2^{n})} \\
&= \mathbb{Z}_{\geq 2^{n}+1-(1+2^{n+1})}
\end{aligned}$$

hold for $n \geq 6$ and for any $v \in \mathcal{V}$ with $v \geq -2^{n-4}$. The inclusions

$$\begin{aligned}
U_{n-1} - 2^{n-2} &\subseteq (\mathbb{Z}_{\leq 2^{n-4}+2^{n-4}-(1+2^{n})} \cup \mathbb{Z}_{\geq 2^{n-2}+2^{n-3}+1-(1+2^{n})}) - 2^{n-2} \\
&= (\mathbb{Z}_{\leq 2^{n-3}-(1+2^{n})} \cup \mathbb{Z}_{\geq 2^{n-2}+2^{n-3}+1-(1+2^{n})}) - 2^{n-2} \\
&= \mathbb{Z}_{\leq 2^{n-3}-(1+2^{n})-2^{n-2}} \cup \mathbb{Z}_{\geq 2^{n-2}+2^{n-3}+1-(1+2^{n})-2^{n-2}} \\
&= \mathbb{Z}_{\leq 2^{n-3}-(1+2^{n})-2^{n-2}} \cup \mathbb{Z}_{\geq 2^{n-3}+1-(1+2^{n})} \\
&= \mathbb{Z}_{\leq 2^{n-1}+2^{n-2}+2^{n-3}-(1+2^{n+1})} \cup \mathbb{Z}_{\geq 2^{n-3}+1-(1+2^{n})}
\end{aligned}$$

$$\subseteq \mathbb{Z}_{\leq 2^{n-1}+2^{n-2}+2^{n-2}-4-(1+2^{n+1})} \cup \mathbb{Z}_{\geq 2^{n-3}+1-(1+2^n)}$$
$$= \mathbb{Z}_{\leq 2^n-4-(1+2^{n+1})} \cup \mathbb{Z}_{\geq 2^n+2^{n-3}+1-(1+2^{n+1})}$$

hold for $n \geq 6$. The inclusions

$$\begin{aligned} U_{n-1} + v &\subseteq \mathbb{Z}_{\leq 2^{n-2}+2^{n-3}+2^{n-5}-(1+2^n)} + v \\ &= \mathbb{Z}_{\leq 2^{n-2}+2^{n-3}+2^{n-5}-(1+2^n)+v} \\ &\subseteq \mathbb{Z}_{\leq 2^{n-2}+2^{n-3}+2^{n-5}-(1+2^n)-2^{n-1}} \\ &= \mathbb{Z}_{\leq 2^{n-1}+2^{n-2}+2^{n-3}+2^{n-5}-(1+2^{n+1})} \\ &\subseteq \mathbb{Z}_{\leq 2^{n-1}+2^{n-2}+2^{n-3}+2^{n-3}-4-(1+2^{n+1})} \\ &= \mathbb{Z}_{\leq 2^n-4-(1+2^{n+1})} \end{aligned}$$

hold for $n \geq 6$ and for any $v \in \mathcal{V}$ with $v \leq -2^{n-1}$. This yields

$$2^n - 2 - (1 + 2^{n+1}) \notin \left(\cup_{m \geq 0, m \neq n-1} U_m + \mathcal{V}\right) \cup (U_{n-1} + (\mathcal{V} \setminus \{-2^{n-3}\})) \quad \text{for } n \geq 6.$$

So Eq. (21) follows. This establishes all of the nine claims, and hence $\mathcal{V}$ is a minimal complement of U. □

Theorem 7 *Let* $U = \{-u_1, -u_2, -u_3, \ldots\}$ *where* $u_1 < u_2 < u_3 < \ldots$. *Define* $U_1 = U$ *and for each positive integer* $i \geq 1$*, define*

$$U_{i+1} := \begin{cases} U_i \setminus \{-u_i\} & \text{if } U_i \setminus \{-u_i\} \text{ is a complement to } \mathcal{V}, \\ U_i & \text{otherwise.} \end{cases}$$

Let $\mathcal{U}$ *denote the subset* $\cap_{i \geq 1} U_i$ *of* $\mathbb{Z}$*. The subsets* $\mathcal{U}$ *and* $\mathcal{V}$ *of* $\mathbb{Z}$ *form a co-minimal pair.*

Proof Note that for any $n \geq 4$ and any $m \geq n$, the inclusions

$$\begin{aligned} U_m + v &\subseteq \mathbb{Z}_{\leq 2^m-(1+2^{m+1})} + v \\ &= \mathbb{Z}_{\leq 2^m-(1+2^{m+1})+v} \\ &\subseteq \mathbb{Z}_{\leq 2^m-(1+2^{m+1})+2^{m-1}} \\ &= \mathbb{Z}_{\leq -(1+2^{m-1})} \\ &\subseteq \mathbb{Z}_{\leq -2^{m-3}-1} \\ &\subseteq \mathbb{Z}_{\leq -2^{n-3}-1} \end{aligned}$$

hold for any $v \leq 2^{m-1}$, the inclusions

$$\begin{aligned}
U_m + 2^m &\subseteq \mathbb{Z}_{\le 2^{m-1}+2^{m-2}+2^{m-4}-(1+2^{m+1})} + 2^m \\
&= \mathbb{Z}_{\le 2^{m-1}+2^{m-2}+2^{m-4}-(1+2^{m+1})+2^m} \\
&\subseteq \mathbb{Z}_{\le 2^{m-1}+2^{m-2}+2^{m-3}-(1+2^{m+1})+2^m} \\
&= \mathbb{Z}_{\le 2^{m-1}+2^{m-2}+2^{m-3}+2^{m-3}-2^{m-3}-(1+2^{m+1})+2^m} \\
&= \mathbb{Z}_{\le -2^{m-3}-1} \\
&\subseteq \mathbb{Z}_{\le -2^{n-3}-1}
\end{aligned}$$

hold, the inclusions

$$\begin{aligned}
U_m + v &\subseteq \mathbb{Z}_{\ge 2^{m-3}+1-(1+2^{m+1})} + v \\
&= \mathbb{Z}_{\ge 2^{m-3}+1-(1+2^{m+1})+v} \\
&\subseteq \mathbb{Z}_{\ge 2^{m-3}+1-(1+2^{m+1})+2^{m+1}} \\
&= \mathbb{Z}_{\ge 2^{m-3}} \\
&\subseteq \mathbb{Z}_{\ge 2^{n-3}}
\end{aligned}$$

hold for any $v \ge 2^{m+1}$. It follows that

$$\{-2^{n-3}, -2^{n-3}+1, -2^{n-3}+2, \dots, -2, -1, 0, 1, 2, 3, \dots, 2^{n-3}-1\} \cap (\cup_{m \ge n} U_m + \mathcal{V}) = \emptyset \tag{24}$$

for any $n \ge 4$. Consequently, for any element $y \in \mathbb{Z}$, the equation $y = u + v$ holds for finitely many pairs $(u, v) \in U \times \mathcal{V}$.

For any $y \in \mathbb{Z}$ and for any $i \ge 1$, there exist elements $u_{y,i} \in U_i$, $v_{y,i} \in \mathcal{V}$ such that $y = u_{y,i} + v_{y,i}$. It follows that for some element $u_y \in U$, the equality $u_y = u_{y,i}$ holds for infinitely many i and for such integers i, we have $v_y = v_{y,i}$ where $v_y := y - u_y \in \mathcal{V}$. Thus $y - v_y = u_{y,i}$ holds for infinitely many i. Hence, for each integer $i \ge 1$, there exists an integer $m_i \ge i$ such that $y - v_y = u_{y,m_i}$, which yields $y \in v_y + U_{m_i} \subseteq v_y + U_i$. Thus y lies in $v_y + \mathcal{U}$, i.e., $y \in \mathcal{U} + \mathcal{V}$. Hence $\mathcal{U}$ is an additive complement of $\mathcal{V}$. It follows that $\mathcal{U}$ is a minimal complement to $\mathcal{V}$. By Theorem 6, it follows that $\mathcal{V}$ is a minimal complement to $\mathcal{U}$. This proves that $(\mathcal{U}, \mathcal{V})$ is a co-minimal pair. □

Thus, Sects. 2 and 3 provide the first explicit constructions of infinite co-minimal pairs in the integers. One may ask the question about how abundant they are. This has been addressed in a subsequent work, see [3].

4 Co-minimal Pairs in the Free Abelian Groups of Higher Rank

Proposition 3 *Let H be a subgroup of an abelian group G. Let (A, B) be a co-minimal pair in H, and C, D be subsets of G whose images form a co-minimal pair in G/H, and the restrictions of the map $G \to G/H$ on C and D induces bijections onto respective images. Then $(A + C, B + D)$ is a co-minimal pair in G.*

Proof Note that $A + C + B + D = G$. Let $a \in A, c \in C$ be such that $(A + C) \setminus \{a + c\}$ is a complement to $B + D$. Let $b \in B, d \in D$ be such that $a + b \notin (A \setminus \{a\}) + B$, $(c + d) \bmod H \notin ((C \setminus \{c\}) + D) \bmod H$. Since $a + c + b + d \in ((A + C) \setminus \{a + c\}) + B + D$, it follows that $a + c + b + d = a' + c' + b' + d'$ for some $a' \in A, c' \in C, b' \in B, d' \in D$ with $a' + c' \neq a + c$. This implies $(c + d) \bmod H = (c' + d') \bmod H$, and hence $c \equiv c' \bmod H$, $d \equiv d' \bmod H$, and thus $c = c', d = d'$, which yields $a + b = a' + b'$. Since $a + b \notin (A \setminus \{a\}) + B$, we obtain $a = a', b = b'$. This implies that $a' + c' = a + c$, which contradicts the fact that $a' + c' \neq a + c$. Hence $A + C$ is a minimal complement to $B + D$. Similarly, it follows that $B + D$ is a minimal complement to $A + C$. □

As an application of the above result, we obtain the following result.

Corollary 1 *Let $\mathcal{S} = \{s_1, s_2, \ldots\}$. For any two sequences $\{x_n\}_{n\geq 1}$ and $\{y_n\}_{n\geq 1}$, each of*

$$(\{(x_i + s_j, s_i) \mid i, j \geq 1\}, \{(y_u + 2^{v-1}, 2^{u-1}) \mid u, v \geq 1\}),$$

$$(\{(x_i + 2^{j-1}, s_i) \mid i, j \geq 1\}, \{(y_u + s_v, 2^{u-1}) \mid u, v \geq 1\})$$

is a co-minimal pair in $\mathbb{Z}^2$.

Proof For any two sequences $\{x_n\}_{n\geq 1}$ and $\{y_n\}_{n\geq 1}$, the subsets

$$(\mathcal{S}, 0) + \{(x_1, s_1), (x_2, s_2), (x_3, s_3), \ldots\} = \{(x_i + s_j, +s_i) \mid i, j \geq 1\},$$

$$(\mathcal{T}, 0) + \{(y_1, 1), (y_2, 2), (y_3, 2^2), \ldots\} = \{(y_u + 2^{v-1}, 2^{u-1}) \mid u, v \geq 1\}$$

of $\mathbb{Z}^2$ form a co-minimal pair in $\mathbb{Z}^2$, and the subsets

$$(\mathcal{T}, 0) + \{(x_1, s_1), (x_2, s_2), (x_3, s_3), \ldots\} = \{(x_i + 2^{j-1}, s_i) \mid i, j \geq 1\},$$

$$(\mathcal{S}, 0) + \{(y_1, 1), (y_2, 2), (y_3, 2^2), \ldots\} = \{(y_u + s_v, 2^{u-1}) \mid u, v \geq 1\}$$

of $\mathbb{Z}^2$ form a co-minimal pair in $\mathbb{Z}^2$ by Proposition 3.

Theorem 8 *Let σ be an automorphism of $\mathbb{Z}^n$ such that there exists an increasing chain of subgroups*

$$0 = M_0 \subsetneq M_1 \subsetneq \cdots \subsetneq M_r = \mathbb{Z}^n$$

of $\mathbb{Z}^n$ such that each of them is stable under the action of σ and the successive quotients M_i/M_{i-1} are free of rank at most two and for any such quotient, the restriction of σ to it is the identity map if the quotient is of rank one, or conjugate to some element of $\mathrm{GL}_2(\mathbb{Z})$ having exactly two nonzero entries, i.e., some of

$$\begin{pmatrix} 1 & 0 \\ 0 & 1 \end{pmatrix}, \begin{pmatrix} 1 & 0 \\ 0 & -1 \end{pmatrix}, \begin{pmatrix} -1 & 0 \\ 0 & 1 \end{pmatrix}, \begin{pmatrix} -1 & 0 \\ 0 & -1 \end{pmatrix}, \begin{pmatrix} 0 & 1 \\ 1 & 0 \end{pmatrix}, \begin{pmatrix} 0 & -1 \\ 1 & 0 \end{pmatrix}, \begin{pmatrix} 0 & 1 \\ -1 & 0 \end{pmatrix}, \begin{pmatrix} 0 & -1 \\ -1 & 0 \end{pmatrix} \tag{25}$$

if the quotient is of rank two.[3] *Then there exists a subset A of* $\mathbb{Z}^n$ *such that* $(A, \sigma(A))$ *is a co-minimal pair in* $\mathbb{Z}^n$.

Proof If $n = 1$, then σ is the identity map, and thus in this case, A can taken to be the subset W of $\mathbb{Z}$ as in [8, Proposition 3].

If $n = 2$, then one of the following conditions hold.

1. M_0 is a subgroup of $\mathbb{Z}^2$ of rank one and σ is conjugate to

$$\begin{pmatrix} 1 & * \\ 0 & 1 \end{pmatrix},$$

2. $M_0 = \mathbb{Z}^2$ and σ is conjugate to one of the matrices in Eq. (25).

For simplicity, we will assume that σ is equal to the above matrix, or equal to one of the matrices in Eq. (25). In the first case, we can take $A = W \times W$. Indeed, $(W \times \{0\}, W \times \{0\})$ is a co-minimal pair in $\mathbb{Z} \times \{0\}$, and the images of $\{0\} \times W$, $\{0\} \times W$ form a co-minimal pair in $(\mathbb{Z} \times \mathbb{Z})/(\mathbb{Z} \times \{0\})$, and the restrictions of the map $\mathbb{Z} \times \mathbb{Z} \to (\mathbb{Z} \times \mathbb{Z})/(\mathbb{Z} \times \{0\})$ on $\{0\} \times W$ induces bijections onto its image. Thus, by Proposition 3, we can take $A = W \times W$ in the first case. In the second case, we can take A to be

- $W \times W$,
- $\{(x, 0) \mid x \in \mathbb{Z}\} \cup \{(0, y) \mid y \in \mathbb{Z}_{\geq 1}\}$,
- $\{(x, 0) \mid x \in \mathbb{Z}_{\geq 1}\} \cup \{(0, y) \mid y \in \mathbb{Z}\}$,
- $\{(x, 0) \mid x \in \mathbb{Z}\} \cup \{(0, y) \mid y \in \mathbb{Z}_{\geq 1}\}$,
- $\mathcal{S} \times \mathcal{T}$,
- $\mathcal{V} \times \mathcal{U}$,
- $\mathcal{U} \times \mathcal{V}$,
- $(-\mathcal{S}) \times \mathcal{T}$

according as σ is equal to the matrices in Eq. (25). Thus Theorem 8 holds for $n = 1, 2$.

Let $n \geq 3$ be an integer such that Theorem 8 holds for free abelian groups of rank $< n$. Let M' be a subgroup of $\mathbb{Z}^n$ such that M' surjects onto M_r/M_{r-1}. Denote the restriction of σ to M_r/M_{r-1} by $\bar{\sigma}$. Since the result holds for $n = 1, 2$ and M_r/M_{r-1} has rank at most two, it follows that there is a subset C of M_r/M_{r-1} such that $(C, \bar{\sigma}(C))$ is a co-minimal pair in M_r/M_{r-1}. Let C' be a subset of M' such that the map $M \to M_r/M_{r-1}$ yields a bijection between C' and C. Note that M_{r-1} is stable under the action of σ. By the induction hypothesis, there is a subset A of M_{r-1} such that $(A, \sigma(A))$ is a co-minimal pair M_{r-1}. By Proposition 3, $(A + C', \sigma(A) + \sigma(C'))$ is a co-minimal pair in $M_r = \mathbb{Z}^n$. Thus Theorem 8 follows. □

[3] Note that the above eight matrices form the subgroup

$$\left\{ \begin{pmatrix} a & b \\ c & d \end{pmatrix} \in \mathrm{GL}_2(\mathbb{Z}) \mid \text{exactly two of } a, b, c, d \text{ are equal to zero} \right\}$$

of $\mathrm{GL}_2(\mathbb{Z})$.

Theorem 9 *Let σ be an automorphism of $\mathbb{Z}^{2d}$ such that its matrix with respect to the standard basis of $\mathbb{Z}^{2d}$ has a block upper triangular form having the matrices*

$$\begin{pmatrix} 0 & 1 \\ 1 & 0 \end{pmatrix}, \begin{pmatrix} 0 & -1 \\ -1 & 0 \end{pmatrix}$$

along the diagonal. Then there exists a subset A of $\mathbb{Z}^{2d}$ contained in a quadrant such that $(A, \sigma(A))$ is a co-minimal pair.

Proof If $d = 1$, then A can be taken to be $\mathcal{S} \times \mathcal{T}, (-\mathcal{S}) \times \mathcal{T}$. Let $d \geq 2$ be an integer and assume that the result holds for free groups of rank $2(d-1)$. Let σ be an automorphism of $\mathbb{Z}^{2d}$ satisfying the given condition. Let M denote the subgroup of $\mathbb{Z}^{2d}$ generated by $e_1, \ldots, e_{2(d-1)}$ and M' denote the subgroup of $\mathbb{Z}^{2d}$ generated by e_{2d-1}, e_{2d}. Note that M' surjects onto $\mathbb{Z}^{2d}/M$. Note that M is stable under the action of σ. By the induction hypothesis, there is a subset A of M contained in a quadrant such that $(A, \sigma(A))$ is a co-minimal pair in M. Since the result holds for $d = 1$, it follows that there exists a subset C of M' contained in a quadrant such that the images of C and $\sigma(C)$ in $\mathbb{Z}^{2d}/M$ form a co-minimal pair. By Proposition 3, $(A + C, \sigma(A) + \sigma(C))$ is a co-minimal pair in $\mathbb{Z}^{2d}$. Since $A + C$ is contained in a quadrant, the result follows by induction. □

Proof (Proof of Theorem* 3 *) The first part follows from Theorem 8.

For $d = 1$, the group $\mathbb{Z}^{2d}$ admits infinitely many automorphisms of the form

$$\begin{pmatrix} 1 & * \\ 0 & 1 \end{pmatrix}$$

with $* \in \mathbb{Z}$. So by Theorem 8, for each such automorphism σ of $\mathbb{Z}^{2d}$, there is a subset A of $\mathbb{Z}^{2d}$ such that $(A, \sigma(A))$ is a co-minimal pair in $\mathbb{Z}^{2d}$. To establish the second part for $d \geq 2$, note that there are infinitely many automorphisms of $\mathbb{Z}^{2d}$ which are block upper triangular where the blocks are of size 2×2 and the matrices lying along the diagonal blocks are equal to some of

$$\begin{pmatrix} 1 & 0 \\ 0 & 1 \end{pmatrix}, \begin{pmatrix} 1 & 0 \\ 0 & -1 \end{pmatrix}, \begin{pmatrix} -1 & 0 \\ 0 & 1 \end{pmatrix}, \begin{pmatrix} -1 & 0 \\ 0 & -1 \end{pmatrix}, \begin{pmatrix} 0 & 1 \\ 1 & 0 \end{pmatrix}, \begin{pmatrix} 0 & -1 \\ 1 & 0 \end{pmatrix}, \begin{pmatrix} 0 & 1 \\ -1 & 0 \end{pmatrix}, \begin{pmatrix} 0 & -1 \\ -1 & 0 \end{pmatrix},$$

then by Theorem 8, for each such automorphism σ of $\mathbb{Z}^{2d}$, there is a subset A of $\mathbb{Z}^{2d}$ such that $(A, \sigma(A))$ is a co-minimal pair in $\mathbb{Z}^{2d}$. □

Acknowledgements We thank the anonymous referee for a careful reading of the manuscript and for a number of helpful comments and suggestions. The first author is supported by the ISF Grant no. 662/15 and the ISF Grant no. 1226/19. He would also like to thank the Department of Mathematics at the Technion where a part of the work was carried out. The second author would like to acknowledge the Initiation Grant from the Indian Institute of Science Education and Research Bhopal, and the INSPIRE Faculty Award from the Department of Science and Technology, Government of India.

References

1. Arindam Biswas and Jyoti Prakash Saha, *Minimal additive complements in finitely generated abelian groups*, Ramanujan J. (accepted).
2. Arindam Biswas and Jyoti Prakash Saha, *Asymptotic complements in the integers*, J. Number Theory **213** (2020), 101–115.
3. Arindam Biswas and Jyoti Prakash Saha, *Infinite co-minimal pairs involving lacunary sequences and generalisations to higher dimensions*, Preprint available at https://arxiv.org/abs/2006.02429, 2020.
4. Arindam Biswas and Jyoti Prakash Saha, *On additive co-minimal pairs*, J. Number Theory (2021), https://doi.org/10.1016/j.jnt.2020.10.010.
5. Arindam Biswas and Jyoti Prakash Saha, *On minimal complements in groups*, Ramanujan J. (2020), https://doi.org/10.1007/s11139-020-00309-3.
6. Paul Erdős, *Some results on additive number theory*, Proc. Amer. Math. Soc. **5** (1954), 847–853. MR 64798
7. Paul Erdős, *Some unsolved problems*, Michigan Math. J. **4** (1957), 291–300. MR 98702
8. Andrew Kwon, *A note on minimal additive complements of integers*, Discrete Math. **342** (2019), no. 7, 1912–1918. MR 3937752
9. G. G. Lorentz, *On a problem of additive number theory*, Proc. Amer. Math. Soc. **5** (1954), 838–841. MR 63389
10. Melvyn B. Nathanson, *Problems in additive number theory, IV: Nets in groups and shortest length g-adic representations*, Int. J. Number Theory **7** (2011), no. 8, 1999–2017. MR 2873139

Rigidity, Graphs and Hausdorff Dimension

Nikolaos Chatzikonstantinou, Alex Iosevich, Sevak Mkrtchyan, and Jonathan Pakianathan

Abstract A set of $k+1$ points in Euclidean space is called a $(k+1)$-point configuration. Two configurations are congruent if they are equal up to an affine isometry. Given a compact subset E of $\mathbb{R}^d$, $d \geq 2$ of Hausdorff dimension greater than $d - \frac{1}{k+1}$ we prove that the Lebesgue measure of noncongruent $(k+1)$-point configurations in E is positive, for $k > d$, complementing the results of [11] for $k \leq d$.

Keywords ■■■

1 Introduction

We study a problem inspired by the Falconer distance conjecture. There the interest is to show that the set of distances of $E \subset \mathbb{R}^d$ has positive Lebesgue measure, when the Hausdorff dimension of E is greater than $d/2$. Since distances may be thought of as line segments identified up to isometry, we obtain the equivalent conjecture that there's "many" unique (up to isometry) line segments in E. More generally, one

---

The second and fourth listed authors were partially supported by the NSA Grant H98230-15-1-0319. The third listed author was partially supported by the Simons Foundation Collaboration Grant No. 422190.

N. Chatzikonstantinou · A. Iosevich (✉) · S. Mkrtchyan · J. Pakianathan
Department of Mathematics, University of Rochester, 500 Joseph C. Wilson Blvd., Rochester, NY 14627, USA
e-mail: alex.iosevich@rochester.edu

N. Chatzikonstantinou
e-mail: nchatzik@ur.rochester.edu

S. Mkrtchyan
e-mail: sevak.mkrtchyan@rochester.edu

J. Pakianathan
e-mail: jonathan.pakianathan@rochester.edu

M. B. Nathanson (ed.), *Combinatorial and Additive Number Theory IV*, Springer Proceedings in Mathematics & Statistics 347,
https://doi.org/10.1007/978-3-030-67996-5_5

can consider $(k+1)$-points in E, and ask if there's "many" of those. For example, one can consider triangles. For various shapes, his problem has been extensively studied in a variety of contexts by Bennett, Bourgain, Chan, Furstenberg, Greenleaf, Katznelson, Łaba, Pramanik, Weiss, Ziegler, the second and fourth named authors and others (see e.g. [2, 3, 5, 7, 10, 11, 15, 17, 26]). Chains were considered in [4] and in [12] necklaces were investigated. More general frameworks were studied in [16].

Our result complements that of [11], where the authors have shown that indeed there are "many" $(k+1)$-point configurations in E for $k \leq d$ when E has large enough Hausdorff dimension. Part of their technique uses the fact that two $(k+1)$-point configurations are equivalent iff they have the same edge distances. Then they measure distances instead of configurations. This only holds when $k \leq d$, since for $k > d$ the dimension of the set of configurations is less than that of distances. For example, consider the 4-point configurations in the plane (see also Fig. 2). It may be shown that the set of such configurations is 5 dimensional, while there's 6 distances between the points, as we determine in Sect. 5.1. This overdetermination does not allow one to measure distances instead of configurations. Note that 4 distances are not enough (there's flexibility in moving the vertices around, hence changing the configuration, while the distances stay fixed). With 5 distances, such flexibility is gone, there are however 2 distinct configurations with the same distance (simply "flip" the fourth vertex about the segment connecting the first and the third).

To the rescue comes the theory of rigidity of graphs, with work by [1, 13, 14, 21, 21] to show that if a graph G is infinitesimally rigid, then the distances determined by its edges are also determining the point configuration, up to finitely many different ones. A uniform bound on the different configurations that have the same distances in the graph is provided by [6], using theory on semialgebraic varieties and their Betti numbers from [19, 20, 23]. With relative ease then the argument of [11] adapts.

The paper is organized as follows. In Sect. 2 we state the main result. In Sect. 3 we introduce the necessary definitions and and provide statements of the distance-type main theorems.

In Sect. 4 we discuss how to understand the moduli space of congruence classes under isometries, explicitly define a Lebesgue measure on it, and show that the main result of Sect. 2 follows from the distance-type results in Sect. 3.

Section 5 gives the proofs of the theorems stated in Sect. 3. Some geometric results are given in Sect. 6. Finally Sect. 7 addresses the issue of sharpness of our dimensional thresholds.

2 Main Result

Consider $(k+1)$-point configurations (for the definition see Sect. 3.2) in $\mathbb{R}^d$. We say two such configurations are equivalent if there is an isometry of $\mathbb{R}^d$ which takes one to the other (see also Definition 3.15). More precisely, for $k+1$ points $x^1, \ldots, x^{k+1} \in \mathbb{R}^d$, we define

$$\boldsymbol{x} = (x^1, \dots, x^{k+1}) \in \mathbb{R}^{d(k+1)},$$

and given $\boldsymbol{x}, \boldsymbol{y} \in \mathbb{R}^{d(k+1)}$ we say $\boldsymbol{x} \sim \boldsymbol{y}$, with $\boldsymbol{y} = (y^1, \dots, y^{k+1}) \in \mathbb{R}^{d(k+1)}$, if there is an isometry $T : \mathbb{R}^d \to \mathbb{R}^d$ such that $Ty^i = x^i$ for all $i = 1, \dots, k+1$.

Given a compact subset $E \subset \mathbb{R}^d$, consider its image $E^{k+1}/\sim$ in the moduli space $\mathbb{R}^{d(k+1)}/\sim$. In other words consider the equivalence classes of $(k+1)$-point configurations of points from the set E. Our goal is to show that there is a dimensional threshold $s_d(k)$, such that if the Hausdorff dimension of the set E is larger than $s_d(k)$, then the Lebesgue measure of the congruence classes of point configurations from E is positive.

Since the moduli space of congruence classes is not a manifold, one needs to define an appropriately modified space together with a natural measure on it to get precise results. We achieve this by showing that by removing from $\mathbb{R}^{d(k+1)}$ a set of Hausdorff codimension at least 1, the image of the remaining set in the moduli space can be explicitly identified with a Euclidean space and thus a Lebesgue measure can be defined on it. We call this space the *reduced moduli space of congruence classes.*

Our main result is the following theorem.

Theorem 2.1 *Let $d \geq 2$ and $k \geq 1$ be arbitrary integers. The dimension of the reduced moduli space of $(k+1)$-point configurations in $\mathbb{R}^d$ is*

$$m = d(k+1) - \binom{d+1}{2}$$

and if $E \subset \mathbb{R}^d$ is any compact set of Hausdorff dimension larger than $s_d(k) = d - \frac{1}{k+1}$, then the m-dimensional Lebesgue measure of the reduced set of congruence classes of $(k+1)$-point configurations of points from E is positive.

As noted in the introduction, we prove this theorem by reducing it to a Falconer distance type problem. One way to identify the congruence class of a $(k+1)$-point configuration is by specifying pairwise distances between the points. However, in general this gives a largely overdetermined system. For example given 4 points in $\mathbb{R}^2$ there are 6 distances that can be specified. However specifying 5 of them already reduces the congruence class to a bounded number of choices, which is good for our purposes. What we need is a way of specifying pairwise distances for only certain pairs of points from the configuration in such a way that its congruence class is identified up to a bounded number of choices. We do this by putting an extra structure on point configurations, which we call a framework. A framework is essentially an ordered collection of points together with a graph which identified pairwise distances that need to be specified. The condition that the distances specified by the graph should identify the congruence class up to a bounded number of choices puts some restrictions on the graph. To address this issue we introduce the notions of rigidity of the graph which allows us to ensure the congruence class is determined, and independence of edges which allows us to ensure the system is not overdetermined.

We solve a Falconer distance type problem for minimally infinitesimally rigid graphs in $\mathbb{R}^d$ and deduce from it our main result. Slightly more work gives us a dimensional threshold for a Falconer distance type problem for an arbitrary graph without isolated vertices. The precise formulations are given in Theorems 3.20 and 3.22.

3 Definitions and Statements of Distance-Type Results

3.1 Notation

Let $k \geq 1$ and let K_{k+1} be the complete graph with vertex set $\{1, \ldots, k+1\}$ and edge set consisting of all the pairs (i, j) where $i < j$ are taken from the vertex set. We order the edge set lexicographically (this will be useful because later we will write down matrices and formulas ranging over the edge set). Let $m \geq 1$ and denote a subgraph of K_{k+1} with $k+1$ vertices and m edges by $G_{k+1,m}$. We define $ij \in G_{k+1,m}$ to mean that $i < j$ and (i, j) is an edge of $G_{k+1,m}$, and when $ij \in G_{k+1,m}$ ranges over all the edges of $G_{k+1,m}$ in an expression, it ascends in the order of the edge set. Let $|\cdot|$ denote the Euclidean distance, $\|\cdot\|_p$ the L^p norm, $\dim(\cdot)$ the Hausdorff dimension, $\mathcal{L}^m(\cdot)$ the Lebesgue measure of measurable subsets of $\mathbb{R}^m$ and for any $s > 0$ let $\mathcal{H}^s(\cdot)$ denote the s-dimensional Hausdorff measure. Let $A \lesssim B$ mean that for some constant $C > 0$, $A \leq C \cdot B$, where C is independent of ϵ, δ and the summation index or integration variable (if the notation is used to bound such expressions). For index sets I, J, if a_{ij} is a matrix, $(i, j) \in I \times J$, then for $B \subseteq I$, $C \subseteq J$, we defined $a_{B,C}$ to be the submatrix a_{ij} with $(i, j) \in B \times C$. We let $GL_d(\mathbb{R})$ denote the general linear group of degree d over $\mathbb{R}$ and $O_d(\mathbb{R})$ be the group of rotations of degree d over $\mathbb{R}$. We also fix a left-invariant metric on $O_d(\mathbb{R})$.

3.2 Configurations, Frameworks and Distances

Definition 3.1 A $(k+1)$-*point configuration* $\boldsymbol{x}$ *in* $\mathbb{R}^d$ is a tuple

$$\boldsymbol{x} = (x^1, x^2, \ldots, x^{k+1}), \ x^j \in \mathbb{R}^d .$$

We may also think of $\boldsymbol{x}$ as an element of $\mathbb{R}^{d(k+1)}$.

Definition 3.2 A *framework of* $G_{k+1,m}$ *in* $\mathbb{R}^d$ is a pair $(G_{k+1,m}, \boldsymbol{x})$, where $\boldsymbol{x}$ is a $(k+1)$-point configuration in $\mathbb{R}^d$.

A convenient way to specify distances is through the distance function which we now define (Fig. 1).

Definition 3.3 Given a graph $G_{k+1,m}$ we define the *distance function*

$$\begin{aligned} f_{G_{k+1,m}} : \mathbb{R}^{d(k+1)} &\to \mathbb{R}^m \\ \boldsymbol{x} &\mapsto \left(|x^i - x^j|\right)_{ij \in G_{k+1,m}} . \end{aligned}$$

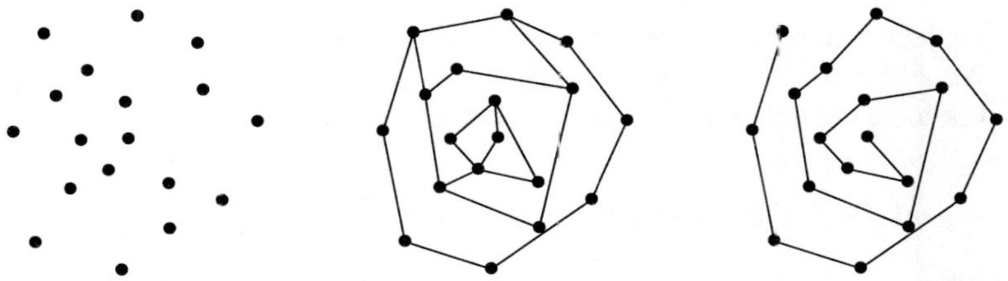

Fig. 1 A configuration followed by two different frameworks on it

We also define the *distance-squared function* $F_{G_{k+1,m}}$ by

$$F_{G_{k+1,m}}(\boldsymbol{x}) = \left(|x^i - x^j|^2\right)_{ij \in G_{k+1,m}} .$$

Definition 3.4 *(Graph distances)* The value $f_{G_{k+1,m}}(\boldsymbol{x})$ is called the $G_{k+1,m}$*-distance of* $\boldsymbol{x}$. When we restrict our domain to some set $\mathfrak{X} \subseteq \mathbb{R}^{d(k+1)}$, we call $f_{G_{k+1,m}}(\boldsymbol{x})$ a $G_{k+1,m}$*-distance on* $\mathfrak{X}$ and we say that $\boldsymbol{x}$ is a realization of this distance in $\mathfrak{X}$. The set of $G_{k+1,m}$-distances on $\mathfrak{X}$ is $f_{G_{k+1,m}}(\mathfrak{X})$ and we denote it by $\Delta(G_{k+1,m}, \mathfrak{X})$.

Remark 3.5 The distance set $\Delta(G_{k+1,m}, \mathfrak{X})$ depends on the numbering of the vertices and the order of the edges. Whereas the order of the edges is superficial, inducing only a permutation in the components of the $G_{k+1,m}$-distances, the numbering of the vertices can significantly change the $G_{k+1,m}$-distance set. Consider $\mathfrak{X} = \{\boldsymbol{x}_0\} \times \mathbb{R}^d \times \cdots \times \mathbb{R}^d$ and a graph $G = G' \cup G'' \cup \{e\}$ where e is a bridge between G' and G''. Then if we number the vertices of G' followed by those of G'', we essentially capture G''-distances only, whereas if we reverse the numbering order of the vertices of G we will capture G'-distances only. In the rest of this paper we take $\mathfrak{X} = E^{k+1}$ for some $E \subset \mathbb{R}^d$, in which case the numbering of the vertices becomes superficial as well. In particular, the dimension of the $G_{k+1,m}$-distance set and its Hausdorff (or Lebesgue) measure are independent of the vertex numbering and edge order.

3.3 Independence and Genericity

We define the notion of independence for subsets of the edge set of K_{k+1} and of maximal independence for subsets of the edge set of $G_{k+1,m}$. We are then ready to define the set of generic configurations as the complement of the zero set of a certain polynomial. We define them in such a way that they are depending only on the dimension d and the number of vertices $k+1$.

Definition 3.6 A subset H of the edge set of K_{k+1} is called *independent in* $\mathbb{R}^d$ *with respect to* $\boldsymbol{x}_0 \in \mathbb{R}^{d(k+1)}$ if the row vectors of $DF_{K_{k+1}}(\boldsymbol{x}_0)$ corresponding to H are linearly independent. We call H *independent in* $\mathbb{R}^d$ if there exists some $\boldsymbol{x}_0$ so that H

is independent with respect to $\boldsymbol{x}_0$, and $\boldsymbol{x}_0$ is said to be a *witness to the independence of* H. We also call H a *maximally independent (in* $\mathbb{R}^d$*) subset of edges of* $G_{k+1,m}$ when it is independent and it is not contained in a larger independent edge set of $G_{k+1,m}$.

For the next definition we employ the previously mentioned submatrix notation.

Definition 3.7 For any nonempty independent in $\mathbb{R}^d$ set H of edges of K_{k+1} we define the polynomial $P_H(\boldsymbol{x})$ to be the sum of squares of $|H| \times |H|$-minors of the submatrix of rows of $DF_{K_{k+1}}$ corresponding to edges of H. Thus,

$$P_H(\boldsymbol{x}) = \sum_{\substack{A \subset \{1,\dots,d(k+1)\} \\ |A|=|H|}} \left|\det(DF_{K_{k+1}}(\boldsymbol{x})_{H,A})\right|^2 .$$

Let X_H denote the zero set of P_H. We define the set of *generic configurations of* $\mathbb{R}^d$ to be the complement of the zero set X of the polynomial $P(\boldsymbol{x})$ defined by

$$P(\boldsymbol{x}) = \prod_{H \text{ independent}} P_H(\boldsymbol{x}) .$$

We call X the set of *singular configurations of* $\mathbb{R}^d$.

Remark 3.8 We have $X = \cup_H X_H$ where the union is taken over all the edge sets H which are independent and the generic configurations are then equal to $\mathbb{R}^{d(k+1)} \setminus X$. Moreover, if a set H of edges is independent then by Definition 3.7 it is generically independent, i.e. independent with respect to any generic $\boldsymbol{x}$. In fact, the set of generic configurations is precisely the set of configurations that simultaneously witness the independence of every independent edge set.

Remark 3.9 The polynomial $P(\boldsymbol{x})$ is nontrivial because every P_H is nontrivial since there is at least one witness $\boldsymbol{x}_H$ for the independence H, which means that $P_H(\boldsymbol{x}_H) \neq 0$. Thus X is a proper algebraic variety of dimension

$$\dim X \leq d(k+1) - 1 . \tag{3.1}$$

Definition 3.10 A framework $(G_{k+1,m}, \boldsymbol{x})$ is called *generic in* $\mathbb{R}^d$ if $\boldsymbol{x}$ is a generic configuration in $\mathbb{R}^d$ and it is called *singular in* $\mathbb{R}^d$ if $\boldsymbol{x}$ is a singular configuration in $\mathbb{R}^d$.

3.4 Structural Rigidity

We are concerned with the Hausdorff measure of the set $\Delta(G_{k+1,m}, E^{k+1})$ and its dimension. An important role is played by properties of the graph $G_{k+1,m}$. In particular it is essential whether the graph is rigid or not.

The key heuristic notion of this paper is that a graph $G_{k+1,m}$ is *rigid in* $\mathbb{R}^d$ if once the m quantities t_{ij} in

$$|x^i - x^j| = t_{ij}, \qquad ij \in G_{k+1,m}$$

are specified, the other distances $|x^i - x^j|$ for $ij \notin G_{k+1,m}$ can only take finitely many different values. For technical reasons, we use a more precise and flexible notion of rigidity described below. A simple example that illustrates the technical obstacles one must contend with is the following. Consider a degenerate quadrilateral in the plane with side lengths $(1, 1, 1, 3)$. This configuration is perfectly rigid in the heuristic sense, but it is not *minimally infinitesimally rigid*, as the reader will see, roughly because the rigidity in this case is not stable under small perturbations of neither the edge lengths nor the vertices.

We now turn to precise definitions, (see also [13] for a complete analysis).

Definition 3.11 An *infinitesimal motion* $\boldsymbol{u} = (u^1, \ldots, u^{k+1})$ *in* $\mathbb{R}^d$ *of* $G_{k+1,m}$ *at* $\boldsymbol{x}$ is a $(k+1)$-tuple $\boldsymbol{u}$ of vectors $u^j \in \mathbb{R}^d$ such that

$$DF_{G_{k+1,m}}(\boldsymbol{x}) \cdot \boldsymbol{u} = 0\,.$$

The set of infinitesimal motions in $\mathbb{R}^d$ of $G_{k+1,m}$ at $\boldsymbol{x}$ is the kernel of $DF_{G_{k+1,m}}(\boldsymbol{x})$. Let us denote by $\mathcal{V}(G_{k+1,m}, \boldsymbol{x})$ the set of infinitesimal motions in $\mathbb{R}^d$ of $G_{k+1,m}$ at $\boldsymbol{x}$. Let $\mathcal{D}(\boldsymbol{x})$ be the set of infinitesimal motions in $\mathbb{R}^d$ of K_{k+1} at $\boldsymbol{x}$.

Remark 3.12 It is evident that $\mathcal{D}(\boldsymbol{x}) \subseteq \mathcal{V}(G_{k+1,m}, \boldsymbol{x})$ since the system of equations $DF_{G_{k+1,m}}(\boldsymbol{x}) \cdot \boldsymbol{u} = 0$ is included in $DF_{K_{k+1}}(\boldsymbol{x}) \cdot \boldsymbol{u} = 0$.

Definition 3.13 A framework $(G_{k+1,m}, \boldsymbol{x})$ is called *infinitesimally rigid in* $\mathbb{R}^d$ when

$$\mathcal{V}(G_{k+1,m}, \boldsymbol{x}) = \mathcal{D}(\boldsymbol{x})\,.$$

It is unnecessarily restrictive to require of a graph to have all its frameworks be infinitesimally rigid, since in fact there may be a small set of singular frameworks which are not rigid in any sense. We shall only require it of generic frameworks.

Definition 3.14 A graph $G_{k+1,m}$ is called *infinitesimally rigid in* $\mathbb{R}^d$ if all its generic frameworks are infinitesimally rigid. It is called *minimally infinitesimally rigid in* $\mathbb{R}^d$ if it is infinitesimally rigid and no proper subgraph (on the same vertex set) is infinitesimally rigid.

We define the notion of congruency for configurations and frameworks.

Definition 3.15 Let $\boldsymbol{x}$ be a $(k+1)$-point configuration in $\mathbb{R}^d$ and define the set of configurations *congruent* to $\boldsymbol{x}$ to be

$$M_{\boldsymbol{x}} = \{(Tx^1, \ldots, Tx^{k+1}) : T \in \mathrm{ISO}(d)\}, \tag{3.2}$$

where $\mathrm{ISO}(d)$ denotes the set of isometries of $\mathbb{R}^d$ to itself.

Definition 3.16 We say that the framework $(G, \boldsymbol{x})$ is *congruent* to the framework $(G', \boldsymbol{x}')$ if $G = G'$ and $\boldsymbol{x}$ is congruent to $\boldsymbol{x}'$ in the sense of Definition 3.15.

Definition 3.17 We say that $\boldsymbol{x} \in \mathbb{R}^{d(k+1)}$ is a *regular configuration of* $F_{G_{k+1,m}}$ if rank $DF_{G_{k+1,m}}$ attains its global maximum at $\boldsymbol{x}$. A framework $(G_{k+1,m}, \boldsymbol{x})$ is a *regular framework* if $\boldsymbol{x}$ is a regular configuration of $F_{G_{k+1,m}}$.

Remark 3.18 It is immediate from the definitions that generic configurations are regular configurations. The other implication does not hold in general, as a regular configuration only needs to witness one maximally independent edge set.

Definition 3.19 The configuration $\boldsymbol{x} = (x^1, \ldots, x^{k+1})$ is said to be in *general position in* $\mathbb{R}^d$ if for every set $J \subseteq \{1, \ldots, k+1\}$ with $|J| \le d+1$ we have that $\{x^j : j \in J\}$ is affinely independent.

3.5 Statements of Distance-Type Results

The first theorem stated, Theorem 3.20, addresses minimally infinitesimally rigid graphs, and says that indeed there is a nontrivial dimensional exponent for which "many" graph-distances may be found. The second theorem, Theorem 3.22 elaborates on the case where the graph is not minimally infinitesimally rigid, essentially by applying Theorem 3.20 at some minimally rigid subgraph. The third theorem, Theorem 3.24, is an attempt in taking advantage of the graph structure, in particular it is shown that degree 1 vertices may be ignored when determining the dimensional exponent, thus for example a 2-chain and a 10-chain can have the same exponent. The last theorem shows that the exponents obtained by Theorem 3.20 are asymptotically correct; that is, the error is of the order $O(k^{-2})$.

Theorem 3.20 *Let $G_{k+1,m}$ be a connected graph that is minimally infinitesimally rigid in $\mathbb{R}^d$, $d \ge 2$ and let E be a compact subset of $\mathbb{R}^d$ with* $\dim E > d - \frac{1}{k+1}$. *Then*

$$\mathcal{L}^m(\Delta(G_{k+1,m}, E^{k+1})) > 0\,.$$

Remark 3.21 We shall see in the proof of Theorem 3.20 in Sect. 5 that if $G_{k+1,m}$ is not connected, the proof naturally breaks down into consideration of the connected components of the graph.

Theorem 3.22 *Let $G_{k+1,m}$ be a graph without isolated vertices and let E be a compact subset of $\mathbb{R}^d$, $d \ge 2$ with* $\dim E > d - \frac{1}{k+1}$. *Let H be a maximally independent subset of edges of $G_{k+1,m}$. Then*

$$\dim \Delta(G_{k+1,m}, E^{k+1}) = |H|$$

and

$$\mathcal{H}^{|H|}\left(\Delta(G_{k+1,m}, E^{k+1})\right) > 0\,.$$

Remark 3.23 It should be pointed out that for our results we only work out the case where $k > d$. Theorem 3.20 for $k \leq d$ was worked out in [11] and the better threshold $\dim E > \frac{dk+1}{k+1}$ was obtained. Theorem 3.22 then follows as a consequence, since when $k \leq d$, the only minimally infinitesimally rigid graph is the complete graph on $k+1$ vertices and the independence condition of Theorem 3.22 is always satisfied (see Theorem 6.5) so H can be taken to be the edge set of $G_{k+1,m}$ and in that case Theorem 3.22 is a consequence of Theorem 3.20 by an application of Fubini's theorem.

Theorem 3.24 (Deforestation) *Let $G_{k+1,m}$, $m \geq 2$ be a graph without isolated vertices with a vertex v of degree 1 and let G_1 be the resulting graph when v is removed from $G_{k+1,m}$. Iterate this process obtaining a sequence $G_1, \ldots, G_n$, until G_n has no more such vertices or when $G_n = K_2$. Then in using Theorem 3.20 or Theorem 3.22 with $G_{k+1,m}$, the dimensional threshold for E obtained may be taken to be*

$$d - \frac{1}{k+1-n}\,.$$

Remark 3.25 Thus trees disjoint from the rest of the graph except for the root can be ignored by applying Theorem 3.24.

We shall also see that our results are fairly sharp in the sense that the sharp dimensional threshold must in general tend to d as the number of vertices grows.

Theorem 3.26 *Let E be a compact subset of $\mathbb{R}^d$, $d \geq 2$ of Hausdorff dimension $s \in (0, d)$. Then the conclusions of Theorems 2.1, 3.20 and 3.22 do not in general hold if $s < d - \frac{\binom{d}{2}}{k}$.*

In dimension $d = 2$, the difference between the thresholds in Theorems 3.20, 3.22 and 3.26 is not very large, $2 - \frac{1}{k+1}$ versus $2 - \frac{1}{k}$. In higher dimensions the gap increases, but Theorem 3.26 still shows that the correct dimensional threshold must in general tend to d as the number of vertices grows.

4 The Lebesgue Measure on the Reduced Moduli Space of Congruence Classes

In this section we describe a smooth manifold and a natural measure on it that functions as the natural moduli space for congruence classes of nonsingular $(k+1)$-point configurations in Euclidean space $\mathbb{R}^d$. We take care to describe all relevant change of coordinates involved and their effect on natural measures. We finally establish that if one has a compact subspace E of the Euclidean space $\mathbb{R}^d$, then E

determines a positive Lebesgue measure of distance types of a given configuration type with $L = \frac{d(d+1)}{2} + d(k-d)$ distances measured implies it determines a positive Lebesgue measure of congruence types.

We will think of a $(k+1)$-point configuration in $\mathbb{R}^d$ as an arbitrary choice of a single point in $\mathbb{R}^{d(k+1)}$. We label this configuration as $(x_1, \dots, x_{k+1})$ with $x_j \in \mathbb{R}^d$. By the Mazur-Ulam theorem isometries of real normed spaces are affine transformations, so we will consider congruence classes under under affine transformations. Throughout this section, $k \geq d$ is assumed and we say that the configuration above is nonsingular if its first $(d+1)$ vectors $\{x_1, \dots x_{d+1}\}$ are affinely independent, or equivalently, if $\{x_2 - x_1, \dots, x_{d+1} - x_1\}$ is a basis of $\mathbb{R}^d$. This condition is equivalent to the determinant of the matrix with these column vectors being nonzero, and hence defines an open subset of $\mathbb{R}^{d(k+1)}$. In fact, it the complement of the proper variety where the determinant quantity mentioned is zero, so only excludes a set of Hausdorff codimension one. We denote the space of these nonsingular $(k+1)$-point configurations in $\mathbb{R}^d$ by $N(k, d)$.

Remark 4.1 Since the dimensional threshold we obtain for E is $s_d(k) = d - \frac{1}{k+1}$, the Hausdorff dimension of the sets $E^{k+1} \subset \mathbb{R}^{d(k+1)}$ we consider will be larger than $d(k+1) - 1$, so for our purposes elimination of a set of Hausdorff codimension 1 will have no negative effect.

Remark 4.2 One could say we are being overly restrictive by requiring the first $d+1$ coordinates to be affinely independent and instead should require that some choice of $d+1$ coordinates out of the $k+1$ available be affinely independent instead. This would cover a larger set of configurations. However for any other choice of $d+1$ coordinates, a similar analysis to what will follow can be done with similar results. The resulting more general configuration set would be a finite union of sets of the form $N(k, d)$ that we are currently analyzing where the union is over all possible choices of $d+1$ coordinates out of the $k+1$ available, and so this more general configuration set would also be understood by the subsequent analysis.

4.1 Step 1: Passage to Origin Pinned Configurations

Given a nonsingular configuration $(x_1, \dots, x_{k+1})$ we define the associated origin-pinned configuration as $(u_1, \dots u_k)$ where $u_j = x_{j+1} - x_1$ for all $1 \leq j \leq k$. The first d-resulting vectors of this process form an invertible matrix whose columns are $u_1, \dots, u_d$ as these vectors were required to be linearly independent and hence are a basis of $\mathbb{R}^d$. Thus we will write the associated origin-pinned configuration as $(\mathbb{A}, u_{d+1}, \dots, u_k)$.

The space for nonsingular origin-pinned configurations is hence identified as $GL_d(\mathbb{R}) \times \mathbb{R}^{d(k-d)}$.

To summarize then, we have a map

$$\pi : N(k, d) \to GL_d(\mathbb{R}) \times \mathbb{R}^{d(k-d)}$$

given by

$$\pi(x_1, \ldots, x_{k+1}) = (\mathbb{A}, u_{d+1}, \ldots, u_k).$$

This map is equivalent to passage to translation classes of nonsingular configurations of $(k+1)$ points in $\mathbb{R}^d$.

Letting $\mathcal{L}^m$ denote the Lebesgue measure on $\mathbb{R}^m$, it is not hard to see that the push forward measure $\pi_*(\mathcal{L}^{d(k+1)})$ is the degenerate infinity measure that assigns every nonempty open set infinite measure. Thus either we have to push forward $\mathcal{L}^{d(k+1)}$ restricted to a compact set or just start with a natural measure on pinned configurations. We will do the latter.

There are some choices of natural measures on $GL_d(\mathbb{R}) \times \mathbb{R}^{d(k-d)}$ as $GL_d(\mathbb{R})$ is an open subset of $\mathbb{R}^{d\times d}$. One is $\mathcal{L}^{d^2} \times \mathcal{L}^{d(k-d)}$ and the other is $H \times \mathcal{L}^{d(k-d)}$ where H is a Haar measure of $GL_d(\mathbb{R})$. As $GL_d(\mathbb{R})$ is unimodular, left invariant Haar measures are also right invariant and vice versa and such a measure is unique up to scaling by a positive real number. It is easy to check that $H = |\det(\mathbb{A})|^{-d}\mathcal{L}^{d^2}$ where $\mathbb{A} \in GL_d(\mathbb{R}) \subset \mathbb{R}^{d\times d}$ serves as a coordinate through its entries.

The two competing measures $H \times \mathcal{L}^{k(d-k)}$ and $\mathcal{L}^{d^2} \times \mathcal{L}^{k(d-k)}$ are hence absolutely continuous with respect to each other with smooth Radon-Nikodym derivatives outside sets of measure zero. When this happens we shall call the measures **nicely related**.

It turns out that for what follows, with push forwards, it will be better for us to use the measure $H \times \mathcal{L}^{k(d-k)}$ for the origin pinned configuration space $GL_d(\mathbb{R}) \times \mathbb{R}^{d(k-d)}$, so we will do so. Using the Lebesgue measure instead would yield nicely related measures but without perhaps as concrete descriptions.

4.2 *Step 2: Analysis of the Orthogonal Group Action on Pinned Configurations and "Moving Frames"*

From the last subsection, we have seen that a nonsingular origin pinned configuration is given by $(\mathbb{A}, u_{d+1} \ldots u_k)$ where $\mathbb{A} \in GL_d(\mathbb{R})$. Congruence classes of such pinned configurations are given by $O_d(\mathbb{R})$-orbits of the action given by $B \in O_d(\mathbb{R})$ acting as

$$B : (\mathbb{A}, u_{d+1}, \ldots, u_d) \mapsto (B\mathbb{A}, Bu_{d+1}, \ldots, Bu_k).$$

Since $O_d(\mathbb{R})$ acts on both the matrix $\mathbb{A}$ and the remaining vectors, to simplify future formulas, we use the method of "moving frames", which we now describe. The columns of $\mathbb{A}$ form a basis of $\mathbb{R}^d$ and we may replace u_j when $j > d$ by their coordinate vectors $c_j = (c_{j1}, \ldots, c_{jd})^T$ in the basis formed by the columns of $\mathbb{A}$. As $\mathbb{A}$ changes, this basis changes, hence the name "moving frames". We have $\mathbb{A}c_j = u_j$.

With respect to these moving frames the action of $B \in O_d(\mathbb{R})$ simplifies, as we have $BAc_j = Bu_j$, which means the coordinate vectors c_j are unchanged by the $O_d(\mathbb{R})$-action. In other words if we re-encode pinned configurations as

$$(u_1, \ldots, u_d, c_{d+1}, \ldots, c_k)$$

then the $O_d(\mathbb{R})$ action only acts on the first d-coordinates and leaves the remaining coordinates unchanged:

$$B \cdot (\mathbb{A}, c_{d+1}, \ldots, c_k) = (B\mathbb{A}, c_{d+1}, \ldots, c_k).$$

The change to the new coordinate system amounts to the transformation

$$(\mathbb{A}, u_{d+1}, \ldots, u_k) \to (\mathbb{A}, \mathbb{A}^{-1}u_{d+1}, \ldots, \mathbb{A}^{-1}u_k).$$

which has a Jacobian determinant $|\det(\mathbb{A})|^{k-d}$, so the Lebesgue measure gets divided by $|\det(\mathbb{A})|^{k-d}$ when switching from u-coordinates to c-coordinates. Thus the original $H \times \mathcal{L}^{k(d-k)}$ measure becomes the $H \times |\det(\mathbb{A})|^{k-d}\mathcal{L}^{k(d-k)}$ measure in this new coordinate system, where H is the Haar measure of $GL_d(\mathbb{R})$.

4.3 Step 3: Quotienting the Action of the Orthogonal Group

Using the moving frame coordinate system, the nonsingular origin-pinned $(k+1)$-point configurations of $\mathbb{R}^d$ are the space $GL_d(\mathbb{R}) \times \mathbb{R}^{d(k-d)}$ with the natural measure

$$H \times |\det(\mathbb{A})|^{k-d}\mathcal{L}^{k(d-k)}.$$

The action of $O_d(\mathbb{R})$ is given by $B \cdot (\mathbb{A}, c_{d+1}, \ldots, c_k) = (B\mathbb{A}, c_{d+1}, \ldots, c_k)$ so the final space for nonsingular congruence classes of configurations of $(k+1)$-points in $\mathbb{R}^d$, which we will call $M(k, d)$, is given by

$$M(k, d) = (O_d(\mathbb{R})\backslash GL_d(\mathbb{R})) \times \mathbb{R}^{d(k-d)}.$$

To make this more explicit, we recall the LU-decomposition of nonsingular matrices that comes from the Gram-Schmidt process. Any $X \in GL_d(\mathbb{R})$ can be written $X = BC$ for unique $B \in O_d(\mathbb{R})$, $C \in L$ where L is the Lie group of upper triangular matrices with positive real entries on the diagonal.

This means as manifolds (but not as groups) $GL_d(\mathbb{R})$ is diffeomorphic to $O(d) \times L$ where the left action of $O(d)$ on $GL_d(\mathbb{R})$ translates to an action on $O(d) \times L$ where $O(d)$ acts only on the left factor by left translation. Thus $O(d)\backslash GL_d(\mathbb{R})$ is naturally diffeomorphic to the Lie group L.

The push forward of the Haar measure H on $GL_d(\mathbb{R})$ is a right Haar measure on L. To see this, let $\pi : GL_d(\mathbb{R}) \to L$ be the map that assigns to input $A = BC$, $B \in O(d)$, $C \in L$, the output L. This is smooth by analysis of the Gram-Schmidt process yielding the LU-decomposition. The push forward measure $m = \pi_*(H)$ can be shown to be the right Haar measure by checking it has the properties uniquely determining these.

Thus the space of congruence classes of nonsingular $(k+1)$-point configurations in $\mathbb{R}^d$ is $L \times \mathbb{R}^{d(k-d)}$ and a natural measure on it is $H' \times |\det(A)|^{k-d}\mathcal{L}^{d(k-d)}$ where H' is the right Haar measure on the Lie group L. Note that when $A = BC$ with $B \in O(d)$, $C \in L$, $|\det(A)| = |\det(C)|$ so the $|\det(A)|$ Jacobian factor is determined by the L-part in the LU-decomposition.

Putting this all together we have:

Theorem 4.3 *The space of congruence classes of nonsingular* $(k+1)$*-point configurations in* $\mathbb{R}^d$ *is* $L \times \mathbb{R}^{d(k-d)}$ *and a natural measure on it is* $H' \times |\det(C)|^{k-d}\mathcal{L}^{d(k-d)}$ *where* L *is the Lie group of upper triangular matrices with positive real entries on the diagonal,* H' *is the right Haar measure on it and* C *is the* L*-part from the* LU*-decomposition.*

4.4 Deduction of the Main Theorem

Any smooth map $F : \mathbb{R}^{d(k+1)} \to \mathbb{R}^m$ that respects congruence induces a smooth map $F : M(k,d) \to \mathbb{R}^m$. This is because all the changes of coordinates we did were diffeomorphisms or modding out Lie group actions and this is a standard fact for those.

When we apply this to a distance type map F, it is clear that there can only be a finite to one correspondence between congruence types and distance types when

$$m = \dim(M(k,d)) = \frac{d(d+1)}{2} + d(k-d) = d(k+1) - \binom{d+1}{2}. \tag{4.1}$$

When this is the case, Sard's theorem says the regular values of F have full measure, and the preimage of them will be discrete (and hence countable) in $M(k,d)$. About each of these preimage points, F is a local diffeomorphism and hence is a diffeomorphism on an neighborhood which preserves the property of having positive measure.

Any result that shows a compact E determines a positive Lebesgue measure worth of distance types in $\mathbb{R}^m$ translates to a positive Lebesgue measure worth of congruence types in $M(k,d)$. (Converse need not hold as the image of a set of positive measure under a smooth map can have measure zero in the codomain). Thus our main theorem follows once we establish Theorem 3.20 since Corollary 6.9 implies a minimally infinitesimally rigid graph must have m edges where m is given by (4.1).

Remark 4.4 It is worth noting that $M(k,d) = M(d,d) \times \mathbb{R}^{d(k-d)}$ where $M(d,d) = L$ is the moduli space for congruence classes of nondegenerate d-simplexes in $\mathbb{R}^d$.

5 Graph Distances of Subsets of $\mathbb{R}^d$

Our goal is to prove that

$$\mathcal{L}^m(\Delta(G_{k+1,m}, E^{k+1})) > 0 \tag{5.1}$$

for some dimensional threshold s_k with $\dim E > s_k$. Here we may assume $G_{k+1,m}$ is connected, since we have

Proposition 5.1 *If $G_1, \ldots, G_n$ are the connected components of $G_{k+1,m}$ on $k_1, \ldots, k_n$ vertices respectively, then for Cartesian products E^{k+1} where $E \subseteq \mathbb{R}^d$ we have*

$$\Delta(G_{k+1,m}, E^{k+1}) = \Delta(G_1, E^{k_1}) \times \cdots \times \Delta(G_n, E^{k_n}) .$$

Proof It is clear (after reordering the vertices if necessary) that

$$f_{G_{k+1,m}} = (f_{G_1}, \ldots, f_{G_n})$$

where $f_{G_{k+1,m}}, f_{G_j}$ are the corresponding distance functions of $G_{k+1,m}, G_j$. The result follows. □

Thus we only need to consider connected graphs, and requiring of $G_{k+1,m}$ to be connected in Theorem 3.20 is not an essential restriction. If (5.1) does not hold, it may be the case that the dimension of the $G_{k+1,m}$-distance set is not full. Theorem 3.22 then provides its Hausdorff dimension and positivity of the Hausdorff measure.

We now describe some examples.

5.1 *Examples of Distance Sets*

In the case where $G_{k+1,m}$ is the complete graph K_2 on 2 vertices and $E \subseteq \mathbb{R}^d$, we recover the distance set of E

$$\Delta(K_2, E^2) = \{|x - y| : x, y \in E\} .$$

Now consider the complete graph K_4. see Fig. 2. Let $d = 2$ and $E = \mathbb{R}^2$. We will directly show that $\mathcal{L}^6(\Delta(K_4, \mathbb{R}^8)) = 0$. This is expected, as we will also show that $\dim \Delta(K_4, \mathbb{R}^8) = 5$ and $\mathcal{H}^5(\Delta(K_4, \mathbb{R}^8)) > 0$. Split the configurations $\boldsymbol{x} \in \mathbb{R}^8$ into two sets A_1 and A_2, $A_j \subset \mathbb{R}^8$, with $\boldsymbol{x} \in A_1$ iff (x^1, x^2, x^3, x^4) are in convex position and A_2 the rest. We will work with A_1 but A_2 is treated similarly.

Let t_{ij} denote the distance from the vertex i to j. By using Euler's theorem for convex quadrilaterals we may obtain the following equation,

$$t_{24}^2 = t_{23}^2 + t_{14}^2 - t_{13}^2 + 2t_{12}t_{34}\cos(\theta - \psi) . \tag{5.2}$$

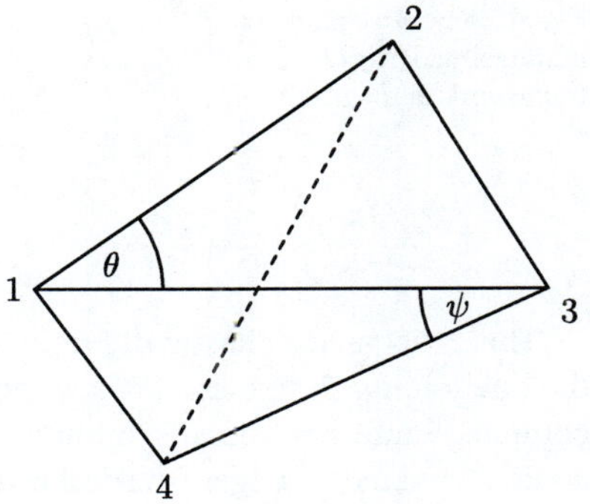

Fig. 2 A framework of K_4 (dashed edge for emphasis)

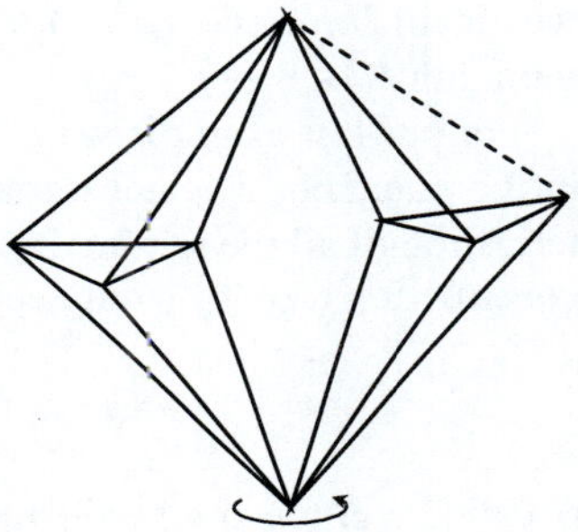

Fig. 3 A framework of the 'double banana' $G_{8,18}$ graph

Here

$$\theta = \cos^{-1}\left(\frac{t_{12}^2 + t_{13}^2 - t_{23}^2}{2t_{12}^2 t_{13}^2}\right), \psi = \cos^{-1}\left(\frac{t_{13}^2 + t_{34}^2 - t_{14}^2}{2t_{13}^2 t_{34}^2}\right).$$

Let $t_{24} = g(\tilde{t})$ where

$$\tilde{t} = (t_{12}, t_{13}, t_{14}, t_{23}, t_{34}).$$

Thus,

$$\mathcal{L}^6(\Delta(K_4, A_1)) \leq \int \int_{t_{24}=g(\tilde{t})} dt_{24} d\tilde{t} = 0.$$

This happens because Eq. (5.2) makes us integrate over the zero-dimensional set (it is a singleton) $t_{24} \in \{g(\tilde{t})\}$. Similarly we may obtain $\mathcal{L}^6(\Delta(K_4, A_2)) = 0$. Thus K_4 in $d = 2$ cannot possibly give us a dimensional threshold since even $E = \mathbb{R}^2$ has a K_4-distance set of zero measure. This happened because the graph had too many edges. Not only $\Delta(K_4, \mathbb{R}^8)$ is a $\mathcal{L}^6$-null set, but in fact its dimension is less than 6. By using Corollary 5.6, we find that its dimension is 5. Then using Theorem 3.22 we see that it has positive $\mathcal{H}^5$-measure.

Now consider the 'double banana' graph $G_{8,18}$ on $\mathbb{R}^3$ (dashed edge for emphasis, but it is in the edge set of $G_{8,18}$) at Fig. 3.

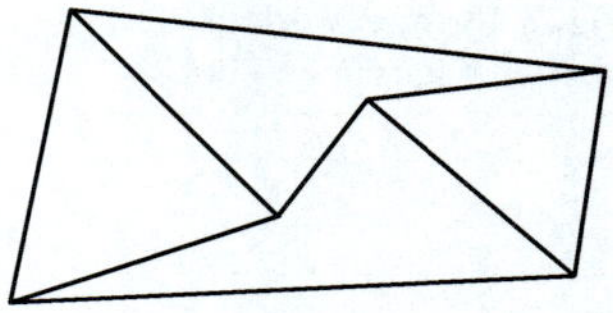

Fig. 4 A non-triangulated infinitesimally rigid framework in the plane

This is not an infinitesimally rigid graph. Each banana may be freely rotated about the line joining the banana ends without altering the edge lengths. Yet it may not be completed into a minimally infinitesimally rigid graph by adding edges because it contains redundant edges (the dashed one, for instance). The solid edges form a maximally independent set H of edges of $G_{8,18}$, thus by an application of Theorem 3.22 we obtain $\dim \Delta(G_{8,18}, E^8) = 17$ and $\mathcal{H}^{17}(\Delta(G_{8,18}, E^8)) > 0$ for $E \subseteq \mathbb{R}^3$ compact with $\dim E > 3 - \frac{1}{9}$.

For our last example we consider a graph that is minimally infinitesimally rigid in the plane but it is not a triangulation. In the plane we have a complete characterization of all the minimally infinitesimally rigid graphs provided by Henneberg's construction (see [14] and Sect. 6 of [18]), which we briefly describe. A sequence

$$K_2 \hookrightarrow G_{3,m(1)} \hookrightarrow \cdots \hookrightarrow G_{k+1,m(k-1))}$$

is called a *Henneberg construction* if each subsequent graph is obtained from the previous by either adding a vertex and joining it to two previous vertices or by subdividing an existing edge with a new vertex which also gets joined to a third vertex. It is the second operation that allows us to produce minimally rigid graphs that are not triangulations, as Fig. 4 shows.

5.2 *A Sharp Upper Bound for the Dimension of the Distance Set*

In this section we determine the Hausdorff dimension of $\Delta(G_{k+1,m}, \mathbb{R}^{d(k+1)})$.

If $G_{k+1,m}$ is a minimally infinitesimally rigid graph in $\mathbb{R}^d$, then from Corollary 6.9, it must have

$$m = d(k+1) - \binom{d+1}{2}. \tag{5.3}$$

We may say that $G_{k+1,m}$ is minimally infinitesimally rigid in $\mathbb{R}^d$ when its edge set is independent and any edge added to $G_{k+1,m}$ turns the rows of $DF_{G_{k+1,m}}$ into a linearly dependent set of vectors, as the next proposition shows.

Proposition 5.2 *Let $G_{k+1,m}$ be a graph. Then the set of edges of $G_{k+1,m}$ is independent in $\mathbb{R}^d$ and may not be enlarged while retaining independence if and only if $G_{k+1,m}$ is minimally infinitesimally rigid.*

Proof If $G_{k+1,m}$ is minimally infinitesimally rigid in $\mathbb{R}^d$, then by definition, for generic configurations $\boldsymbol{x} \in \mathbb{R}^{d(k+1)}$ the kernel of $DF_{G_{k+1,m}}(\boldsymbol{x})$ has the smallest dimension possible (in view of Theorem 6.6 and the dimension of the rotation group). Thus the edge set can not be enlarged while retaining independence, since a larger independent set of edges would produce an even smaller kernel.

On the other hand, assume $G_{k+1,m}$ has an independent in $\mathbb{R}^d$ edge set that may not be enlarged while retaining independence. We have just argued that $G_{k+1,m}$ cannot contain a minimally infinitesimally rigid proper subgraph. Assuming then that $G_{k+1,m}$ is not minimally infinitesimally rigid itself, for generic configurations $\boldsymbol{x} \in \mathbb{R}^{d(k+1)}$ we know that $\mathcal{V}(G_{k+1,m}, \boldsymbol{x})$ properly contains $\mathcal{D}(\boldsymbol{x})$. Thus there must be a set of edges $H \subset K_{k+1}$ disjoint from those of $G_{k+1,m}$ with

$$\mathcal{V}(H \cup G_{k+1,m}, \boldsymbol{x}) = \mathcal{D}(\boldsymbol{x}) .$$

But that implies $\operatorname{rank} D_{F_{H \cup G_{k+1,m}}}(\boldsymbol{x}) > \operatorname{rank} D_{F_{G_{k+1,m}}}(\boldsymbol{x})$, a contradiction to the assumption that $G_{k+1,m}$ has an edge set that may not be enlarged while retaining independence. □

Proposition 5.3 *If the edge set of $G_{k+1,m}$ is independent in $\mathbb{R}^d$, then a minimally infinitesimally rigid (in $\mathbb{R}^d$) graph $\overline{G}_{k+1,\overline{m}}$ containing $G_{k+1,m}$ exists.*

Proof If $k \leq d$, we just complete $G_{k+1,m}$ to the complete graph K_{k+1} since that is the only rigid graph on $k + 1$ vertices in $\mathbb{R}^d$ (see Theorem 6.5).

Otherwise we pick $\boldsymbol{x}$ at random from a continuous distribution (say, the Gaussian distribution) on $\mathbb{R}^{d(k+1)}$. Since the set of singular frameworks is a proper algebraic variety, it has Lebesgue measure zero and we have almost certainly (that is, with probability 1) picked a generic framework.

As long as the property of independence with respect to $\boldsymbol{x}$ is retained, we keep adding edges to $G_{k-1,m}$ until no more edges may be added. We then end up with a graph that is minimally infinitesimally rigid. □

Remark 5.4 The graph $\overline{G}_{k+1,\overline{m}}$ need not be unique. For instance, if $G_{k+1,m}$ is a tree, for large enough k there are many different minimally infinitesimally rigid graphs it may complete to.

Theorem 5.5 *Let $G_{k+1,m}$ be a connected graph and H a maximally independent in $\mathbb{R}^d$ subset of edges of $G_{k+1,m}$. Then*

$$\dim \Delta(G_{k+1,m}, \mathbb{R}^{d(k+1)}) = |H| .$$

Proof Let W_{ij} denote the plane $\{\boldsymbol{x} : x^i = x^j\}$. The map $f_{G_{k+1,m}}$ is smooth away from $W = \bigcup_{ij \in G_{k+1,m}} W_{ij}$ and the rank of its total derivative does not exceed $|H|$

there. Thus, since $f_{G_{k+1,m}}$ has regular configurations away from W, we see that $f_{G_{k+1,m}}(\mathbb{R}^{d(k+1)} \setminus W)$ has dimension $|H|$. Now consider $f_{G_{k+1,m}}$ restricted to W_{ij}. There the function is smooth away from $W_{ij} \cap W_{kl}$ for $kl \in G_{k+1,m}$ with $kl \neq ij$. The rank of the derivative is less than or equal to $|H|$, so inductively $f_{G_{k+1,m}}(W_{ij})$ has dimension less than or equal to $|H|$. □

Corollary 5.6 *If $G_{k+1,m}$ is a connected graph that contains a minimally infinitesimally rigid (in $\mathbb{R}^d$) subgraph G^*_{k+1,m^*}, then*

$$\dim \Delta(G_{k+1,m}, \mathbb{R}^{d(k+1)}) = m^* .$$

5.3 Bounds on the Number of Noncongruent Realizations

Let $G_{k+1,m}$ be minimally infinitesimally rigid in $\mathbb{R}^d$ and let $\boldsymbol{x} \in \mathbb{R}^{d(k+1)}$.

We consider the set of preimages of $f_{G_{k+1,m}}(\boldsymbol{x})$:

$$N_{\boldsymbol{x}} := \{\boldsymbol{y} : f_{G_{k+1,m}}(\boldsymbol{y}) = f_{G_{k+1,m}}(\boldsymbol{x})\} .$$

Define the equivalence relation $\boldsymbol{y} \sim \boldsymbol{z}$ by $\boldsymbol{y} \in M_{\boldsymbol{z}}$, where $M_{\boldsymbol{z}}$ is defined in (3.2) to be the set of configurations congruent to $\boldsymbol{z}$. The set $N_{\boldsymbol{x}}$ is defined by the system of quadratic equations

$$|y^i - y^j|^2 = |x^i - x^j|^2, \quad ij \in G_{k+1,m} ,$$

and results on bounds of the Betti numbers of semi-algebraic varieties by Oleinik and Petrovskii, Thom and Milnor (see [19, 20, 23]) allow us to conclude that $N_{\boldsymbol{x}}$, hence (since ISO(d) has two connected components), $N_{\boldsymbol{x}}/\sim$, has less than $C_{d,k} \cdot 2^{dk}$ connected components, for some $C_{d,k} > 0$. For a better bound see [6]. In particular, when $\boldsymbol{x}$ is regular valued, we may conclude that there are at most $C_{d,k} \cdot 2^{dk}$ preimages of $f_{G_{k+1,m}}(\boldsymbol{x})$ up to congruences by Proposition 6.10. If $\boldsymbol{x}$ is not regular valued, it is possible for noncongruent preimages to lie in the same connected component, but in the argument to follow we will avoid singular frameworks. For our purposes we only need the fact that if the singular configurations are removed, then the preimage set $N_{\boldsymbol{x}}$ is finite up to congruences, with the bound independent of $\boldsymbol{x}$.

5.4 The Proof of the Dimensional Threshold

We prove Theorem 3.20.

Proof Let $G = G_{k+1,m}$ to ease subscript use. Fix $E \subset \mathbb{R}^d$ compact, and let $\mu = \mu_s$ be a Borel probability measure supported on E, with Frostman exponent s. Thus there exists some constant $C_\mu > 0$ with $\mu(B(x,r)) \le C_\mu r^s$ for all balls $B(x,r)$ of radius $r > 0$, and we may choose $s < \dim E$ arbitrarily close to $\dim E$. (See [25], Chap. 8 for the existence and properties of such measures). In particular, we may choose

$$s > d - \frac{1}{k+1} . \tag{5.4}$$

Let us first prove that the set of singular frameworks X is a null set for μ^{k+1}:

$$\mu^{k+1}(X) = 0 .$$

Observe that μ^{k+1} is a Frostman measure of exponent $s(k+1)$. This follows easily from the fact that any ball $B(\boldsymbol{x}, r)$ is contained in a concentric cube $Q = Q_1 \times \cdots \times Q_{k+1}$ of side $2r$, where each Q_j in turn is contained in a ball $B(x^j, 2r)$. Since $\mu^{k+1}(Q) = \mu(Q_1) \cdots \mu(Q_{k+1}) \lesssim r^{s(k+1)}$ and $s(k+1) > \dim(X)$, which we may assume since the dimension of E is big enough (by (3.1) and (5.4)) and using the fact that sets of positive measure of a Frostman measure have dimension greater or equal to the Frostman exponent (see Lemma 6.14), we conclude that X is a null set for μ^{k+1}.

Let $\delta > 0$ be such that

$$\mu^{k+1}(E^{k+1} \setminus X_\delta) > 1/2$$

for X_δ the δ-neighborhood of X defined by $X_\delta = \{\boldsymbol{y} \in \mathbb{R}^{d(k+1)} : |\boldsymbol{y} - \boldsymbol{x}| < \delta\}$. Such a δ exists since X is closed and $X = \bigcap_{\delta>0} X_\delta$. We want to avoid getting close to X because our named constants in the arguments to follow blow up near it. Let $A = E^{k+1} \setminus X_\delta$ and let $\nu(\boldsymbol{t})$ be the pushforward of $\mu^{k+1}(\boldsymbol{x})$ by $f_G|_A$, that is, for any measurable function $g(\boldsymbol{t})$ the integral $\int g d\nu$ is defined by

$$\int g(\boldsymbol{t}) d\nu(\boldsymbol{t}) = \int_A g\left((|x^i - x^j|)_{ij\in G}\right) d\mu(x^1) \cdots d\mu(x^{k+1}) .$$

From now on we shall write $d\mu^{k+1}(\boldsymbol{x})$ for $d\mu(x^1) \cdots d\mu(x^{k+1})$. We shall show that $\nu(\mathbb{R}^m) > 0$ and $\nu \in L^2(\mathbb{R}^m)$ implying $\mathcal{L}^m(\operatorname{supp} \nu) > 0$ which concludes the proof since $\operatorname{supp} \nu \subset \Delta(G, E^{k+1})$. For the first claim,

$$\nu(\mathbb{R}^m) = \mu^{k+1}(E^{k+1} \setminus X_\delta) > 1/2 .$$

Now we will prove that $\nu \in L^2(\mathbb{R}^m)$. Let $\nu^\epsilon = \phi^\epsilon * \nu$, where $\phi^\epsilon(\boldsymbol{t}) = \epsilon^{-m}\phi(\epsilon^{-1}\boldsymbol{t})$ and $\phi \in C_c^\infty(\mathbb{R}^m)$ is a nonnegative radial function with $\int \phi = 1$, $\phi \le 1$ and $\operatorname{supp} \phi \subset B(0, 2)$. Here we denote by χ the characteristic function of a set. We have,

$$\begin{aligned} \nu^\epsilon(t) &= \int_A \epsilon^{-m}\phi\left(\epsilon^{-1}(f_G(\boldsymbol{x}) - \boldsymbol{t})\right) d\mu^{k+1}(\boldsymbol{x}) \\ &\le \int_A \epsilon^{-m}\chi\left\{\left|f_G(\boldsymbol{x}) - \boldsymbol{t}\right| < 2\epsilon\right\} d\mu^{k+1}(\boldsymbol{x}) \,. \end{aligned} \tag{5.5}$$

By Lemma 6.12, as $\epsilon \to 0$ we have $\nu^\epsilon \to \nu$ in the weak topology of the dual of $C_0(\mathbb{R}^m)$. We conclude that $\liminf \|\nu^\epsilon\|_2 \ge \|\nu\|_2$ from Lemma 6.13, and thus it suffices to bound $\liminf \|\nu^\epsilon\|_2$. By an application of the triangle inequality on (5.5) we have,

$$\begin{aligned} &\|\nu^\epsilon\|_2^2 \\ &\le \epsilon^{-2m}\int\int_A\int_A \chi\left\{\left|f_G(\boldsymbol{x}) - \boldsymbol{t}\right| < 2\epsilon\right\}\chi\left\{\left|f_G(\boldsymbol{y}) - \boldsymbol{t}\right| < 2\epsilon\right\} d\mu^{k+1}(\boldsymbol{x})d\mu^{k+1}(\boldsymbol{y})d\boldsymbol{t} \\ &\le \epsilon^{-2m}\int_{|\boldsymbol{t}|<2\epsilon} d\boldsymbol{t}\cdot\int_A\int_A \chi\left\{\left|f_G(\boldsymbol{x}) - f_G(\boldsymbol{y})\right| < 4\epsilon\right\} d\mu^{k+1}(\boldsymbol{x})d\mu^{k+1}(\boldsymbol{y}) \\ &\lesssim \epsilon^{-m}\int_A\int_A \chi\left\{\left|f_G(\boldsymbol{x}) - f_G(\boldsymbol{y})\right| < 4\epsilon\right\} d\mu^{k+1}(\boldsymbol{x})d\mu^{k+1}(\boldsymbol{y}) \,. \end{aligned}$$

Consider $\boldsymbol{y}$ to be fixed now. Note that it is a regular configuration since it belongs to A. The condition that the images of $\boldsymbol{x}$ and $\boldsymbol{y}$ are ϵ-close is giving us $\lesssim 2^{dk}$ open sets of $\mathbb{R}^{d(k+1)}$ where $\boldsymbol{x}$ may lie (as explained in Sect. 5.3). Let $U_1, \ldots, U_l$ be those open sets, and let $Z = \{\boldsymbol{z}_1, \ldots, \boldsymbol{z}_l\}$ be such that each $\boldsymbol{z}_j \in U_j \cap A$ (potentially picking less than l configurations, if some intersections are empty). Cover the compact Riemannian manifold $O_d(\mathbb{R})$ by ϵ-balls $T_1^\epsilon, \ldots, T_{K(\epsilon)}^\epsilon$ of finite (uniformly in ϵ) overlap with centers $g_1, \ldots, g_{K(\epsilon)}$.

Then the set $\{\boldsymbol{x} \in A : |f_G(\boldsymbol{x}) - f_G(\boldsymbol{y})| < \epsilon\}$ is a subset of the set

$$\bigcup_{\boldsymbol{z}\in Z}\bigcup_{k=1}^{K(\epsilon)}\{\boldsymbol{x} \in A : |(x^i - x^j) - g_k(z^i - z^j)| < c\epsilon, \quad \forall ij \in G\}$$

for some $c > 0$ that depends continuously on $\boldsymbol{y}$ (as Proposition 6.10 shows, f_G is bi-Lipschitz in $U_1, \ldots, U_l$, once congruences are identified). Since A is a compact set, c attains a maximum value, so pick such c to lift the dependence on $\boldsymbol{y}$.

Since $|Z| \lesssim 2^{dk}$, it follows that,

$$\begin{aligned} &\|\nu^\epsilon\|_2^2 \\ &\lesssim \epsilon^{-m}\int_A\sum_{\boldsymbol{z}\in Z}\sum_{k=1}^{K(\epsilon)}\int_A \chi\{|(x^i - x^j) - g_k(z^i - z^j)| < c\epsilon, \ \forall ij \in G\}d\mu^{k+1}(\boldsymbol{x})d\mu^{k+1}(\boldsymbol{y}) \\ &\lesssim 2^{dk}\epsilon^{-m}\sum_{k=1}^{K(\epsilon)}\int_A\int_A \chi\{|(x^i - x^j) - g_k(y^i - y^j)| < c\epsilon, \ \forall ij \in G\}d\mu^{k+1}(\boldsymbol{x})d\mu^{k+1}(\boldsymbol{y}) \,. \end{aligned}$$

The volume of the ϵ-balls of $O_d(\mathbb{R})$ is $c_d \cdot \epsilon^{d(d-1)/2}$ for some $c_d > 0$. In what follows, we estimate the value of a function at a point by twice the average of that function around an ϵ-ball and use the fact that they cover $O_d(\mathbb{R})$ with finite overlap to obtain,

$$\begin{aligned}
&\|\nu^\epsilon\|_2^2 \\
&\lesssim \epsilon^{-m} \sum_{k=1}^{K(\epsilon)} \frac{1}{\epsilon^{d(d-1)/2}} \int_{T_k^\epsilon} \int_A \int_A \chi\{|(x^i - x^j) - g(y^i - y^j)| < c\epsilon,\ \forall ij \in G\} \\
&d\mu^{k+1}(x) d\mu^{k+1}(y) dg \\
&\lesssim \epsilon^{-dk} \int_{O_d(\mathbb{R})} \int_A \int_A \chi\{|(x^i - x^j) - g(y^i - y^j)| < c\epsilon,\ \forall ij \in G\} d\mu^{k+1}(x) d\mu^{k+1}(y) dg\,.
\end{aligned} \tag{5.6}$$

Here we used (5.3) to get $m + d(d-1)/2 = dk$. For $g \in O_d(\mathbb{R})$, define ν_g by

$$\int f(x) d\nu_g(x) = \int \int f(u - gv) d\mu(u) d\mu(v)\,.$$

Let $G' \subset G$ be a spanning tree. We continue (5.6) with

$$\begin{aligned}
&\|\nu^\epsilon\|_2^2 \\
&\lesssim \epsilon^{-dk} \int_{O_d(\mathbb{R})} \int_A \int_A \chi\{|(x^i - x^j) - g(y^i - y^j)| < c\epsilon,\ \forall ij \in G'\} d\mu^{k+1}(x) d\mu^{k+1}(y) dg\,.
\end{aligned} \tag{5.7}$$

Using Lemma 5.10 (to be proved below) on (5.7) we obtain

$$\|\nu^\epsilon\|_2^2 \lesssim \int \int \nu_g^{k+1}(x) dx dg\,. \tag{5.8}$$

Theorem 5.8 shows this integral to be finite for $s > d - \frac{1}{k+1}$, concluding the proof that $\nu \in L^2(\mathbb{R}^m)$ and thus showing that $\mathcal{L}^m(\Delta(G, E^{k+1})) > 0$. □

We may now go a step further and prove Theorem 3.22.

Proof If $G_{k+1,m}$ is any connected graph, and H is a maximally independent subset of the edge set of $G_{k+1,m}$, we may complete H to a minimally infinitesimally rigid graph $\overline{H}_{k+1,\overline{m}}$ (see Proposition 5.3). By using Theorem 3.20, we obtain the nontrivial threshold $d - \frac{1}{k+1}$ for $\Delta(\overline{H}_{k+1,\overline{m}}, E^{k+1})$ to have positive Lebesgue measure. We project $\Delta(\overline{H}_{k+1,\overline{m}}, E^{k+1}) \to \Delta(H, E^{k+1})$ by $(t_{ij})_{ij \in \overline{H}_{k+1,\overline{m}}} \mapsto (t_{ij})_{ij \in H}$ to show that $\Delta(H, E^{k+1})$ has positive Lebesgue measure by Fubini.

Now, projecting $\Delta(G_{k+1,m}, E^{k+1}) \to \Delta(H, E^{k+1})$ by $(t_{ij})_{ij \in G_{k+1,m}} \mapsto (t_{ij})_{ij \in H}$ shows that $\Delta(G_{k+1,m}, E^{k+1})$ has positive $\mathcal{H}^{|H|}$-measure, because the projection is Lipschitz. Lastly, Theorem 5.5 shows that $\dim \Delta(G_{k+1,m}, E^{k+1}) = |H|$.

Moreover, if $G_{k+1,m}$ has connected components $G_1, \dots G_n$, we will obtain a maximally independent subset $H = H_1 \cup \dots \cup H_n$ of $G_{k+1,m}$ where each H_j is a maximally independent subset of G_j for $j = 1, \dots, n$. Using Proposition 5.1 and what we just argued for connected graphs, we again obtain a positive $|H|$-Hausdorff measures worth of distances. □

We now prove Theorem 3.24.

Proof As before let $G = G_{k+1,m}$ to avoid notational clutter. Let σ_t denote the surface measure of the sphere $tS^{d-1} \subset \mathbb{R}^d$ of radius $t > 0$ centered at 0. Let $\phi^\epsilon(x) = \epsilon^{-d}\phi(\epsilon^{-1}x)$ where $\phi \in C_c^\infty(\mathbb{R}^d)$ is a nonnegative radial function with $\phi = 1$ on $B(0, 1)$, $\phi \leq 1$ and $\operatorname{supp} \phi \subset B(0, 2)$. Let $\sigma_t^\epsilon = \phi^\epsilon * \sigma_t$. We note that

$$c \cdot \chi_{\{y:||y|-t|<\epsilon\}}(x) \leq \epsilon\sigma_t^\epsilon(x) \leq C \cdot \chi_{\{y:||y|-t|<2\epsilon\}}(x) \tag{5.9}$$

for some constants $c > 0$ and $C > 0$ depending on ϕ only. Without loss of generality assume $v = k + 1$ and that $\{k, k + 1\}$ is an edge of G. Let $\boldsymbol{t} = (t_{ij})_{ij\in G}$, $t_{ij} \in (0, +\infty)$, and consider the function (on the domain of $\boldsymbol{t}$ just mentioned),

$$\Lambda_{G,\mu}^\epsilon(\boldsymbol{t}) = \int \prod_{ij\in G} \sigma_{t_{ij}}^\epsilon(x^i - x^j)d\mu(x^1)\cdots d\mu(x^{k+1}) .$$

Using (5.9) we see that obtaining a bound $\|\Lambda_{G,\mu}^\epsilon\|_2 \leq M$ with M independent of ϵ is equivalent to obtaining a bound for $\|\nu^\epsilon\|_2$ independent of ϵ, in particular showing that $\nu \in L^2(\mathbb{R}^m)$. Write then

$$\begin{aligned}\Lambda_\mu^\epsilon(\boldsymbol{t}) = \int \prod_{ij\in G, ij\neq k(k+1)} & \sigma_{t_{ij}}^\epsilon(x^i - x^j)d\mu(x^1)\cdots d\mu(x^{k-1}) \\ & \cdot \int \sigma_{t_{k(k+1)}}^\epsilon(x^k - x^{k+1})d\mu(x^k)d\mu(x^{k+1}) \\ = \int \Bigg(\prod_{ij\in G, ij\neq k(k+1)} & \sigma_{t_{ij}}^\epsilon(x^i - x^j)d\mu(x^1)\cdots d\mu(x^{k-1})\Bigg) (\sigma_{t_{k(k+1)}}^\epsilon * \mu)(x^k)d\mu(x^k)\end{aligned}$$

From Lemma 6.15 we know that $\|\sigma_t^\epsilon * \mu\|_{L^1(\mu)} \lesssim 1$ and so using Chebyshev's inequality we may obtain a compact subset $E' \subset E$ with $\mu(E') > 0$ and $\|\sigma_t^\epsilon * \mu\|_{L^\infty(E',\mu)} \lesssim 1$. Denoting by μ' the restriction of μ to E', we obtain a Frostman measure of the same exponent. Denote by G' the graph G with the vertex v removed. Thus we have now

$$\begin{aligned}\Lambda_{G,\mu'}^\epsilon(\boldsymbol{t}) &\lesssim \int \prod_{ij\in G'} \sigma_{t_{ij}}^\epsilon(x^i - x^j)d\mu'(x^1)\cdots d\mu'(x^{k+1}) \\ &= \Lambda_{G',\mu'}^\epsilon(\boldsymbol{t})\end{aligned}$$

The rest of the proof proceeds as in Theorem 3.20, with μ' in place of μ and G' in place of G. □

5.5 The Natural Measure ν_g on $E - gE$

We denote by S^{d-1} the $(d-1)$-dimensional unit sphere centered at 0 in $\mathbb{R}^d$. We will need the following result.

Theorem 5.7 (Wolff-Erdoğan Theorem) *Let μ be a compactly supported Borel measure in $\mathbb{R}^d$. Then, for $s \geq d/2$ and $\epsilon > 0$,*

$$\int_{S^{d-1}} |\hat{\mu}(t\omega)|^2 d\omega \leq C_\epsilon I_s(\mu) t^{\epsilon-\gamma(s,d)} ,$$

with $\gamma(s,d) = (d+2s-2)/4$ if $d/2 \leq s \leq (d+2)/2$ and $\gamma(s,d) = s-1$ for $s \geq (d+2)/2$ where $I_s(u)$ is the s-energy of μ, $I_s(\mu) = \iint |x-y|^{-s} d\mu(x) d\mu(y)$.

For $s \leq (d+2)/2$, see Wolff [24] for $d = 2$, and Erdoğan [8] for $d \geq 3$. For the case $s \geq (d+2)/2$, see Sjölin [22]). See [25] Chap. 8 for the definition and relevant properties of s-energy, we are only interested in the fact that it will be a finite number.

Theorem 5.8 (Natural measure on $E - gE$) *Let $k \geq 2$ and let $E \subset \mathbb{R}^d$, $d \geq 2$ be a compact set with $\dim E > \frac{d}{2}$. Let μ be a Borel probability measure on E of Frostman exponent $s < \dim E$ with s satisfying*

$$\begin{cases} s > \frac{d(4k-1)+2}{4k+2} & \text{for} \quad \frac{d}{2} < \dim E \leq \frac{d+2}{2} \\ s > \frac{4kd-1}{4k+1} & \text{for} \quad \frac{d+2}{2} < \dim E . \end{cases}$$

Let g be a rotation, $g \in O_d(\mathbb{R})$ and define the measure ν_g by

$$\int f d\nu_g = \int\int f(u - gv) d\mu(u) d\mu(v) . \tag{5.10}$$

Then the integral

$$\int\int \nu_g^{k+1}(x) dx dg$$

is a finite quantity and in particular ν_g is absolutely continuous for g-a.e.

Remark 5.9 The threshold for s when $\frac{d}{2} < \dim E \leq \dim \frac{d}{2} + 1$ is not useful unless $d = 2$ or $d = 3$, $k = 1, 2$ and $k = 1$, since in that case

$$\frac{d(4k-1)+2}{4k+2} \geq \frac{d+2}{2} \geq \dim E .$$

In particular, below is a table for the readers convenience.

Exponents for s when $\frac{d}{2} < \dim E \le \frac{d+2}{2}$	
$d = 2$	$s > \frac{4k}{2k+1}$
$d = 3, k = 1, 2$	$s > \frac{12k-1}{4k+2}$
$d > 3, k = 1$	$s > \frac{d}{2} + \frac{1}{3}$

If the values of d, k are not listed on this table then we have

$$\dim E > s > \frac{4kd-1}{4k+1}.$$

Proof Let $\psi \ge 0$ be a smooth radial function supported in $\{\xi \in \mathbb{R}^d : 1/2 \le |\xi| \le 4\}$, identically equal to 1 in $\{\xi \in \mathbb{R}^d : 1 \le |\xi| \le 2\}$ with $\sqrt{\psi}$ also smooth. Let $\psi_j(\xi) = \psi(2^{-j}\xi)$. Moreover, we require $\sum_{j=-\infty}^{+\infty} \psi_j(\xi) = c$, for a suitable constant $c > 0$, for all $\xi \neq 0$ (see [25] §7 for existence of such ψ). Let $\nu_{g,j}$ denote the j-th Littlewood-Paley piece of ν_g defined by $\hat{\nu}_{g,j}(\xi) = \hat{\nu}_g(\xi)\psi_j(\xi)$. Since ν_g is a finite measure, in bounding the pieces, we may assume that j is bounded from below. Using the Littlewood-Paley decomposition of ν_g, we may write $\int \nu_g^{k+1}(x)dx$ as

$$\int \sum_{j_1,\ldots,j_{k+1}} \nu_{g,j_1}(x) \cdots \nu_{g,j_{k+1}}(x)dx \,.$$

This is bounded above by

$$(k+1)! \int \sum_{j_1 \ge j_2 \ge \cdots \ge j_{k+1}} \nu_{g,j_1}(x) \cdots \nu_{g,j_{k+1}}(x)dx \,.$$

Now using Plancherel, we see that since $\hat{\nu}_{g,j_2} * \cdots * \hat{\nu}_{g,j_{k+1}}$ is supported on scale $2^{j_2} + \cdots + 2^{j_{k+1}} \le 2^{j_2+1}$ while $\hat{\nu}_{g,j_1}$ is supported on an annulus of scale 2^{j_1}, the sum vanishes if $j_1 - j_2 > 2$ for j_2 large. Thus it suffices to consider the case $j_1 = j_2$ (the case $j_1 = j_2 + 1$ is similarly treated) and to look at the sum

$$\sum_{j_1 = j_2 \ge j_3 \ge \cdots \ge j_{k+1}} \int \nu_{g,j_1}^2(x)\nu_{g,j_3}(x) \cdots \nu_{g,j_{k+1}}(x)dx \,. \tag{5.11}$$

From the definition of $\nu_{g,j}$ it follows that $\nu_{g,j} = \mu_j(-\cdot) * \mu_j(g\cdot)$, where $\hat{\mu}_j = \hat{\mu}\sqrt{\psi_j}$.

By Young's inequality,

$$\|\nu_{g,j}\|_\infty \le \|\mu_j\|_1 \cdot \|\mu_j\|_\infty \,.$$

Trivially $\|\mu_j\|_1 \leq 1$ since μ is a probability measure. Also

$$|\mu_j(x)| = 2^{dj}|\mu * \widehat{\sqrt{\psi}}(-2^j x)| \leq C_N 2^{dj} \int (1 + 2^j|x - y|)^{-N} d\mu(y) \leq C_N' 2^{j(d-s)}$$

for any $N > 1$ since μ is a Frostman measure on E. Using this estimate on the terms corresponding to the indices $j_3, \ldots, j_{k+1}$ we can bound (5.11) by a constant multiple of

$$\int \sum_j \left(\sum_{j \geq j_3 \geq \cdots \geq j_{k+1}} 2^{(j_3 + \cdots + j_{k+1})(d-s)} \right) \nu_{g,j}^2(x)dx \lesssim \int \sum_j 2^{j(k-1)(d-s)} \nu_{g,j}^2(x)dx .$$

It follows that

$$\int \int \nu_g^{k+1}(x)dxdg \lesssim \sum_j 2^{j(k-1)(d-s)} \cdot \int \int \nu_{g,j}^2(x)dxdg .$$

We will show that $\int \int \nu_{g,j}^2(x)dxdg \lesssim 2^{j(d-s)}2^{-j\gamma(s,d)}$ where the quantity $\gamma(s,d)$ is defined in Theorem 5.7, which completes the proof.

Since we have $\hat{\nu}_{g,j} = \hat{\mu}_j(\xi)\hat{\mu}_j(g\xi)$, via Plancherel we obtain

$$\begin{aligned}\int \int \nu_{g,j}^2(x)dxdg &= \int \int |\hat{\mu}_j(\xi)|^2 |\hat{\mu}_j(g\xi)|^2 d\xi dg \\ &= \int_0^\infty \int_{S^{d-1}} |\hat{\mu}_j(t\omega)|^2 \left(\int |\hat{\mu}_j(gt\omega)|^2 dg \right) t^{d-1} d\omega dt .\end{aligned}$$

Since $O_d(\mathbb{R})$ acts transitively on the sphere, the quantity in the parentheses is constant in ω, and in particular it is a constant multiple of $\int |\hat{\mu}_j(t\omega')|^2 d\omega'$. Thus we have that

$$\begin{aligned}\int \int \nu_{g,j}^2(x)dxdg &= C \int \left(\int_{S^{d-1}} |\hat{\mu}_j(t\omega)|^2 d\omega \right)^2 t^{d-1} dt \\ &= C \int \left(\int_{S^{d-1}} |\hat{\mu}(t\omega)|^2 \psi(2^{-j}t\omega) d\omega \right)^2 t^{d-1} dt \\ &= C' \int_{2^{j-1}}^{2^{j+2}} \left(\int_{S^{d-1}} |\hat{\mu}(t\omega)|^2 d\omega \right)^2 t^{d-1} dt .\end{aligned}$$

Since we are summing over j and the intervals $[2^{j-1}, 2^{j+2}]$ have finite overlap with each other, we may as well bound $\sum \int \int \nu_{g,j}^2(x)dxdg$ by a constant multiple of

$$\int_{2^j}^{2^{j+1}} \left(\int_{S^{d-1}} |\hat{\mu}(t\omega)|^2 d\omega \right)^2 t^{d-1} dt .$$

The proof is finished by using Theorem 5.7, showing that $\int\int \nu_g^2(x) dxdg < \infty$. □

The proof is now complete up the proof of Lemma 5.10 and the geometric results in Sect. 4. We prove the lemma below. The geometric results are established in Sect. 6.

Lemma 5.10 *Let $G'_{k+1,m}$ be a tree. Then for small enough ϵ*

$$\epsilon^{-dk}\int_{O_d(\mathbb{R})}\int_A\int_A \chi\{|(x^i - x^j) - g(y^i - y^j)| < c\epsilon : ij \in G'_{k+1,m}\} d\mu^{k+1}(\boldsymbol{x}) d\mu^{k+1}(\boldsymbol{y}) dg \tag{5.12}$$

is bounded by a constant multiple of

$$\int\int \nu_g^{k+1}(x) dxdg\ .$$

Proof First we bound (5.12) by

$$\epsilon^{-dk}\int\int\int \chi\{|(x^1 - x^j) - g(y^1 - y^j)| < (k+1)c\epsilon : j > 1\} d\mu^{k+1}(\boldsymbol{x}) d\mu^{k+1}(\boldsymbol{y}) dg. \tag{5.13}$$

This is accomplished as follows: Fix $ij \in G'_{k+1,m}$ and let $(x^1, x^{l_2}, \ldots, x^{l_p}, x^i, x^j)$ be a path of length $p+1$, from x^1 to x^j in $G'_{k+1,m}$. Set $l_1 = 1$ and $l_{p+1} = i$, $l_{p+2} = j$. Using the triangle inequality we see that the set

$$\{(\boldsymbol{x}, \boldsymbol{y}) : |(x^i - x^j) - g(y^i - y^j)| < c\epsilon\}$$

is contained in the set

$$\{(\boldsymbol{x}, \boldsymbol{y}) : \sum_{f=1}^{p+1} |(x^{l_f} - x^{l_{f+1}}) - g(y^{l_f} - y^{l_{f+1}})| < (p+1)c\epsilon\}$$

which is contained in

$$\{(\boldsymbol{x}, \boldsymbol{y}) : |(x^1 - x^j) - g(y^1 - y^j)| < (p+1)c\epsilon\}\ .$$

Since $k \geq p$ we conclude that (5.12) is bounded by (5.13). Using now the natural measure ν_g we write (5.13) as

$$\epsilon^{-dk}\int\int\cdots\int \chi\{|z^1 - z^j| < c\epsilon : j > 1\} d\nu_g(z^1)\cdots d\nu_g(z^{k+1}) dg\ .$$

Now it is obvious that taking $\epsilon \to 0$ and using the absolute continuity of ν_g and the dominated convergence theorem, we may bound (5.13) by a constant multiple of

$$\int\int \nu_g^{k+1}(z)dzdg$$

finishing the proof. □

6 Geometric Results

For $d \geq 2$ and each $1 \leq q \leq d$ we show that the edge set of K_{q+1} is independent in $\mathbb{R}^d$. We show that infinitesimal rigidity of a fixed graph $G_{k+1,m}$ is a generic property, either holding for all generic frameworks, or none of them. We count the number of edges a minimally infinitesimally rigid graph must have. The behavior of the distance function near regular configurations is investigated. Our approach follows closely that of [13]. See also [21] for motivation and examples.

6.1 Generic Frameworks

In this section we prove various results for generic frameworks in $\mathbb{R}^d$. Some of the statements are for regular frameworks, but as noted in Remark 3.18, generic frameworks are regular.

The following lemma, while technically obvious, serves to remind the reader of the form that $DF_{G_{k+1,m}}$ takes, which will be useful in subsequent proofs in this section.

Lemma 6.1 *Fix $d \geq 2$. We have $DF_{G_{k+1,m}}(\boldsymbol{x}) \cdot \boldsymbol{u} = 0$ if and only if $\boldsymbol{u}$ is a solution to the following system of m equations in $d(k+1)$ variables:*

$$(x^i - x^j) \cdot (u^i - u^j) = 0, \ \textit{for}\ ij \in G_{k+1\,m}\,. \tag{6.1}$$

Proof Since $F_{G_{k+1,m}}$ is a function $\mathbb{R}^{d(k+1)} \to \mathbb{R}^m$, $DF_{G_{k+1,m}}$ is a $m \times d(k+1)$ matrix with rows corresponding to edges $ij \in G_{k+1,m}$ and columns corresponding to the scalar components of $\boldsymbol{x}$. The ij-th row is equal to the following vector

$$(0,\ldots,0,\overbrace{2(x^i - x^j)}^{d(i-1)+1 \text{ to } di},0,\ldots,0,\overbrace{-2(x^i - x^j)}^{d(j-1)+1 \text{ to } dj},0,\ldots,0)\,.$$

Here every component $x^i - x^j$ is also a vector ($x^1,\ldots,x^{k+1} \in \mathbb{R}^d$ are vectors). Thus we can see that $DF_{G_{k+1,m}}(\boldsymbol{x}) \cdot \boldsymbol{u} = 0$ is equivalent to (6.1). □

Proposition 6.2 *If $H \subseteq K_{k+1}$ is an independent (in $\mathbb{R}^d$) set of edges then $\boldsymbol{x} \in X_H$ if and only if the rows of $DF_{K_{k+1}}(\boldsymbol{x})$ corresponding to edges of H are linearly dependent. Moreover if $H \subset H'$ (H, H' independent), then $X_H \subseteq X_{H'}$.*

Proof Since the rank of the matrix is less than $|H|$, the $|H|$ row vectors are linearly dependent. Conversely, if the rows are linearly dependent every minor has to be zero since all the submatrices will satisfy the same dependence.

Moreover, if $H \subset H'$ then the matrix corresponding to H' will have rank at least that of the one for H. □

Theorem 6.3 *The set of generic configurations in $\mathbb{R}^d$ is an open dense set of full Lebesgue measure. Moreover every independent (in $\mathbb{R}^d$) set H is in fact independent in $\mathbb{R}^d$ with respect to any generic configuration in $\mathbb{R}^d$.*

Proof Note that to each polynomial P_H corresponds at least one configuration $\boldsymbol{x}_0$ for which P_H is nonzero, thus the zero sets X_H are proper algebraic varieties. Thus the set of generic configurations of G is nonempty, and in particular open dense of full measure (since the complement X is of codimension at least 1, as a proper algebraic variety).

The independence of H for any generic configuration follows from the definition of genericity. In particular if $\boldsymbol{x}$ is generic then $\boldsymbol{x} \notin X$, thus $\boldsymbol{x} \notin X_H$. By Proposition 6.2 it follows that H is independent with respect to $\boldsymbol{x}$. □

Lemma 6.4 *If A is an invertible affine transformation of $\mathbb{R}^d$ and we set $A\boldsymbol{x} = (Ax^1, \ldots, Ax^{k+1})$, then if X is the set of singular configurations of $\mathbb{R}^d$ (see Definition 3.7) we have that $AX = X$. Thus invertible affine transformations preserve genericity.*

Proof If $Ax = Bx + b$ where B is an invertible linear transformation and b a vector, then since the row vectors of $DF_{K_{k+1}}(\boldsymbol{x})$ contain the entries $\pm(x^i - x^j)$, we see that the row vectors of $DF_{K_{k+1}}(A\boldsymbol{x})$ contain the entries $\pm(Ax^i - Ax^j) = \pm B(x^i - x^j)$. In particular we see that the rows of $DF_{K_{k+1}}(\boldsymbol{x})$ corresponding to H are linearly independent if and only if those of $DF_{K_{k+1}}(A\boldsymbol{x})$ are. Using Proposition 6.2 we conclude that affine transformations preserve genericity. □

Theorem 6.5 *Assume $q \leq d$. The edge set of K_{q+1} is then independent in $\mathbb{R}^d$, in fact with respect to any configuration in general position.*

Proof Let $\boldsymbol{x}_0$ be in general position and assume that the rows of $DF_{K_{q+1}}(\boldsymbol{x}_0)$ are linearly dependent, say

$$\sum_{1 \leq i < j \leq q} s_{ij} r_{ij} = 0$$

where r_{ij} is the row corresponding to the edge $ij \in K_{q+1}$. Suppose $a < b$ is such that $s_{ab} \neq 0$. Then if we focus on the column corresponding to x^a we get the nontrivial equation

$$\sum_{j<a} s_{ja}(x^a - x^j) + \sum_{j>a} s_{aj}(x^a - x^j) = 0$$

which contradicts the fact that $\boldsymbol{x}_0$ is in general position, i.e. that $\{x^a - x^j : j \neq a\}$ is a linearly independent set of vectors. □

Theorem 6.6 *Let x be a generic configuration in $\mathbb{R}^d$. Then x is in general position in $\mathbb{R}^d$.*

Proof Assume x is not in general position. Thus for some $1 \leq q \leq d$ without loss of generality we may assume $\{x^1, \ldots, x^{q+1}\}$ are affinely dependent with $\{x^1, \ldots, x^q\}$ affinely independent. As proven in Theorem 6.5, the edge set of K_{q+1} is independent. Let A be the affine transformation taking each x^j, $1 \leq j \leq q$ to e_j, the standard j-th basis vector (using Lemma 6.4). Then by affine dependence we must have $Ax^{q+1} = t_1 e_1 + \cdots + t_q e_q$ with $t_1 + \cdots + t_q = 0$. Setting $s_{ij} = -t_i t_j$ for $1 \leq i < j \leq q$ and $s_{i(q+1)} = t_i$ for $1 \leq i \leq q$ we easily check that for any $1 \leq a \leq q+1$,

$$\sum_{ia \in K_{q+1}} s_{ia}(x^a - x^i) + \sum_{aj \in K_{q+1}} s_{aj}(x^a - x^j) = 0\,.$$

Thus the edge set of K_{q+1} is not independent with respect to $(x^1, \ldots, x^{q+1})$ and so by Theorem 6.3 we conclude that $(x^1, \ldots, x^{q+1})$ is not a generic configuration. We now note that if x were a generic configuration then $(x^1, \ldots, x^{q+1})$ would also be a generic configuration, simply because we are removing $d(k+1) - d(q+1)$ column vectors from $DF_{K_{k+1}}$ to test for the genericity of $(x^1, \ldots, x^{q+1})$. This contradicts what we have found therefore x is not generic. □

Lemma 6.7 *If x is in general position in $\mathbb{R}^d$ then* $\dim \mathcal{D}(x) = \binom{d+1}{2}$.

Proof First let us show that if $\gamma(t) : (-1, 1) \to M_x$ is a smooth curve with $\gamma(0) = x$, then $\gamma'(0) \in \mathcal{D}(x)$. This follows from the fact that the composition $F(\gamma(t))$ is constant, so by the chain rule $DF(\gamma(0)) \cdot \gamma'(0) = 0$. Note that $\dim M_x = \binom{d+1}{2}$, giving us

$$\dim \mathcal{D}(x) \geq \binom{d+1}{2}.$$

For the reverse inequality, we will show that any infinitesimal motion $u \in \mathcal{D}(x)$ projects injectively to an infinitesimal motion $\tilde{u}$ of $\mathcal{V}(K_{d+1}, \tilde{x})$ where $\tilde{u} = (u^1, \ldots, u^{d+1})$ and $\tilde{x} = (x^1, \ldots, x^{d+1})$. By the rank-nullity theorem and Theorem 6.5,

$$\dim \mathcal{V}(K_{d+1}, \tilde{x}) = d(d+1) - \binom{d+1}{2} = \binom{d+1}{2},$$

establishing the reverse inequality and completing the proof.

Thus it remains to show injectivity. Let $u, v \in \mathcal{D}(x)$ and assume $\tilde{u} = \tilde{v}$, that is, $u^i = v^i$ for $1 \leq i \leq d+1$. Since the space $\mathcal{D}(x)$ is a vector space, for $w = u - v$ we have

$$(w^i - w^j) \cdot (x^i - x^j) = 0, \text{ for } ij \in K_{k-1}\,.$$

Now if $i = 1, \ldots, d$ we have $w^i = 0$, so that for any $j > d$ we have

$$w^j \cdot (x^i - x^j) = 0, \text{ for } i = 1, \ldots, d\,.$$

But that means w^j is perpendicular to d linearly independent vectors, since $\boldsymbol{x}$ is in general position, thus $w^j = 0$ as well, and $\boldsymbol{u} = \boldsymbol{v}$. □

Theorem 6.8 *Let $(G_{k+1,m}, \boldsymbol{x}_0)$ be an infinitesimally rigid framework in $\mathbb{R}^d$ with $\boldsymbol{x}_0$ generic. Then for all generic configurations $\boldsymbol{x}$, the frameworks $(G_{k+1,m}, \boldsymbol{x})$ are infinitesimally rigid in $\mathbb{R}^d$.*

Proof Since $\mathcal{V}(G_{k+1,m}, \boldsymbol{x}_0) = \mathcal{D}(\boldsymbol{x}_0)$, combine Lemma 6.7 and Theorem 6.6 to obtain

$$\dim \mathcal{V}(G_{k+1,m}, \boldsymbol{x}_0) = \binom{d+1}{2}.$$

Since $\mathcal{V}(G_{k+1,m}, \boldsymbol{x}_0) = \ker DF_{G_{k+1,m}}(\boldsymbol{x}_0)$ by the rank-nullity theorem we obtain

$$\dim \mathcal{V}(G_{k+1,m}, \boldsymbol{x}_0) = d(k+1) - \operatorname{rank} DF_{G_{k+1,m}}(\boldsymbol{x}_0) . \tag{6.2}$$

Combining these two equations we find that

$$\operatorname{rank} DF_{G_{k+1,m}}(\boldsymbol{x}_0) = d(k+1) - \binom{d+1}{2}.$$

Since generic configurations have the same rank (by Theorem 6.3), by using Eq. (6.2) with $\boldsymbol{x}$ in place of $\boldsymbol{x}_0$ we see that $\dim \mathcal{V}(G_{k+1,m}, \boldsymbol{x}) = \binom{d+1}{2}$, implying that $\mathcal{V}(G_{k+1,m}, \boldsymbol{x}) = \mathcal{D}(\boldsymbol{x})$, so that $(G_{k+1,m}, \boldsymbol{x})$ is infinitesimally rigid. □

Corollary 6.9 *A minimally infinitesimally rigid graph $G_{k+1,m}$ in $\mathbb{R}^d$ satisfies*

$$m = d(k+1) - \binom{d+1}{2}.$$

Proof Let $G_{k+1,m}$ be minimally infinitesimally rigid. Then $m \geq \operatorname{rank} DF_{G_{k+1,m}}(\boldsymbol{x}_0)$ $= d(k+1) - \binom{d+1}{2}$, as the proof of Theorem 6.8 shows. Let $(G_{k+1,m}, \boldsymbol{x}_0)$ be a regular framework. If we assume $m > d(k+1) - \binom{d+1}{2}$, there must be a subset H of edges of $G_{k+1,m}$ such that

$$\operatorname{rank} DF_H = d(k+1) - \binom{d+1}{2} = |H|$$

about $\boldsymbol{x}_0$, therefore H is infinitesimally rigid about $\boldsymbol{x}_0$. Since that is an open set, by Theorem 6.8 the generic behavior of H is determined, thus H is infinitesimally rigid, which is a contradiction since H has less edges than $G_{k+1,m}$. □

Proposition 6.10 *Let $G_{k+1,m}$ be a minimally infinitesimally rigid graph in $\mathbb{R}^d$ and $(G_{k+1,m}, \boldsymbol{x}_0)$ be a regular framework. Then there exists some open neighborhood U of $\boldsymbol{x}_0$ and an embedded m-dimensional submanifold $M \subset U$ that contains $\boldsymbol{x}_0$, with $F_{G_{k+1,m}}$ restricted on M a diffeomorphism onto its image. Moreover if $\boldsymbol{x} \in U$, letting $N_{\boldsymbol{x}} = \{\boldsymbol{y} : F_G(\boldsymbol{y}) = F_G(\boldsymbol{x})\}$ denote the level curves, we have*

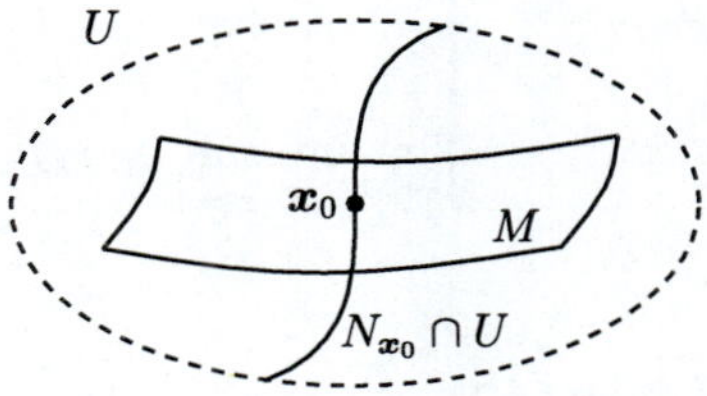

Fig. 5 The local behavior of the distance function at a regular configuration

$$N_x \cap U = \{(Tx^1, \ldots, Tx^{k+1}) : T \in \mathrm{ISO}(d)\} \cap U \ .$$

The same also hold for $f_{G_{k+1,m}}$.

Proof Since rank $DF_{G_{k+1,m}}(\boldsymbol{x}_0) = m$ is maximal, it stays maximal around an open neighborhood U of $\boldsymbol{x}_0$. The Inverse Function Theorem yields local coordinates $(\boldsymbol{p}, \boldsymbol{q})$ at $\boldsymbol{x}_0$ such that $F_{G_{k+1,m}}(\boldsymbol{p}, \boldsymbol{q}) = \boldsymbol{p}$. The manifold M is the image of $(\boldsymbol{p}, 0)$. It clearly is of dimension m. Using Corollary 6.9 and the fact that dim ISO$(d) = \binom{d+1}{2}$, the other claims follow. □

Remark 6.11 We justly say that the regular frameworks of a minimally infinitesimally rigid graph are locally uniquely realizable, in the sense that modulo isometries the distance function is a local diffeomorphism as Proposition 6.10 shows.

6.2 Useful Lemmas

Here we collect the rest of lemmas that were used, that are not related to graph rigidity (Fig. 5).

Lemma 6.12 *Let* $\phi \in C_c^\infty(\mathbb{R}^m)$ *be a nonnegative radial function with* $\int \phi = 1$, $\phi \leq 1$ *and* supp $\phi \subset \{t : |t| \leq 2\}$ *and for* $\epsilon > 0$ *set* $\phi^\epsilon(t) = \epsilon^{-m}\phi(\epsilon^{-1}t)$. *Let* $\nu(t)$ *be a Borel probability measure. Then* $\nu^\epsilon = \phi^\epsilon * \nu$ *converges weakly-* to* ν.

Proof Let $f \in C_0(\mathbb{R}^m)$ be a nonnegative function that vanishes at infinity. Then

$$\int f d\nu^\epsilon = \int (f * \phi^\epsilon) d\nu \ .$$

It is a well known result of mollifiers that $f * \phi^\epsilon \to f$ pointwise and by the Dominated Convergence Theorem we may conclude that $\int f d\nu^\epsilon \to \int f d\nu$. □

Lemma 6.13 *Let* V *be a normed vector space and* V^* *its dual equipped with the operator norm. If* $\nu_n \to \nu$ *weakly-* in* V^* *then* $\liminf \|\nu_n\| \geq \|\nu\|$.

Proof

$$\liminf_{n\to\infty} \inf_{k\geq n} \sup_{\|x\|=1} \langle x, \nu_k\rangle \geq \lim_{n\to\infty} \sup_{\|x\|=1} \inf_{k\geq n} \langle x, \nu_k\rangle \geq \sup_{\|x\|=1} \langle x, \nu\rangle .$$

□

Lemma 6.14 *If μ is a Borel measure on $\mathbb{R}^n$ and $C_\mu > 0$ a constant with $\mu(B(x,r)) \leq C_\mu r^s$ for some $s > 0$ and all $r > 0$ and $x \in \mathbb{R}^n$, then for any measurable subset A of $\mathbb{R}^n$ with $\mu(A) > 0$ we have* $\dim A \geq s$.

Proof Let U_j be open balls of radius r_j covering A. We then have

$$0 < \mu(A) \leq \sum_{j=1}^{\infty} \mu(U_j) \leq C_\mu \sum_{j=1}^{\infty} r_j^s .$$

Taking the infimum over all such collections U_j we obtain that $\mathcal{H}^s(A) > 0$. □

Lemma 6.15 *Let σ_t denote the surface measure of the sphere $tS^{d-1} \subset \mathbb{R}^d$ of radius t centered at 0. Let $\phi^\epsilon(x) = \epsilon^{-d}\phi(\epsilon^{-1}x)$ and $\phi \in C_c^\infty(\mathbb{R}^d)$ is a nonnegative radial function with $\int \phi = 1$, $\phi \leq 1$ and $\operatorname{supp}\phi \subset B(0,2)$. Let $\sigma_t^\epsilon = \phi^\epsilon * \sigma_t$. Let μ be a Frostman measure on $E \subset \mathbb{R}^d$, E compact, with Frostman exponent $s > \frac{d+1}{2}$. Then there exists a constant $C_t > 0$ independent of ϵ with*

$$\|\sigma_t^\epsilon * \mu\|_{L^1(\mu)} < C_t .$$

Proof We use Plancherel and the stationary phase of the sphere, (see [25]), that tells us that for ξ of large norm and some $c > 0$ we have

$$|\widehat{\sigma_t^\epsilon}(\xi)| \leq c t^{\frac{d-1}{2}} |\xi|^{-\frac{d-1}{2}} ,$$

to obtain that for some $C > 0$ depending on the diameter of E,

$$\begin{aligned}
\int \sigma_t^\epsilon * \mu(x) d\mu(x) &= \int \widehat{\sigma_t}(\xi)\widehat{\phi}(\epsilon\xi)|\widehat{\mu}|^2(\xi)d\xi \\
&\lesssim 1 + \int |\xi|^{-\frac{d-1}{2}} |\widehat{\mu}|^2(\xi) d\xi \\
&\lesssim 1 + \int\int |x-y|^{-\frac{d+1}{2}} d\mu(x)d\mu(y) \\
&\lesssim 1 + \int\int_C^{+\infty} \mu\left\{x : |x-y| < \lambda^{-\frac{2}{d+1}}\right\} d\lambda d\mu(y) \\
&\lesssim 1 + \int_C^{+\infty} \lambda^{-\frac{2}{d+1}s} d\lambda .
\end{aligned}$$

The last integral is finite by assumption. □

7 Proof of Theorem 3.26

Let q be a positive integer and define E_q to be the $q^{-\frac{d}{s}}$-neighborhood of $\frac{1}{q}\left\{\mathbb{Z}^d \cap [0, q]^d\right\}$ with $s \in \left(\frac{d}{2}, d\right)$ to be determined later. It is known (see e.g. [9]) that if we choose $q_1 = 2, q_{i+1} > q_i^i$, then the Hausdorff dimension of $E = \cap_i E_{q_i}$ is equal to s.

Lemma 7.1 *The number of congruence classes of frameworks with $k+1$ vertices in $\mathbb{Z}^d \cap [0, q]^d$ is bounded above by Cq^{dk}.*

To prove the lemma, fix one of the vertices at the origin, which we may do since $\mathbb{Z}^d$ is translation invariant. The number of the remaining k-point configurations is $\leq q^{dk}$ by construction. This proves the lemma.

We now consider an infinitesimally rigid framework on $k+1$ vertices in $\mathbb{Z}^d \cap [0, q]^d$ described by the graph $G_{k+1,m}$. By Corollary 6.9, the number of edges is $m = d(k+1) - \binom{d+1}{2} = dk - \binom{d}{2}$. It follows that

$$\mathcal{H}^{dk-\binom{d}{2}}(\Delta(G_{k+1,m}, E_q^{k+1})) \leq C(q^{-\frac{d}{s}})^{dk-\binom{d}{2}} \cdot q^{dk}.$$

This quantity tends to 0 as $q \to \infty$ if $s < d - \frac{\binom{d}{2}}{k}$ and Theorem 3.26 is proved.

References

1. L. Asimow and B. Roth. The rigidity of graphs. *Trans. Amer. Math. Soc.*, 245:279–289, 1978.
2. Michael Bennett, Derrick Hart, Alex Iosevich, Jonathan Pakianathan, and Misha Rudnev. Group actions and geometric combinatorics in $\mathbb{F}_q^d$. *Forum Math.*, 29(1):91–110, 2017.
3. Mike Bennett, Alex Iosevich, and Jonathan Pakianathan. Three-point configurations determined by subsets of $\mathbb{F}_q^2$ via the Elekes-Sharir paradigm. *Combinatorica*, 34(6):689–706, 2014.
4. Michael Bennett, Alexander Iosevich, and Krystal Taylor. Finite chains inside thin subsets of $\mathbb{R}^d$. *Anal. PDE*, 9(3):597–614, 2016.
5. J. Bourgain. A Szemerédi type theorem for sets of positive density in $\mathbf{R}^k$. *Israel J. Math.*, 54(3):307–316, 1986.
6. Ciprian Borcea and Ileana Streinu. The number of embeddings of minimally rigid graphs. *Discrete Comput. Geom.*, 31(2):287–303, 2004.
7. Vincent Chan, Izabella Ł aba, and Malabika Pramanik. Finite configurations in sparse sets. *J. Anal. Math.*, 128:289–335, 2016.
8. M. Burak Erdŏ gan. A bilinear Fourier extension theorem and applications to the distance set problem. *Int. Math. Res. Not.*, (23):1411–1425, 2005.
9. K. J. Falconer. *The geometry of fractal sets*, volume 85 of *Cambridge Tracts in Mathematics*. Cambridge University Press, Cambridge, 1986.
10. Hillel Furstenberg, Yitzchak Katznelson, and Benjamin Weiss. Ergodic theory and configurations in sets of positive density. In *Mathematics of Ramsey theory*, volume 5 of *Algorithms Combin.*, pages 184–198. Springer, Berlin, 1990.
11. Allan Greenleaf, Alex Iosevich, Bochen Liu, and Eyvindur Palsson. A group-theoretic viewpoint on Erdős-Falconer problems and the Mattila integral. *Rev. Mat. Iberoam.*, 31(3):799–810, 2015.

12. Allan Greenleaf, Alex Iosevich, and Malabika Pramanik. On necklaces inside thin subsets of $\mathbb{R}^d$. *Math. Res. Lett.*, 24(2):347–362, 2017.
13. Jack Graver, Brigitte Servatius, and Herman Servatius. *Combinatorial rigidity*, volume 2 of *Graduate Studies in Mathematics*. American Mathematical Society, Providence, RI, 1993.
14. L. Henneberg. Die graphische statik der starren systeme. *Leipzig*, 1911.
15. Derrick Hart and Alex Iosevich. Ubiquity of simplices in subsets of vector spaces over finite fields. *Anal. Math.*, 34(1):29–38, 2008.
16. Kevin Henriot, Izabella Ł aba, and Malabika Pramanik. On polynomial configurations in fractal sets. *Anal. PDE*, 9(5):1153–1184, 2016.
17. Alex Iosevich and Bochen Liu. Falconer distance problem, additive energy and Cartesian products. *Ann. Acad. Sci. Fenn. Math.*, 41(2):579–585, 2016.
18. G. Laman. On graphs and rigidity of plane skeletal structures. *J. Engrg. Math.*, 4:331–340, 1970.
19. J. Milnor. On the Betti numbers of real varieties. *Proc. Amer. Math. Soc.*, 15:275–280, 1964.
20. I. G. Petrovskii and O. A. Oleinik. On the topology of real algebraic surfaces. *Amer. Math. Soc. Translation*, 1952(70):20, 1952.
21. B. Roth. Rigid and flexible frameworks. *Amer. Math. Monthly*, 88(1):6–21, 1981.
22. Per Sjölin. Estimates of spherical averages of Fourier transforms and dimensions of sets. *Mathematika*, 40(2):322–330, 1993.
23. René Thom. Sur l'homologie des variétés algébriques réelles. In *Differential and Combinatorial Topology (A Symposium in Honor of Marston Morse)*, pages 255–265. Princeton Univ. Press, Princeton, N.J., 1965.
24. Thomas Wolff. Decay of circular means of Fourier transforms of measures. *Internat. Math. Res. Notices*, (10):547–567, 1999.
25. Thomas H. Wolff. *Lectures on harmonic analysis*, volume 29 of *University Lecture Series*. American Mathematical Society, Providence, RI, 2003. With a foreword by Charles Fefferman and preface by Izabella Ł aba, Edited by Ł aba and Carol Shubin.
26. Tamar Ziegler. Nilfactors of $\mathbb{R}^m$-actions and configurations in sets of positive upper density in $\mathbb{R}^m$. *J. Anal. Math.*, 99:249–266, 2006.

On Generalized Harmonic Numbers

Yong-Gao Chen and Bing-Ling Wu

Abstract For three positive integers a, b and n, let $H_{a,b}(n)$ be the sum of the reciprocals of the first n terms of arithmetic progression $\{ak + b : k = 0, 1, \ldots\}$ and let $v_{a,b}(n)$ be the denominator of $H_{a,b}(n)$. In this paper, we prove that for two coprime positive integers a and b, (i) if p is a prime with $p \nmid a$, then the set of positive integers n with $p \mid v_{a,b}(n)$ has asymptotic density one; (ii) the set of positive integers n with $v_{a,b}(n) = v_{a,b}(n + 1)$ has asymptotic density one.

Keywords Harmonic number · Arithmetic progression · Eswarathasan-Levine conjecture · Asymptotic density

1 Introduction

For any positive integer n, the nth harmonic number is defined by

$$H_n = 1 + \frac{1}{2} + \frac{1}{3} + \cdots + \frac{1}{n}.$$

Harmonic numbers have a long history and are important in many fields. In 1730, Stirling found the asymptotic formula for factorial $n!$ by using the digamma psi function $\psi(n) = H_n - \log n$. In 1737, Euler used harmonic numbers to prove that the sum of the reciprocals of the primes is divergent (see [4, Theorem 19 and its note]). It is well known that $\psi(n) \to \gamma$ as $n \to +\infty$, where γ is Euler's constant. A classic result on harmonic numbers is Wolstenholme's theorem (see [8]): If $p \geq 5$ is

Y.-G. Chen (✉)
School of Mathematical Sciences and Institute of Mathematics, Nanjing Normal University, Nanjing 210023, People's Republic of China
e-mail: ygchen@njnu.edu.cn

B.-L. Wu
School of Science, Nanjing University of Posts and Telecommunications, Nanjing 210023, People's Republic of China
e-mail: wubingling911219@163.com

M. B. Nathanson (ed.), *Combinatorial and Additive Number Theory IV*, Springer Proceedings in Mathematics & Statistics 347,
https://doi.org/10.1007/978-3-030-67996-5_6

a prime, then the numerator of H_{p-1} is divisible by p^2. Another well known result is that none of H_n $(n = 2, 3, \ldots)$ is an integer which is proved by Theisinger [7].

We write H_n as a fraction

$$H_n = \frac{u_n}{v_n}, \quad (u_n, v_n) = 1,\ v_n \geq 1.$$

For any prime p, let J_p be the set of all positive integers n with $p|u_n$. In 1991, Eswarathasan and Levine [2] conjectured that J_p is finite for any prime number p. Boyd [1] conjectured that $|J_p| = O(p^2(\log\log p)^{2+\varepsilon})$, where $|J_p|$ denotes the cardinality of the set J_p. An important progress on the Eswarathasan–Levine conjecture is due to Sanna [5]. Sanna [5] proved that the set J_p has asymptotic density zero by proving $J_p(x) \ll x^{0.765}$. This has been improved by Wu and Chen [9] to

$$J_p(x) \leq 3x^{\frac{2}{3}+\frac{1}{25\log p}}. \tag{1}$$

Shiu [6] proved that $v(n) = v(n+1)$ for infinite many positive integers n. Recently, Wu and Chen [11] proved that the set of positive integers n with $v(n) = v(n+1)$ has asymptotic density one.

In this paper, we consider general harmonic numbers

$$H_{a,b}(n) = \sum_{k=0}^{n-1} \frac{1}{ak+b},$$

where a, b are two positive integers. It is clear that $H_{1,1}(n) = H_n$. We appoint $H_{a,b}(0) = 0$. We write $H_{a,b}(n)$ as a fraction

$$H_{a,b}(n) = \frac{u_{a,b}(n)}{v_{a,b}(n)}, \quad (u_{a,b}(n), v_{a,b}(n)) = 1,\ v_{a,b}(n) \geq 1.$$

Then $u_{a,b}(0) = 0$ and $v_{a,b}(0) = 1$. For any two positive integers a, b and any prime p, let $J_{a,b,p}$ be the set of nonnegative integers n with $p \mid u_{a,b}(n)$ and let $I_{a,b,p}$ be the set of nonnegative integers n with $p \nmid v_{a,b}(n)$. It is clear that $0 \in J_{a,b,p}$, $1 \notin J_{a,b,p}$, and $J_{a,b,p} \subseteq I_{a,b,p}$. For any set S of nonnegative integers and any positive real number x, let $S(x)$ be the number of integers in S which are less than or equal to x. If $\lim_{x\to+\infty} S(x)/x$ exists, then the limit is called the asymptotic density of S. For any positive integer n and any prime p, let $\nu_p(n)$ denote the p-adic valuation of n, that is, $\nu_p(n)$ is the largest nonnegative integer α with $p^\alpha|n$.

In this paper, the following results are proved.

Theorem 1 *Let a and b be two positive integers and let p be a prime.*

(i) If $\nu_p(a) \leq \nu_p(b)$, then both $J_{a,b,p}$ and $I_{a,b,p}$ have asymptotic densities zero.

(ii) If $\nu_p(a) > \nu_p(b) \geq 1$, then both $J_{a,b,p}$ and $I_{a,b,p}$ have positive asymptotic densities which is between $1/(ab)$ and $(ab-1)/(ab)$.

(iii) If $v_p(a) > v_p(b) = 0$, then the set $J_{a,b,p}$ has positive asymptotic density which is between $1/(ab)$ and $(ab-1)/(ab)$ and the set $I_{a,b,p}$ is equal to the set of all nonnegative integers.

From Theorem 1 we have the following corollary immediately.

Corollary 1 *Let a and b be two coprime positive integers and let p be a prime.*

(i) If $p \nmid a$, then the set of positive integers n with $p \mid v_{a,b}(n)$ has asymptotic density one;

(ii) If $p \mid a$, then there are no nonnegative integers n with $p \mid v_{a,b}(n)$.

For two positive integers a and b, let

$$G_{a,b} = \left(b, \frac{a}{(a,b)}\right).$$

Theorem 2 *Let a and b be two positive integers.*

(i) If $G_{a,b} = 1$, then the set of positive integers n with $v_{a,b}(n) = v_{a,b}(n+1)$ has asymptotic density one, and there are infinitely many positive integers n such that $v_{a,b}(n) < v_{a,b}(n+1)$;

(ii) If $G_{a,b}$ is even, then each of sets of positive integers n with $v_{a,b}(n) < v_{a,b}(n+1)$ and $v_{a,b}(n) > v_{a,b}(n+1)$, respectively, has positive asymptotic density and there are no nonnegative integers n with $v_{a,b}(n) = v_{a,b}(n+1)$;

(iii) If $G_{a,b} > 1$ and $G_{a,b}$ is odd, then each of sets of positive integers n with $v_{a,b}(n) < v_{a,b}(n+1)$, $v_{a,b}(n) = v_{a,b}(n+1)$ and $v_{a,b}(n) > v_{a,b}(n+1)$, respectively, has positive asymptotic density.

By Theorem 2(i) we have the following corollary.

Corollary 2 *Let a and b be two positive integers with $(a,b) = 1$. Then the set of positive integers n with $v_{a,b}(n) = v_{a,b}(n+1)$ has asymptotic density one.*

For Theorem 2(i), we pose a problem here.

Problem 1 Are there two positive integers a and b with $(a,b) = 1$ such that $v_{a,b}(n) > v_{a,b}(n+1)$ for at most finitely many positive integers n?

2 Preliminary Lemmas

Firstly we give the following lemmas.

Lemma 1 ([9, Lemma 2.3]) *Let y be a real number with $y \geq 8/3$ and let $c_1, \ldots, c_l$ be integers with $0 \leq c_k \leq k-1$ $(1 \leq k \leq l)$ such that*

$$\sum_{k=1}^{l} kc_k \le y.$$

Then

$$1 + \sum_{k=1}^{l} c_k \le \left(\frac{9}{8}\right)^{\frac{1}{3}} y^{\frac{2}{3}}.$$

Lemma 2 *Let p be an odd prime and let x, y be two real numbers with $8/3 \le y < p$. Then, for any positive integers a, b with $p \nmid a$, we have*

$$|J_{a,b,p} \cap [x, x+y]| \le \left(\frac{9}{8}\right)^{\frac{1}{3}} y^{\frac{2}{3}}.$$

Proof If $|J_{a,b,p} \cap [x, x+y]| \le 2$, then the lemma is trivial. Now we assume that $|J_{a,b,p} \cap [x, x+y]| \ge 3$. Let

$$J_{a,b,p} \cap [x, x+y] = \{n_1 < n_2 < \cdots < n_l\}.$$

For any integer $1 \le k \le y$, let $c_k = |\{j : n_j - n_{j-1} = k\}|$. Now we prove that $c_k \le k-1$. Let

$$f_k(x) = (ax+b)(ax+a+b)\cdots(ax+(k-1)a+b)$$

and

$$g_k(x) = \frac{f_k(x)}{ax+b} + \frac{f_k(x)}{ax+a+b} + \cdots + \frac{f_k(x)}{ax+(k-1)a+b}.$$

Then $g_k(x)$ is an integral coefficient polynomial of degree $k-1$ and its leading coefficient ka^{k-1} is indivisible by p. For $1 \le i \le l$, we have $p \nmid v_{a,b}(n_i)$ since $p \mid u_{a,b}(n_i)$. Thus, for $i \in \{j : n_j - n_{j-1} = k\}$, noting that $p|u_{a,b}(n_i)$ and $p|u_{a,b}(n_{i-1})$, we have

$$\begin{aligned} &g_k(n_{i-1})v_{a,b}(n_i)v_{a,b}(n_{i-1}) \\ &= (H_{a,b}(n_i) - H_{a,b}(n_{i-1}))f_k(n_{i-1})v_{a,b}(n_i)v_{a,b}(n_{i-1}) \\ &= (u_{a,b}(n_i)v_{a,b}(n_{i-1}) - v_{a,b}(n_i)u_{a,b}(n_{i-1}))f_k(n_{i-1}) \end{aligned}$$

is divisible by p. It follows that $g_k(n_{i-1})$ is divisible by p. That is, n_{i-1} is a solution of $g_k(x) \equiv 0 \pmod{p}$ for each $i \in \{j : n_j - n_{j-1} = k\}$. Hence $c_k \le \deg(g_k(x)) = k-1$.

Since $y \ge 8/3$ and

$$\sum_k kc_k = \sum_{i=2}^{l}(n_i - n_{i-1}) = n_l - n_1 \le y,$$

it follows from Lemma 1 that

$$|J_{a,b,p} \cap [x, x+y]| = l = 1 + \sum_k c_k \le \left(\frac{9}{8}\right)^{\frac{1}{3}} y^{\frac{2}{3}}.$$

This completes the proof of Lemma 2. □

3 Proof of Theorem 1

Firstly, we prove the following theorem which is also interesting in a sense.

Theorem 3 *Let a and b be two positive integers and let p be a prime. If $\nu_p(a) \le \nu_p(b)$, then*

$$I_{a,b,p}(x) \le 2p^{\frac{1}{3}} x^{\frac{2}{3}+\frac{1}{25\log p}}, \quad J_{a,b,p}(x) \le 2x^{\frac{2}{3}+\frac{1}{25\log p}}, \quad x \ge 1.$$

Proof We assume that a and b are two positive integers with $\nu_p(a) \le \nu_p(b)$. Let

$$a = p^w a', \quad b = p^w b', \quad p \nmid a', \quad a', b', w \in \mathbb{Z}_{\ge 0}.$$

By the definitions of $I_{a,b,p}$ and $J_{a,b,p}$, we have $I_{a,b,p} \subseteq I_{a',b',p}$ and $J_{a,b,p} \subseteq J_{a',b',p}$. So $I_{a,b,p}(x) \le I_{a',b',p}(x)$ and $J_{a,b,p}(x) \le J_{a',b',p}(x)$. Thus, without loss of generality, we may further assume that $p \nmid a$.

If $1 \le x \le p$, then

$$I_{a,b,p}(x) \le x + 1 \le 2x \le 2p^{\frac{1}{3}} x^{\frac{2}{3}} \le 2p^{\frac{1}{3}} x^{\frac{2}{3}+\frac{1}{25\log p}}.$$

If $1 \le x \le 3$, then

$$J_{a,b,p}(x) \le x + 1 \le 2x^{\frac{2}{3}} \le 2x^{\frac{2}{3}+\frac{1}{25\log p}}.$$

If $3 < x < p$, then by Lemma 2,

$$J_{a,b,p}(x) \le \left(\frac{9}{8}\right)^{\frac{1}{3}} x^{\frac{2}{3}} \le 2x^{\frac{2}{3}+\frac{1}{25\log p}}.$$

In the following, we assume that $x \ge p$.

Let $b_0 = b$. We define $m_0, b_1, m_1, b_2, \ldots$ in order as follows: for $i = 0, 1, \ldots$, let m_i be the least nonnegative integer with $p \mid am_i + b_i$ and let $b_{i+1} = (am_i + b_i)/p$. It is clear that $0 \le m_i \le p - 1$ and b_i are positive integers $(i = 0, 1, \ldots)$. Let $m \in I_{a,b_i,p}$ and $m \ge m_i + 1$. It is clear that

$$H_{a,b_i}(m) = \sum_{\substack{l=0\\ p\nmid al+b_i}}^{m-1} \frac{1}{al+b_i} + \sum_{\substack{l=0\\ p\mid al+b_i}}^{m-1} \frac{1}{al+b_i}.$$

Write

$$\sum_{\substack{l=0\\ p\nmid al+b_i}}^{m-1} \frac{1}{al+b_i} = \frac{d}{c},$$

where c, d are two positive integers with $p \nmid c$. Let $m = pn + m_i - r$ with $0 \le r < p$. Then $n \ge 1$ and $m - 1 - m_i = pn - (r+1)$ with $1 \le r+1 \le p$. It follows that

$$n - 1 \le \frac{m-1-m_i}{p} < n.$$

Thus,

$$\begin{aligned}
\sum_{\substack{l=0\\ p\mid al+b_i}}^{m-1} \frac{1}{al+b_i} &= \sum_{\substack{l=0\\ p\mid al+am_i+b_i}}^{m-1-m_i} \frac{1}{al+am_i+b_i}\\
&= \sum_{\substack{l=0\\ p\mid al+pb_{i+1}}}^{m-1-m_i} \frac{1}{al+pb_{i+1}}\\
&= \frac{1}{p}\sum_{l=0}^{n-1} \frac{1}{al+b_{i+1}}\\
&= \frac{1}{p}\frac{u_{a,b_{i+1}}(n)}{v_{a,b_{i+1}}(n)}.
\end{aligned}$$

Hence,

$$H_{a,b_i}(m) = \frac{pdv_{a,b_{i+1}}(n) + cu_{a,b_{i+1}}(n)}{pcv_{a,b_{i+1}}(n)}.$$

By $m \in I_{a,b_i,p}$, we have $p \nmid v_{a,b_i}(m)$. It follows that

$$p \mid pdv_{a,b_{i+1}}(n) + cu_{a,b_{i+1}}(n).$$

Noting that $p \nmid c$, we have $p \mid u_{a,b_{i+1}}(n)$. This means that $n \in J_{a,b_{i+1},p}$.

We divide into two cases according to $p = 2$ and $p \ge 3$.

Case 1: $p = 2$. We will prove that for every nonnegative integer i, $I_{a,b_i,2} \subseteq \{0, 1\}$. Suppose that for some i, $I_{a,b_i,2} \nsubseteq \{0, 1\}$. Let m be the least integer with $m \ge 2$ for which there exists an i with $m \in I_{a,b_i,2}$. Recall that m_i is the least nonnegative integer with $2 \mid am_i + b_i$, we have $m_i = 0, 1$. Write m as $m = 2n + m_i - r$ with $0 \le r < 2$. Then $1 \le n < m$. By the previous arguments, we have $n \in J_{a,b_{i+1},2}$. Noting that

$J_{a,b_{i+1},2} \subseteq I_{a,b_{i+1},2}$, we have $n \in I_{a,b_{i+1},2}$. By $1 \le n < m$ and the definition of m, we have $n = 1$, a contradiction with $1 \notin J_{a,b_{i+1},2}$. Hence, $I_{a,b_i,2} \subseteq \{0, 1\}$ for all $i \ge 0$. Since $J_{a,b_i,2} \subseteq I_{a,b_i,2}$ and $1 \notin J_{a,b_i,2}$, it follows that $J_{a,b_i,2} = \{0\}$. Noting that $b_0 = b$, we have

$$I_{a,b,2}(x) \le 2 \le 2x^{\frac{2}{3}}, \quad J_{a,b,2}(x) = 1 \le x^{\frac{2}{3}}, \quad x \ge 1.$$

Case 2: $p \ge 3$. Let $x \ge 1$ and let k be the integer with $p^k \le x < p^{k+1}$. By $x \ge p$, we have $k \ge 1$. Let $m \in I_{a,b,p} \cap [0, x]$ and $m \ge m_0 + 1$. Write m as $m = pn + m_0 - r$ with $0 \le r < p$. Then $1 \le n \le (x + p - 1)/p$ and $n \in J_{a,b_1,p}$. Thus, by $0 \le m_0 \le p - 1$,

$$\begin{aligned} I_{a,b,p}(x) &\le |I_{a,b,p} \cap [0, m_0]| + \sum_{n \in J_{a,b_1,p} \cap [1,(x+p-1)/p]} |I_{a,b,p} \cap [pn + m_0 - p + 1, pn + m_0]| \\ &\le p + p \left| J_{a,b_1,p} \cap \left[1, \frac{x}{p} + \frac{p-1}{p}\right] \right| \\ &= p \left| J_{a,b_1,p} \cap \left[0, \frac{x}{p} + \frac{p-1}{p}\right] \right|. \end{aligned}$$

Let

$$A_p = \left(\frac{9}{8}\right)^{\frac{1}{3}} p^{\frac{2}{3}}.$$

Since $J_{a,b,p}(x) \subseteq I_{c,b,p}(x)$, it follows from Lemma 2 and $0 \le m_0 \le p - 1$ that

$$\begin{aligned} J_{a,b,p}(x) &\le |J_{a,b,p} \cap [0, m_0]| \\ &\quad + \sum_{n \in J_{a,b_1,p} \cap [1,(x+p-1)/p]} |J_{a,b,p} \cap [pn + m_0 - p + 1, pn + m_0]| \\ &\le |J_{a,b,p} \cap [0, p-1]| \\ &\quad + \sum_{n \in J_{a,b_1,p} \cap [1,(x+p-1)/p]} |J_{a,b,p} \cap [pn + m_0 - p + 1, pn + m_0]| \\ &\le A_p + A_p \left| J_{a,b_1,p} \cap \left[1, \frac{x}{p} + \frac{p-1}{p}\right] \right| \\ &= A_p \left| J_{a,b_1,p} \cap \left[0, \frac{x}{p} + \frac{p-1}{p}\right] \right|. \end{aligned}$$

Similarly, we have

$$\begin{aligned} &\left| J_{a,b_1,p} \cap \left[0, \frac{x}{p} + \frac{p-1}{p}\right] \right| \\ &\le A_p \left| J_{a,b_2,p} \cap \left[0, \frac{x}{p^2} + \frac{p-1}{p^2} + \frac{p-1}{p}\right] \right| \\ &\le \cdots \end{aligned}$$

$$\le A_p^{k-1}\left|J_{a,b_k,p}\cap\left[0,\frac{x}{p^k}+\frac{p-1}{p^k}+\cdots+\frac{p-1}{p}\right]\right|$$
$$\le A_p^{k-1}\left|J_{a,b_k,p}\cap\left[0,\frac{x}{p^k}+1\right)\right|.$$

If
$$\frac{x}{p^k}+1<\frac{8}{3},$$
then, by $1\notin J_{a,b_k,p}$,
$$\left|J_{a,b_k,p}\cap\left[0,\frac{x}{p^k}+1\right)\right|\le\left|J_{a,b_k,p}\cap[0,2]\right|\le 2\le 2\left(\frac{x}{p^k}\right)^{\frac{2}{3}}.$$

If
$$\frac{x}{p^k}+1\ge\frac{8}{3}$$
and $x<(p-1)p^k$, then by Lemma 2 we have
$$\begin{aligned}\left|J_{a,b_k,p}\cap\left[0,\frac{x}{p^k}+1\right)\right|&\le\left(\frac{9}{8}\right)^{\frac{1}{3}}\left(\frac{x}{p^k}+1\right)^{\frac{2}{3}}\\&\le\left(\frac{9}{8}\right)^{\frac{1}{3}}\left(\frac{2x}{p^k}\right)^{\frac{2}{3}}\\&\le 2\left(\frac{x}{p^k}\right)^{\frac{2}{3}}.\end{aligned}$$

If $x\ge(p-1)p^k$, then by Lemma 2 we have
$$\begin{aligned}\left|J_{a,b_k,p}\cap\left[0,\frac{x}{p^k}+1\right)\right|&\le\left|J_{a,b_k,p}\cap[0,p]\right|\\&\le\left(\frac{9}{8}\right)^{\frac{1}{3}}(p-1)^{\frac{2}{3}}+1\\&\le 2(p-1)^{\frac{2}{3}}\\&\le 2\left(\frac{x}{p^k}\right)^{\frac{2}{3}}.\end{aligned}$$

Therefore,
$$I_{a,b,p}(x)\le 2pA_p^{k-1}\left(\frac{x}{p^k}\right)^{\frac{2}{3}}=2\frac{p}{A_p}\left(\frac{9}{8}\right)^{\frac{k}{3}}x^{\frac{2}{3}}<2p^{\frac{1}{3}}\left(\frac{9}{8}\right)^{\frac{k}{3}}x^{\frac{2}{3}}$$

and

$$J_{a,b,p}(x) \le 2A_p^k\left(\frac{x}{p^k}\right)^{\frac{2}{3}} = 2\left(\frac{9}{8}\right)^{\frac{k}{3}} x^{\frac{2}{3}}.$$

Since

$$\log\left(\frac{9}{8}\right)^{\frac{k}{3}} = \frac{k}{3}\log\frac{9}{8} \le \left(\frac{1}{3}\log\frac{9}{8}\right)\frac{\log x}{\log p} < \frac{\log x}{25\log p},$$

it follows that

$$I_{a,b,p}(x) \le 2p^{\frac{1}{3}}x^{\frac{2}{3}+\frac{1}{25\log p}}, \quad J_{a,b,p}(x) \le 2x^{\frac{2}{3}-\frac{1}{25\log p}}.$$

This completes the proof of Theorem 3. □

Remark 1 With a slight adjustment of the proof of Theorem 3, we can prove that for any prime $p \ge 5$,

$$J_{1,1,p}(x) \le \left(\frac{9}{8}\right)^{\frac{1}{3}} x^{\frac{2}{3}+\frac{1}{25\log p}}, \quad x \ge 1. \tag{2}$$

Proof (*Proof of Theorem* 1)

Theorem 1(i) follows from Theorems 3 immediately.

Now we prove Theorem 1(ii) and (iii). We assume that a and b are two positive integers with $\nu_p(a) > \nu_p(b)$.

For any positive integer n, we have

$$\begin{aligned}
H_{a,b}(n+ab) &= H_{a,b}(n) + \sum_{k=n}^{n+ab-1}\frac{1}{ak+b} \\
&= H_{a,b}(n) + \sum_{r=0}^{b-1}\sum_{n\le bl+r\le n+ab-1}\frac{1}{abl+ar+b} \\
&= H_{a,b}(n) + \sum_{r=0}^{b-1}\sum_{n\le bl+r\le n+ab-1}\left(\frac{1}{abl+ar+b}-\frac{1}{ar+b}\right) \\
&\quad + \sum_{r=0}^{b-1}\sum_{n\le bl+r\le n+ab-1}\frac{1}{ar+b} \\
&= H_{a,b}(n) - \sum_{r=0}^{b-1}\sum_{n\le bl+r\le n+ab-1}\frac{abl}{(abl+ar+b)(ar+b)} \\
&\quad + \sum_{r=0}^{b-1}\frac{a}{ar+b}.
\end{aligned}$$

By $\nu_p(a) > \nu_p(b)$, we have

$$\nu_p(abl) \geq \nu_p(a) + \nu_p(b) > \nu_p(b) + \nu_p(b) = \nu_p(abl + ar + b) + \nu_p(ar + b)$$

and $\nu_p(a) > \nu_p(b) = \nu_p(ar + b)$. It follows that

$$H_{a,b}(n + ab) = H_{a,b}(n) + \frac{pd}{c}, \tag{3}$$

where c, d are two integers with $p \nmid c$.

By (3), $p \mid u_{a,b}(n)$ if and only if $p \mid u_{a,b}(n + ab)$. Thus $n \in J_{a,b,p}$ if and only if $n + ab \in J_{a,b,p}$. Noting that $0 \in J_{a,b,p}$ and $1 \notin J_{a,b,p}$, the set $J_{a,b,p}$ has positive asymptotic density which is between $1/(ab)$ and $(ab - 1)/(ab)$.

Again, by (3), $p \nmid v_{a,b}(n + ab)$ if and only if $p \nmid v_{a,b}(n)$. That is, $n + ab \in I_{a,b,p}$ if and only if $n \in I_{a,b,p}$. Since $0 \in I_{a,b,p}$, it follows that $I_{a,b,p}$ has positive asymptotic density at least $1/ab$. If $p \mid b$, then $1 \notin I_{a,b,p}$ and so $I_{a,b,p}$ has positive asymptotic density at most $(ab - 1)/ab$. If $p \nmid b$, then by $\nu_p(a) > \nu_p(b) = 0$ we see that $p \nmid ak + b$ for all integers k. It follows that $p \nmid v_{a,b}(n)$ for all $n \geq 0$. Therefore, the set $I_{a,b,p}$ is equal to the set of all nonnegative integers.

This completes the proof of Theorem 1(ii) and (iii).

□

4 Proof of Theorem 2

Given a positive integer M. For any positive integers n, let $n_{\overline{M}}$ be the largest factor m of n with $(m, M) = 1$ and let $n_M = n/n_{\overline{M}}$. We may call n_M the M-part and $n_{\overline{M}}$ the non M-part of n.

Let $\sim$ be one of symbols $<, =, >$. Define

$$N_{ab}^{\sim} = |\{n : v_{a,b}(n)_M \sim v_{a,b}(n + 1)_M, 0 \leq n \leq ab - 1\}|,$$

where $M = G_{a,b}$ and $v_{a,b}(m)_M = (v_{a,b}(m))_M$ for any $m \geq 1$.

Firstly, we prove the following theorem.

Theorem 4 *Let a and b be two positive integers and $M = G_{a,b}$.*

(i) The set of positive integers n with $v_{a,b}(n)_{\overline{M}} \neq v_{a,b}(n + 1)_{\overline{M}}$ has asymptotic density zero.

Furthermore, there exist two positive constants c_1 and c_2 such that

$$\frac{c_1 x}{\log x} \leq |\{n \leq x : v_{a,b}(n)_{\overline{M}} \neq v_{a,b}(n + 1)_{\overline{M}}\}| \leq \frac{c_2 x}{\log x}$$

for all sufficiently large x.

(ii) Let $\sim$ be arbitrary one of symbols $<, =, >$. Then the set of positive integers n with $v_{a,b}(n) \sim v_{a,b}(n+1)$ has positive asymptotic density $N_{ab}^{\sim}/(ab)$.

Proof Let $T_{a,b}$ be the set of positive integers n such that $v_{a,b}(n+1)_{\overline{M}} \neq v_{a,b}(n)_{\overline{M}}$.

Let $(a, b) = d$, $a = da'$ and $b = db'$. Then $(a', b') = 1$. By the prime number theorem for arithmetic progressions (see [3, Chap. 22]), the number of primes not exceeding x which are congruent to b' modulo a' is

$$\frac{1}{\phi(a')} \int_2^x \frac{1}{\log t} dt + O(x \exp(-c\sqrt{\log x})$$

for a positive constant c. So the number of $n \le x$ such that $a'n + b'$ is prime is greater than

$$\frac{c_1 x}{\log x}$$

for a positive constant c_1. Let n be an integer such that $a'n + b'$ is prime and $a'n + b' > (a, b) = d$. Then $a'n + b' \nmid M$ and $a'n + b' \nmid d(a'm + b')$ for all $0 \le m \le n - 1$. It follows that

$$a'n + b' \nmid v_{a,b}(n), \quad a'n + b' \mid v_{a,b}(n+1).$$

So $v_{a,b}(n+1)_{\overline{M}} \neq v_{a,b}(n)_{\overline{M}}$. Therefore,

$$T_{a,b}(x) > \frac{c_1 x}{\log x}$$

for all sufficiently large x.

Now we prove that

$$T_{a,b}(x) \ll \frac{x}{\log x}.$$

Suppose that $n \in T_{a,b}$, that is,

$$v_{a,b}(n+1)_{\overline{M}} \neq v_{a,b}(n)_{\overline{M}}.$$

Then there exists a prime $p \nmid M$ such that

$$\nu_p(v_{a,b}(n+1)) \neq \nu_p(v_{a,b}(n)). \tag{4}$$

It is clear that if (4) holds for some $p \nmid M$, then $n \in T_{a,b}$. Hence, $T_{a,b}$ is equal to the set of positive integers n such that (4) holds for some $p \nmid M$.

Firstly we consider those $n \in T_{a,b}$ with (4) for some prime p with $p \nmid a$.

Since

$$\frac{u_{a,b}(n+1)}{v_{a,b}(n+1)} = \frac{u_{a,b}(n)}{v_{a,b}(n)} + \frac{1}{an+b},$$

it follows from (4) that

$$p \mid an+b, \quad \nu_p(an+b) \geq \nu_p(v_{a,b}(n)). \tag{5}$$

Let l_p be the least integer such that $p \mid al_p+b$. In view of $p \nmid a$, $p \mid al_p+b$ and $p \mid an+b$, we have $p \mid n-l_p$. Let $n-l_p=pk$ and $b'_p=(al_p+b)/p$. Suppose that

$$p \nmid (ak+b'_p)u_{a,b'_p}(k+1)u_{a,b'_p}(k).$$

Then, by $p \nmid ak+b'_p$ and

$$\frac{u_{a,b'_p}(k+1)}{v_{a,b'_p}(k+1)} = \frac{u_{a,b'_p}(k)}{v_{a,b'_p}(k)} + \frac{1}{ak+b'_p},$$

we have

$$\nu_p(v_{a,b'_p}(k+1)) = \nu_p(v_{a,b'_p}(k)).$$

Write

$$H_{a,b}(n+1) = \sum_{\substack{l=0 \\ p\nmid al+b}}^{n} \frac{1}{al+b} + \sum_{\substack{l=0 \\ p\mid al+b}}^{n} \frac{1}{al+b}$$

and

$$H_{a,b}(n) = \sum_{\substack{l=0 \\ p\nmid al+b}}^{n-1} \frac{1}{al+b} + \sum_{\substack{l=0 \\ p\mid al+b}}^{n-1} \frac{1}{al+b}.$$

Noting that $p \mid an+b$, we have

$$\sum_{\substack{l=0 \\ p\nmid al+b}}^{n} \frac{1}{al+b} = \sum_{\substack{l=0 \\ p\nmid al+b}}^{n-1} \frac{1}{al+b} = \frac{t}{s},$$

where s, t are integers with $p \nmid s$. Since

$$\begin{aligned}\sum_{\substack{l=0 \\ p\mid al+b}}^{n} \frac{1}{al+b} &= \sum_{\substack{l=0 \\ p\mid al+al_p+b}}^{n-l_p} \frac{1}{al+al_p+b} = \sum_{\substack{l=0 \\ p\mid al+pb'_p}}^{n-l_p} \frac{1}{al+pb'_p} \\ &= \frac{1}{p}\sum_{l'=0}^{k} \frac{1}{al'+b'_p} = \frac{1}{p}\frac{u_{a,b'_p}(k+1)}{v_{a,b'_p}(k+1)}\end{aligned}$$

and

$$\sum_{\substack{l=0\\ p\nmid al+b}}^{n-1}\frac{1}{al+b}=\sum_{\substack{l=0\\ p\mid al+al_p+b}}^{n-1-l_p}\frac{1}{al+al_p+b}=\sum_{\substack{l=0\\ p\mid al+pb'_p}}^{n-1-l_p}\frac{1}{al+pb'_p}$$
$$=\frac{1}{p}\sum_{l'=0}^{k-1}\frac{1}{al'+b'_p}=\frac{1}{p}\frac{u_{a,b'_p}(k)}{v_{a,b'_p}(k)},$$

it follows that

$$H_{a,b}(n+1)=\frac{ptv_{a,b'_p}(k+1)+su_{a,b'_p}(k+1)}{psv_{a,b'_p}(k+1)}$$

and

$$H_{a,b}(n)=\frac{ptv_{a,b'_p}(k)+su_{a,b'_p}(k)}{psv_{a,b'_p}(k)}.$$

In view of $p\nmid u_{a,b'_p}(k+1)u_{a,b'_p}(k)$ and $p\nmid s$, we have

$$p\nmid ptv_{a,b'_p}(k+1)+su_{a,b'_p}(k+1),\quad p\nmid ptv_{a,b'_p}(k)+su_{a,b'_p}(k).$$

Thus,

$$\begin{aligned}\nu_p(v_{a,b}(n+1))&=\nu_p(psv_{a,b'_p}(k+1))=1+\nu_p(v_{a,b'_p}(k+1))\\&=1+\nu_p(v_{a,b'_p}(k))=\nu_p(psv_{a,b'_p}(k))=\nu_p(v_{a,b}(n)),\end{aligned}$$

a contradiction with (4). Hence

$$p\mid(ak+b'_p)u_{a,b'_p}(k+1)u_{a,b'_p}(k). \tag{6}$$

If $p\mid ak+b'_p$, then

$$an+b=a(n-l_p)+al_p+b=apk+pb'_p=p(ak+b'_p)$$

is divisible by p^2.

Noting that $n-l_p=pk$, we write $k=k_{n,p}$ in the following arguments. By (5) and (6), the set of $n\in T_{a,b}$ with (4) for some prime $p\nmid a$ is a subset of

$$\begin{aligned}&\{n: p^2\mid an+b,\ \nu_p(an+b)\ge\nu_p(v_{a,b}(n))\text{ for some prime }p\nmid a\}\\&\cup\{n: p\mid an+b,\ \nu_p(an+b)\ge\nu_p(v_{a,b}(n)),\ p\mid u_{a,b'_p}(k_{n,p}+1)\text{ for some prime }p\nmid a\}\\&\cup\{n: p\mid an+b,\ \nu_p(an+b)\ge\nu_p(v_{a,b}(n)),\ p\mid u_{a,b'_p}(k_{n,p})\text{ for some prime }p\nmid a\}.\end{aligned}$$

Now we consider those $n \in T_{a,b}$ with (4) for some prime p with $p \mid a$ and $p \nmid M$.

If $p \nmid b$, then by $p \mid a$ we have $p \nmid al + b$ for all integers l. It follows that $\nu_p(v_{a,b}(m)) = 0$ for every positive integer m, a contradiction with (4).

If $p \mid b$, then by $p \nmid M$, we have $\nu_p(a) \le \nu_p(b)$. Let

$$a = p^w a', \quad b = p^w b', \quad a', b', w \in \mathbb{Z}_{\ge 0}, \ p \nmid a'.$$

By Theorem 1, positive integers n with

$$p \mid u_{a',b'}(n) u_{a',b'}(n+1)$$

contribute to $T_{a,b}$ at most

$$4x^{\frac{2}{3}+\frac{1}{25 \log p}} \le 4x^{\frac{2}{3}+\frac{1}{25 \log 2}} \le 4x^{\frac{3}{4}}.$$

Thus, considering all prime factors p of a, positive integers n with

$$p \mid u_{a',b'}(n) u_{a',b'}(n+1)$$

contribute to $T_{a,b}$ at most

$$4\omega(a) x^{\frac{3}{4}},$$

where $\omega(a)$ denotes the number of distinct prime factors of a.

Now we consider those $n \in T_{a,b}$ with (4) for some prime $p \mid a$, $p \nmid M$ and $p \nmid u_{a',b'}(n) u_{a',b'}(n+1)$.

Since

$$H_{a,b}(n+1) = \frac{u_{a,b}(n+1)}{v_{a,b}(n+1)} = \frac{1}{p^w} \frac{u_{a',b'}(n+1)}{v_{a',b'}(n+1)}$$

and

$$H_{a,b}(n) = \frac{u_{a,b}(n)}{v_{a,b}(n)} = \frac{1}{p^w} \frac{u_{a',b'}(n)}{v_{a',b'}(n)},$$

it follows from $p \nmid u_{a',b'}(n) u_{a',b'}(n+1)$ and (4) that

$$\nu_p(v_{a',b'}(n+1)) \ne \nu_p(v_{a',b'}(n)). \tag{7}$$

Let l'_p be the least integer such that $p \mid a' l'_p + b'$. Since

$$\frac{u_{a',b'}(n+1)}{v_{a',b'}(n+1)} = \frac{u_{a',b'}(n)}{v_{a',b'}(n)} + \frac{1}{a'n + b'},$$

it follows from (7) that

$$p \mid a'n + b', \quad \nu_p(a'n + b') \ge \nu_p(v_{a',b'}(n)).$$

Let x be a sufficiently large number. For any prime p, let $\beta_{p,x}$ be the integer with

$$p^{\beta_{p,x}} \le x^{\frac{1}{8}} < p^{\beta_{p,x}+1}.$$

Then

$$\begin{aligned}
T_{a,b}(x) &\le \sum_{p\nmid M} |\{n \le x : \nu_p(v_{a,b}(n+1)) \ne \nu_p(v_{a,b}(n))\}| \\
&= \sum_{p\nmid M, p\nmid a} |\{n \le x : \nu_p(v_{a,b}(n+1)) \ne \nu_p(v_{a,b}(n))\}| \\
&\quad + \sum_{p\nmid M, p\mid a} |\{n \le x : \nu_p(v_{a,b}(n+1)) \ne \nu_p(v_{a,b}(n))\}| \\
&= \sum_{p\nmid a} |\{n \le x : \nu_p(v_{a,b}(n+1)) \ne \nu_p(v_{a,b}(n))\}| \\
&\quad + \sum_{p\nmid M, p\mid a} |\{n \le x : \nu_p(v_{a,b}(n+1)) \ne \nu_p(v_{a,b}(n))\}| \\
&\le \sum_{p\nmid a} |\{n \le x : p^2 \mid an+b, \nu_p(an+b) \ge \nu_p(v_{a,b}(n))\}| \\
&\quad + \sum_{p\nmid a} |\{n \le x : p \mid an+b, \nu_p(an+b) \ge \nu_p(v_{a,b}(n)), p \mid u_{a,b'_p}(k_{n,p}+1)\}| \\
&\quad + \sum_{p\nmid a} |\{n \le x : p \mid an+b, \nu_p(an+b) \ge \nu_p(v_{a,b}(n)), p \mid u_{a,b'_p}(k_{n,p})\}| \\
&\quad + 4\omega(a)x^{\frac{3}{4}} + \sum_{p\mid(a,b), p\nmid a'} |\{n \le x : \nu_p(a'n+b') \ge \nu_p(v_{a',b'}(n))\}| \\
&\le \Delta_1 + \Delta_1' + \Delta_2 + \Delta_2' + \Delta_3 + \Delta_4 + \Delta_5 + 4\omega(a)x^{\frac{3}{4}},
\end{aligned}$$

where

$$\Delta_1 = \sum_{p\le x^{1/16}, p\nmid a} |\{n \le x : \nu_p(v_{a,b}(n)) \le \beta_{p,x}\}|,$$

$$\Delta_1' = \sum_{p\mid(a,b), p\nmid a'} |\{n \le x : \nu_p(v_{a',b'}(n)) \le \beta_{p,x}\}|,$$

$$\Delta_2 = \sum_{p\le x^{1/16}, p\nmid a} |\{n \le x : \nu_p(an+b) > \beta_{p,x}\}|,$$

$$\Delta_2' = \sum_{p\mid(a,b), p\nmid a'} |\{n \le x : \nu_p(a'n+b') > \beta_{p,x}\}|,$$

$$\Delta_3 = \sum_{p>x^{1/16}, p\nmid a} |\{n \le x : p^2 \mid an+b\}|,$$

$$\Delta_4 = \sum_{p>x^{1/16}, p\nmid a} |\{n \le x : p \mid an+b,\ p \mid u_{a,b'_p}(k_{n,p})\}|,$$

and

$$\Delta_5 = \sum_{p>x^{1/16}, p\nmid a} |\{n \le x : p \mid an+b,\ p \mid u_{a,b'_p}(k_{n,p}+1)\}|.$$

In the following, we will prove that

$$\Delta_i \ll \frac{x}{\log x}, \quad i = 1, 2, 3, 4, 5.$$

and

$$\Delta'_i \ll \frac{x}{\log x}, \quad i = 1, 2.$$

□

4.1 The Estimates for Δ_1 and Δ'_1

For any prime p with $p \nmid a$ and $p < x^{1/16}$ and any positive integer $n \le x$ with $\nu_p(v_{a,b}(n)) \le \beta_{p,x}$, we have

$$p^{\beta_{p,x}+1} \nmid v_{a,b}(n).$$

Let $l_{p,x}$ be the least nonnegative integer with

$$p^{\beta_{p,x}+1} \mid al_{p,x} + b$$

and let

$$b_{p,x} = \frac{al_{p,x}+b}{p^{\beta_{p,x}+1}}.$$

It is clear that $l_{p,x} \le p^{\beta_{p,x}+1} - 1$. Write $n = p^{\beta_{p,x}+1}m + l_{p,x} - r$ with $0 \le r < p^{\beta_{p,x}+1}$. Then

$$m - 1 \le \frac{n - l_{p,x} - 1}{p^{\beta_{p,x}+1}} < m \le \frac{x}{p^{\beta_{p,x}+1}} + 1.$$

Write

$$H_{a,b}(n) = \sum_{\substack{l=0 \\ p^{\beta_{p,x}+1} \nmid al+b}}^{n-1} \frac{1}{al+b} + \sum_{\substack{l=0 \\ p^{\beta_{p,x}+1} \mid al+b}}^{n-1} \frac{1}{al+b}.$$

It is clear that

$$\sum_{\substack{l=0 \\ p^{\beta_{p,x}+1}\nmid al+b}}^{n-1} \frac{1}{al+b} = \frac{1}{p^{\beta_{p,x}}}\frac{r'}{s'},$$

where r', s' are integers with $p \nmid s'$. Since

$$\begin{aligned}
\sum_{\substack{l=0 \\ p^{\beta_{p,x}+1}|al+b}}^{n-1} \frac{1}{al+b} &= \sum_{\substack{l=0 \\ p^{\beta_{p,x}+1}|al+b}}^{n-1} \frac{1}{a(l-l_{p,x})+al_{p,x}+b} \\
&= \sum_{\substack{l=0 \\ p^{\beta_{p,x}+1}|al+al_{p,x}+b}}^{n-l_{p,x}-1} \frac{1}{al+al_{p,x}+b} \\
&= \sum_{\substack{l=0 \\ p^{\beta_{p,x}+1}|al+p^{\beta_{p,x}+1}b_{p,x}}}^{n-l_{p,x}-1} \frac{1}{al+p^{\beta_{p,x}+1}b_{p,x}} \\
&= \frac{1}{p^{\beta_{p,x}+1}}\sum_{l'=0}^{m-1}\frac{1}{al'+b_{p,x}},
\end{aligned}$$

it follows that

$$H_{a,b}(n) = \frac{1}{p^{\beta_{p,x}}}\frac{r'}{s'} + \frac{1}{p^{\beta_{p,x}+1}}\frac{u_{a,b_{p,x}}(m)}{v_{a,b_{p,x}}(m)} = \frac{pr'v_{a,b_{p,x}}(m)+s'u_{a,b_{p,x}}(m)}{p^{\beta_{p,x}+1}s'v_{a,b_{p,x}}(m)}. \tag{8}$$

In view of (8) and

$$p^{\beta_{p,x}+1} \nmid v_{a,b}(n), \quad p \nmid s',$$

we have $p \mid u_{a,b_{p,x}}(m)$. That is, $m \in J_{a,b_{p,x},p}$. Hence,

$$\begin{aligned}
\Delta_1 &= \sum_{p\le x^{1/16},\, p\nmid a} |\{n \le x : \nu_p(v_{a,b}(n)) \le \beta_{p,x}\}| \\
&\le \sum_{p\le x^{1/16},\, p\nmid a}\ \sum_{r=0}^{p^{\beta_{p,x}+1}-1}\ \sum_{\substack{m\le (x/p^{\beta_{p,x}+1})+1 \\ m\in J_{a,b_{p,x},p}}} 1 \\
&\le \sum_{p\le x^{1/16},\, p\nmid a} p^{\beta_{p,x}+1} J_{a,b_{p,x},p}\left(\frac{x}{p^{\beta_{p,x}+1}}+1\right).
\end{aligned}$$

By Theorem 3, we have

$$J_{a,b_{p,x},p}\left(\frac{x}{p^{\beta_{p,x}+1}}+1\right) \ll \left(\frac{x}{p^{\beta_{p,x}+1}}+1\right)^{\frac{2}{3}+\frac{1}{25\log p}} \ll p^{-\frac{2}{3}(\beta_{p,x}+1)}x^{\frac{2}{3}+\frac{1}{25\log p}}.$$

It follows that

$$\begin{aligned}\Delta_1 &\ll \sum_{p\le x^{1/16}} p^{\frac{1}{3}(\beta_{p,x}+1)}x^{\frac{2}{3}+\frac{1}{25\log p}}\\ &\ll 2^{\frac{1}{3}}x^{\frac{17}{24}+\frac{1}{25\log 2}} + \sum_{2<p\le x^{1/16}} p^{\frac{1}{3}}x^{\frac{1}{24}}x^{\frac{2}{3}+\frac{1}{25\log 3}}\\ &\ll x^{\frac{17}{24}+\frac{1}{25\log 2}} + x^{\frac{1}{16}}x^{\frac{1}{48}}x^{\frac{17}{24}+\frac{1}{25}}\\ &\ll x^{\frac{19}{24}+\frac{1}{25}} \ll \frac{x}{\log x}.\end{aligned}$$

Similarly, we have

$$\Delta_1' \ll \frac{x}{\log x}.$$

4.2 The Estimates for Δ_2, Δ_2', and Δ_3

For Δ_2, we have

$$\begin{aligned}\Delta_2 &\le \sum_{p\le x^{1/16},\, p\nmid a} |\{n\le x : \nu_p(an+b) > \beta_{p,x}\}|\\ &= \sum_{p\le x^{1/16},\, p\nmid a} |\{n\le x : p^{\beta_{p,x}+1} \mid an+b\}|\\ &\ll \sum_{p\le x^{1/16}} \frac{x}{p^{\beta_{p,x}+1}}\\ &\ll \sum_{p\le x^{1/16}} x^{\frac{7}{8}} \ll x^{\frac{15}{16}} \ll \frac{x}{\log x}.\end{aligned}$$

Similarly, we have

$$\Delta_2' \ll \frac{x}{\log x}.$$

For Δ_3, we have

$$\begin{aligned}\Delta_3 &= \sum_{p>x^{1/16}, p\nmid a} |\{n \le x : p^2 \mid an+b\}| \\ &\ll \sum_{x^{1/16}<p\le\sqrt{ax+b}} \frac{ax+b}{p^2} \\ &\ll x^{\frac{15}{16}} \ll \frac{x}{\log x}.\end{aligned}$$

4.3 The Estimates for Δ_4 and Δ_5

If $p > x$ with $p \nmid a$ and n is a nonnegative integer with $n \le x$ and $p \mid an+b$, then $p \le ax+b$, $l_p \le x$ and $p \mid n - l_p$ by $p \nmid a$. Since $0 \le n - l_p \le n \le x < p$, it follows that $n - l_p = 0$. Hence, by the prime number theorem (see [3]),

$$\sum_{p>x, p\nmid a} |\{n \le x : p \mid an+b\}| \le \sum_{x<p\le ax+b} 1 \ll \frac{x}{\log x}.$$

By Theorem 3, we have

$$\begin{aligned}\Delta_4 &= \sum_{p>x^{1/16}, p\nmid a} |\{n \le x : p \mid an+b,\ p \mid u_{a,b'_p}(k_{n,p})\}| \\ &\ll \sum_{x^{1/16}<p\le x, p\nmid a} |\{k \le \frac{x}{p} : p \mid u_{a,b'_p}(k)\}| + \frac{x}{\log x} \\ &\ll \sum_{x^{1/16}<p\le x} \left(\frac{x}{p}\right)^{\frac{2}{3}+\frac{1}{25\log p}} + \frac{x}{\log x} \\ &\ll \sum_{x^{1/16}<p\le x} \left(\frac{x}{p}\right)^{\frac{2}{3}} \left(\frac{p^{16}}{p}\right)^{\frac{1}{25\log p}} + \frac{x}{\log x} \\ &\ll x^{\frac{2}{3}} \sum_{x^{1/16}<p\le x} p^{-\frac{2}{3}} + \frac{x}{\log x}.\end{aligned}$$

In view of the prime number theorem (see [3]),

$$\begin{aligned}\sum_{x^{1/16}<p\le x} p^{-\frac{2}{3}} &= \int_{x^{1/16}}^{x} y^{-\frac{2}{3}} d\pi(y) \ll \int_{x^{1/16}}^{x} y^{-\frac{2}{3}} \frac{1}{\log y} dy \\ &\ll \frac{1}{\log x} \int_{x^{1/16}}^{x} y^{-\frac{2}{3}} dy \ll \frac{x^{\frac{1}{3}}}{\log x}.\end{aligned}$$

Hence

$$\Delta_4 \ll x^{\frac{2}{3}} \sum_{x^{1/16}<p\le x} p^{-\frac{2}{3}} + \frac{x}{\log x} \ll \frac{x}{\log x}.$$

Similarly, we have

$$\Delta_5 \ll \frac{x}{\log x}.$$

Finally, we have

$$T_{a,b}(x) \le \Delta_1 + \Delta_1' + \Delta_2 + \Delta_2' + \Delta_3 + \Delta_4 + \Delta_5 + 4\omega(a)x^{\frac{3}{4}} \ll \frac{x}{\log x}.$$

This completes the proof of Theorem 4(i).

Now we prove Theorem 4(ii).

In view of Theorem 4(i), it is enough to prove that the set of positive integers n with $v_{a,b}(n)_M \sim v_{a,b}(n+1)_M$ has positive asymptotic density $N_{ab}^{\sim}/(ab)$, where $M = G_{a,b}$. If $M = 1$, then the conclusion is trivial. Now we assume that $M > 1$.

Let p be a prime factor of M. Then $\nu_p(a) > \nu_p(b) \ge 1$. By (3),

$$H_{a,b}(n+ab) = H_{a,b}(n) + \frac{pd}{c}, \tag{9}$$

where c, d are two integers with $p \nmid c$. So

$$\nu_p(v_{a,b}(n+ab)) = \nu_p(v_{a,b}(n)).$$

Hence

$$v_{a,b}(n+ab)_M = v_{a,b}(n)_M.$$

It follows that the set of positive integers n with $v_{a,b}(n)_M \sim v_{a,b}(n+1)_M$ has positive asymptotic density $N_{ab}^{\sim}/(ab)$, where $M = G_{a,b}$. This completes the proof of Theorem 4(ii). □

Proof (*Proof of Theorem* 2) Let $M = G_{a,b}$ and let $d = (a, b)$, $a = da'$ and $b = db'$. Then $(a', b') = 1$.

Firstly, we prove Theorem 2(i).

Since $M = 1$, it follows that $v_{a,b}(n)_M = 1 = v_{a,b}(n+1)_M$. By Theorem 4(i), the set of positive integers n with $v_{a,b}(n)_{\overline{M}} \ne v_{a,b}(n+1)_{\overline{M}}$ has asymptotic density zero. So, the set of positive integers n with $v_{a,b}(n)_{\overline{M}} = v_{a,b}(n+1)_{\overline{M}}$ has asymptotic density one. Therefore, the set of positive integers n with $v_{a,b}(n) = v_{a,b}(n+1)$ has asymptotic density one.

By Dirichlet's theorem for arithmetic progressions, there are infinitely many primes of the form $a'n + b'$. Let $a'n + b'$ be a sufficiently large prime. Since

$$H_{a,b}(n) = \frac{1}{d}\frac{u_{a',b'}(n)}{v_{a',b'}(n)},$$

$$\begin{aligned} H_{a,b}(n+1) &= \frac{1}{d}\frac{u_{a',b'}(n+1)}{v_{a',b'}(n+1)} \\ &= \frac{1}{d}\frac{u_{a',b'}(n)}{v_{a',b'}(n)} + \frac{1}{d}\frac{1}{a'n+b'} \\ &= \frac{1}{d}\frac{u_{a',b'}(n)(a'n+b') + v_{a',b'}(n)}{v_{a',b'}(n)(a'n+b')} \end{aligned}$$

and

$$(u_{a',b'}(n)(a'n+b') + v_{a',b'}(n), v_{a',b'}(n)(a'n+b')) = 1,$$

it follows that

$$v_{a,b}(n+1) \geq v_{a',b'}(n)(a'n+b') > dv_{a',b'}(n) \geq v_{a,b}(n).$$

Therefore, there are infinitely many positive integers n such that

$$v_{a,b}(n+1) > v_{a,b}(n).$$

Now, we prove Theorem 2(ii) and (iii).

Since $M = G_{a,b} > 1$, we have $v_{a,b}(0)_M = 1 < b_M = v_{a,b}(1)_M$. So $N^{<}_{a,b} \geq 1$. By Theorem 4, the set of positive integers n with $v_{a,b}(n) < v_{a,b}(n+1)$ has positive asymptotic density.

Let p be the least prime factor of $M = G_{a,b}$. Then $p \mid d$, $p \mid a'$ and $p \nmid b'$. Since

$$H_{a,b}(n) = \frac{1}{d}H_{a',b'}(n)$$

and

$$\begin{aligned} H_{a',b'}(n) &= \sum_{k=0}^{n-1}\left(\frac{1}{b'+ka'} - \frac{1}{b'}\right) + \frac{n}{b'} \\ &= -\sum_{k=0}^{n-1}\frac{ka'}{b'(b'+ka')} + \frac{n}{b'} \\ &= -\frac{a't_n}{s_n} + \frac{n}{b'} \\ &= \frac{-a't_nb' + ns_n}{b's_n}, \end{aligned}$$

where s_n and t_n are integers with $(a', s_n) = 1$, it follows that

$$\nu_p(v_{a,b}(p)) \le \nu_p(d) - 1 < \nu_p(d) = \nu_p(v_{a,b}(p-1)).$$

For any prime factor q of $M = G_{a,b}$ with $q > p$, we have $q \mid a'$ and $q \nmid b'$. So $q \nmid b's_n$ for all n. It follows that $q \nmid -a't_nb' + ns_n$ for $n \le p$. Hence, for $n \le p$, $\nu_q(v_{a,b}(n)) = \nu_q(d)$. Thus,

$$\nu_q(v_{a,b}(p)) = \nu_q(d) = \nu_q(v_{a,b}(p-1)).$$

Combining the above arguments, we have

$$v_{a,b}(p-1)_M > v_{a,b}(p)_M.$$

So $N^{>}_{a,b} \ge 1$. By Theorem 4, the set of positive integers n with $v_{a,b}(n) > v_{a,b}(n+1)$ has positive asymptotic density.

To complete the proof, now we divide into two cases:

Case 1: $2 \mid G_{a,b}$. Then $2 \mid d$, $2 \mid a'$ and $2 \nmid b'$.

Let n be a nonnegative integer. It is clear that

$$\begin{aligned} H_{a,b}(n+1) &= H_{a,b}(n) + \frac{1}{an+b} \\ &= \frac{1}{d}\frac{u_{a',b'}(n)}{v_{a',b'}(n)} + \frac{1}{d}\frac{1}{a'n+b'} \\ &= \frac{1}{d}\frac{u_{a',b'}(n)(a'n+b') + v_{a',b'}(n)}{v_{a',b'}(n)(a'n+b')}. \end{aligned}$$

Since $2 \mid a'$ and $2 \nmid b'$, it follows that $v_{a',b'}(n)$ and $a'n+b'$ are both odd. Thus, exactly one of $u_{a',b'}(n)$ and $u_{a',b'}(n)(a'n+b') + v_{a',b'}(n)$ is even. Noting that $2 \mid d$, we have

$$\nu_2(v_{a,b}(n+1)) \ne \nu_2(v_{a,b}(n)).$$

Therefore,

$$v_{a,b}(n+1) \ne v_{a,b}(n).$$

Case 2: $2 \nmid G_{a,b}$ and $G_{a,b} > 1$.

For any prime factor p of $M = G_{a,b}$, we have $p \mid a'$ and $p \nmid b'$. So $(M, b') = 1$ and $(M, b' + a') = 1$. Since

$$H_{a,b}(1) = \frac{1}{b} = \frac{1}{db'}$$

and

$$H_{a,b}(2) = \frac{1}{b} + \frac{1}{b+a} = \frac{2b'+a'}{db'(b'+a')},$$

it follows that

$$v_{a,b}(1)_M = (db')_M = d_M = (db'(b'+a'))_M = v_{a,b}(2)_M.$$

So $N^{=}_{a,b} \geq 1$. It follows from Theorem 4 that the set of positive integers n with $v_{a,b}(n) = v_{a,b}(n+1)$ has positive asymptotic density.

This completes the proof of Theorem 2.

□

Acknowledgements This work was supported by the National Natural Science Foundation of China, Grant No. 11771211 and a project funded by the Priority Academic Program Development of Jiangsu Higher Education Institutions.

References

1. Boyd, D.W.: A p-adic study of the partial sums of the harmonic series. Exp. Math. **3 (4)**, 287–302 (1994)
2. Eswarathasan, A., Levine, E.: p-integral harmonic sums. Discrete Math. **91(3)**, 249–257 (1991)
3. Davenport, H.: Multiplicative number theory. 2nd Ed., Springer (1980)
4. Hardy, G.H., Wright, E.M.: An introduction to the theory of numbers. 5th ed., Oxford Univ. Press (1979)
5. Sanna, C.: On the p-adic valuation of harmonic numbers. J. Number Theory **166**, 41–46 (2016)
6. Shiu, P.: The denominators of harmonic numbers. arXiv:1607.02863v1
7. Theisinger, L.: Bemerkung über die harmonische Reihe. Monatsh. für Math. **26**, 132–134 (1915)
8. Wolstenholme, J.: On certain properties of prime numbers. Quart. J. Pure Appl. Math. **5**, 35–39 (1862)
9. Wu, B.L., Chen, Y.G.: On certain properties of harmonic numbers. J. Number Theory **175**, 66–86 (2017)
10. Wu, B.L., Chen,Y.G.: On the denominators of harmonic numbers. C. R. Acad. Sci. Paris, Ser I **356**, 129–132 (2018)
11. Wu, B.L., Chen, Y.G.: On the denominators of harmonic numbers, II. J. Number Theory **200**, 397–406 (2019)

Partitions for Semi-magic Squares of Size Three

Robert W. Donley

Abstract In the theory of Clebsch-Gordan coefficients, one may recognize the domain space as the set of weakly semi-magic squares of size three. Two partitions on this set are considered: a triangle-hexagon model based on top lines, and one based on the orbits under a finite group action. In addition to giving another proof of McMahon's formula, we give a generating function that counts the so-called trivial zeros of Clebsch-Gordan coefficients and its associated quasi-polynomial.

Keywords Clebsch-Gordan coefficient · Semi-magic square · Stochastic matrix

1 Introduction

A basic open problem of elementary representation theory concerns the classification of zeros for Clebsch-Gordan coefficients of $SU(2)$. Regge made the remarkable observation that the domain space for Clebsch-Gordan coefficients corresponds precisely to weakly semi-magic of squares of size three and that, suitably normalized, the associated group of symmetries of the determinant acts upon values of Clebsch-Gordan coefficients by sign changes. In particular, these symmetries preserve the zero locus for Clebsch-Gordan coefficients.

In the notation of [3], a parametrization for the domain of $c_{m,n,k}(i, j)$ is given by the set of semi-magic squares

$$\begin{Vmatrix} n-k & m-k & k \\ i & j & m+n-i-j-k \\ m-i & n-j & i+j-k \end{Vmatrix} \tag{1}$$

with non-negative integer entries. In particular, the Clebsch-Gordan coefficients for a fixed top line record all data for the projection

R. W. Donley (✉)
Queensborough Community College, Bayside, New York, NY, USA
e-mail: RDonley@qcc.cuny.edu

M. B. Nathanson (ed.), *Combinatorial and Additive Number Theory IV*,
Springer Proceedings in Mathematics & Statistics 347,
https://doi.org/10.1007/978-3-030-67996-5_7

$$V(m) \otimes V(n) \to V(m+n-2k).$$

The bottom rows sum according to the projection and may be considered as exponent pairs. With respect to the natural invariant bilinear form, the weight vectors

$$f^i \phi_m \quad \text{and} \quad f^{m-i} \phi_m$$

pair non-trivially.

The classification of zeros for Clebsch-Gordan coefficients typically starts by extracting the so-called trivial zeros; these zeros arise from certain fixed points of Regge symmetries, and, as we will show, display a measurable regularity. The remaining zeros may be further stratified by magic number J, the minimal entry of the magic square (the degree), and the order; see [9] for a definition of the latter item and for results on non-trivial zeros with $J \leq 900$ and beyond.

Our strategy for enumerating non-trivial zeros is as follows:

- for a fixed J, elementary MAPLE programming creates the list of all zeros as magic squares. Output consists of seven integers - the square (five parameters), the minimal entry in the square, and the determinant of the square,
- for odd J, there are trivial zeros; the associated magic squares have determinant zero, but non-trivial zeros may also have this property,
- such non-trivial zeros are detected by comparing counts for determinant zero with the generating function that counts trivial zeros; these zeros are sparse and readily found, and
- to both enumerate orbits and count orbit sizes, representatives for non-trivial zeros are narrowed down by sorting the list in EXCEL using determinant and smallest value.

In this work, we consider results on enumerating trivial zeros as a proof of concept for techniques associated to two partitions. Sections 3–5 develop a triangle-hexagon model; another proof of McMahon's enumeration of semi-magic squares of size three is given in Sect. 3 (Proposition 3), and Secs. 4 and 5 provide further examples. Sections 6–8 consider orbit-based partitioning. The main results are the generating function that counts trivial zeros in Sect. 7 (Theorem 1), and the corresponding quasi-polynomial for trivial zeros in Sect. 8 (Theorem 2). Section 9 gives a detailed application of the theory to the example of $J = 15$.

2 Magic Squares and Clebsch-Gordan Coefficients

Definition 1 Let $\mathbb{M}_3$ denote the monoid of weakly semi-magic squares of size three; that is, each element of $\mathbb{M}_3$ is of the form

$$M = \begin{Vmatrix} a & b & k \\ r & * & * \\ * & * & c \end{Vmatrix} \tag{2}$$

such that

- all entries are nonnegative integers, and
- all line sums along rows and columns are equal.

This sum, the **magic number**, equals $J = a + b + k$. Such matrices are also called **integer doubly-stochastic matrices**.

Definition 2 Fix $J \geq 0$. Let $\mathbb{M}_3(J)$ denote the subset in $\mathbb{M}_3$ of all semi-magic squares with magic number J. Define $H_3(J)$ to be the cardinality of $\mathbb{M}_3(J)$.

Of course, when $J > 0$, any M in $\mathbb{M}_3(J)$ becomes a proper doubly-stochastic matrix with rational entries upon division by J. The earliest formula for $H_3(J)$ is due to MacMahon [7]; see [12], and Proposition 3 below. For a general overview of properties of $H_n(J)$, see Chap. 1 of [11]. An extensive presentation of the theory of doubly-stochastic matrices with respect to composite quantum systems may be found in [6].

With this parametrization, the corresponding Clebsch-Gordan coefficients belong to the tensor product

$$V(b+k) \otimes V(a+k) \to V(a+b);$$

each highest weight is the pairwise sum of elements in the top line.

Definition 3 The Clebsch-Gordan function

$$C : \mathbb{M}_3 \to \mathbb{Z}$$

is the map $M = \begin{Vmatrix} a & b & k \\ r & s & * \\ * & * & c \end{Vmatrix} \mapsto$

$$C(M) = \sum_{l=0}^{k} (-1)^l \binom{c}{r-l} \binom{b+k-l}{b} \binom{a+l}{a}.$$

Here $s = c - r + k$, and $C(M)$ is the coefficient of

$$x^s y^r$$

in the power series expansion of

$$\frac{(x+y)^c}{(1-x)^{b+1}(1+y)^{a+1}}.$$

3 A Partition for Semi-magic Squares of Size Three

For square matrices of size three, let G be the group of determinantal symmetries; that is,

- G is generated by row switches, column switches, and transpose,
- every element g of G may be expressed uniquely as

$$g = R(\sigma)C(\tau)T^{\varepsilon} \quad \text{with} \quad \sigma, \tau \in S_3, \ \varepsilon \in \{0, 1\} \tag{3}$$

  and
- $|G| = 72$.

In the above factorization, $R(\sigma)$ denotes a permutation of rows, $C(\tau)$ a permutation of columns, and T the transpose operation. For basic group theoretic properties of these symmetries, such as conjugacy classes, normal subgroups, and character table, see [8].

These symmetries preserve

1. the semi-magic square property,
2. the magic number J and thus each $M_3(J)$, and
3. the zero locus for Clebsch-Gordan coefficients.

The classification of zeros of Clebsch-Gordan coefficients has traditionally been a subject of intense study, largely focused in the mathematical physics literature, and remains an open problem. In the present work and preceding issues [3–5], one goal has been to reconstruct parts of the general theory using combinatorial methods. A goal of this work is to give a natural visual representation of the domain space. That is, we parametrize this subset of semi-magic squares as

$$\left\| \begin{matrix} a & b & k \\ r & * & * \\ * & * & c \end{matrix} \right\| \subset \mathbb{N}^5 \tag{4}$$

in terms of a triangle-hexagon model; generally, each lattice point in a three-dimensional triangular cone represents a hexagonal family of magic squares.

The partitioning of this set occurs in three steps:

- fix a magic number $J \geq 0$,
- organize the compositions of J with three parts into an equilateral triangle of size $J + 1$, and
- organize all magic squares with a fixed top line, as a composition of J, into a possibly degenerate hexagon.

In this way, the data attached to

$$V(b + k) \otimes V(a + k) \to V(a + b)$$

is indexed by all magic squares with top line (a, b, k).

To organize the top lines for a fixed $J \geq 0$, we index the first two entries of the triple as usual for matrices, but starting at index zero. These triples may then be arranged as an equilateral triangle as noted. For any top line in the triangle, the entries denote the number of rows above, the number of spaces to the left, and the number of spaces to the right, respectively,

To further analyze the set of magic squares associated to a top line, we consider (r, c) as coordinates in a square matrix of size $J + 1$ in which the first row and column are indexed by zero. By the magic square property, r and c are subject to the inequalities

$$0 \leq r \leq b + k, \qquad 0 \leq c \leq a + b, \qquad -k \leq c - r \leq a. \tag{5}$$

The solution set of these inequalities may be interpreted as a rectangle with two congruent right triangles removed from the upper right and lower left corners. Since $(r, c) = (0, 0)$ is always a solution of these inequalities, the solution set always has one vertex in the upper left corner.

Furthermore the shape of the solution set is determined by the following side lengths:

- vertical side: $k + 1$,
- diagonal side: $b - 1$,
- horizontal side: $a + 1$.

If exactly one of a, b, or k is zero, the solution set is a parallelogram, and we obtain a line segment if two parameters equal zero; otherwise, the solution set is a hexagon with parallel opposing sides. Proposition 2 below follows directly.

Considering the characterization by side lengths, we see that a permutation of columns permutes the entries of a given top line, with the effect that the corresponding hexagon has the same shape in a new orientation.

Finally, we note formulas for enumerating the various objects of interest:

Proposition 1 *The number of top lines associated to a fixed J is given by*

$$1 + 2 + 3 + \cdots + (J + 1) = \frac{(J + 1)(J + 2)}{2}. \tag{6}$$

Proposition 2 *Let $(a, b, k)|$ denote the number of semi-magic squares with top line (a, b, k). Then*

$$|(a, b, k)| = 1 + (a + b + k) + (ab + ak + bk). \tag{7}$$

In particular, $|(a, b, k)|$ is invariant under permutations of a, b and k, a fact consistent with our characterization of column permutations as isometries of the corresponding hexagon. See Sect. 5 for an algorithm that computes the array of all $|(a, b, k)|$ for a given J.

Proposition 3 ([10], A002817) *Let $H_3(J)$ be the number of semi-magic squares with magic number J. Then*

$$H_3(J) = \frac{(J+1)(J+2)(J^2+3J+4)}{8}. \tag{8}$$

Remark 1 For alternative expressions to (8), we have

$$H_3(J) = 3\binom{J+3}{4} + \binom{J+2}{2} \tag{9}$$

$$= \binom{J+5}{5} - \binom{J+2}{5} \tag{10}$$

$$= \binom{J+4}{4} + \binom{J+3}{4} + \binom{J+2}{4}. \tag{11}$$

Equation (9) is given in [7], Art. 407. In [12], one finds (10) and (11) in an exercise (Chap. 2, Ex. 15, pp. 225, 236); see also [2]. From the third equality, we have the generating function

$$\sum_{J\geq 0} H_3(J)x^J = \frac{1+x+x^2}{(1-x)^5},$$

also found in Sect. 4.6.1 of [12]. For general properties of $H_n(J)$, see [11], Chap. 1.1–1.9.

Proof (Proposition 3) From the partition triangle, we have that

$$H_3(J) = \sum_{k=0}^{J}\sum_{b=0}^{J-k} |(J-b-k, b, k)|. \tag{12}$$

Using Proposition 2 and the identities

$$\sum_{k=0}^{N} k = \frac{N(N+1)}{2}, \quad \sum_{k=0}^{N} k^2 = \frac{N(N+1)(2N+1)}{6}, \quad \sum_{k=0}^{N} k^3 = \frac{N^2(N+1)^2}{4},$$

the result follows from direct computation. Alternatively, the identities imply that $H_3(J)$ is a polynomial in J of degree 4, and the coefficients may be simply computed using the first five values of $H_3(J)$:

$$H_3(0) = 1, \;\; H_3(1) = 6, \;\; H_3(2) = 21, \;\; H_3(3) = 55, \;\; H_3(4) = 120. \qquad \square$$

Remark 2 An outline for this proof is noted in Sect. 6.2 of [6]; Proposition 2 provides the missing step in (6.13).

$$\begin{array}{c} (0,0,4),(0,1,3),(0,2,2),(0,3,1),\boxed{(0,4,0)} \\ (1,0,3),(1,1,2),\boxed{(1,2,1)},(1,3,0) \\ (2,0,2),(2,1,1),(2,2,0) \\ \boxed{(3,0,1)},(3,1,0) \\ (4,0,0) \end{array} \qquad \begin{array}{ccccccccc} 5 & & 8 & & 9 & & 8 & & \boxed{5} \\ & 8 & & 10 & & \boxed{10} & & 8 & \\ & & 9 & & 10 & & 9 & & \\ & & & \boxed{8} & & 8 & & & \\ & & & & 5 & & & & \end{array}$$

Fig. 1 $J = 4$: **a** Top lines and **b** magic square counts

4 Examples

We consider $\mathbb{M}_3(4)$, the subset of semi-magic squares with $J = 4$. The triangle of top lines and corresponding magic square counts are given as Fig. 1. Here we have boxed entries to be used below.

Applying MacMahon's formula, we obtain 120 magic squares, confirmed by the counts above, as are the 15 top lines.

Now consider the boxed top line $(0, 4, 0)$ with 5 magic squares:

$$\left\|\begin{matrix} 0 & 4 & 0 \\ r & 0 & * \\ * & 0 & c \end{matrix}\right\| \mapsto \begin{bmatrix} * & & & & \\ & * & & & \\ & & * & & \\ & & & \boxed{*} & \\ & & & & * \end{bmatrix}.$$

The starred entries on the right correspond to (r, c) that yield a valid magic square; the boxed entry corresponds to $(r, c) = (3, 3)$, which gives the magic square

$$\left\|\begin{matrix} 0 & 4 & 0 \\ 3 & 0 & 1 \\ 1 & 0 & 3 \end{matrix}\right\|.$$

On the other hand, the pair $(r, c) = (1, 0)$ does not produce a valid magic square, signified by the blank space below the first star.

Next we have the boxed top line $(3, 0, 1)$ with 8 magic squares:

$$\left\|\begin{matrix} 3 & 0 & 1 \\ r & * & * \\ * & * & c \end{matrix}\right\| \mapsto \begin{bmatrix} * & * & * & * \\ * & * & \boxed{*} & * \\ & & & \\ & & & \end{bmatrix}.$$

The boxed entry corresponds to $(r, c) = (1, 2)$, corresponding to the magic square

$$\left\| \begin{matrix} 3 & 0 & 1 \\ 1 & 2 & 1 \\ 0 & 2 & 2 \end{matrix} \right\| .$$

Finally consider the boxed top line $(1, 2, 1)$ with 10 magic squares:

$$\left\| \begin{matrix} 1 & 2 & 1 \\ r & * & * \\ * & * & c \end{matrix} \right\| \mapsto \begin{bmatrix} * & * & & \\ * & * & * & \\ & * & \boxed{*} & * \\ & & * & * \end{bmatrix} .$$

The boxed entry corresponds to $(r, c) = (2, 2)$, giving the magic square

$$\left\| \begin{matrix} 1 & 2 & 1 \\ 2 & 1 & 1 \\ 1 & 1 & 2 \end{matrix} \right\| .$$

Of course, if we evaluate $C(M)$ for all M in each array, we obtain, respectively,

$$\begin{bmatrix} 1 & & & & \\ & 1 & & & \\ & & 1 & & \\ & & & 1 & \\ & & & & 1 \end{bmatrix}, \quad \begin{bmatrix} 1 & 1 & 1 & 1 \\ -4 & -3 & -2 & -1 \end{bmatrix}, \quad \begin{bmatrix} 3 & 3 & & \\ -2 & 1 & 4 & \\ & -2 & -1 & 3 \\ & & -2 & -3 \end{bmatrix} .$$

5 Examples of Zeros

We now apply the partition of Sect. 3 to organize the zeros of $C(M)$ in each $\mathbb{M}_3(J)$. For instance, if $J = 8$, Fig. 2 gives an enumeration by top lines of magic squares and zeros of $C(M)$.

$$\begin{array}{c} 9 \quad 16 \quad 21 \quad 24 \quad 25 \quad 24 \quad 21 \quad 16 \quad 9 \\ 16 \quad 22 \quad 26 \quad 28 \quad 28 \quad 26 \quad 22 \quad 16 \\ 21 \quad 26 \quad 29 \quad 30 \quad 29 \quad \boxed{26} \quad 21 \\ 24 \quad 28 \quad 30 \quad 30 \quad 28 \quad 24 \\ 25 \quad 28 \quad 29 \quad 28 \quad 25 \\ 24 \quad 26 \quad 26 \quad 24 \\ 21 \quad 22 \quad 21 \\ 16 \quad 16 \\ 9 \end{array} \qquad \begin{array}{c} \circ \quad \circ \quad \circ \quad \circ \quad \circ \quad \circ \quad \circ \quad \circ \quad \circ \\ \circ \quad \cdot \quad 2 \quad \cdot \quad \cdot \quad 2 \quad \cdot \quad \circ \\ \circ \quad 2 \quad 2 \quad \cdot \quad 2 \quad \boxed{2} \quad \circ \\ \circ \quad \cdot \quad \cdot \quad \cdot \quad \cdot \quad \circ \\ \circ \quad \cdot \quad 2 \quad \cdot \quad \circ \\ \circ \quad 2 \quad 2 \quad \circ \\ \circ \quad \cdot \quad \circ \\ \circ \quad \circ \\ \circ \end{array}$$

Fig. 2 $J = 8$: **a** Magic square counts and **b** zero counts

For the figure on the left, we note that the magic square counts may be generated algorithmically using (7). That is,

- the vertices have value $J+1$,
- the differences along rows progress arithmetically, decreasing by increments of 2, and
- the starting differences for each row begin with $J-1$ in the top row and decrease by increments of 1.

One notes that values are preserved under the symmetries induced by permutations of columns. Top line orbits have orders 1, 3, or 6. Any fixed point is of the form (a, a, a) and corresponds to the central entry in the triangle when $J = 3a$.

On the right in Fig. 2, we see that there are 18 zeros of $C(M)$ for $J = 8$. These zeros correspond to a single orbit of magic squares under the full determinantal group, with representative noted below.

The boxed entries correspond to the top line (2, 5, 1), confirmed by counting rows above and spaces to each side of the box. The hexagon corresponding to top line (2, 5, 1) is given by

$$\begin{bmatrix} 6 & 6 & 6 & & & & & \\ -3 & 3 & 9 & 15 & & & & \\ & -3 & \boxed{0} & 9 & 24 & & & \\ & & -3 & -3 & 6 & 30 & & \\ & & & -3 & -6 & \boxed{0} & 30 & \\ & & & & -3 & -9 & -9 & 21 \\ & & & & & -3 & -12 & -21 \end{bmatrix}, \tag{13}$$

from which one readily counts 26 magic squares and two zeros. In particular, for the boxed zeros, we have coordinates $(r, c) = (2, 2),\ (4, 5)$, yielding the magic squares

$$\left\| \begin{matrix} 2 & 5 & 1 \\ 2 & 1 & 5 \\ 4 & 2 & 2 \end{matrix} \right\|, \quad \left\| \begin{matrix} 2 & 5 & 1 \\ 4 & 2 & 2 \\ 2 & 1 & 5 \end{matrix} \right\|.$$

These squares are related by the Weyl group symmetry, which switches the lower rows; the effect on the hexagon is to rotate by 180 degrees, while the Clebsch-Gordan coefficients transform according to formula (8.5) in [3]. Additionally, we recall that this hexagon records all data for the tensor product

$$V(6) \otimes V(3) \to V(7).$$

In general, we will consider only the triangle for the zero locus of $C(M)$ in $\mathbb{M}_3(J)$ and triangles for the orbits of zeros under the full symmetry group. For example, the zero locus corresponding to $J = 24$ contains six orbits contributing 252 zeros. Figure 3 gives the full portrait of zero counts, and Fig. 4 gives the portrait of zero counts for the orbit containing the twelve zeros for top line (8, 8, 8).

Fig. 3 The 252 zeros for $J = 24$

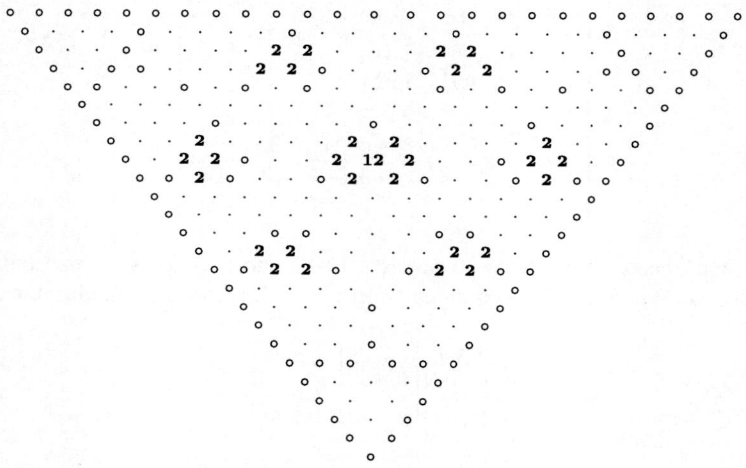

Fig. 4 The 72 zeros for the orbit of $M_{(8,8,8)}$

The hexagon for top line (8, 8, 8) is preserved under a symmetry group of type D_{12}, the dihedral group of order 12, and all zeros for (8, 8, 8) consist of a single orbit under D_{12}. A representative of this orbit is given by

$$M_{(8,8,8)} = \left\| \begin{matrix} 8 & 8 & 8 \\ 2 & 9 & 13 \\ 14 & 7 & 3 \end{matrix} \right\| .$$

A necessary condition to have an orbit of order 12 is that the hexagon be regular, or equivalently that the top line is (a, a, a), the center of the triangle when $J = 3a$. As of this writing, the author knows of no other examples; see Fig. 2 in [4].

6 Orbits and a Second Partition

For another approach to Proposition 3, we consider a different partitioning of each $\mathbb{M}_3(J)$, closer in spirit to Sect. 4 of [1], which uses convex geometry and specific representatives to construct generating functions for related classes of magic squares.

Consider the following definitions:

Definition 4 Let U denote the element of $\mathbb{M}_3(3)$ with all entries having value 1.

Definition 5 We say that an element M of $\mathbb{M}_3$ is a **reduced** if 0 occurs as an entry in M. For a fixed $J \geq 0$, we denote the set of all reduced squares in $\mathbb{M}_3(J)$ by $N_3(J)$.

If any entry of a top line equals zero, then the corresponding polygon consists entirely of reduced magic squares; otherwise, the boundary of the corresponding hexagon consists precisely of the reduced magic squares with this top line.

Definition 6 Let $\mathbb{M}_3^s$ denote the subset of $\mathbb{M}_3$ of squares with smallest entry s, and let $\mathbb{M}_3^s(J) = \mathbb{M}_3^s \cap \mathbb{M}_3(J)$.

The partitioning in question is rooted in the following observation:

Lemma 1 *Let M be an element of $\mathbb{M}_3^s(J)$. Then there exists a unique decomposition of M as*

$$M = sU + N$$

with N a reduced square in $\mathbb{M}_3(J - 3s)$.

Of course,

$$\mathbb{M}_3^s(J) = sU + N_3(J - 3s). \tag{14}$$

When $J \geq 3$, we may partition $\mathbb{M}_3(J)$ as

$$\begin{aligned} \mathbb{M}_3(J) &= N_3(J) \cup (U + \mathbb{M}_3(J-3)) \\ &= N_3(J) \cup (U + N_3(J-3)) \cup (2U + N_3(J-6)) \cup \ldots. \end{aligned} \tag{15}$$

The first equation partitions $\mathbb{M}_3(J)$ into reduced and non-reduced squares, while the latter partitions by smallest value. In any case, the cardinality of $\mathbb{M}_3(J)$ may be obtained from the cardinalities of $N_3(l)$ with $l \leq J$.

Now U is unchanged under the group of determinantal symmetries, so it is enough to decompose each $N_3(J)$ into orbits under the group G and count orbits using a

Table 1 Orbit counts for reduced squares

Orbit size	# Orbits of reduced squares		Stabilizer
1	1	$J = 0$	G
6	1	J odd	$\langle R(\sigma)C(\sigma), T\rangle \cong D_{12}$
	2	J even	
9	1	$J = 4t + 2$	$\langle R(23), C(23), T\rangle \cong D_8$
	2	$J = 4t$	
12	$t - 1$	$J = 2t$	$\langle R(123)C(132), T\rangle \cong S_3$
	t	$J = 2t + 1$	
18	$2t$	$J = 2t + 1$	$\langle R(23)C(23), T\rangle \cong \mathbb{Z}/2 \times \mathbb{Z}/2$
	$5t - 3$	$J = 4t$	
	$5t$	$J = 4t + 2$	
36	$(t - 1)(3t - 4)/2$	$J = 2t$	$\langle T\rangle \cong \mathbb{Z}/2$
	$3t(t - 1)/2$	$J = 2t + 1$	
36	$t - 1$	$J = 4t$	$\langle R(23)\rangle \cong \mathbb{Z}/2$
	t	$J = 4t + 2$	
72	$t(t - 1)(t - 2)/6$	$J = 2t + 1 \geq 7$	$\langle e\rangle$
	$t(t - 1)(4t - 5)/3$	$J = 4t \geq 8$	
	$t(t - 1)(4t + 1)/3$	$J = 4t + 2 \geq 10$	

consistent set of representatives. Omitting a representative for orbit size 72, the following representatives correspond to the data in Table 1:

$$\left\|\begin{matrix} a & a & a \\ a & a & a \\ a & a & a \end{matrix}\right\|, \left\|\begin{matrix} b & a & a \\ a & b & a \\ a & a & b \end{matrix}\right\|, \left\|\begin{matrix} b & c & c \\ c & a & a \\ c & a & a \end{matrix}\right\|, \left\|\begin{matrix} a & c & b \\ c & b & a \\ b & a & c \end{matrix}\right\|,$$
$$\left\|\begin{matrix} b & a & a \\ a & c & d \\ a & d & c \end{matrix}\right\|, \left\|\begin{matrix} a & d & f \\ d & b & e \\ f & e & c \end{matrix}\right\|, \left\|\begin{matrix} d & e & f \\ a & b & c \\ a & b & c \end{matrix}\right\|. \tag{16}$$

Stabilizer subgroups D_{2n} of G in Table 1 are dihedral groups with $2n$ elements.

To verify Table 1, the counts for orbits of constant or linear size may be computed directly. For the orbit type of size 36 with symmetric representative, we note that two cases occur: either two zeros may be placed with $f = 0$ or a single zero with $a = 0$. Then M is the determined by the choice of d and e. The range of allowed values in either case forms one or two isosceles right triangles. Counting over arithmetic progressions, we obtain a quadratic polynomial in t. Initial conditions determine the coefficients.

For the orbits of size 72, we note that

$$|N_3(J)| = H_3(J) - H_3(J-3) = \frac{3J(J^2+3)}{2}; \tag{17}$$

the formulas follow by removing counts for orbits of smaller size in each case.

Of course, if we calculate the latter orbit counts directly, one could give another proof of Proposition 3. Instead, we illustrate the method for this proof to count trivial zeros in $\mathbb{M}_3(J)$ in the next section.

7 Trivial Zeros

In this section, we define the class of trivial zeros of $C(M)$ and give generating functions for both the number of orbits and number of trivial zeros in $\mathbb{M}_3(J)$ for any J. In this work, we apply the notion of trivial only to elements in $\mathbb{M}_3(J)$. Traditionally, the domain for Clebsch-Gordan coefficients is extended by zero in an appropriate manner.

Definition 7 Suppose M is in $\mathbb{M}_3(J)$ with all positive entries. Then M is called a **trivial zero** of $C(M)$ if and only if J is odd and M has at least one pair of matching rows or columns.

With the matching condition and even J, $C(M)$ is non-zero and may be computed explicitly. An orbit of zeros for $C(M)$ under G corresponding to a trivial zero consists of trivial zeros. We call such an orbit a **trivial orbit** of zeros.

Definition 8 For $J \geq 0$, let $O_3^T(J)$ be the number of trivial orbits in $\mathbb{M}_3(J)$, and let $H_3^T(J)$ be the number of trivial zeros in $\mathbb{M}_3(J)$.

Theorem 1 *The generating function that counts the number of trivial orbits $O_3^T(J)$ in $\mathbb{M}_3(J)$ is given by*

$$\sum_{J\geq 0} O_3^T(J)\, x^J = \frac{x^3(1+2x^6-x^{10})}{(1-x^4)^2(1-x^6)}. \tag{18}$$

The generating function that counts the number of trivial zeros $H_3^T(J)$ in $\mathbb{M}_3(J)$ is given by

$$\sum_{J\geq 0} H_3^T(J)\, x^J = \frac{x^3(1+9x^2+16x^4+27x^6+19x^8)}{(1-x^4)^2(1-x^6)}. \tag{19}$$

Proof To count orbits, we first consider those orbits in Table 1 corresponding to semi-magic squares with matching rows or columns. Only three types contribute; these orbits have sizes 1, 9, or 36.

For example, consider the number of trivial orbits for stabilizer of type $\langle R(23)\rangle$ in $\mathbb{M}_3(J)$ with J odd. Reduced squares of this type only occur for even J; thus none contribute directly to $O_3^T(J)$.

Suppose such a reduced square M occurs in $\mathbb{M}_3(J')$ with $J' = 4t$. Addition of U to M once gives a trivial orbit for $J = 4t + 3$; contributions to odd J follow now by repeatedly adding $2U$. To the generating function that counts trivial orbits, the contribution for reduced squares of this type in $\mathbb{M}_3(4t)$ is

$$(t-1)x^{4t+3}(1 + x^6 + x^{12} + \ldots) = \frac{(t-1)x^{4t+3}}{1-x^6}. \tag{20}$$

Summing over positive t and noting

$$1 + 2x + 3x^2 + \cdots = \frac{1}{(1-x)^2},$$

we obtain

$$\frac{x^{11}}{(1-x^6)(1-x^4)^2}.$$

Likewise, for reduced squares associated to $J' = 4t + 2$, the contribution is

$$\frac{x^9}{(1-x^6)(1-x^4)^2}.$$

Thus the generating function that counts trivial orbits of this type is

$$\frac{x^9 + x^{11}}{(1-x^6)(1-x^4)^2}. \tag{21}$$

A similar calculation shows that the orbit counts for sizes 1 and 9 are given by

$$\frac{x^3}{1-x^6}, \quad \frac{x^5 + 2x^7}{(1-x^4)(1-x^6)}, \tag{22}$$

respectively. Both parts of the theorem now follow. □

8 Triangles for Trivial Zeros

Given a fixed J, the top line partition for trivial zeros admit a uniform description, as do the orbits under G. There are three components to the top line partition:

- in general, for each top line (a, b, k) with all odd entries, values of 1 occur in alternating rows and then alternate along these rows,
- when $J \neq 3a$, the bisectors from each vertex to the opposite edge have values that progress from 1 to $\frac{J-1}{2}$; if $J = 3a$, a similar progression occurs, but the value of the middle entry is replaced by $J - 2$, and
- the triangle with vertices at edge midpoints consists of values of 2, except with values of 3 occurring from the first item and values on the bisectors given by the second item.

For example, consider the triangle of zeros for $J = 13$, all of which are trivial (Fig. 5).

To explain these counts, it may also help to consult the next section for the example of $J = 15$. First, the pattern for value 1 is Proposition 15 of [4]; the Weyl group symmetry on a hexagon can only have a fixed point at the center, and this center exists and has $C(M) = 0$ exactly when all entries of the top line are odd and the lower rows of M match. For other trivial zeros, any pair of columns or one of the other row pairs match, and these zeros occur as doublets under the Weyl group symmetry.

For the bisector progressions, see Propositions 11, 13, and 14 of Sect. 4 in [4]; in these cases, the top line has repeat values, and the hexagon has additional symmetries. When the top line has a matching pair and distinct third value, the hexagon has four sides of equal length and a diagonal of trivial zeros occur; the values along a bisector change with the width of the hexagon. For top line (a, a, a), the hexagon is regular, and three diagonals of zeros of equal length intersect at the center.

Finally, for the central triangle with midpoint vertices, the values of 2 arise when the upper row of M matches only one of the other rows. Off the bisectors, values of 3 in the triangle occur when the top line has all odd entries.

Fig. 5 All zeros for $J = 13$ are trivial

```
o o o o o o o o o o o o o o
 o 1 · 1 · 1 6 1 · 1 · 1 o
  o · 2 · · 2 2 · · 2 · o
   o 1 · 3 2 5 2 3 · 1 o
    o · · 2 4 4 2 · · o
     o 1 2 5 4 5 2 1 o
      o 6 2 2 2 2 6 o
       o 1 · 3 · 1 o
        o · · · · o
         o 1 2 1 o
          o · · o
           o 1 o
            o o
             o
```

With this enumeration of trivial zeros, we may directly determine $H_3^T(J)$ as a quasi-polynomial, implied by Theorem 1; that is, $H_3^T(J)$ is a polynomial on each remainder class modulo 12.

Theorem 2 *Let $H_3^T(J)$ denote the number of trivial zeros of $C(M)$ with magic number J. For $J \geq 0$,*

$$H_3^T(J) = \begin{cases} 0 & J\ even \\ \frac{3}{4}(J-1)(J-2) & J = 1, 5\ (mod\ 12) \\ \frac{3}{4}(J-1)(J-2)+4 & J = 9\ (mod\ 12) \\ \frac{3}{4}(J+1)(J-4) & J = 7, 11\ (mod\ 12) \\ \frac{3}{4}(J+1)(J-4)+4 & J = 3\ (mod\ 12). \end{cases}$$

Proof Summing over arithmetic progressions, the contributions of 1 and 2 from the first and third bullet points are given by

$$\frac{J^2-1}{8} \quad \text{and} \quad \frac{(J+1)(J+3)}{4},$$

respectively. The bisector counts are given by

$$\frac{3(J^2-1)}{8}$$

unless $J = 3a$, in which case this count is off by a linear error. The total bisector error in the full count, which is dependent on the parity of the bisector length and multiplicity 3, is linear with quasi-period 12. Thus, for each odd residue class, the leading term is $\frac{3}{4}J^2$, and values for $1 \leq J < 24$ determine the remaining coefficients. □

We may further decompose the triangle of trivial zeros into trivial orbits. Again see the next section for the complete case when $J = 15$. In this case, non-zero entries in the triangle may take values 1, 2, or 6. A sextet of trivial zeros occur at the midpoint (a, a, a) only if $J = 3a$; these six zeros are equidistant from the center along the three diagonals in the corresponding hexagon. Doublets of trivial zeros occur in two cases. If the zero doublet occurs on a bisector, a representative has matching columns; otherwise, another row must match to the first row.

In general, for a trivial orbit of size 1, we must have $J = 3a$, and the triangle for this unique orbit has a single 1 in the middle entry. For orbit size 9 (resp. 36), the convex hull of the non-zero entries in the triangle form an equilateral triangle (resp. hexagon) with a 1 at each vertex and a 2 at the midpoint of each edge.

The hexagon case is filled out with three parallelograms based at the midpoints of the edges and inside the triangle formed by these midpoints of the long edges. The entries adjacent to the midpoints are given by intersecting the equilateral triangles formed from the midpoints; these entries always have value 2. The remaining vertices may meet at the center with value 6 only if $J = 3a$; otherwise, these vertices have value 2.

9 Example: Triangles for $J = 15$

To illustrate the various portraits described in the previous section, the case of $J = 15$ is large enough to exhibit most of the phenomena described in the previous section. Here we have eight orbits, seven trivial and one non-trivial, and trivial orbits of size 36 that both contain and do not contain the center. Figure 6 displays counts for all zeros and all trivial zeros. Figures 7, 8 and 9 give the trivial orbits, and Fig. 10 shows the only non-trivial orbit. We omit the triangle for the trivial orbit with top line $(5, 5, 5)$, which consists of a single 1 at the center of the triangle.

Finally, we recall the 36-pointed star of Theorem 2 in [4]. In the case of $J = 15$, this star is composed of 73 trivial zeros, represented as the union of trivial orbits corresponding to top lines $(1, 5, 9)$, $(3, 5, 7)$, and $(5, 5, 5)$. See Fig. 9.

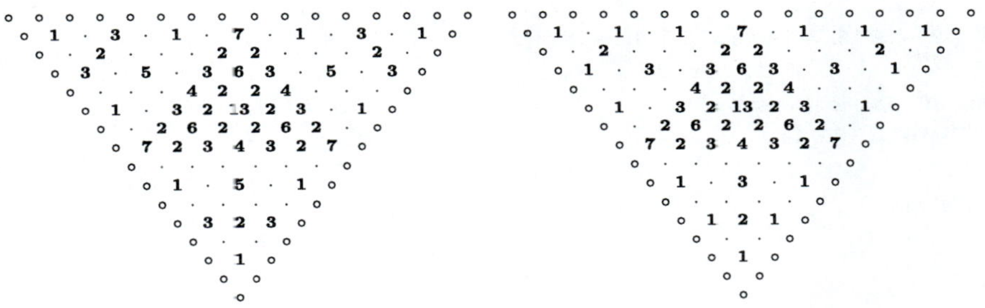

Fig. 6 $J = 15$: **a** All zeros and **b** trivial zeros

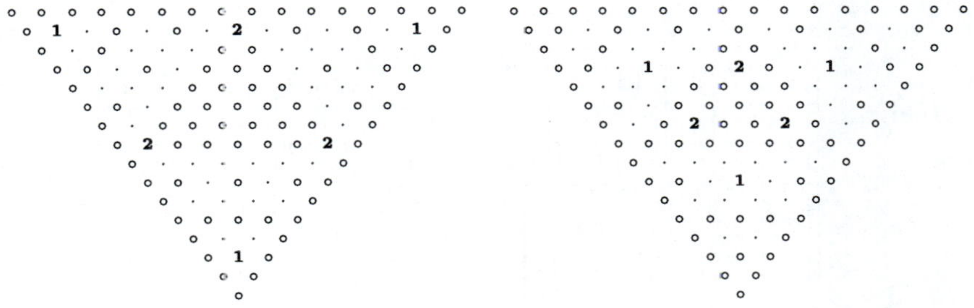

Fig. 7 $J = 15$: Top lines **a** $(13, 1, 1)$ and **b** $(9, 3, 3)$

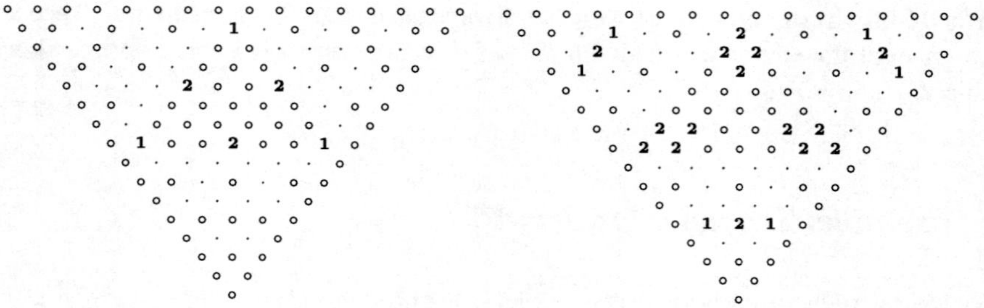

Fig. 8 $J = 15$: Top lines **a** $(1, 7, 7)$ and **b** $(1, 3, 11)$

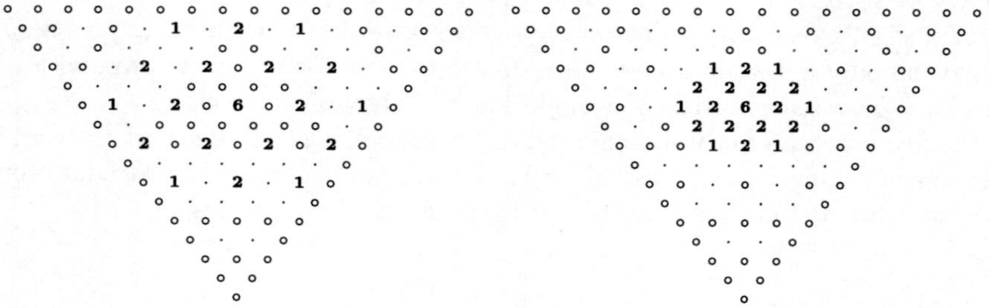

Fig. 9 $J = 15$: Top lines **a** $(1, 5, 9)$ and **b** $(3, 5, 7)$

Fig. 10 $J = 15$: Non-trivial orbit with top line $(9, 3, 3)$

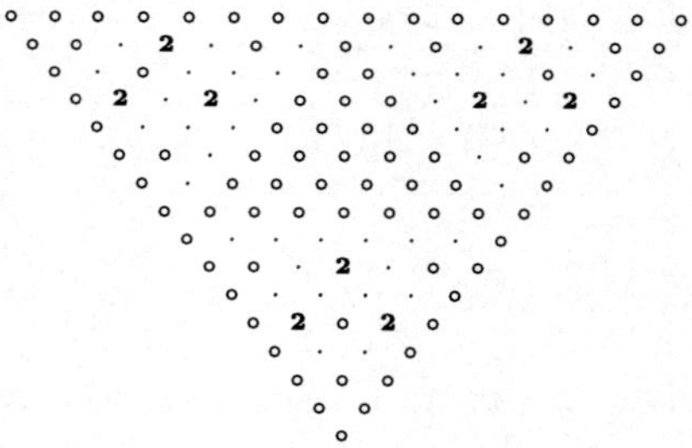

$$\textit{Orbits}: \left\|\begin{matrix} 5 & 5 & 5 \\ 5 & 5 & 5 \\ 5 & 5 & 5 \end{matrix}\right\|, \left\|\begin{matrix} 13 & 1 & 1 \\ 1 & 7 & 7 \\ 1 & 7 & 7 \end{matrix}\right\|, \left\|\begin{matrix} 9 & 3 & 3 \\ 3 & 6 & 6 \\ 3 & 6 & 6 \end{matrix}\right\|, \left\|\begin{matrix} 1 & 7 & 7 \\ 7 & 4 & 4 \\ 7 & 4 & 4 \end{matrix}\right\|, \left\|\begin{matrix} 1 & 3 & 11 \\ 7 & 6 & 2 \\ 7 & 6 & 2 \end{matrix}\right\|,$$
$$\left\|\begin{matrix} 1 & 5 & 9 \\ 7 & 5 & 3 \\ 7 & 5 & 3 \end{matrix}\right\|, \left\|\begin{matrix} 3 & 5 & 7 \\ 6 & 5 & 4 \\ 6 & 5 & 4 \end{matrix}\right\|, \left\|\begin{matrix} 9 & 3 & 3 \\ 3 & 11 & 1 \\ 3 & 1 & 11 \end{matrix}\right\|.$$

References

1. Beck, M., Zaslavsky, T.: Six little squares and how their numbers grow. J. Integer Seq. **13**, 1–45 (2010)
2. Bóna, M.: A new proof of the formula for the number of 3 × 3 magic squares. Math. Mag. **70**, 201–203 (1997)
3. Donley, R.W., Jr., Kim, W.G.: A rational theory of Clebsch-Gordan coefficients. In: Representation theory and harmonic analysis on symmetric spaces. American Mathematical Society, Providence. Contemp. Math., Vol. 714, 115–130 (2018)
4. Donley, R. W., Jr.: Central values of Clebsch-Gordan coefficients. In: Combinatorial and additive number theory III. Springer, New York. Springer Proc. in Math. and Stat., Vol. 297, 26 pp. (2019)
5. Donley, R. W., Jr.: Binomial arrays and generalized Vandermonde identities. Preprint, 28 pp (2019)
6. Louck, J. D.: Applications of unitary symmetry and combinatorics. World Scientific, Teaneck, N.J. (2011)
7. MacMahon, P. A.: Combinatory Analysis, Vols. 1 and 2. Cambridge University Press, (1916); reprinted by Chelsea, New York (1960) and Dover, New York (2004)
8. Rao, K. S., Van der Jeugt, J., Raynal, J., Jagannathan, R., Rajeswari, V.: Group theoretic basis for the terminating ${}_3F_2(1)$ series. J. Phys. A **25**, 861–876 (1992)
9. Raynal, J., Van der Jeugt, J., Rao, K.S., Rajeswari, V.: On the zeros of 3j coefficients: Polynomial degree versus recurrence order. J. Phys. A **26**, 2607–2623 (1993)
10. Sloane, N. J. A.: The On-Line Encyclopedia of Integer Sequences. Published electronically at http://oeis.org/.
11. Stanley, R.: Combinatorics and commutative algebra. Second edition. Birkhäuser, Boston (1996)
12. Stanley, R.: Enumerative combinatorics, Vol. 1. Second edition. Cambridge University Press, New York (2011)

A Sum of Negative Degrees of the Gaps Values in Two-Generated Numerical Semigroups and Identities for the Hurwitz Zeta Function

Leonid G. Fel, Takao Komatsu, and Ade Irma Suriajaya

Abstract We derive an explicit expression for an inverse power series over the gaps values of numerical semigroups generated by two integers. It implies the multiplication theorem for the Hurwitz zeta function $\zeta(n, q)$.

Keywords Numerical semigroups · Gaps and non-gaps · Hurwitz zeta function

2010 Mathematics Subject Classification Primary—20M14 · Secondary—11P81

1 Introduction

A sum of integer powers of gaps values in numerical semigroups $S_m = \langle d_1, \ldots, d_m \rangle$, where $\gcd(d_1, \ldots, d_m) = 1$, is referred often as semigroup series

$$g_n(S_m) = \sum_{s \in \mathbb{N} \setminus S_m} s^n, \qquad n \in \mathbb{Z}, \tag{1}$$

where $\mathbb{N} \setminus S_m$ is known as a set of gaps of S_m and $g_0(S_m)$ is called a genus of S_m. For $n \geq 0$ an explicit expression of $g_n(S_2)$ and implicit expression of $g_n(S_3)$ were

L. G. Fel (✉)
Department of Civil Engineering, 32000 Technion, Haifa, Israel
e-mail: lfel@technion.ac.il

T. Komatsu
School of Science, Zhejiang Sci-Tech University, Hangzhou 310018, China
e-mail: komatsu@zstu.edu.cn

A. I. Suriajaya
Faculty of Mathematics, Kyushu University, Fukuoka 819-0395, Japan
e-mail: adeirmasuriajaya@math.kyushu-u.ac.jp

M. B. Nathanson (ed.), *Combinatorial and Additive Number Theory IV*,
Springer Proceedings in Mathematics & Statistics 347,
https://doi.org/10.1007/978-3-030-67996-5_8

given in [2, 4], respectively. In this paper we derive a formula for semigroup series $g_{-n}(S_2) = \sum_{s\in\mathbb{N}\setminus S_2} s^{-n}$ $(n \geq 1)$.

Consider a numerical semigroup $S_2 = \langle d_1, d_2\rangle$, generated by two integers $d_1, d_2 \geq 2$, $\gcd(d_1, d_2) = 1$, with the Hilbert series $H(z; S_2)$ and the gaps generating function $\Phi(z; S_2)$ given as follows.

$$H(z; S_2) = \sum_{s\in S_2} z^s, \qquad \Phi(z; S_2) = \sum_{s\in\mathbb{N}\setminus S_2} z^s,$$
$$H(z; S_2) + \Phi(z; S_2) = \frac{1}{1-z}, \quad z < 1, \tag{2}$$

where $\min\{\mathbb{N} \setminus S_2\} = 1$ and $\max\{\mathbb{N} \setminus S_2\} = d_1d_2 - d_1 - d_2$ is called the Frobenius number and denoted by F_2. A rational representation (Rep) of $H(z; S_2)$ is given by

$$H(z; S_2) = \frac{1 - z^{d_1d_2}}{(1 - z^{d_1})(1 - z^{d_2})}. \tag{3}$$

Introduce a new generating function $\Psi_1(z; S_2)$ by

$$\Psi_1(z; S_2) = \int_0^z \frac{\Phi(t; S_2)}{t}dt = \sum_{s\in\mathbb{N}\setminus S_2} \frac{z^s}{s}, \qquad \Psi_1(1; S_2) = g_{-1}(S_2). \tag{4}$$

Plugging (2) into (4), we obtain

$$\Psi_1(z; S_2) = \int_0^z \left(\frac{1}{1-t} - H(t; S_2)\right)\frac{dt}{t}. \tag{5}$$

Keeping in mind $(1 - t^{d_i})^{-1} = \sum_{k_i=0}^{\infty} t^{k_id_i}$, substitute the last into (3) and obtain

$$H(t; S_2) = \sum_{k_1,k_2=0}^{\infty} t^{k_1d_1+k_2d_2} - \sum_{k_1,k_2=0}^{\infty} t^{k_1d_1+k_2d_2+d_1d_2}. \tag{6}$$

The expression (6) presents a difference between two infinite series and coincides with expression for $H(t; S_2)$ in (2). In (6) the degree $s = k_1d_1 + k_2d_2$ is running over all nodes in the sublattice $\mathbb{K}$ of the integer lattice $\mathbb{Z}_2$,

$$\mathbb{K} = \{0, 0\} \cup \mathbb{K}_1 \cup \mathbb{K}_2, \qquad \begin{cases} \mathbb{K}_1 = \{1 \leq k_1 \leq d_2 - 1, \;\; k_2 = 0\}, \\ \mathbb{K}_2 = \{0 \leq k_1 \leq d_2 - 1, \;\; 1 \leq k_2 \leq \infty\}. \end{cases} \tag{7}$$

In Fig. 1, we present, as an example, a part of the integer lattice $\mathbb{K}$ for the numerical semigroup $\langle 5, 8\rangle = \{0, 5, 8, 10, 13, 15, 16, 18, 20, 21, 23, 24, 25, 26, 28, \longmapsto\}$, where a symbol $\longmapsto$ denotes an infinite set of natural numbers exceeding 28.

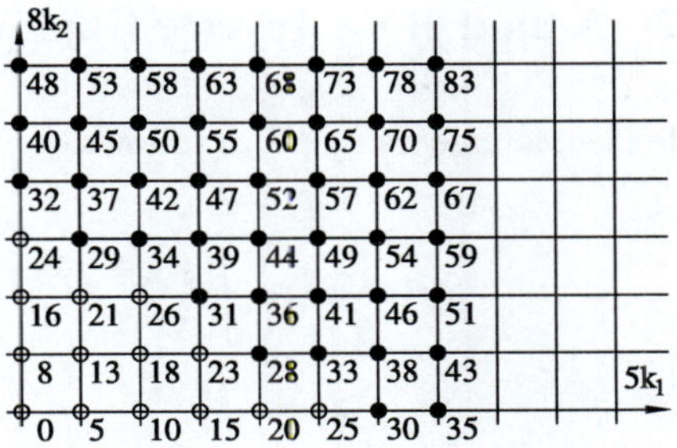

Fig. 1 A part of the integer lattice $\mathbb{K} \subset \mathbb{Z}_2$ for the numerical semigroup $\langle 5, 8\rangle$. The nodes mark the non-gaps of semigroup: the values, assigned to the black and white nodes, exceed and precede $F_2 = 27$, respectively

Lemma 1 *There exists a bijection between the infinite set of nodes in the integer lattice $\mathbb{K}$ and an infinite set of non-gaps of the semigroup $\langle d_1, d_2\rangle$.*

Proof We have to prove two statements of existence and uniqueness:

1. Every $s \in \langle d_1, d_2\rangle$ has its Rep node in $\mathbb{K}$,
2. All $s \in \langle d_1, d_2\rangle$ have their Rep nodes in $\mathbb{K}$ only once.

1. Let $s \in \langle d_1, d_2\rangle$ be given. Then by definition of $\langle d_1, d_2\rangle$ an integer s has Rep,

$$s = k_1 d_1 + k_2 d_2, \quad 0 \le k_1, k_2 \le \infty. \tag{8}$$

Choose s such that $k_1 = p d_2 + q$, where $p = \lfloor k_1/d_2 \rfloor$, i.e., $0 \le q \le d_2 - 1$, and $\lfloor r \rfloor$ denotes an integer part of a real number r. Then Rep (8) reads,

$$s = q d_1 + (k_2 + p d_1) d_2,$$

and s has its Rep node in $\mathbb{K}$.

2. By way of contradiction, assume that there exist two nodes $\{k_1, k_2\} \in \mathbb{K}$ and $\{l_1, l_2\} \in \mathbb{K}$ such that

$$\begin{aligned} &k_1 d_1 + k_2 d_2 = l_1 d_1 + l_2 d_2, \\ &0 \le k_1, l_1 \le d_2 - 1, \quad 0 \le k_2, l_2 \le \infty, \quad k_1 > l_1, \quad k_2 < l_2, \end{aligned} \tag{9}$$

namely, there exists such $s \in \langle d_1, d_2\rangle$ which has two different Rep nodes in $\mathbb{K}$. Rewrite equality (9) as follows,

$$(k_1 - l_1) d_1 = (l_2 - k_2) d_2. \tag{10}$$

Since $\gcd(d_1, d_2) = 1$, the equality (10) implies that

$$k_1 - l_1 = b d_2, \quad b \ge 1 \quad \longrightarrow \quad k_1 = l_1 + b d_2 \quad \longrightarrow \quad k_1 \ge d_2,$$

that contradict the assumption $\{k_1, k_2\} \in \mathbb{K}$. □

2 A Sum of the Inverse Gaps Values $g_{-1}(S_2)$

Present an integral in (5) as follows.

$$\Psi_1(z;S_2)=\int_0^z\left(\sum_{k=0}^{\infty}t^{k-1}-\frac{H(t;S_2)}{t}\right)dt,\qquad \text{where} \tag{11}$$

$$\frac{H(t;S_2)}{t}=\sum_{j=0}^{2}h_j(t;S_2),\qquad h_0(t;S_2)=\frac{1}{t},$$

$$h_1(t;S_2)=\sum_{k_1=1}^{d_2-1}t^{k_1d_1-1},\qquad h_2(t;S_2)=\sum_{k_1,k_2\in\mathbb{K}_2}t^{k_1d_1+k_2d_2-1}.$$

Perform an integration in (11),

$$\Psi_1(z;S_2)=\sum_{k=1}^{\infty}\frac{z^k}{k}-\frac{1}{d_1}\sum_{k_1=1}^{d_2-1}\frac{z^{k_1d_1}}{k_1}-\sum_{k_1,k_2\in\mathbb{K}_2}\frac{z^{k_1d_1+k_2d_2}}{k_1d_1+k_2d_2}, \tag{12}$$

and obtain by (4) and (7),

$$g_{-1}(S_2)=\sum_{k=1}^{\infty}\frac{1}{k}-\sum_{k_1,k_2\in\mathbb{K}_2}\frac{1}{k_1d_1+k_2d_2}-\frac{1}{d_1}\sum_{k_1=1}^{d_2-1}\frac{1}{k_1}. \tag{13}$$

By Lemma 1, after subtraction in (13) there are left a finite number of terms, since all terms, which exceed F_2 in the two first infinite series in (13), are cancelled. To emphasize that fact, we represent formula (13) as follows,

$$g_{-1}(S_2)=\sum_{k=1}^{c_2}\frac{1}{k}-\sum_{k_1,k_2\in\mathbb{K}_2}^{k_1d_1+k_2d_2\le c_2}\frac{1}{k_1d_1+k_2d_2}-\frac{1}{d_1}\sum_{k_1=1}^{d_2-1}\frac{1}{k_1},$$

where $c_2=F_2+1$ is called a conductor of semigroup S_2.

3 A Sum of the Negative Degrees of Gaps Values $g_{-n}(S_2)$

Generalize formula (12) and introduce a new generating function $\Psi_n(z;S_2)$ $(n\ge 2)$

$$\Psi_n(z;S_2)=\int_0^z\frac{dt_1}{t_1}\int_0^{t_1}\frac{dt_2}{t_2}\ldots\int_0^{t_{n-1}}\Phi(t_n;S_2)\frac{dt_n}{t_n}=\sum_{s\in\mathbb{N}\setminus S_2}\frac{z^s}{s^n}, \tag{14}$$

where $\Psi_n(1; S_2) = g_{-n}(S_2)$ and satisfies the following recursive relation,

$$\Psi_{k+1}(t_{n-k-1}; S_2) = \int_0^{t_{n-k-1}} \frac{dt_{n-k}}{t_{n-k}} \Psi_k(t_{n-k}; S_2), \quad k \geq 0,$$

$$\Psi_0(t_n; S_2) = \Phi(t_n; S_2), \quad t_0 = z,$$

namely,

$$\Psi_1(t_{n-1}; S_2) = \int_0^{t_{n-1}} \frac{dt_n}{t_n} \Psi_0(t_n; S_2),$$

$$\Psi_2(t_{n-2}; S_2) = \int_0^{t_{n-2}} \frac{dt_{n-1}}{t_{n-1}} \Psi_1(t_{n-1}; S_2), \quad \ldots$$

Performing integration in (14), we obtain

$$\Psi_n(z; S_2) = \sum_{k=1}^{\infty} \frac{z^k}{k^n} - \frac{1}{d_1^n} \sum_{k_1=1}^{d_2-1} \frac{z^{k_1 d_1}}{k_1^n} - \sum_{k_1, k_2 \in \mathbb{K}_2} \frac{z^{k_1 d_1 + k_2 d_2}}{(k_1 d_1 + k_2 d_2)^n}.$$

Thus, for $z = 1$ we have

$$g_{-n}(S_2) = \sum_{k=1}^{\infty} \frac{1}{k^n} - \sum_{k_1=0}^{d_2-1} \sum_{k_2=1}^{\infty} \frac{1}{(k_1 d_1 + k_2 d_2)^n} - \frac{1}{d_1^n} \sum_{k_1=1}^{d_2-1} \frac{1}{k_1^n}, \quad n \geq 2. \tag{15}$$

Define a ratio $\delta = d_1/d_2$ and represent (15) as follows,

$$g_{-n}(S_2) = \sum_{k=1}^{\infty} \frac{1}{k^n} - \frac{1}{d_2^n} \sum_{k_2=1}^{\infty} \frac{1}{k_2^n} - \frac{1}{d_2^n} \sum_{k_1=1}^{d_2-1} \sum_{k_2=1}^{\infty} \frac{1}{(k_1 \delta + k_2)^n} - \frac{1}{d_1^n} \sum_{k_1=1}^{d_2-1} \frac{1}{k_1^n}.$$

Making use of the Hurwitz $\zeta(n, q) = \sum_{k=0}^{\infty} (k + q)^{-n}$ and the Riemann zeta functions $\zeta(n) = \zeta(n, 1)$, we represent the last formula as follows,

$$g_{-n}(S_2) = \left(1 - \frac{1}{d_2^n}\right) \zeta(n) - \frac{1}{d_2^n} \sum_{k_1=1}^{d_2-1} \zeta(n, k_1 \delta), \quad n \geq 2. \tag{16}$$

Interchanging the generators d_1 and d_2 in (16), we get an alternative expression for $g_{-n}(S_2)$:

$$g_{-n}(S_2) = \left(1 - \frac{1}{d_1^n}\right)\zeta(n) - \frac{1}{d_1^n}\sum_{k_2=1}^{d_1-1}\zeta\left(n, \frac{k_2}{\delta}\right). \tag{17}$$

4 An Application to the Hurwitz Zeta Function

Finally we introduce an implication of the identity (16) for $g_{-n}(S_2)$ to zeta functions. The Hurwitz zeta function is defined to be the Dirichlet series

$$\zeta(s, \mathfrak{a}) := \sum_{n=0}^{\infty} \frac{1}{(n+\mathfrak{a})^s}$$

for $0 < \mathfrak{a} \le 1$, in the half plane $\mathrm{Re}(s) > 1$ which can be analytically continued to $\mathbb{C} \setminus \{1\}$. This is a generalization of the well-known Riemann zeta function

$$\zeta(s) := \sum_{n=1}^{\infty} \frac{1}{n^s}, \qquad \mathrm{Re}(s) > 1$$

which is also analytic on $\mathbb{C} \setminus \{1\}$. The Hurwitz zeta function $\zeta(s, \mathfrak{a})$ obeys the multiplication theorem

$$\sum_{r=1}^{d} \zeta\left(n, \frac{r}{d}\right) = d^n \zeta(n) \tag{18}$$

(see [5, p. 226]) which decomposes values of the Riemann zeta function $\zeta(s)$ at positive integers into sums of values of $\zeta(s, \mathfrak{a})$. This multiplication theorem is well-known and easy to prove. We found however that this is also an implication of the identity (16) for $g_{-n}(S_2)$ with $n \ge 2$. In this section we show how to deduce (18) from (16). We further remark that (16) and (18) are indeed equivalent expressions. This is an interesting implication since (18) is an elementary identity known in the theory of zeta functions and (16) gives its interpretation in the study of numerical semigroups.

We prove (18) for one of the generators of the numerical semigroup $\langle d_1, d_2 \rangle$. However, due to arbitrariness of the chosen semigroup, the proof for any positive integer d follows.

We begin by presenting two Lemmas 2, 3 on Rep of semigroup gaps $s \in \mathbb{N} \setminus S_2$ and one technical identity (23). Denote by l the integer part of kd_1/d_2,

$$l = \left\lfloor \frac{kd_1}{d_2} \right\rfloor. \tag{19}$$

Without loss of generality, we assume $d_1 > d_2$ in the rest of this section, hence $l \geq 1$ for $k \geq 1$.

Recall lemma [1] on the unique Rep of semigroup gaps $s \in \mathbb{N} \setminus S_2$, adapted to our assumption $d_1 > d_2$.

Lemma 2 ([1, Lemma 2]) *Let a numerical semigroup* $S_2 = \langle d_1, d_2 \rangle$ *be given, then* $s \in \mathbb{N} \setminus S_2$ *if and only if*

$$s = d_1 d_2 - p d_1 - q d_2,$$
$$1 \leq p \leq d_2 - 1, \quad 1 \leq q \leq \left\lfloor d_1 - \frac{d_1}{d_2} \right\rfloor, \quad 1 < d_2 < d_1. \tag{20}$$

Corollary 1 *Each gap* $s \in \mathbb{N} \setminus S_2$ *has a unique Rep*

$$s = k d_1 - j d_2, \qquad k \in \{1, 2, \ldots, d_2 - 1\}, \quad j \in \{1, 2, \ldots, l\}. \tag{21}$$

Proof Consider Rep (20) and let p be given. Then by Lemma 2 and $s \geq 0$, the variable q varies in the range which depends on k and is determined by $[1, l]$, where l is defined in (19). Replacing $p = d_2 - k$ and $q = j$, we conclude that a gap $s \in \mathbb{N} \setminus S_2$ has a Rep (21). Finally, we show the uniqueness of Rep (21) for every gap $s \in \mathbb{N} \setminus S_2$. Suppose $k' d_1 - j' d_2 = k'' d_1 - j'' d_2$ for some $k', k'' \in \{1, 2, \ldots, d_2 - 1\}$ and $j', j'' \in \{1, 2, \ldots, l\}$. Then

$$(k' - k'') d_1 = (j' - j'') d_2.$$

Since $\gcd(d_1, d_2) = 1$, we have $d_1 | (j' - j'')$ and $d_2 | (k' - k'')$. Since $k', k'' \in \{1, 2, \ldots, d_2 - 1\}$, we have $2 - d_2 \leq k' - k'' \leq d_2 - 2$. Hence $k' = k''$ and since $d_2 \neq 0$, we have $j' = j''$. □

Lemma 3 *The following set equality holds:*

$$\{k d_1 - l d_2\}_{k=1}^{d_2 - 1} = \{1, 2, \ldots, d_2 - 1\}. \tag{22}$$

Proof Equation (31) in [1] states that

$$1 \leq k d_1 - l d_2 \leq d_2 - 1, \qquad k \in \{1, 2, \ldots, d_2 - 1\}.$$

Combining the last inequality with Corollary 1 we arrive at (22). □

Next we show that the Hurwitz zeta function satisfies the following identity:

$$\zeta\left(n, k\frac{d_1}{d_2}\right) = \zeta\left(n, \frac{k d_1 - l d_2}{d_2}\right) - \sum_{j=0}^{l-1} \left(j + \frac{k d_1 - l d_2}{d_2}\right)^{-n}. \tag{23}$$

Expanding $\zeta(n, k d_1 / d_2)$, we get

$$\zeta\left(n,k\frac{d_1}{d_2}\right)=\zeta\left(n,l+\frac{kd_1-ld_2}{d_2}\right)=\sum_{m=0}^{\infty}\left(m+l+\frac{kd_1-ld_2}{d_2}\right)^{-n}$$
$$=\sum_{j=l}^{\infty}\left(j+\frac{kd_1-ld_2}{d_2}\right)^{-n}$$
$$=\sum_{j=0}^{\infty}\left(j+\frac{kd_1-ld_2}{d_2}\right)^{-n}-\sum_{j=0}^{l-1}\left(j+\frac{kd_1-ld_2}{d_2}\right)^{-n},$$

and identity (23) follows. Now we are ready to prove the multiplication theorem (18).

Proof (*Proof of* (18)) Combining formula (16) for $g_{-n}(S_2)$ and identity (23), we obtain

$$g_{-n}(S_2)=\left(1-\frac{1}{d_2^n}\right)\zeta(n)$$
$$-\frac{1}{d_2^n}\sum_{k=1}^{d_2-1}\left(\zeta\left(n,\frac{kd_1-ld_2}{d_2}\right)-\sum_{j=0}^{l-1}\left(j+\frac{kd_1-ld_2}{d_2}\right)^{-n}\right). \quad (24)$$

Calculating the double sum in (24) and applying Corollary 1 at the last step, we get

$$\sum_{k=1}^{d_2-1}\sum_{j=0}^{l-1}\left(j+\frac{kd_1-ld_2}{d_2}\right)^{-n}=d_2^n\sum_{k=1}^{d_2-1}\sum_{j=1}^{l}(kd_1-jd_2)^{-n}=d_2^n\sum_{s\in\mathbb{N}\setminus S_2}s^{-n}.$$

Recalling the definition $g_{-n}(S_2)=\sum_{s\in\mathbb{N}\setminus S_2}s^{-n}$ and substituting the last equality into (24), we obtain

$$\left(1-\frac{1}{d_2^n}\right)\zeta(n)=\frac{1}{d_2^n}\sum_{k=1}^{d_2-1}\zeta\left(n,\frac{kd_1-ld_2}{d_2}\right). \quad (25)$$

By Lemma 3, the sum on the right hand side of (25) reads

$$\sum_{k=1}^{d_2-1}\zeta\left(n,\frac{kd_1-ld_2}{d_2}\right)=\sum_{r=1}^{d_2-1}\zeta\left(n,\frac{r}{d_2}\right).$$

Substituting the last equality into (25), and multiplying both sides by d_2^n and recalling $\zeta(n)=\zeta(n,1)$, we obtain

$$d_2^n\zeta(n)=\sum_{r=1}^{d_2-1}\zeta\left(n,\frac{r}{d_2}\right)+\zeta(n,1)=\sum_{r=1}^{d_2}\zeta\left(n,\frac{r}{d_2}\right).$$

Thus, the multiplication theorem (18) for the Hurwitz zeta function with $d = d_2$ is proven. A generalization of the proof for any positive integer d follows immediately from the arbitrariness of the chosen semigroup. □

We conclude with an explicit example which originates from identities (16) and (17) for semigroup series $g_{-n}(S_2)$ and is implicitly embedded in the multiplication theorem (18) of the Hurwitz zeta function. This gives explicit identities for the Hurwitz zeta function.

Example 1 Consider two semigroups $R_1 = \langle 2, 3\rangle$, $R_2 = \langle 3, 4\rangle$ with sets of gaps $\mathbb{N} \setminus R_1 = \{1\}$, $\mathbb{N} \setminus R_2 = \{1, 2, 5\}$, respectively. Substituting the gaps values into (16, 17), we get identity (26a) for R_1 and identity (26b) for R_2.

$$\begin{aligned}
&\text{a)}\quad \zeta\left(n, \frac{2}{3}\right) + \zeta\left(n, \frac{4}{3}\right) = (3^n - 1)\zeta(n) - 3^n,\\
&\text{b)}\quad \zeta\left(n, \frac{4}{3}\right) + \zeta\left(n, \frac{8}{3}\right) = (3^n - 1)\zeta(n) - \left(3^n + \left(\frac{3}{2}\right)^n + \left(\frac{3}{5}\right)^n\right).
\end{aligned} \tag{26}$$

Equalities (26a) and (26b) may be obtained by multiplication theorem (18) with $d = 3$ and the following identities,

$$\zeta\left(n, \frac{1}{3}\right) + \zeta\left(n, \frac{2}{3}\right) = (3^n - 1)\zeta(n), \qquad \zeta\left(n, \frac{4}{3}\right) = \zeta\left(n, \frac{1}{3}\right) - 3^n,$$
$$\zeta\left(n, \frac{8}{3}\right) = \zeta\left(n, \frac{2}{3}\right) - \left(\frac{3}{2}\right)^n - \left(\frac{3}{5}\right)^n.$$

Combining equality (26a) and (26b), we arrive at another identity:

$$\zeta\left(n, \frac{2}{3}\right) - \zeta\left(n, \frac{8}{3}\right) = \left(\frac{3}{2}\right)^n + \left(\frac{3}{5}\right)^n.$$

Making use of semigroups S_2 with more complex set of gaps, we can deduce other explicit identities.

Acknowledgements The main part of the paper was written during the stay of one of the authors (LGF) at the School of Mathematics and Statistics of Wuhan University and its hospitality is highly appreciated. The research was supported in part (LGF) by the Kamea Fellowship and JSPS KAKENHI Grant Number 18K13400, and was in part conducted (AIS) under RIKEN Special Postdoctoral Researcher program.
The present paper is an extended version of the preprint [6] posted on arXiv.org.

References

1. L.G. Fel, *Frobenius problem for* $S(d_1, d_2, d_3)$, Funct. Anal. Other Math. **1**, 119–157 (2006).
2. L.G. Fel and B.Y. Rubinstein, *Power sums related to semigroups* $S(d_1, d_2, d_3)$, Semigroup Forum, **74**, 93–98 (2007).
3. R. L. Graham, D. E. Knuth and O. Patashnik, *Concrete Mathematics*, Addison-Wesley, Reading MA, 1988.
4. Ö.J. Rödseth, *A note on Brown and Shiues paper on a remark related to the Frobenius problem*, Fibonacci Quart., **32** (5), 407–408 (1994).
5. J. Steuding, *Value-distribution of L-functions*, Lecture Notes in Mathematics, 1877. Springer, Berlin, 2007.
6. L.G.Fel and T. Komatsu, *Semigroup's series for negative degrees of the gaps values in numerical semigroups generated by two integers and identities for the Hurwitz zeta function*, https://arxiv.org/pdf/1711.00353.pdf (2017)

Widely Digitally Stable Numbers

Michael Filaseta, Jacob Juillerat, and Jeremiah Southwick

Abstract We show that there are infinitely many composite numbers N, relatively prime to 10, that remain composite if you insert any digit anywhere in its base 10 representation, including between two of the infinitely many leading zeros of N and to the right of the units digit of N.

Keywords Composite · Prime · Digitally delicate · Digitally stable

1 Introduction

In 1979, Erdős [2], answering a problem posed by Klamkin [8], showed that there are infinitely many prime numbers N, with the property that you can change any digit in the base 10 representation of N, and the resulting number will be composite. The first such prime, called a *digitally delicate* prime in [6], is 294001. Later, Tao [10] showed a positive proportion of the primes are digitally delicate; and Hopper and Pollack [6] resolved a question of Tao's allowing for an arbitrary but fixed number of digit changes to the beginning and end of the prime.

As Konyagin [9] has pointed out, the methods of the others above imply that a positive proportion of composite numbers N, coprime to 10, satisfy the property that if any digit in the base 10 representation of N is changed, then the resulting number remains composite. For example, the number $N = 212159$ satisfies this property. That is, every number in the set

$$\{d12159, 2d2159, 21d159, 212d59, 2121d9, 21215d : d \in \{0, 1, 2, \ldots, 9\}\}$$

M. Filaseta (✉) · J. Juillerat · J. Southwick
Department of Mathematics, University of South Carolina, Columbia, SC 29208, USA
e-mail: filaseta@math.sc.edu

J. Juillerat
e-mail: juillerj@math.sc.edu

J. Southwick
e-mail: soutjt14@alumni.wfu.edu

M. B. Nathanson (ed.), *Combinatorial and Additive Number Theory IV*,
Springer Proceedings in Mathematics & Statistics 347,
https://doi.org/10.1007/978-3-030-67996-5_9

is composite. Note that if we remove the requirement that N be coprime to 10, then the results are less intriguing. For example, for m a positive integer, the number

$$N = (11 \cdot 13 \cdot 17 \cdot 19 \cdot 10)m + 15$$

is composite and remains composite when replacing any digit with $d \in \{0, 1, \ldots, 9\}$. This can be easily seen since changing any digit besides the right-most digit will result in a number divisible by 5. A similar conclusion can be obtained by looking at the number

$$N = (11 \cdot 13 \cdot 17 \cdot 19 \cdot 10)m + 12.$$

When starting with a composite number N, Filaseta et al. [4] noted that you can get an existence result for inserting a digit instead of replacing a digit. In other words, they showed that there are infinitely many composite numbers N, coprime to 10, with the property that if you insert any digit into the base 10 representation of N, then the resulting number remains composite. The first composite number with this property is $N = 25011$; that is, each number in the set

$$\{d25011, 2d5011, 25d011, 250d11, 2501d1, 25011d : d \in \{0, 1, 2, \ldots, 9\}\}$$

is composite. A similar result is not currently known to hold for prime numbers N.

In 2020, Filaseta and Southwick [5] showed that a positive proportion of the primes N, are such that you can replace any digit in the base 10 representation of N, *including one of the infinitely many leading zeros*, and the resulting number will be composite. They called such numbers *widely digitally delicate* primes. Notice that the first digitally delicate prime, 294001, is not widely digitally delicate since 10294001 is prime. In fact, even though a positive proportion of the primes are widely digitally delicate, no specific examples of widely digitally delicate primes are known.

Our goal is to consider the case of composite numbers N, coprime to 10, that remain composite when any digit is inserted in the decimal expansion of N, including between two of the infinitely many leading zeros of N and to the right of the units digit of N. We call such a number a *widely digitally stable* composite number. Our main theorem is the following:

Theorem 1 *There are infinitely many widely digitally stable composite numbers.*

Providing an explicit example of a widely digitally stable composite number does not appear to be easy. Our proof of Theorem 1 is nevertheless constructive, so in theory one can construct such a number from our proof. Furthermore, we emphasize that unlike the prior results which give that a positive proportion of the primes or numbers satisfy a certain property, we do not know if a positive proportion of the composite numbers are widely digitally stable.

The analogous result for Theorem 1 in bases $b \in \{2, 3, \ldots, 9\}$ has also been obtained by the second author with details to be provided in [7]. We note that the number

$$6135559 = (10111011001111100000111)_2$$

is an example of a number N in base 2 which remains composite when any binary digit 0 or 1 is inserted in the number including between two of the infinitely many leading binary zeros of N and to the right of the units digit of N. This provides an explicit example of a widely digitally stable number in base 2. This example was found by looking through a list of Sierpiński numbers, which easily provides other examples in base 2, including the base 2 representations of the base 10 numbers 7134623, 8629967, 9454129, 16010419, 16907749, 34158143, and 34629797.

2 Initial Ideas Behind the Proof

We write the decimal expansion of a natural number N, as

$$N = d_{n-1}d_{n-2}\ldots d_1d_0, \quad d_i \in \{0, 1, \ldots, 9\}, \quad n \geq 1, \quad d_{n-1} = 7.$$

Then N can also be written as

$$N = 7 \cdot \frac{10^n - 1}{9} + M, \tag{1}$$

where n and M are large natural numbers to be determined and n is large enough that the left-most digit of N is 7. Notice that N can be written as

$$N = 00\ldots00\underbrace{77\ldots77}_{n \text{ total } 7's}+M,$$

where $77\ldots77$ is a string of n digits that are all 7's.

To prove Theorem 1, we will determine a positive integer M and a set of primes $\mathcal{P}$ such that for infinitely many choices of the positive integer n, when we insert any $x \in \{0, 1, \ldots, 9\}$ to the right of any digit in the decimal expansion of N, including a leading zero, the resulting number is divisible by at least one prime in $\mathcal{P}$. This goal will be achieved by forming covering systems of the integers. The role of 7 above is motivated by the idea of taking $7 \in \mathcal{P}$ and $M \equiv 0 \pmod 7$ so that any insertion of the digit 0 or the digit 7 into the leading 0's or leading 7's of N will produce a number divisible by 7. Thus, we do not need to worry about these particular insertions. We will also require N to be congruent to one of 1, 3, 7 and 9 modulo 10 to ensure $\gcd(N, 10) = 1$; observe that this condition on N is equivalent to the condition that M is congruent to one of 0, 2, 4 and 6 modulo 10.

For a nonnegative integer k, let $N^{(k)}(x)$ denote the result of inserting a digit $x \in \{0, 1, \ldots, 9\}$ to the right of the $(k+1)^{\text{st}}$ digit of N, that is to the right of d_k. For example,

$$N^{(2)}(x) = 00\ldots00d_{n-1}d_{n-2}\ldots d_2xd_1d_0.$$

We set K to be one more than the number of digits of M; thus, initially, K is unknown to us. Observe that $d_k = 7$ for $K \le k \le n-1$. Shortly, we will break up the argument into three cases: $k \ge n$ corresponding to the leading zeros, $K \le k \le n-1$ corresponding to the leading sevens, and $0 \le k < K$ corresponding to M. We can formulate $N^{(k)}(x)$ above nicely for the first two cases. Specifically, with $x \in \{0, 1, \dots, 9\}$, for $k \ge n$, the value of $N^{(k)}(x)$ takes the form

$$N_0^{(k)}(x) = 7 \cdot \frac{10^n - 1}{9} + M + x \cdot 10^k\,; \tag{2}$$

and for $K \le k < n$, the value of $N^{(k)}(x)$ takes the form

$$N_7^{(k)}(x) = 7 \cdot 10^n + 7 \cdot \frac{10^n - 1}{9} + M + (x-7) \cdot 10^k. \tag{3}$$

For $k \ge n$, we see that $N_0^{(k)}(x)$ is the result of inserting the digit x to the right of a leading zero; for $K \le k < n$, we see that $N_7^{(k)}(x)$ is the result of inserting the digit x to the right of a leading seven.

Recall that by taking $M \equiv 0 \pmod 7$, we have $N^{(k)}(0)$ and $N^{(k)}(7)$ are divisible by the prime $7 \in \mathcal{P}$ for all $k \ge K$. We now take $3 \in \mathcal{P}$ and $M \equiv 1 \pmod 3$. If in addition $n \equiv 0 \pmod 3$, then one checks that

$$N^{(k)}(2) \equiv N^{(k)}(5) \equiv N^{(k)}(8) \equiv 0 \pmod 3$$

for every nonnegative integer k. Thus, inserting an $x \in \{2, 5, 8\}$ into the number N results in a number divisible by 3. At this point, we have that inserting an $x \in \{0, 2, 5, 7, 8\}$ into the number N produces a number divisible by a prime in $\{3, 7\} \subseteq \mathcal{P}$.

To handle $x \in \{1, 3, 4, 6, 9\}$, we will do a bit more work, and this is where we will consider $N_0^{(k)}(x)$ and $N_7^{(k)}(x)$ separately, as well as inserting x in the right-most K digits of N. We begin with the following.

Definition. Let t be a positive integer. For $1 \le i \le t$, let a_i and m_i be integers with m_i positive. The finite system of congruences $x \equiv a_i \pmod{m_i}$, for $1 \le i \le t$, is called a *covering system* (or simply a *covering*) *of the integers* if every integer satisfies at least one of the congruences.

For example, the congruences

$$x \equiv 0 \pmod 2, \quad x \equiv 1 \pmod 6, \quad x \equiv 5 \pmod 6,$$
$$x \equiv 3 \pmod{12}, \quad x \equiv 9 \pmod{12}$$

form a covering of the integers. This covering, as well as the coverings used in this paper, will not require that the moduli be distinct.

The following lemma is similar to [4, Lemma 1] and clarifies our use of covering systems. This lemma will be used for inserting a digit in the leading zeros. By way of notation, we define $c(p)$ to be the multiplicative order of 10 modulo a prime with $\gcd(10, p) = 1$.

Lemma 1 *Let N and M be natural numbers such that*

$$N = 7 \cdot \frac{10^n - 1}{9} + M,$$

where N has the decimal expansion

$$N = d_{n-1}d_{n-2}\dots d_1d_0, \quad d_i \in \{0, 1, \dots, 9\}, \quad n \geq 1, \quad d_{n-1} = 7.$$

Let K be a nonnegative integer such that $d_k = 7$ for $K \leq k \leq n-1$, and set $d_k = 0$ for $k \geq n$. For a fixed $x \in \{0, 1, \dots, 9\}$, suppose we have distinct primes $p_1, \dots, p_t$, each > 5, for which

(i) there exists a covering of the integers

$$k \equiv b_i \pmod{c(p_i)}, \quad 1 \leq i \leq t,$$

(ii) $n \equiv 0 \pmod{lcm(c(p_1), \dots, c(p_t))}$,
(iii) M is a solution to the system of congruences

$$M \equiv -x \cdot 10^{b_i} \pmod{p_i}, \quad 1 \leq i \leq t.$$

Then, for all nonnegative integers k, we have

$$N_0^{(k)}(x) = 7 \cdot \frac{10^n - 1}{9} + M + x \cdot 10^k$$

is divisible by at least one of the primes p_i where $1 \leq i \leq t$.

We will not give a proof of Lemma 1 as it is fairly straight forward (see Lemma 1 in [4]). Although the conclusion holds for all nonnegative integers k, we are interested only in the case $k \geq n$ of the lemma. We make some observations. For each $x \in \{1, 3, 4, 6, 9\}$, we want to create a covering system to use with Lemma 1. The covering system will be of the form given in (i) corresponding to inserting x to the right of the digit d_k. This lemma is specifically to address the insertion of x into the leading 0's; we will want a similar but different lemma to address inserting a digit into the leading 7's and yet a different argument for inserting a digit to the right of d_k for $k < K$. For the covering in (i), we do not need $c(p_i)$ to be the order of 10 modulo p_i but rather that $10^{c(p_i)} \equiv 1 \pmod{p_i}$. In other words, it is enough to have $c(p_i)$ divides the order of 10 modulo p_i. However, in practice, we want to use the order of 10 modulo p_i provided it can be easily computed. To obtain the primes p of a given order $c = c(p)$, one merely needs to look at the primes p which divide $\Phi_c(10)$, where $\Phi_c(x)$ is the cth cyclotomic polynomial. Every prime divisor p of $\Phi_c(10)$ either will be such that the order of 10 modulo p is c or will be equal to the largest prime divisor of c. In fact, it is also known that the largest prime divisor of c can only divide $\Phi_c(10)$ to the first power. There are also no other primes besides those primes dividing $\Phi_c(10)$ for which the order of 10 modulo p is c. It follows that for the covering system given in (i), we can use the modulus c multiple times, where the multiplicity is equal to the number of distinct prime divisors of $\Phi_c(10)$ other than the largest prime divisor of c. What complicates the covering systems as we proceed is that the primes considered in general cannot be used for different $x \in \{1, 3, 4, 6, 9\}$. More precisely, a congruence involving a prime p_i in (i) can only be used for two different values of x if the congruence condition on M in (iii) is the same for those values of x. In practice, this rarely happens, but we will definitely take advantage of some instances where this is the case.

To keep track of the number of admissible different congruences for (i) involving a specific value of c, Table 1 provides a lower bound $L = L(c)$ on the number of distinct primes with 10 of a given order c, that is the number of distinct prime divisors of $\Phi_c(10)$ excluding the largest prime factor of c. Thus,

$$L(c) \leq \sum_{\substack{p \text{ prime} \\ p \mid \Phi_c(10),\ p \nmid c}} 1.$$

The lower bounds $L(c)$ were obtained through factorizations or partial factorizations largely using Magma V2.22 and checked using Maple 2019 and are tabulated in Table 1. Many of the specific primes will appear in our covering systems to follow, and a complete list of the known primes in these factorizations can be found in [3] or [7] (also, see [1]). To clarify, for each c, we attempted to factor $\Phi_c(10)$, and $L(c)$ is the number of distinct primes found, minus 1 if the largest prime factor of c divides $\Phi_c(10)$, plus τ, where $\tau = 0$ if the prime factorization we obtained was a complete factorization, $\tau = 1$ if the prime factorization we obtained was incomplete and the remaining factor is relatively prime to the product of the known prime divisors of $\Phi_c(10)$ and c and the remaining factor is a probable prime, and $\tau = 2$ if the prime factorization we obtained was incomplete and the remaining factor is relatively prime to the product of the known prime divisors of $\Phi_c(10)$ and c and the remaining factor is neither a probable prime nor a prime power (Table 1).

3 The Coverings for the Leading Zeros

Set $\mathcal{P}_2 = \mathcal{P}_5 = \mathcal{P}_8 = \{3\}$ and $\mathcal{P}_0 = \mathcal{P}_7 = \{7\}$. For $x \in \{1, 3, 4, 6, 9\}$, let $\mathcal{P}_x$ be a set of primes, to be determined, as in Lemma 1. For this section, we make use of covering systems developed in [4, 5], with some minor modifications. Thus, the tables given in this section are not identical to the tables in [4, 5]. Some minor edits to the congruences were made so that some of the same moduli can be used for more congruences when we look at inserting a digit into the leading sevens in the next section.

Table 1 Part I, lower bounds on the number $L = L(c)$ of prime factors of $\Phi_c(10)$ relatively prime to c

c	L	c	L	c	L	c	L	c	L	c	L	c	L
2	1	49	2	114	2	195	3	297	6	456	5	672	5
3	1	50	3	115	5	196	1	300	6	459	3	675	3
4	1	51	4	116	4	198	2	306	6	464	6	680	4
5	2	52	2	117	4	200	4	308	2	465	4	684	2
6	1	54	2	119	4	203	4	310	4	468	3	696	4
7	2	55	4	120	1	204	6	312	5	476	5	700	2
8	2	56	2	121	4	207	3	315	3	480	3	714	2
9	1	57	3	124	2	208	4	322	5	483	5	715	2
10	1	58	2	125	4	210	3	330	3	484	6	726	5
11	2	60	3	126	2	217	3	333	3	496	3	728	5
12	1	62	1	128	3	220	6	336	3	500	5	740	10
13	3	63	5	130	2	222	4	340	5	504	2	744	7
14	1	64	4	132	3	224	4	342	5	510	5	748	4
15	2	65	2	133	3	225	5	345	6	512	4	750	3
16	2	66	2	135	5	228	7	348	4	518	2	765	2
17	2	68	3	136	2	231	3	350	3	522	4	768	2
18	2	69	3	138	3	232	5	357	6	525	5	777	5
19	1	70	2	140	5	234	3	360	3	528	3	780	5
20	2	72	3	143	4	238	3	363	4	532	5	783	2
21	3	74	3	145	4	240	3	364	3	540	4	784	4
22	3	75	3	147	5	242	5	370	2	544	3	792	2
23	1	76	2	148	6	245	5	372	3	546	2	798	3
24	1	77	4	150	1	248	5	374	4	552	2	810	4
25	3	78	4	152	4	250	4	375	5	555	5	814	3
26	2	84	2	153	6	252	3	380	4	558	3	816	4
27	2	85	3	154	6	255	4	384	5	560	5	819	2
28	3	87	3	155	1	256	4	390	3	561	3	828	2
29	5	88	2	156	3	259	5	392	2	570	4	833	4
30	3	90	3	161	5	260	4	396	3	572	5	840	5
31	3	91	7	165	3	261	7	399	5	580	3	850	3
32	5	92	3	168	3	264	7	405	6	588	2	858	5
33	2	93	1	170	3	266	3	406	5	594	5	868	6
34	3	95	5	171	3	270	4	407	6	595	4	870	2
35	3	96	4	174	3	272	2	408	5	600	3	888	4
36	1	98	2	175	3	273	5	414	5	605	3	896	3
37	3	99	4	180	3	275	5	420	5	608	3	900	7
38	1	100	4	182	2	276	5	425	3	609	4	910	2
39	1	102	2	184	2	279	3	429	3	612	8	930	3
40	2	104	2	185	4	285	3	434	5	616	2	931	7
42	3	105	3	186	4	286	5	435	4	620	2	935	3
44	2	108	3	187	3	288	3	444	3	624	5	936	5
45	2	110	4	189	3	290	6	448	2	630	4	952	5
46	4	111	3	190	3	294	1	450	2	665	2	1000	2
48	1	112	3	192	4	296	2	455	5	666	3	1020	7

Table 1 Part II, lower bounds on the number $L = L(c)$ of prime factors of $\Phi_c(10)$ relatively prime to c

c	L	c	L	c	L	c	L	c	L	c	L	c	L
1036	3	1200	3	1360	4	1628	5	1872	3	2432	2	4050	5
1044	4	1210	4	1428	4	1632	3	1904	2	2442	3	4356	4
1050	5	1215	3	1430	4	1638	2	1995	3	2464	4	4464	2
1064	3	1216	7	1452	1	1666	4	2016	3	2496	6	4725	3
1071	3	1218	2	1456	5	1674	4	2025	3	2499	3	4752	4
1092	3	1221	5	1488	3	1680	5	2040	4	2660	4	4788	3
1116	4	1224	4	1500	5	1700	5	2100	4	2664	2	5320	2
1120	5	1232	4	1530	6	1716	2	2128	4	2730	4	5808	4
1144	2	1248	5	1560	4	1776	5	2145	3	2904	2	6120	7
1170	3	1275	3	1566	2	1785	3	2184	4	3060	8		
1176	6	1330	2	1575	2	1792	5	2232	1	3120	2		
1188	2	1332	5	1584	3	1824	3	2340	3	3192	2		
1190	5	1344	6	1596	4	1860	7	2394	2	3256	2		
1197	2	1350	3	1620	9	1870	2	2420	6	3672	7		

Table 2 Covering used in Lemma 1 for $x = 9$

Congruence	Prime	Congruence	Prime
$k \equiv 0 \pmod 2$	11	$k \equiv 1 \pmod 8$	73
$k \equiv 3 \pmod 4$	101	$k \equiv 5 \pmod 8$	137

When $x = 9$, the covering system used is given in Table 2. To clarify, the covering system consists of the congruences in the left columns. One can check that this is a covering system by noting that every integer k can be written in the form $k = 8q + r$ where q and r are integers with $0 \le r \le 7$. If $r \in \{0, 2, 4, 6\}$, then $k \equiv 0 \pmod 2$. If $r \in \{3, 7\}$, then $k \equiv 3 \pmod 4$. For the remaining possibilities $r = 1$ and $r = 5$, one sees that $k \equiv 1 \pmod 8$ and $k \equiv 5 \pmod 8$, respectively. One can also see that

$$2 = c(11) = \mathrm{ord}_{11}(10), \qquad 4 = c(101) = \mathrm{ord}_{101}(10),$$
$$8 = c(73) = \mathrm{ord}_{73}(10), \qquad 8 = c(137) = \mathrm{ord}_{137}(10).$$

What Table 2 is indicating is that

$$N_0^{(k)}(9) = 7 \cdot \frac{10^n - 1}{9} + M + 9 \cdot 10^k \equiv 0 \pmod p,$$

for some $p \in \{11, 101, 73, 137\}$, depending on which congruence k satisfies from the covering. Note that the prime 11 corresponds to the one prime that exists with

Table 3 Covering used in Lemma 1 for $x = 3$

Congruence	Prime	Congruence	Prime
$k \equiv 0 \pmod{3}$	37	$k \equiv 53 \pmod{54}$	70541929
$k \equiv 1 \pmod{6}$	13	$k \equiv 26 \pmod{54}$	14175966169
$k \equiv 2 \pmod{9}$	333667	$k \equiv 4 \pmod{12}$	9901
$k \equiv 14 \pmod{18}$	19	$k \equiv 10 \pmod{24}$	99990001
$k \equiv 5 \pmod{18}$	52579	$k \equiv 22 \pmod{72}$	3169
$k \equiv 17 \pmod{27}$	757	$k \equiv 46 \pmod{72}$	98641
$k \equiv 8 \pmod{27}$	$p_2(27)$	$k \equiv 70 \pmod{72}$	$p_3(72)$

$c(10) = 2$ indicated in Table 1, the prime 101 corresponds to the one prime with $c(10) = 4$ from Table 1, and the primes 73 and 137 correspond to the two primes with $c(10) = 8$ from Table 1. The significance of Table 1 is that the specific values of these primes are not important; only the order of 10 modulo these primes help determine the covering system.

With regard to Lemma 1, we take

$$t = 4, \quad p_i \in \mathcal{P}_9 = \{11, 73, 101, 137\}, \quad n \equiv 0 \pmod{8},$$

and M satisfying all the congruences

$$M \equiv -9 \cdot 10^0 \pmod{11}, \qquad M \equiv -9 \cdot 10^3 \pmod{101}$$
$$M \equiv -9 \cdot 10^1 \pmod{73}, \qquad M \equiv -9 \cdot 10^5 \pmod{137}.$$

Since we have different prime moduli above, the Chinese Remainder Theorem ensures us that there is a single congruence for M modulo $11 \cdot 101 \cdot 73 \cdot 137$ that is equivalent to the above congruences on M. Thus, Lemma 1 says that under these conditions, the number $N_0^{(k)}(9)$ is composite for all $k \geq n$.

Combining the information so far, we see that if $n \equiv 0 \pmod{24}$ and M satisfies all of the congruences on M modulo 11, 101, 73 and 137 above as well as $M \equiv 0 \pmod{7}$ and $M \equiv 1 \pmod{3}$, then $N_0^{(k)}(x)$ will be divisible by at least one prime in $\{3, 7, 11, 73, 101, 137\}$ for each $x \in \{0, 2, 5, 7, 8, 9\}$ and $k \geq n$.

For $x = 3$, we use the primes found in [4, Table 1]. These primes correspond to the primes indicated here in Table 1 for the entries $c \in \{3, 6, 9, 12, 18, 24, 27, 54, 72\}$. The details of the congruences are provided in Table 3. Note that Table 1 indicates that there are two primes with $c(10) = 27$. Table 3 tells us that one of them is 757. The other, indicated by $p_2(27)$, is 440334654777631. Similarly, from Table 1, there are three primes with $c(10) = 72$. In Table 3, these primes are given as 3169, 98641 and $p_3(72)$, where the latter is 3199044596370769. As noted above, the values of $p_2(27)$ and $p_3(72)$ are not important to our proof of Theorem 1; we only need to know that these primes exist.

The set $\mathcal{P}_3$ is the set of 14 primes in the right columns. One can check that the congruences in the left columns form a covering of the integers. The least common multiple of the moduli found in Table 3 is 216, so verifying that the congruneces in Table 3 form a covering amounts to verifying that each of 0, 1, ..., 215 satisfies at least one of the congruences. One further checks that each prime p listed in the second columns of Table 3 has $c(p)$ equal to the corresponding modulus in that row. We can then apply Lemma 1 as done previously to see that the number $N_0^{(k)}(3)$ is composite for all $k \geq n$.

We use the same technique for $x \in \{1, 4, 6\}$. For $x \in \{1, 6\}$, we use similar coverings to those found in [4], and for $x = 4$ we use a similar covering to that found in Tables 6 and 7 from [5]. We provide our tables for $x \in \{1, 4, 6\}$ in the appendix. In general, one can verify that the congruences given in a table for a given x is a covering by setting ℓ to be the least common multiple of the moduli in the set of congruences and verifying that each integer in $[0, \ell - 1]$ satisfies one of the congruences. Then it is not difficult to see that every integer will satisfy one of the congruences. In cases where ℓ is large, a slightly modified approach can be used. For example, when $x = 4$, we have

$$\ell = 18295200.$$

One can check that $\ell = 30 \cdot 609840$ and that all but a set S of 29722 integers in $[1, 609840]$ satisfy a congruence in the table for $x = 4$ where the modulus divides 609840. Then one can verify that each number of the form $609840u + v$ where $0 \leq u \leq 29$ and $v \in S$ satisfies one of the congruences in the table for $x = 4$, from which it follows that every integer in $[0, \ell - 1]$ satisfies one of the congruences.

For $x \in \{1, 3, 4, 6, 9\}$, the set of primes $\mathcal{P}_x$ corresponding to a given x for Lemma 1 is the set of primes appearing in the last column of the table given for x. For example, as noted earlier, $\mathcal{P}_9 = \{11, 73, 101, 137\}$. We now have a congruence condition on M of the form given in Lemma 1(iii) for each prime in $\mathcal{P}_1 \cup \mathcal{P}_2 \cup \cdots \cup \mathcal{P}_9$. In the next section, we will develop more congruence conditions on M to handle inserting a digit into one of the leading sevens of N.

4 The Coverings for the Leading Sevens

We now work on an analogous argument for inserting a digit x in the string of leading sevens. Recall that the case $x \in \{0, 2, 5, 7, 8\}$ has already been handled; given that $M \equiv 7 \pmod{21}$, inserting any one of the digits in $\{0, 2, 5, 7, 8\}$ anywhere in N as in (1) will produce a number divisible by 3 or 7. So now we are interested in the case where one of $x \in \{1, 3, 4, 6, 9\}$ is inserted in the leading sevens. We start with a Lemma similar to Lemma 1.

Lemma 2 *Let N and M be natural numbers such that*

$$N = 7 \cdot \frac{10^n - 1}{9} + M,$$

where N has the decimal expansion

$$N = d_{n-1}d_{n-2}\dots d_1d_0, \quad d_i \in \{0, 1, \dots, 9\}, \quad n \geq 1, \quad d_{n-1} = 7.$$

Let K be a nonnegative integer such that $d_k = 7$ for $K \leq k \leq n-1$, and set $d_k = 0$ for $k \geq n$. For a fixed $d \in \{0, 1, \dots, 9\}$, suppose we have distinct primes $p_1, \dots, p_t$, each > 5, for which

(i) there exists a covering of the integers

$$k \equiv b_i \pmod{c(p_i)}, \quad 1 \leq i \leq t,$$

(ii) $n \equiv 0 \pmod{lcm(c(p_1), \dots, c(p_t))}$,
(iii) M is a solution to the system of congruences

$$M \equiv -7 - (d-7) \cdot 10^{b_i} \pmod{p_i}, \quad 1 \leq i \leq t.$$

Then, for all nonnegative integers k, we have

$$N_7^{(k)}(d) = 7 \cdot 10^n + 7 \cdot \frac{10^n - 1}{9} + M + (d-7) \cdot 10^k$$

is divisible by at least one of the primes p_i where $1 \leq i \leq t$.

We omit the details of the proof as it is similar to the proof of Lemma 1 which in turn is similar to Lemma 1 in [4]. We are interested in the conclusion of the lemma for $K \leq k \leq n-1$, though as indicated the conclusion follows for all nonnegative integers k. Notice that in Lemma 2, we refer to the digit to be inserted as d rather than x. From now forward, this will help clarify when we are talking about tables where the digit d is inserted into the leading sevens (for applying Lemma 2) and when we are talking about tables where the digit x is inserted into the leading zeros (for applying Lemma 1).

In the previous section, when making our coverings for $x \in \{1, 3, 4, 6, 9\}$, we wanted the primes considered to be distinct so that we could apply the Chinese Remainder Theorem to justify the existence of M satsifying the congruences simultaneously that arise in Lemma 1(iii). Due to the fact that $N_7^{(k)}(d)$ has the additional $7 \cdot 10^n$ term and $d - 7$ instead of x, we can reuse some primes for $d \in \{1, 3, 4, 6, 9\}$ as long as we use them in a specific way. For example, consider the prime 101. We used this prime in Table 2 for inserting a 9 into the leading zeros via the congruence $k \equiv 3 \pmod 4$. From this, we get that

$$n \equiv 0 \pmod 4 \quad \text{and} \quad M \equiv -9 \cdot 10^3 \equiv 90 \pmod{101}.$$

When this happens, $N_0^{(k)}(9) \equiv 0 \pmod{101}$. Looking at $N_7^{(k)}(d)$, if

$$k \equiv 2 \pmod 4, \quad n \equiv 0 \pmod 4, \quad \text{and} \quad d = 3,$$

then using $M \equiv 90 \pmod{101}$, we have

Table 4 Reusable primes for inserting a digit in the leading string of sevens

Prime p_i	x	x congruence	$M \pmod{p_i}$	d	d congruence
11	9	$k \equiv 0 \pmod{2}$	$M \equiv 2 \pmod{11}$	9	$k \equiv 0 \pmod{2}$
13	3	$k \equiv 1 \pmod{6}$	$M \equiv 9 \pmod{13}$	3	$k \equiv 5 \pmod{6}$
13	3	$k \equiv 1 \pmod{6}$	$M \equiv 9 \pmod{13}$	4	$k \equiv 0 \pmod{6}$
19	3	$k \equiv 14 \pmod{18}$	$M \equiv 9 \pmod{19}$	4	$k \equiv 9 \pmod{18}$
29	1	$k \equiv 10 \pmod{28}$	$M \equiv 23 \pmod{29}$	1	$k \equiv 18 \pmod{28}$
31	6	$k \equiv 9 \pmod{15}$	$M \equiv 28 \pmod{31}$	6	$k \equiv 12 \pmod{15}$
37	3	$k \equiv 0 \pmod{3}$	$M \equiv 34 \pmod{37}$	3	$k \equiv 0 \pmod{3}$
37	3	$k \equiv 0 \pmod{3}$	$M \equiv 34 \pmod{37}$	4	$k \equiv 2 \pmod{3}$
41	6	$k \equiv 0 \pmod{5}$	$M \equiv 35 \pmod{41}$	6	$k \equiv 0 \pmod{5}$
43	1	$k \equiv 9 \pmod{21}$	$M \equiv 2 \pmod{43}$	1	$k \equiv 10 \pmod{21}$
61	6	$k \equiv 18 \pmod{60}$	$M \equiv 59 \pmod{61}$	3	$k \equiv 40 \pmod{60}$
71	1	$k \equiv 4 \pmod{35}$	$M \equiv 11 \pmod{71}$	1	$k \equiv 9 \pmod{35}$
101	9	$k \equiv 3 \pmod{4}$	$M \equiv 90 \pmod{101}$	3	$k \equiv 2 \pmod{4}$
137	9	$k \equiv 5 \pmod{8}$	$M \equiv 90 \pmod{137}$	3	$k \equiv 5 \pmod{8}$
211	6	$k \equiv 7 \pmod{30}$	$M \equiv 171 \pmod{211}$	3	$k \equiv 26 \pmod{30}$
239	1	$k \equiv 0 \pmod{7}$	$M \equiv 238 \pmod{239}$	1	$k \equiv 0 \pmod{7}$

$$N_7^{(k)}(3) \equiv 7 + 0 + 90 + (3 - 7) \cdot 10^2 \equiv 97 - 400 \equiv 0 \pmod{101}.$$

Thus, we can reuse the prime 101 assuming we use it via the congruence $k \equiv 2 \pmod{4}$ for $d = 3$. Table 4 lists the primes that we will use again and how we will use them when making coverings for various values of d.

We put the coverings for inserting the digit $d \in \{1, 3, 4, 6, 9\}$ in the leading sevens, making use of Lemma 2, in the appendix. In the way of some details, we note that in Table 8 for $d = 1$, the least common multiple of the moduli is

$$78218300160 = 2^8 \cdot 3^3 \cdot 5 \cdot 7^2 \cdot 11 \cdot 13 \cdot 17 \cdot 19.$$

Checking directly that every integer in the interval [0, 78218300160) satisfies one of the 358 congruences in Table 8 is unreasonable. A more indirect approach is as follows. The covering was developed by addressing k separately depending on its residue class modulo 7. Let $\mathcal{C}$ be the complete set of 358 congruences $k \equiv b \pmod{c}$ in Table 8. We split $\mathcal{C}$ into 8 disjoint sets defined by

$$\mathcal{C}' = \{k \equiv b \pmod{c} : k \equiv b \pmod{c} \text{ is in } \mathcal{C}, 7 \nmid c\},$$

and, for $r \in \{0, 1, \ldots, 6\}$,

$$\mathcal{C}_r = \{k \equiv b \pmod{c} : k \equiv b \pmod{c} \text{ is in } \mathcal{C}, 7|c \text{ and } b \equiv r \pmod{7}\}.$$

Observe that an integer congruent to a fixed $r \in \{0, 1, \ldots, 6\}$ modulo 7 can only satisfy those congruences in $\mathcal{C}$ from $\mathcal{C}'$ and $\mathcal{C}_r$. For each $r \in \{0, 1, \ldots, 6\}$, we computed the least common multiple ℓ_r of the moduli in $\mathcal{C}' \cup \mathcal{C}_r$. The goal now is to determine whether, for each r, every integer in the interval $[0, \ell_r)$ that is r modulo 7 satisfies one of the congruences in $\mathcal{C}' \cup \mathcal{C}_r$. The largest value of ℓ_r is $\ell_4 = 286513920$. We only need to look at the integers in $[0, \ell_4)$ that are 4 modulo 7, so we are left with 40930560 integers to verify satisfy one of the congruences in $\mathcal{C}' \cup \mathcal{C}_4$ and then to do a similar but smaller computation with the other values of $r \in \{0, 1, \ldots, 6\}$.

This latter process too can be sped up as follows. To simplify matters, we note that each of the values of ℓ_r is divisible by $102 = 2 \cdot 3 \cdot 17$ as well as by 7; we can use a different factor for each ℓ_r, but the common factor 102 will suffice. Let $\ell'_r = \ell_r/102$. We first determine the set of integers $\mathcal{S}_r$ in $[0, \ell'_r)$ that are r mod 7 and satisfy at least one congruence from $\mathcal{C}' \cup \mathcal{C}_r$ where we only consider congruences with moduli dividing ℓ'_r. Suppose $k_0 \in \mathcal{S}_r$ and an integer k is congruent to k_0 modulo ℓ'_r. Then $k \equiv k_0 \pmod{\ell'_r}$ implies that k satisfies whatever congruences k_0 satisfies from $\mathcal{C}' \cup \mathcal{C}_r$ with moduli dividing ℓ'_r. We are left then with only verifying that integers in $[0, \ell_r)$ that are r modulo 7 and are *not* congruent to an element of $\mathcal{S}_r$ modulo ℓ'_r satisfy some congruence in $\mathcal{C}' \cup \mathcal{C}_r$. In the way of an example, we see that

$$\ell'_4 = \frac{286513920}{102} = 2808960,$$

and a computation gives that there are 5680 integers in [0, 2808960) that are 4 modulo 7 and are *not* in $\mathcal{S}_4$. Thus, with $r = 4$, we are left with checking if the integers

$$2808960i + j, \quad 0 \le i < 102 \text{ and } j \in [0, 2808960) \cap \mathbb{Z} \setminus \mathcal{S}_4$$

all satisfy one of the congruences in $\mathcal{C}' \cup \mathcal{C}_4$. As j comes from a set of 5680 integers, this is a simple computation. After the analogous computation for each $r \in \{0, 1, \ldots, 6\}$, the verification that the congruences in Table 8 form a covering is justified.

For the case $d = 3$ in Table 9, there are 74 congruences and the least common multiple of the moduli for these congruences is 299520. A direct analysis as done with inserting a digit x into the leading zeros can be done in this case. This is also true for $d = 4$ and Table 10, where there are 126 congruences and the least common multiple of the moduli is 8648640. For $d = 6$ and Table 11, there are 291 congruences and the least common multiple of the moduli for these congruences is 1272348000. An analysis similar to that above for $d = 1$ can be done by replacing the role of 7 above with 5 and keeping the role of 102 the same. For $d = 9$ and Table 12, there are 329 congruences and the least common multiple of the moduli for these congruences is 127242949680. For this case, we found the easiest verification of the covering came by viewing the integers as lying in the congruence classes

$$x \equiv 0 \pmod{2}, \quad x \equiv 1 \pmod{6}, \quad x \equiv 5 \pmod{6},$$
$$x \equiv 3 \pmod{12}, \quad \text{and} \quad x \equiv 9 \pmod{12}.$$

The five congruences above, themselves, form a covering system, so every integer satisfies one of these five congruences. Observe that $k \equiv 0 \pmod{2}$ is the first congruence listed in Table 12, so every integer satisfying the first of the five congruences above (the even integers) satisfy a congruence in Table 12. Now, we want to consider each of the four remaining congruences above and show that the integers satisfying each of these also satisfy a congruence in Table 12. One can accomplish this as follows. First, we check that every integer satisfying $x \equiv 1 \pmod{6}$ satisfies one of the 58 congruences in Table 12 with moduli dividing 57960. Observe that 57960 is divisible by 6 and since every integer congruent to 1 modulo 6 in $[0, 57960)$ satisfies one of these 58 congruences, we can deduce every integer congruent to 1 modulo 6 not in $[0, 57960)$ also does. Indeed, if $k \equiv k_0 \pmod{57960}$ and $k_0 \equiv b \pmod{c}$ for some c dividing 57960, then $k \equiv b \pmod{c}$. Thus, every integer satisfying $x \equiv 1 \pmod{6}$ also satisfies a congruence in Table 12. Similarly, one can check that every integer satisfying $x \equiv 5 \pmod{6}$ satisfies one of the 88 congruences in Table 12 with moduli dividing 438480; every integer satisfying $x \equiv 3 \pmod{12}$ satisfies one of the 82 congruences in Table 12 with moduli dividing 468720; and every integer satisfying $x \equiv 9 \pmod{12}$ satisfies one of the 104 congruences in Table 12 with moduli dividing 2051280.

5 The Right-Most Digits

Define B to be the least common multiple of the moduli appearing in the congruences $k \equiv b \pmod{c}$ in the tables obtained from using Lemma 1(i) in Sect. 3 and from using Lemma 2(i) in Sect. 4. Note that from Table 3, we see that 6 divides B; in particular, the order of 10 modulo 3 and 7 also divides B. For Lemmas 1(ii) and 2(ii), we take $n \equiv 0 \pmod{B}$. Let $\mathcal{P}'_d$ be the set of primes appearing in the tables associated with $d \in \{1, 3, 4, 6, 9\}$ in the previous section. Recalling the sets $\mathcal{P}_j$ in Sect. 3, we set

$$\mathcal{P} = \{3, 7\} \cup \mathcal{P}_1 \cup \mathcal{P}_3 \cup \mathcal{P}_4 \cup \mathcal{P}_6 \cup \mathcal{P}_9 \cup \mathcal{P}'_1 \cup \mathcal{P}'_3 \cup \mathcal{P}'_4 \cup \mathcal{P}'_6 \cup \mathcal{P}'_9,$$

and let P denote the product of the primes in $\mathcal{P}$. From Lemma 1(iii) to Lemma 2(iii), we have congruences conditions on M where the moduli are the primes $p \in \mathcal{P}$. We deduce from these lemmas that if any digit is inserted in any of the leading zeros or leading sevens, then N is divisible by a prime in $\mathcal{P}$. We impose the further condition on M that $M \equiv 0 \pmod{10}$. From the definition of N in Lemmas 1 and 2, we see then that $N \equiv 7 \pmod{10}$ so that N is coprime to 10.

We fix M as above, and modify n, using an argument similar to that used in [4], to obtain an N for which inserting a digit in the remaining right-most digits of N also results in a composite number. The basic strategy we describe next is to consider primes, possibly not in $\mathcal{P}$, that divide the number we obtain after inserting a digit in the remaining right-most digits of N. Each insertion will correspond to a prime, though these primes need not be distinct. As M is fixed, there are a fixed number of such insertions and hence a fixed number of these primes to consider.

Recall that $N^{(k)}(x)$ is the number obtained by inserting a digit $x \in \{0, 1, \ldots, 9\}$ to the right of the $(k+1)$st digit of N. We are now interested in the case that $k \in \{0, \ldots, K-1\}$, where M is fixed as above and K is one more than the number of digits of M. Fix $n_0 \equiv 0 \pmod{B}$ satisfying $10^{n_0-2} > M$. We let n vary with $n \equiv 0 \pmod{B}$ and $n \geq n_0$. For any such n, the number N has a string of leading sevens. Since M is fixed, the natural number K is also fixed and does not vary as n varies. Therefore, there are finitely many ways to insert a digit in M; that is, there are finitely many choices of $k \in \{0, 1, \ldots, K-1\}$ and $x \in \{0, 1, \ldots, 9\}$ independent of n.

Recall M and n_0 are fixed. Momentarily, fix also $k \in \{0, 1, \ldots, K-1\}$ and $x \in \{0, 1, \ldots, 9\}$. Set

$$\overline{N} = 7 \cdot \frac{10^{n_0} - 1}{9} + M.$$

Observe that $\overline{N}$ is coprime to 10 since, as noted above, we take $M \equiv 0 \pmod{10}$. We denote by $\overline{N}^{(k)}(x)$ the number obtained by inserting $x \in \{0, 1, \ldots, 9\}$ to the right of the $(k+1)^{\text{st}}$ digit of $\overline{N}$. Let $q = q(k, x)$ denote the least prime $q = q(k, x)$ dividing $\overline{N}^{(k)}(x)$. If $q \notin \{2, 5\}$ and $n \equiv n_0 \pmod{c(q)}$, then

$$N^{(k)}(x) - \overline{N}^{(k)}(x) = 7 \cdot 10^{n_0} \cdot \frac{10^{n-n_0} - 1}{9} \equiv 0 \pmod{q}.$$

Thus, we also have q divides $N^{(k)}(x)$ whenever $n \equiv n_0 \pmod{c(q)}$. If $q \in \{2, 5\}$, then the units digit of $\overline{N}^{(k)}(x)$ will be even or 5, which implies that, for $n \geq n_0$, the units digit of $N^{(k)}(x)$ is even or 5. Thus, in this case, $N^{(k)}(x)$ is composite.

We now let $k \in \{0, 1, \ldots, K-1\}$ and $x \in \{0, 1, \ldots, 9\}$ vary and justify that we can take $n \equiv n_0 \pmod{c(q(k, x))}$ for all such k and x and $n \equiv 0 \pmod{B}$. In fact, since $n_0 \equiv 0 \pmod{B}$, we see that these simultaneous congruences have the solution $n = n_0$ and hence infinitely many solutions. Therefore, for such n, we see that after inserting any digit anywhere in N, including the leading zeros and to the right of the units digit, the resulting number will be divisible by one of the primes $q(k, x)$ above or one of the primes $p \in \mathcal{P}$. Taking n sufficiently large satisfying $n \equiv n_0$

(mod $c(q(k, x))$) for all such k and x and $n \equiv 0 \pmod{B}$, we obtain Theorem 1 holds.

Appendix

The following lists the congruences for the covering systems involving Lemma 1 for $x \in \{1, 4, 6\}$ and Lemma 2 for $x \in \{1, 3, 4, 6, 9\}$. Only primes with ≤ 14 digits are listed. The congruences take the form $k \equiv b \pmod{c}$ in the first column, where the order of 10 modulo the prime p given in the second column is c. Recall that Table 1 indicates a lower bound on the number of distinct primes p, excluding the largest prime factor of c, with 10 of order c modulo p. We denote these primes $p_1(c), p_2(c), \ldots$. Thus, for example, Table 1 tells us that there are 5 primes p for which $c(p) = 140$. In Table 5 below, there are correspondingly 5 congruences involving a modulus $c = 140$. The first 4 listed, give an explicit prime needed for the congruence condition on M in (iii) of Lemma 1. The fifth prime is not given explicitly in the table, due to its size, but it is

Table 5 Covering used in Lemma 1 for $x = 1$

Congruence	Prime
$k \equiv 0 \pmod{7}$	239
$k \equiv 1 \pmod{7}$	4649
$k \equiv 9 \pmod{21}$	43
$k \equiv 2 \pmod{21}$	1933
$k \equiv 16 \pmod{21}$	10838689
$k \equiv 3 \pmod{14}$	909091
$k \equiv 10 \pmod{28}$	29
$k \equiv 24 \pmod{28}$	281
$k \equiv 4 \pmod{35}$	71
$k \equiv 11 \pmod{35}$	123551
$k \equiv 18 \pmod{35}$	$p_3(35)$
$k \equiv 25 \pmod{70}$	4147571
$k \equiv 60 \pmod{70}$	$p_2(70)$
$k \equiv 32 \pmod{105}$	30703738801
$k \equiv 67 \pmod{105}$	625437743071
$k \equiv 102 \pmod{105}$	$p_3(105)$
$k \equiv 5 \pmod{42}$	127
$k \equiv 26 \pmod{42}$	2689
$k \equiv 12 \pmod{42}$	459691

Congruence	Prime
$k \equiv 33 \pmod{84}$	226549
$k \equiv 75 \pmod{84}$	$p_2(84)$
$k \equiv 19 \pmod{63}$	10837
$k \equiv 40 \pmod{63}$	23311
$k \equiv 61 \pmod{63}$	45613
$k \equiv 6 \pmod{28}$	121499449
$k \equiv 13 \pmod{56}$	7841
$k \equiv 41 \pmod{56}$	$p_2(56)$
$k \equiv 20 \pmod{140}$	421
$k \equiv 48 \pmod{140}$	3471301
$k \equiv 76 \pmod{140}$	13489841
$k \equiv 104 \pmod{140}$	60368344121
$k \equiv 132 \pmod{140}$	$p_5(140)$
$k \equiv 27 \pmod{112}$	113
$k \equiv 83 \pmod{112}$	$p_2(112)$
$k \equiv 55 \pmod{168}$	11189053009
$k \equiv 111 \pmod{168}$	$p_2(168)$
$k \equiv 167 \pmod{168}$	$p_3(168)$

Table 6 Covering used in Lemma 1 for $x = 4$, Part 1

Congruence	Prime
$k \equiv 16 \pmod{66}$	183411838171
$k \equiv 136 \pmod{264}$	$p_1(264)$
$k \equiv 202 \pmod{264}$	$p_2(264)$
$k \equiv 70 \pmod{528}$	$p_1(528)$
$k \equiv 268 \pmod{528}$	$p_2(528)$
$k \equiv 334 \pmod{528}$	$p_3(528)$
$k \equiv 532 \pmod{1584}$	$p_1(1584)$
$k \equiv 4 \pmod{4752}$	19009
$k \equiv 1588 \pmod{4752}$	1109703258577
$k \equiv 3 \pmod{16}$	5882353
$k \equiv 11 \pmod{32}$	353
$k \equiv 27 \pmod{32}$	449
$k \equiv 0 \pmod{11}$	21649
$k \equiv 1 \pmod{11}$	513239
$k \equiv 10 \pmod{16}$	17
$k \equiv 2 \pmod{22}$	23
$k \equiv 13 \pmod{22}$	4093
$k \equiv 3 \pmod{22}$	8779
$k \equiv 14 \pmod{44}$	89
$k \equiv 36 \pmod{44}$	1052788969
$k \equiv 15 \pmod{33}$	67
$k \equiv 26 \pmod{33}$	$p_2(33)$
$k \equiv 37 \pmod{66}$	599144041
$k \equiv 38 \pmod{132}$	5419170769
$k \equiv 126 \pmod{132}$	$p_2(132)$
$k \equiv 60 \pmod{132}$	$p_3(132)$
$k \equiv 5 \pmod{88}$	617
$k \equiv 49 \pmod{88}$	$p_2(88)$
$k \equiv 104 \pmod{264}$	2377
$k \equiv 236 \pmod{264}$	16369
$k \equiv 71 \pmod{264}$	432961
$k \equiv 159 \pmod{264}$	6796152793
$k \equiv 247 \pmod{264}$	24387741577
$k \equiv 6 \pmod{55}$	1321
$k \equiv 17 \pmod{55}$	62921
$k \equiv 28 \pmod{55}$	83251631
$k \equiv 39 \pmod{55}$	$p_4(55)$
$k \equiv 50 \pmod{110}$	331
$k \equiv 105 \pmod{110}$	5171
$k \equiv 7 \pmod{110}$	20163494891
$k \equiv 62 \pmod{110}$	$p_4(110)$
$k \equiv 18 \pmod{220}$	661
$k \equiv 73 \pmod{220}$	18041
$k \equiv 128 \pmod{220}$	148721

Congruence	Prime
$k \equiv 183 \pmod{220}$	1121407321
$k \equiv 40 \pmod{275}$	7151
$k \equiv 95 \pmod{275}$	15401
$k \equiv 150 \pmod{275}$	59951
$k \equiv 29 \pmod{165}$	471241
$k \equiv 84 \pmod{165}$	$p_2(165)$
$k \equiv 139 \pmod{165}$	$p_3(165)$
$k \equiv 51 \pmod{220}$	$p_5(220)$
$k \equiv 161 \pmod{220}$	$p_6(220)$
$k \equiv 106 \pmod{330}$	4124507971
$k \equiv 216 \pmod{330}$	$p_2(330)$
$k \equiv 326 \pmod{330}$	$p_3(330)$
$k \equiv 8 \pmod{77}$	5237
$k \equiv 19 \pmod{77}$	42043
$k \equiv 30 \pmod{77}$	29920507
$k \equiv 41 \pmod{77}$	$p_4(77)$
$k \equiv 52 \pmod{154}$	463
$k \equiv 129 \pmod{154}$	24179
$k \equiv 63 \pmod{154}$	590437
$k \equiv 140 \pmod{154}$	7444361
$k \equiv 74 \pmod{154}$	$p_5(154)$
$k \equiv 151 \pmod{154}$	$p_6(154)$
$k \equiv 9 \pmod{99}$	199
$k \equiv 20 \pmod{99}$	397
$k \equiv 31 \pmod{99}$	34849
$k \equiv 42 \pmod{99}$	$p_4(99)$
$k \equiv 53 \pmod{198}$	7093127053
$k \equiv 152 \pmod{198}$	$p_2(198)$
$k \equiv 64 \pmod{396}$	79082656489
$k \equiv 163 \pmod{396}$	$p_2(396)$
$k \equiv 262 \pmod{396}$	$p_3(396)$
$k \equiv 361 \pmod{792}$	761113
$k \equiv 757 \pmod{792}$	440718109921
$k \equiv 75 \pmod{297}$	55243
$k \equiv 174 \pmod{297}$	198397
$k \equiv 273 \pmod{297}$	1981560241
$k \equiv 86 \pmod{297}$	31600574312077
$k \equiv 185 \pmod{297}$	$p_5(297)$
$k \equiv 284 \pmod{297}$	$p_6(297)$
$k \equiv 97 \pmod{594}$	7129
$k \equiv 196 \pmod{594}$	$p_2(594)$
$k \equiv 295 \pmod{594}$	$p_3(594)$
$k \equiv 394 \pmod{594}$	$p_4(594)$
$k \equiv 493 \pmod{594}$	$p_5(594)$

Table 6 Covering used in Lemma 1 for $x = 4$, Part 2

Congruence	Prime
$k \equiv 592 \pmod{1188}$	765144469
$k \equiv 1186 \pmod{1188}$	$p_2(1188)$
$k \equiv 10 \pmod{121}$	15973
$k \equiv 21 \pmod{121}$	38237
$k \equiv 32 \pmod{121}$	274187
$k \equiv 43 \pmod{121}$	$p_4(121)$
$k \equiv 54 \pmod{242}$	4357
$k \equiv 175 \pmod{242}$	25169
$k \equiv 65 \pmod{242}$	1485397
$k \equiv 186 \pmod{242}$	$p_4(242)$
$k \equiv 76 \pmod{242}$	$p_5(242)$
$k \equiv 197 \pmod{484}$	56629
$k \equiv 439 \pmod{484}$	170369
$k \equiv 87 \pmod{484}$	29606281
$k \equiv 208 \pmod{484}$	$p_4(484)$
$k \equiv 98 \pmod{363}$	622001227
$k \equiv 219 \pmod{363}$	$p_2(363)$
$k \equiv 109 \pmod{726}$	727
$k \equiv 230 \pmod{726}$	1453
$k \equiv 351 \pmod{726}$	3481311540961
$k \equiv 593 \pmod{1452}$	$p_1(1452)$
$k \equiv 1319 \pmod{4356}$	949609
$k \equiv 2771 \pmod{4356}$	384538969
$k \equiv 2166 \pmod{5808}$	21582529
$k \equiv 5070 \pmod{5808}$	114690577

Congruence	Prime
$k \equiv 2892 \pmod{5808}$	48098483178241
$k \equiv 5796 \pmod{5808}$	$p_4(5808)$
$k \equiv 120 \pmod{605}$	9666954991
$k \equiv 362 \pmod{1210}$	10891
$k \equiv 967 \pmod{1210}$	131891
$k \equiv 1088 \pmod{2420}$	20142648596621
$k \equiv 2298 \pmod{2420}$	$p_2(2420)$
$k \equiv 604 \pmod{2420}$	1006721
$k \equiv 1209 \pmod{2420}$	2323201
$k \equiv 1814 \pmod{2420}$	1754328181
$k \equiv 2419 \pmod{2420}$	151620139001
$k \equiv 1060 \pmod{1584}$	$p_2(1584)$
$k \equiv 3172 \pmod{4752}$	$p_3(4752)$
$k \equiv 205 \pmod{275}$	$p_4(275)$
$k \equiv 260 \pmod{275}$	$p_5(275)$
$k \equiv 329 \pmod{484}$	$p_5(484)$
$k \equiv 450 \pmod{484}$	$p_6(484)$
$k \equiv 340 \pmod{363}$	$p_3(363)$
$k \equiv 472 \pmod{726}$	$p_4(726)$
$k \equiv 4223 \pmod{4356}$	$p_3(4356)$
$k \equiv 714 \pmod{2904}$	$p_1(2904)$
$k \equiv 1440 \pmod{2904}$	$p_2(2904)$
$k \equiv 241 \pmod{605}$	$p_2(605)$
$k \equiv 483 \pmod{1210}$	$p_3(1210)$

Table 7 Covering used in Lemma 1 for $x = 6$

Congruence	Prime
$k \equiv 0 \pmod{5}$	41
$k \equiv 1 \pmod{5}$	271
$k \equiv 2 \pmod{10}$	9091
$k \equiv 3 \pmod{20}$	3541
$k \equiv 13 \pmod{20}$	27961
$k \equiv 7 \pmod{30}$	211
$k \equiv 17 \pmod{30}$	241
$k \equiv 27 \pmod{30}$	2161
$k \equiv 8 \pmod{40}$	5964848081
$k \equiv 28 \pmod{40}$	1676321

Congruence	Prime
$k \equiv 18 \pmod{60}$	61
$k \equiv 58 \pmod{60}$	4188901
$k \equiv 38 \pmod{60}$	39526741
$k \equiv 9 \pmod{15}$	31
$k \equiv 4 \pmod{15}$	2906161
$k \equiv 29 \pmod{45}$	238681
$k \equiv 14 \pmod{45}$	$p_2(45)$
$k \equiv 89 \pmod{90}$	29611
$k \equiv 44 \pmod{90}$	3762091

Table 8 Covering used in Lemma 2 for $d = 1$, Part 1

Congruence	Prime
$k \equiv 0 \pmod{7}$	239
$k \equiv 10 \pmod{21}$	43
$k \equiv 59 \pmod{63}$	45121231
$k \equiv 45 \pmod{63}$	$p_2(63)$
$k \equiv 38 \pmod{126}$	5274739
$k \equiv 101 \pmod{126}$	$p_2(126)$
$k \equiv 213 \pmod{252}$	1009
$k \equiv 150 \pmod{252}$	43266855241
$k \equiv 87 \pmod{252}$	$p_3(252)$
$k \equiv 24 \pmod{504}$	41593295521
$k \equiv 276 \pmod{504}$	$p_2(504)$
$k \equiv 17 \pmod{189}$	$p_1(189)$
$k \equiv 255 \pmod{315}$	631
$k \equiv 192 \pmod{315}$	142809770881
$k \equiv 129 \pmod{315}$	$p_3(315)$
$k \equiv 381 \pmod{630}$	131041
$k \equiv 66 \pmod{630}$	525326613841
$k \equiv 48 \pmod{49}$	505885997
$k \equiv 41 \pmod{49}$	$p_2(49)$
$k \equiv 34 \pmod{98}$	197
$k \equiv 83 \pmod{98}$	$p_2(98)$
$k \equiv 27 \pmod{147}$	63799
$k \equiv 76 \pmod{147}$	4715467
$k \equiv 125 \pmod{147}$	$p_3(147)$
$k \equiv 118 \pmod{147}$	$p_4(147)$
$k \equiv 69 \pmod{147}$	$p_5(147)$
$k \equiv 167 \pmod{294}$	$p_1(294)$
$k \equiv 209 \pmod{245}$	336737801
$k \equiv 160 \pmod{245}$	23609565631
$k \equiv 111 \pmod{245}$	$p_3(245)$
$k \equiv 153 \pmod{196}$	$p_1(196)$
$k \equiv 251 \pmod{784}$	54881
$k \equiv 447 \pmod{784}$	$p_2(784)$
$k \equiv 986 \pmod{1176}$	737353
$k \equiv 790 \pmod{1176}$	1482937
$k \equiv 594 \pmod{1176}$	70251889
$k \equiv 398 \pmod{1176}$	14498348929
$k \equiv 202 \pmod{1176}$	3343677109249
$k \equiv 6 \pmod{1176}$	$p_6(1176)$
$k \equiv 89 \pmod{91}$	547
$k \equiv 82 \pmod{91}$	14197
$k \equiv 75 \pmod{91}$	17837
$k \equiv 68 \pmod{91}$	4262077
$k \equiv 61 \pmod{91}$	43442141653
$k \equiv 54 \pmod{91}$	$p_6(91)$
$k \equiv 47 \pmod{91}$	$p_7(91)$
$k \equiv 40 \pmod{182}$	21705503
$k \equiv 131 \pmod{182}$	$p_2(182)$
$k \equiv 33 \pmod{273}$	32786209
$k \equiv 124 \pmod{273}$	$p_2(273)$
$k \equiv 215 \pmod{273}$	1093
$k \equiv 572 \pmod{819}$	31123
$k \equiv 299 \pmod{819}$	$p_2(819)$
$k \equiv 845 \pmod{1638}$	24571
$k \equiv 26 \pmod{1638}$	280099
$k \equiv 19 \pmod{455}$	175631
$k \equiv 110 \pmod{455}$	$p_2(455)$
$k \equiv 201 \pmod{455}$	911
$k \equiv 285 \pmod{364}$	654721485601
$k \equiv 642 \pmod{728}$	574393
$k \equiv 551 \pmod{728}$	10727809
$k \equiv 460 \pmod{728}$	286569193
$k \equiv 733 \pmod{1092}$	37275014869
$k \equiv 369 \pmod{1092}$	428085457801
$k \equiv 5 \pmod{1092}$	$p_3(1092)$
$k \equiv 278 \pmod{1456}$	10193
$k \equiv 1006 \pmod{1456}$	436801
$k \equiv 187 \pmod{1456}$	64348368449
$k \equiv 1552 \pmod{2184}$	3624275167969
$k \equiv 824 \pmod{2184}$	64365563809
$k \equiv 18 \pmod{28}$	29
$k \equiv 32 \pmod{112}$	$p_1(112)$
$k \equiv 81 \pmod{420}$	1384194841
$k \equiv 165 \pmod{420}$	42681134161
$k \equiv 249 \pmod{420}$	424451728681
$k \equiv 333 \pmod{420}$	$p_4(420)$
$k \equiv 417 \pmod{420}$	$p_5(420)$
$k \equiv 179 \pmod{210}$	29970369241
$k \equiv 95 \pmod{210}$	$p_2(210)$
$k \equiv 11 \pmod{210}$	$p_3(210)$
$k \equiv 137 \pmod{840}$	134401
$k \equiv 557 \pmod{840}$	575794801
$k \equiv 53 \pmod{840}$	$p_3(840)$
$k \equiv 109 \pmod{672}$	1454209
$k \equiv 277 \pmod{672}$	9396577
$k \equiv 445 \pmod{672}$	$p_3(672)$
$k \equiv 25 \pmod{1680}$	3361
$k \equiv 361 \pmod{1680}$	20161

Table 8 Covering used in Lemma 2 for $d = 1$, Part 2

Congruence	Prime	Congruence	Prime
$k \equiv 697 \pmod{1680}$	75500478859681	$k \equiv 64 \pmod{1904}$	46502552676497
$k \equiv 193 \pmod{2016}$	2017	$k \equiv 1135 \pmod{1190}$	42841
$k \equiv 865 \pmod{2016}$	518113	$k \equiv 897 \pmod{1190}$	14281
$k \equiv 1537 \pmod{2016}$	$p_3(2016)$	$k \equiv 659 \pmod{1190}$	674731
$k \equiv 95 \pmod{224}$	673	$k \equiv 295 \pmod{357}$	4999
$k \equiv 207 \pmod{224}$	43735845217	$k \equiv 176 \pmod{357}$	143556841
$k \equiv 39 \pmod{336}$	337	$k \equiv 57 \pmod{357}$	3329808692827
$k \equiv 291 \pmod{1344}$	205633	$k \equiv 288 \pmod{357}$	$p_4(357)$
$k \equiv 739 \pmod{1344}$	4546753	$k \equiv 638 \pmod{714}$	17851
$k \equiv 1187 \pmod{1344}$	1342657	$k \equiv 281 \pmod{714}$	$p_2(714)$
$k \equiv 403 \pmod{1344}$	10418689	$k \equiv 876 \pmod{1071}$	43834194307
$k \equiv 60 \pmod{560}$	1378721	$k \equiv 519 \pmod{1071}$	$p_2(1071)$
$k \equiv 172 \pmod{560}$	5758943337281	$k \equiv 162 \pmod{1071}$	6427
$k \equiv 284 \pmod{560}$	$p_3(560)$	$k \equiv 1114 \pmod{1428}$	898827469
$k \equiv 459 \pmod{896}$	105528193	$k \equiv 757 \pmod{1428}$	1429
$k \equiv 123 \pmod{1120}$	48102033281	$k \equiv 512 \pmod{595}$	10711
$k \equiv 347 \pmod{1120}$	555204874561	$k \equiv 393 \pmod{595}$	9521
$k \equiv 571 \pmod{1120}$	$p_3(1120)$	$k \equiv 1226 \pmod{1785}$	$p_1(1785)$
$k \equiv 235 \pmod{1792}$	7710977	$k \equiv 743 \pmod{833}$	17241400681
$k \equiv 1131 \pmod{1792}$	12939800833	$k \equiv 624 \pmod{833}$	1667
$k \equiv 907 \pmod{1792}$	45501964033	$k \equiv 1100 \pmod{1666}$	609757
$k \equiv 1124 \pmod{1232}$	30357713	$k \equiv 267 \pmod{1666}$	$p_2(1666)$
$k \equiv 1012 \pmod{1232}$	3697	$k \equiv 1695 \pmod{2499}$	2962284613
$k \equiv 60 \pmod{308}$	39511854229	$k \equiv 5 \pmod{17}$	5363222357
$k \equiv 256 \pmod{308}$	$p_2(308)$	$k \equiv 15 \pmod{17}$	2071723
$k \equiv 228 \pmod{231}$	2311	$k \equiv 8 \pmod{34}$	103
$k \equiv 74 \pmod{231}$	22187551	$k \equiv 25 \pmod{34}$	21993833369
$k \equiv 151 \pmod{231}$	$p_3(231)$	$k \equiv 18 \pmod{34}$	4013
$k \equiv 1348 \pmod{2464}$	29569	$k \equiv 35 \pmod{68}$	$p_1(68)$
$k \equiv 116 \pmod{2464}$	232722337	$k \equiv 1 \pmod{68}$	1491383821
$k \equiv 1236 \pmod{2464}$	$p_3(2464)$	$k \equiv 9 \pmod{35}$	71
$k \equiv 4 \pmod{2464}$	$p_4(2464)$	$k \equiv 128 \pmod{133}$	1597
$k \equiv 113 \pmod{119}$	923441	$k \equiv 121 \pmod{133}$	$p_2(133)$
$k \equiv 106 \pmod{119}$	3924966376871	$k \equiv 114 \pmod{133}$	$p_3(133)$
$k \equiv 85 \pmod{238}$	1868879293	$k \equiv 107 \pmod{266}$	$p_1(266)$
$k \equiv 204 \pmod{238}$	$p_2(238)$	$k \equiv 100 \pmod{798}$	$p_1(798)$
$k \equiv 197 \pmod{238}$	$p_3(238)$	$k \equiv 93 \pmod{399}$	6316969
$k \equiv 428 \pmod{476}$	2381	$k \equiv 226 \pmod{399}$	34282879
$k \equiv 309 \pmod{476}$	$p_2(476)$	$k \equiv 359 \pmod{399}$	1473464802559
$k \equiv 190 \pmod{476}$	$p_3(476)$	$k \equiv 79 \pmod{532}$	14691812549
$k \equiv 302 \pmod{952}$	371281	$k \equiv 212 \pmod{532}$	2129
$k \equiv 540 \pmod{952}$	6514537	$k \equiv 345 \pmod{532}$	$p_3(532)$
$k \equiv 778 \pmod{952}$	953	$k \equiv 205 \pmod{1596}$	3301756921
$k \equiv 1016 \pmod{1904}$	7024789118033	$k \equiv 1668 \pmod{2128}$	57457

Table 8 Covering used in Lemma 2 for $d = 1$, Part 3

Congruence	Prime
$k \equiv 1136 \pmod{2128}$	$p_2(2128)$
$k \equiv 863 \pmod{931}$	16759
$k \equiv 730 \pmod{931}$	83791
$k \equiv 597 \pmod{931}$	13071241
$k \equiv 464 \pmod{931}$	7645033117
$k \equiv 331 \pmod{931}$	$p_5(931)$
$k \equiv 1472 \pmod{1995}$	496774951
$k \equiv 1 \pmod{19}$	$p_1(19)$
$k \equiv 32 \pmod{38}$	$p_1(38)$
$k \equiv 51 \pmod{76}$	$p_1(76)$
$k \equiv 13 \pmod{76}$	$p_2(76)$
$k \equiv 44 \pmod{57}$	21319
$k \equiv 6 \pmod{57}$	10749631
$k \equiv 25 \pmod{57}$	$p_3(57)$
$k \equiv 18 \pmod{95}$	191
$k \equiv 37 \pmod{95}$	59281
$k \equiv 56 \pmod{95}$	63841
$k \equiv 75 \pmod{95}$	$p_4(95)$
$k \equiv 94 \pmod{95}$	$p_5(95)$
$k \equiv 87 \pmod{114}$	1458973
$k \equiv 49 \pmod{114}$	$p_2(114)$
$k \equiv 11 \pmod{228}$	$p_1(228)$
$k \equiv 125 \pmod{228}$	$p_2(228)$
$k \equiv 30 \pmod{152}$	$p_1(152)$
$k \equiv 68 \pmod{152}$	$p_2(152)$
$k \equiv 106 \pmod{152}$	1403417
$k \equiv 144 \pmod{152}$	457
$k \equiv 156 \pmod{190}$	1812604116731
$k \equiv 118 \pmod{190}$	$p_2(190)$
$k \equiv 80 \pmod{190}$	$p_3(190)$
$k \equiv 4 \pmod{380}$	417332341
$k \equiv 190 \pmod{380}$	74861
$k \equiv 213 \pmod{228}$	229
$k \equiv 175 \pmod{228}$	2281
$k \equiv 137 \pmod{228}$	4789
$k \equiv 99 \pmod{228}$	304077901
$k \equiv 61 \pmod{228}$	52875286008709
$k \equiv 54 \pmod{171}$	$p_1(171)$
$k \equiv 263 \pmod{285}$	$p_1(285)$
$k \equiv 35 \pmod{570}$	571
$k \equiv 320 \pmod{570}$	$p_2(570)$
$k \equiv 92 \pmod{570}$	$p_3(570)$
$k \equiv 377 \pmod{570}$	$p_4(570)$
$k \equiv 301 \pmod{342}$	2053
$k \equiv 244 \pmod{342}$	4410785971
$k \equiv 187 \pmod{342}$	2911579215499
$k \equiv 130 \pmod{342}$	$p_4(342)$
$k \equiv 73 \pmod{342}$	$p_5(342)$
$k \equiv 2529 \pmod{2660}$	85121
$k \equiv 1997 \pmod{2660}$	73700621
$k \equiv 1332 \pmod{1596}$	539449
$k \equiv 268 \pmod{4788}$	71821
$k \equiv 667 \pmod{1064}$	$p_1(1064)$
$k \equiv 135 \pmod{456}$	27817
$k \equiv 287 \pmod{456}$	376102873
$k \equiv 439 \pmod{456}$	36120612721
$k \equiv 534 \pmod{608}$	3031489
$k \equiv 306 \pmod{1216}$	665153
$k \equiv 914 \pmod{1216}$	1601473
$k \equiv 230 \pmod{1216}$	65384321
$k \equiv 838 \pmod{1216}$	$p_4(1216)$
$k \equiv 154 \pmod{1216}$	1217
$k \equiv 1218 \pmod{1824}$	1252017313
$k \equiv 80 \pmod{189}$	$p_2(189)$
$k \equiv 143 \pmod{189}$	$p_3(189)$
$k \equiv 318 \pmod{630}$	$p_3(630)$
$k \equiv 3 \pmod{630}$	$p_4(630)$
$k \equiv 240 \pmod{266}$	$p_2(266)$
$k \equiv 233 \pmod{266}$	$p_3(266)$
$k \equiv 366 \pmod{798}$	$p_2(798)$
$k \equiv 632 \pmod{798}$	$p_3(798)$
$k \equiv 352 \pmod{399}$	$p_4(399)$
$k \equiv 219 \pmod{399}$	$p_5(399)$
$k \equiv 485 \pmod{1197}$	$p_1(1197)$
$k \equiv 884 \pmod{1197}$	$p_2(1197)$
$k \equiv 86 \pmod{2394}$	$p_1(2394)$
$k \equiv 1283 \pmod{2394}$	$p_2(2394)$
$k \equiv 478 \pmod{532}$	$p_4(532)$
$k \equiv 471 \pmod{532}$	$p_5(532)$
$k \equiv 338 \pmod{1064}$	$p_2(1064)$
$k \equiv 870 \pmod{1064}$	$p_3(1064)$
$k \equiv 737 \pmod{1596}$	$p_3(1596)$
$k \equiv 1269 \pmod{1596}$	$p_4(1596)$
$k \equiv 604 \pmod{2128}$	$p_3(2128)$
$k \equiv 72 \pmod{2128}$	$p_4(2128)$
$k \equiv 198 \pmod{931}$	$p_6(931)$
$k \equiv 65 \pmod{931}$	$p_7(931)$
$k \equiv 541 \pmod{665}$	$p_1(665)$

Table 8 Covering used in Lemma 2 for $d = 1$, Part 4

Congruence	Prime	Congruence	Prime
$k \equiv 408 \pmod{665}$	$p_2(665)$	$k \equiv 103 \pmod{364}$	$p_3(364)$
$k \equiv 275 \pmod{1330}$	$p_1(1330)$	$k \equiv 12 \pmod{728}$	$p_4(728)$
$k \equiv 940 \pmod{1330}$	$p_2(1330)$	$k \equiv 376 \pmod{728}$	$p_5(728)$
$k \equiv 807 \pmod{1995}$	$p_2(1995)$	$k \equiv 915 \pmod{1456}$	$p_4(1456)$
$k \equiv 142 \pmod{1995}$	$p_3(1995)$	$k \equiv 96 \pmod{2184}$	$p_3(2184)$
$k \equiv 42 \pmod{380}$	$p_3(380)$	$k \equiv 473 \pmod{840}$	$p_4(840)$
$k \equiv 232 \pmod{380}$	$p_4(380)$	$k \equiv 613 \pmod{672}$	$p_4(672)$
$k \equiv 23 \pmod{456}$	$p_4(456)$	$k \equiv 529 \pmod{672}$	$p_5(672)$
$k \equiv 251 \pmod{456}$	$p_5(456)$	$k \equiv 1033 \pmod{1680}$	$p_4(1680)$
$k \equiv 111 \pmod{171}$	$p_2(171)$	$k \equiv 1369 \pmod{1680}$	$p_5(1680)$
$k \equiv 168 \pmod{171}$	$p_3(171)$	$k \equiv 88 \pmod{224}$	$p_3(224)$
$k \equiv 206 \pmod{285}$	$p_2(285)$	$k \equiv 200 \pmod{224}$	$p_4(224)$
$k \equiv 149 \pmod{285}$	$p_3(285)$	$k \equiv 151 \pmod{336}$	$p_2(336)$
$k \equiv 16 \pmod{684}$	$p_1(684)$	$k \equiv 263 \pmod{336}$	$p_3(336)$
$k \equiv 358 \pmod{684}$	$p_2(684)$	$k \equiv 67 \pmod{448}$	$p_1(448)$
$k \equiv 1465 \pmod{2660}$	$p_3(2660)$	$k \equiv 179 \pmod{448}$	$p_2(448)$
$k \equiv 933 \pmod{2660}$	$p_4(2660)$	$k \equiv 851 \pmod{1344}$	$p_5(1344)$
$k \equiv 401 \pmod{5320}$	$p_1(5320)$	$k \equiv 1299 \pmod{1344}$	$p_6(1344)$
$k \equiv 3061 \pmod{5320}$	$p_2(5320)$	$k \equiv 396 \pmod{560}$	$p_4(560)$
$k \equiv 800 \pmod{3192}$	$p_1(3192)$	$k \equiv 508 \pmod{560}$	$p_5(560)$
$k \equiv 2396 \pmod{3192}$	$p_2(3192)$	$k \equiv 683 \pmod{896}$	$p_2(896)$
$k \equiv 1864 \pmod{4788}$	$p_2(4788)$	$k \equiv 795 \pmod{1120}$	$p_4(1120)$
$k \equiv 3460 \pmod{4788}$	$p_3(4788)$	$k \equiv 1019 \pmod{1120}$	$p_5(1120)$
$k \equiv 458 \pmod{608}$	$p_2(608)$	$k \equiv 11 \pmod{1792}$	$p_4(1792)$
$k \equiv 382 \pmod{608}$	$p_3(608)$	$k \equiv 900 \pmod{1232}$	$p_3(1232)$
$k \equiv 762 \pmod{1216}$	$p_6(1216)$	$k \equiv 788 \pmod{1232}$	$p_4(1232)$
$k \equiv 686 \pmod{1216}$	$p_7(1216)$	$k \equiv 452 \pmod{616}$	$p_1(616)$
$k \equiv 1294 \pmod{2432}$	$p_1(2432)$	$k \equiv 340 \pmod{616}$	$p_2(616)$
$k \equiv 78 \pmod{2432}$	$p_2(2432)$	$k \equiv 99 \pmod{119}$	$p_3(119)$
$k \equiv 610 \pmod{1824}$	$p_2(1824)$	$k \equiv 92 \pmod{119}$	$p_4(119)$
$k \equiv 2 \pmod{1824}$	$p_3(1824)$	$k \equiv 316 \pmod{476}$	$p_4(476)$
$k \equiv 314 \pmod{588}$	$p_1(588)$	$k \equiv 78 \pmod{476}$	$p_5(476)$
$k \equiv 20 \pmod{588}$	$p_2(588)$	$k \equiv 547 \pmod{952}$	$p_4(952)$
$k \equiv 62 \pmod{245}$	$p_4(245)$	$k \equiv 71 \pmod{952}$	$p_5(952)$
$k \equiv 13 \pmod{245}$	$p_5(245)$	$k \equiv 421 \pmod{1190}$	$p_4(1190)$
$k \equiv 104 \pmod{392}$	$p_1(392)$	$k \equiv 183 \pmod{1190}$	$p_5(1190)$
$k \equiv 300 \pmod{392}$	$p_2(392)$	$k \equiv 169 \pmod{357}$	$p_5(357)$
$k \equiv 643 \pmod{784}$	$p_3(784)$	$k \equiv 50 \pmod{357}$	$p_6(357)$
$k \equiv 55 \pmod{784}$	$p_4(784)$	$k \equiv 400 \pmod{1428}$	$p_3(1428)$
$k \equiv 117 \pmod{273}$	$p_4(273)$	$k \equiv 43 \pmod{1428}$	$p_4(1428)$
$k \equiv 208 \pmod{273}$	$p_5(273)$	$k \equiv 274 \pmod{595}$	$p_3(595)$
$k \equiv 292 \pmod{455}$	$p_4(455)$	$k \equiv 155 \pmod{595}$	$p_4(595)$
$k \equiv 383 \pmod{455}$	$p_5(455)$	$k \equiv 631 \pmod{1785}$	$p_2(1785)$
$k \equiv 194 \pmod{364}$	$p_2(364)$	$k \equiv 36 \pmod{1785}$	$p_3(1785)$

Table 8 Covering used in Lemma 2 for $d = 1$, Part 5

Congruence	Prime	Congruence	Prime
$k \equiv 505 \pmod{833}$	$p_3(833)$	$k \equiv 148 \pmod{1666}$	$p_4(1666)$
$k \equiv 386 \pmod{833}$	$p_4(833)$	$k \equiv 862 \pmod{2499}$	$p_2(2499)$
$k \equiv 981 \pmod{1666}$	$p_3(1666)$	$k \equiv 29 \pmod{2499}$	$p_3(2499)$

Table 9 Covering used in Lemma 2 for $d = 3$

Congruence	Prime	Congruence	Prime
$k \equiv 2 \pmod{4}$	101	$k \equiv 24 \pmod{64}$	6187457
$k \equiv 5 \pmod{8}$	137	$k \equiv 56 \pmod{64}$	19841
$k \equiv 0 \pmod{3}$	37	$k \equiv 3115 \pmod{3120}$	38609381516161
$k \equiv 5 \pmod{6}$	13	$k \equiv 1555 \pmod{3120}$	$p_2(3120)$
$k \equiv 26 \pmod{30}$	211	$k \equiv 307 \pmod{1560}$	66365521
$k \equiv 40 \pmod{60}$	61	$k \equiv 619 \pmod{1560}$	$p_2(1560)$
$k \equiv 1 \pmod{96}$	97	$k \equiv 907 \pmod{1872}$	20593
$k \equiv 25 \pmod{96}$	206209	$k \equiv 595 \pmod{936}$	937
$k \equiv 49 \pmod{96}$	66554101249	$k \equiv 283 \pmod{936}$	$p_2(936)$
$k \equiv 73 \pmod{96}$	75118313082913	$k \equiv 883 \pmod{936}$	$p_3(936)$
$k \equiv 7 \pmod{48}$	$p_1(48)$	$k \equiv 571 \pmod{936}$	$p_4(936)$
$k \equiv 31 \pmod{120}$	$p_1(120)$	$k \equiv 259 \pmod{936}$	$p_5(936)$
$k \equiv 79 \pmod{240}$	$p_1(240)$	$k \equiv 235 \pmod{624}$	148068337
$k \equiv 127 \pmod{240}$	$p_2(240)$	$k \equiv 523 \pmod{624}$	$p_2(624)$
$k \equiv 175 \pmod{240}$	1132716961	$k \equiv 211 \pmod{624}$	10662171313
$k \equiv 223 \pmod{480}$	177601	$k \equiv 19 \pmod{156}$	3121
$k \equiv 116 \pmod{128}$	1265011073	$k \equiv 43 \pmod{156}$	$p_2(156)$
$k \equiv 100 \pmod{128}$	$p_2(128)$	$k \equiv 67 \pmod{156}$	53397071018461
$k \equiv 84 \pmod{128}$	$p_3(128)$	$k \equiv 91 \pmod{312}$	$p_1(312)$
$k \equiv 196 \pmod{256}$	257	$k \equiv 163 \pmod{312}$	$p_2(312)$
$k \equiv 68 \pmod{256}$	15361	$k \equiv 115 \pmod{312}$	$p_3(312)$
$k \equiv 180 \pmod{256}$	453377	$k \equiv 187 \pmod{312}$	1101673
$k \equiv 52 \pmod{256}$	$p_4(256)$	$k \equiv 139 \pmod{312}$	313
$k \equiv 420 \pmod{512}$	10753	$k \equiv 0 \pmod{192}$	193
$k \equiv 292 \pmod{512}$	8253953	$k \equiv 96 \pmod{192}$	769
$k \equiv 164 \pmod{512}$	9524994049	$k \equiv 32 \pmod{192}$	$p_3(192)$
$k \equiv 36 \pmod{512}$	73171503617	$k \equiv 128 \pmod{192}$	$p_4(192)$
$k \equiv 276 \pmod{384}$	3457	$k \equiv 64 \pmod{288}$	13249
$k \equiv 148 \pmod{384}$	12289	$k \equiv 160 \pmod{288}$	1067329
$k \equiv 20 \pmod{384}$	418725889	$k \equiv 256 \pmod{288}$	$p_3(288)$
$k \equiv 388 \pmod{768}$	434689	$k \equiv 463 \pmod{480}$	$p_2(480)$
$k \equiv 4 \pmod{768}$	859393	$k \equiv 260 \pmod{384}$	$p_4(384)$
$k \equiv 16 \pmod{32}$	641	$k \equiv 132 \pmod{384}$	$p_5(384)$
$k \equiv 28 \pmod{32}$	1409	$k \equiv 931 \pmod{1560}$	$p_3(1560)$
$k \equiv 12 \pmod{32}$	69857	$k \equiv 1243 \pmod{1560}$	$p_4(1560)$
$k \equiv 8 \pmod{64}$	$p_1(64)$	$k \equiv 1843 \pmod{1872}$	$p_2(1872)$
$k \equiv 40 \pmod{64}$	976193	$k \equiv 547 \pmod{624}$	$p_4(624)$

Table 10 Covering used in Lemma 2 for $d = 4$, Part 1

Congruence	Prime
$k \equiv 2 \pmod{3}$	37
$k \equiv 0 \pmod{6}$	13
$k \equiv 9 \pmod{18}$	19
$k \equiv 11 \pmod{13}$	53
$k \equiv 8 \pmod{13}$	79
$k \equiv 5 \pmod{13}$	265371653
$k \equiv 28 \pmod{39}$	$p_1(39)$
$k \equiv 12 \pmod{26}$	859
$k \equiv 25 \pmod{26}$	1058313049
$k \equiv 139 \pmod{143}$	2823679
$k \equiv 126 \pmod{143}$	180523201
$k \equiv 87 \pmod{286}$	51767
$k \equiv 230 \pmod{286}$	22144088539
$k \equiv 74 \pmod{286}$	264752347289
$k \equiv 217 \pmod{286}$	$p_4(286)$
$k \equiv 204 \pmod{286}$	$p_5(286)$
$k \equiv 347 \pmod{572}$	17576772101461
$k \equiv 633 \pmod{1144}$	119424449
$k \equiv 61 \pmod{1144}$	$p_2(1144)$
$k \equiv 477 \pmod{572}$	$p_2(572)$
$k \equiv 334 \pmod{572}$	$p_3(572)$
$k \equiv 321 \pmod{429}$	50184916627561
$k \equiv 737 \pmod{858}$	$p_1(858)$
$k \equiv 594 \pmod{858}$	$p_2(858)$
$k \equiv 451 \pmod{858}$	6007
$k \equiv 880 \pmod{1716}$	1063921
$k \equiv 22 \pmod{1716}$	152980598629
$k \equiv 1010 \pmod{1430}$	708433441
$k \equiv 295 \pmod{1430}$	$p_2(1430)$
$k \equiv 1439 \pmod{2145}$	51481
$k \equiv 724 \pmod{2145}$	6224791
$k \equiv 9 \pmod{2145}$	2832665551
$k \equiv 331 \pmod{390}$	15601
$k \equiv 253 \pmod{390}$	925081
$k \equiv 175 \pmod{390}$	$p_3(390)$
$k \equiv 97 \pmod{1170}$	75417940111411
$k \equiv 175 \pmod{546}$	102103
$k \equiv 331 \pmod{546}$	$p_2(546)$
$k \equiv 1579 \pmod{2730}$	5528251
$k \equiv 1189 \pmod{2730}$	6097697971
$k \equiv 474 \pmod{910}$	2475034612051
$k \equiv 19 \pmod{910}$	$p_2(910)$
$k \equiv 58 \pmod{78}$	157
$k \equiv 16 \pmod{78}$	388847808493
$k \equiv 55 \pmod{78}$	216451
$k \equiv 52 \pmod{78}$	6397
$k \equiv 1183 \pmod{1248}$	57653857
$k \equiv 1027 \pmod{1248}$	4694971009
$k \equiv 871 \pmod{1248}$	$p_3(1248)$
$k \equiv 1651 \pmod{2496}$	192193
$k \equiv 403 \pmod{2496}$	965953
$k \equiv 1495 \pmod{2496}$	22389121
$k \equiv 247 \pmod{2496}$	199369365313
$k \equiv 1339 \pmod{2496}$	324200647681
$k \equiv 91 \pmod{2496}$	$p_6(2496)$
$k \equiv 169 \pmod{780}$	19501
$k \equiv 325 \pmod{780}$	$p_2(780)$
$k \equiv 481 \pmod{780}$	$p_3(780)$
$k \equiv 13 \pmod{2340}$	25741
$k \equiv 793 \pmod{2340}$	257764562641
$k \equiv 1573 \pmod{2340}$	6253963297921
$k \equiv 10 \pmod{117}$	$p_1(117)$
$k \equiv 49 \pmod{117}$	$p_2(117)$
$k \equiv 88 \pmod{117}$	$p_3(117)$
$k \equiv 85 \pmod{117}$	$p_4(117)$
$k \equiv 46 \pmod{234}$	461917
$k \equiv 163 \pmod{234}$	60034573
$k \equiv 124 \pmod{234}$	$p_3(234)$
$k \equiv 7 \pmod{468}$	$p_1(468)$
$k \equiv 30 \pmod{52}$	521
$k \equiv 43 \pmod{52}$	$p_2(52)$
$k \equiv 17 \pmod{104}$	$p_1(104)$
$k \equiv 69 \pmod{104}$	1580801
$k \equiv 4 \pmod{208}$	1249
$k \equiv 56 \pmod{208}$	49297
$k \equiv 108 \pmod{208}$	300977
$k \equiv 160 \pmod{208}$	648961
$k \equiv 40 \pmod{65}$	162503518711
$k \equiv 53 \pmod{65}$	$p_2(65)$
$k \equiv 27 \pmod{130}$	131
$k \equiv 92 \pmod{130}$	$p_2(130)$
$k \equiv 14 \pmod{260}$	2311921
$k \equiv 79 \pmod{260}$	$p_2(260)$
$k \equiv 144 \pmod{260}$	$p_3(260)$
$k \equiv 209 \pmod{260}$	$p_4(260)$
$k \equiv 1 \pmod{195}$	1951
$k \equiv 66 \pmod{195}$	35081393881
$k \equiv 131 \pmod{195}$	$p_3(195)$

Table 10 Covering used in Lemma 2 for $d = 4$, Part 2

Congruence	Prime	Congruence	Prime
$k \equiv 21 \pmod{36}$	999999000001	$k \equiv 191 \pmod{572}$	$p_4(572)$
$k \equiv 3 \pmod{108}$	109	$k \equiv 48 \pmod{572}$	$p_5(572)$
$k \equiv 39 \pmod{108}$	$p_2(108)$	$k \equiv 178 \pmod{429}$	$p_2(429)$
$k \equiv 75 \pmod{108}$	153469	$k \equiv 35 \pmod{429}$	$p_3(429)$
$k \equiv 87 \pmod{90}$	8985695684401	$k \equiv 308 \pmod{858}$	$p_4(858)$
$k \equiv 69 \pmod{180}$	181	$k \equiv 165 \pmod{858}$	$p_5(858)$
$k \equiv 159 \pmod{180}$	$p_2(180)$	$k \equiv 581 \pmod{715}$	$p_1(715)$
$k \equiv 141 \pmod{180}$	$p_3(180)$	$k \equiv 438 \pmod{715}$	$p_2(715)$
$k \equiv 51 \pmod{360}$	265183201	$k \equiv 867 \pmod{1430}$	$p_3(1430)$
$k \equiv 33 \pmod{270}$	6481	$k \equiv 152 \pmod{1430}$	$p_4(1430)$
$k \equiv 123 \pmod{270}$	577603663291	$k \equiv 487 \pmod{1170}$	$p_2(1170)$
$k \equiv 213 \pmod{270}$	31023833790241	$k \equiv 877 \pmod{1170}$	$p_3(1170)$
$k \equiv 195 \pmod{270}$	$p_4(270)$	$k \equiv 799 \pmod{2730}$	$p_3(2730)$
$k \equiv 105 \pmod{540}$	329941	$k \equiv 409 \pmod{2730}$	$p_4(2730)$
$k \equiv 375 \pmod{540}$	68189581	$k \equiv 715 \pmod{1248}$	$p_4(1248)$
$k \equiv 15 \pmod{540}$	49229101	$k \equiv 559 \pmod{1248}$	$p_5(1248)$
$k \equiv 285 \pmod{540}$	$p_4(540)$	$k \equiv 637 \pmod{780}$	$p_4(780)$
$k \equiv 113 \pmod{143}$	$p_3(143)$	$k \equiv 241 \pmod{468}$	$p_2(468)$
$k \equiv 100 \pmod{143}$	$p_4(143)$	$k \equiv 231 \pmod{360}$	$p_2(360)$

Table 11 Covering used in Lemma 2 for $d = 6$, Part 1

Congruence	Prime	Congruence	Prime
$k \equiv 0 \pmod{5}$	41	$k \equiv 68 \pmod{125}$	$p_{11}(125)$
$k \equiv 12 \pmod{15}$	31	$k \equiv 43 \pmod{125}$	$p_{12}(125)$
$k \equiv 21 \pmod{25}$	21401	$k \equiv 143 \pmod{250}$	21001
$k \equiv 16 \pmod{25}$	25601	$k \equiv 18 \pmod{250}$	162251
$k \equiv 11 \pmod{25}$	182521213001	$k \equiv 208 \pmod{225}$	2002877551
$k \equiv 6 \pmod{75}$	151	$k \equiv 133 \pmod{225}$	$p_8(225)$
$k \equiv 31 \pmod{75}$	4201	$k \equiv 58 \pmod{225}$	$p_9(225)$
$k \equiv 56 \pmod{75}$	$p_9(75)$	$k \equiv 483 \pmod{675}$	40588518151
$k \equiv 76 \pmod{100}$	60101	$k \equiv 308 \pmod{375}$	21751
$k \equiv 51 \pmod{100}$	7019801	$k \equiv 233 \pmod{375}$	47001751
$k \equiv 26 \pmod{100}$	14103673319201	$k \equiv 158 \pmod{375}$	4471031976001
$k \equiv 1 \pmod{100}$	$p_{12}(100)$	$k \equiv 153 \pmod{175}$	$p_3(175)$
$k \equiv 23 \pmod{50}$	251	$k \equiv 228 \pmod{350}$	3612546001
$k \equiv 48 \pmod{50}$	5051	$k \equiv 253 \pmod{350}$	299547376801
$k \equiv 13 \pmod{50}$	78875943472201	$k \equiv 78 \pmod{350}$	$p_9(350)$
$k \equiv 188 \pmod{200}$	401	$k \equiv 403 \pmod{700}$	422100001
$k \equiv 138 \pmod{200}$	1201	$k \equiv 53 \pmod{700}$	701
$k \equiv 88 \pmod{200}$	1601	$k \equiv 378 \pmod{525}$	79801
$k \equiv 38 \pmod{200}$	$p_{12}(200)$	$k \equiv 203 \pmod{525}$	35120401
$k \equiv 118 \pmod{125}$	751	$k \equiv 28 \pmod{525}$	435288001
$k \equiv 93 \pmod{125}$	1797655751	$k \equiv 3153 \pmod{4725}$	286826401

Table 11 Covering used in Lemma 2 for $d = 6$, Part 2

Congruence	Prime	Congruence	Prime
$k \equiv 1578 \pmod{4725}$	918984151	$k \equiv 647 \pmod{900}$	139501
$k \equiv 3 \pmod{4725}$	7650670784401	$k \equiv 497 \pmod{900}$	33301
$k \equiv 127 \pmod{135}$	1577071	$k \equiv 347 \pmod{900}$	560701
$k \equiv 112 \pmod{135}$	16357951	$k \equiv 197 \pmod{900}$	5030101
$k \equiv 97 \pmod{135}$	310362841	$k \equiv 332 \pmod{450}$	270001
$k \equiv 82 \pmod{135}$	258360989311	$k \equiv 182 \pmod{450}$	$p_2(450)$
$k \equiv 67 \pmod{135}$	$p_5(135)$	$k \equiv 932 \pmod{1350}$	371251
$k \equiv 52 \pmod{405}$	21871	$k \equiv 917 \pmod{1050}$	1051
$k \equiv 187 \pmod{405}$	61561	$k \equiv 767 \pmod{1050}$	$p_2(1050)$
$k \equiv 322 \pmod{405}$	5222071	$k \equiv 617 \pmod{1050}$	$p_3(1050)$
$k \equiv 37 \pmod{405}$	40435201	$k \equiv 1517 \pmod{2100}$	1194901
$k \equiv 697 \pmod{810}$	811	$k \equiv 467 \pmod{2100}$	62134801
$k \equiv 562 \pmod{810}$	213872067091	$k \equiv 1367 \pmod{2100}$	219345525001
$k \equiv 967 \pmod{1620}$	1621	$k \equiv 317 \pmod{2100}$	5455568729101
$k \equiv 157 \pmod{1620}$	19441	$k \equiv 452 \pmod{600}$	32401
$k \equiv 832 \pmod{1620}$	27541	$k \equiv 602 \pmod{1200}$	32233075296001
$k \equiv 22 \pmod{1620}$	119881	$k \equiv 84 \pmod{85}$	262533041
$k \equiv 1492 \pmod{1620}$	68041	$k \equiv 79 \pmod{85}$	8119594779271
$k \equiv 1087 \pmod{1620}$	$p_6(1620)$	$k \equiv 74 \pmod{85}$	$p_3(85)$
$k \equiv 682 \pmod{1620}$	8101	$k \equiv 69 \pmod{170}$	87211
$k \equiv 952 \pmod{1215}$	369361	$k \equiv 154 \pmod{170}$	787223761
$k \equiv 1627 \pmod{2025}$	180860878351	$k \equiv 149 \pmod{170}$	$p_3(170)$
$k \equiv 2437 \pmod{4050}$	4051	$k \equiv 574 \pmod{680}$	1361
$k \equiv 412 \pmod{4050}$	1931851	$k \equiv 404 \pmod{680}$	787131281
$k \equiv 2032 \pmod{4050}$	$p_3(4050)$	$k \equiv 25 \pmod{68}$	28559389
$k \equiv 237 \pmod{250}$	10893295001	$k \equiv 127 \pmod{136}$	152533657
$k \equiv 162 \pmod{250}$	$p_8(250)$	$k \equiv 59 \pmod{136}$	$p_2(136)$
$k \equiv 87 \pmod{500}$	4001	$k \equiv 178 \pmod{204}$	409
$k \equiv 337 \pmod{500}$	76001	$k \equiv 144 \pmod{204}$	3061
$k \equiv 12 \pmod{500}$	1610501	$k \equiv 110 \pmod{204}$	5969449
$k \equiv 362 \pmod{750}$	$p_1(750)$	$k \equiv 76 \pmod{204}$	134703241
$k \equiv 687 \pmod{1000}$	24001	$k \equiv 42 \pmod{204}$	$p_5(204)$
$k \equiv 187 \pmod{1000}$	1378001	$k \equiv 8 \pmod{204}$	$p_6(204)$
$k \equiv 887 \pmod{1500}$	3001	$k \equiv 37 \pmod{51}$	613
$k \equiv 137 \pmod{1500}$	$p_2(1500)$	$k \equiv 20 \pmod{51}$	210631
$k \equiv 812 \pmod{1500}$	$p_3(1500)$	$k \equiv 3 \pmod{51}$	52986961
$k \equiv 122 \pmod{150}$	$p_1(150)$	$k \equiv 49 \pmod{51}$	$p_4(51)$
$k \equiv 257 \pmod{300}$	601	$k \equiv 83 \pmod{102}$	291078844423
$k \equiv 107 \pmod{300}$	261301	$k \equiv 32 \pmod{102}$	$p_2(102)$
$k \equiv 242 \pmod{300}$	3903901	$k \equiv 270 \pmod{408}$	2857
$k \equiv 92 \pmod{300}$	168290119201	$k \equiv 219 \pmod{408}$	13266937
$k \equiv 227 \pmod{300}$	$p_5(300)$	$k \equiv 168 \pmod{408}$	$p_3(408)$
$k \equiv 77 \pmod{300}$	$p_6(300)$	$k \equiv 423 \pmod{816}$	54673
$k \equiv 797 \pmod{900}$	1801	$k \equiv 15 \pmod{816}$	5637065089

Table 11 Covering used in Lemma 2 for $d = 6$, Part 3

Congruence	Prime
$k \equiv 372 \pmod{1224}$	282803977
$k \equiv 780 \pmod{1224}$	$p_2(1224)$
$k \equiv 3177 \pmod{3672}$	3673
$k \equiv 1953 \pmod{3672}$	1813969
$k \equiv 729 \pmod{3672}$	161569
$k \equiv 2769 \pmod{3672}$	87140233
$k \equiv 1545 \pmod{3672}$	322941468457
$k \equiv 146 \pmod{153}$	307
$k \equiv 129 \pmod{153}$	18973
$k \equiv 112 \pmod{153}$	11910133
$k \equiv 95 \pmod{153}$	25332185271529
$k \equiv 78 \pmod{153}$	$p_5(153)$
$k \equiv 61 \pmod{153}$	$p_6(153)$
$k \equiv 350 \pmod{459}$	12853
$k \equiv 197 \pmod{459}$	919
$k \equiv 44 \pmod{459}$	2735641
$k \equiv 486 \pmod{612}$	54469
$k \equiv 333 \pmod{612}$	158963941
$k \equiv 180 \pmod{612}$	2709009355501
$k \equiv 27 \pmod{612}$	3626707988341
$k \equiv 469 \pmod{612}$	$p_5(612)$
$k \equiv 316 \pmod{612}$	$p_6(612)$
$k \equiv 163 \pmod{612}$	$p_7(612)$
$k \equiv 10 \pmod{612}$	$p_8(612)$
$k \equiv 209 \pmod{255}$	77967508765681
$k \equiv 124 \pmod{255}$	$p_2(255)$
$k \equiv 289 \pmod{306}$	2142001
$k \equiv 238 \pmod{306}$	5364487
$k \equiv 187 \pmod{306}$	832339891
$k \equiv 136 \pmod{306}$	276402747619
$k \equiv 85 \pmod{306}$	2405782797823
$k \equiv 34 \pmod{306}$	$p_6(306)$
$k \equiv 284 \pmod{340}$	26861
$k \equiv 199 \pmod{340}$	7568346838961
$k \equiv 114 \pmod{340}$	$p_3(340)$
$k \equiv 1554 \pmod{1700}$	5101
$k \equiv 1214 \pmod{1700}$	3221501
$k \equiv 874 \pmod{1700}$	$p_3(1700)$
$k \equiv 1129 \pmod{1360}$	72547841
$k \equiv 789 \pmod{1360}$	$p_2(1360)$
$k \equiv 449 \pmod{1360}$	$p_3(1360)$
$k \equiv 109 \pmod{1360}$	$p_4(1360)$
$k \equiv 228 \pmod{272}$	13355595217
$k \equiv 160 \pmod{272}$	$p_2(272)$
$k \equiv 364 \pmod{544}$	58627959
$k \equiv 840 \pmod{1632}$	3156663361
$k \equiv 359 \pmod{425}$	2551
$k \equiv 274 \pmod{425}$	1076120401
$k \equiv 189 \pmod{425}$	$p_3(425)$
$k \equiv 529 \pmod{850}$	2254201
$k \equiv 869 \pmod{1275}$	122401
$k \equiv 444 \pmod{1275}$	4029001
$k \equiv 19 \pmod{1275}$	10947151
$k \equiv 439 \pmod{510}$	102001
$k \equiv 354 \pmod{510}$	5516286288241
$k \equiv 269 \pmod{510}$	$p_3(510)$
$k \equiv 524 \pmod{1020}$	1021
$k \equiv 14 \pmod{1020}$	855781
$k \equiv 944 \pmod{1020}$	2586721
$k \equiv 774 \pmod{1020}$	5071197096181
$k \equiv 604 \pmod{1020}$	8161
$k \equiv 1284 \pmod{2040}$	$p_1(2040)$
$k \equiv 264 \pmod{2040}$	$p_2(2040)$
$k \equiv 1369 \pmod{1530}$	1531
$k \equiv 1199 \pmod{1530}$	7947357582331
$k \equiv 1029 \pmod{1530}$	40984651817371
$k \equiv 859 \pmod{1530}$	12804651623971
$k \equiv 2219 \pmod{3060}$	9181
$k \equiv 689 \pmod{3060}$	302941
$k \equiv 2049 \pmod{3060}$	49281384122461
$k \equiv 519 \pmod{3060}$	$p_4(3060)$
$k \equiv 1879 \pmod{3060}$	$p_5(3060)$
$k \equiv 349 \pmod{3060}$	6121
$k \equiv 3239 \pmod{6120}$	36721
$k \equiv 179 \pmod{6120}$	24481
$k \equiv 4599 \pmod{6120}$	12241
$k \equiv 3069 \pmod{6120}$	1211761
$k \equiv 1539 \pmod{6120}$	10527797306161
$k \equiv 106 \pmod{187}$	143899867
$k \equiv 225 \pmod{374}$	192611
$k \equiv 38 \pmod{374}$	1284680342573
$k \equiv 327 \pmod{374}$	$p_3(374)$
$k \equiv 140 \pmod{374}$	$p_4(374)$
$k \equiv 242 \pmod{561}$	1123
$k \equiv 55 \pmod{561}$	51613
$k \equiv 429 \pmod{561}$	$p_3(561)$
$k \equiv 718 \pmod{748}$	155623169021
$k \equiv 531 \pmod{748}$	7481

Table 11 Covering used in Lemma 2 for $d = 6$, Part 4

Congruence	Prime
$k \equiv 259 \pmod{935}$	1871
$k \equiv 183 \pmod{225}$	$p_{10}(225)$
$k \equiv 108 \pmod{225}$	$p_{11}(225)$
$k \equiv 258 \pmod{675}$	$p_4(675)$
$k \equiv 33 \pmod{675}$	$p_5(675)$
$k \equiv 83 \pmod{375}$	$p_{10}(375)$
$k \equiv 8 \pmod{375}$	$p_{11}(375)$
$k \equiv 128 \pmod{175}$	$p_4(175)$
$k \equiv 103 \pmod{175}$	$p_5(175)$
$k \equiv 353 \pmod{525}$	$p_4(525)$
$k \equiv 178 \pmod{525}$	$p_5(525)$
$k \equiv 1053 \pmod{1575}$	$p_1(1575)$
$k \equiv 528 \pmod{1575}$	$p_2(1575)$
$k \equiv 172 \pmod{405}$	$p_5(405)$
$k \equiv 307 \pmod{405}$	$p_6(405)$
$k \equiv 427 \pmod{810}$	$p_3(810)$
$k \equiv 292 \pmod{810}$	$p_4(810)$
$k \equiv 277 \pmod{1620}$	$p_8(1620)$
$k \equiv 547 \pmod{1215}$	$p_2(1215)$
$k \equiv 142 \pmod{1215}$	$p_3(1215)$
$k \equiv 1222 \pmod{2025}$	$p_2(2025)$
$k \equiv 817 \pmod{2025}$	$p_3(2025)$
$k \equiv 7 \pmod{4050}$	$p_4(4050)$
$k \equiv 262 \pmod{500}$	$p_4(500)$
$k \equiv 437 \pmod{500}$	$p_5(500)$
$k \equiv 287 \pmod{750}$	$p_2(750)$
$k \equiv 212 \pmod{750}$	$p_3(750)$
$k \equiv 62 \pmod{1500}$	$p_4(1500)$
$k \equiv 47 \pmod{900}$	$p_6(900)$
$k \equiv 482 \pmod{1350}$	$p_2(1350)$
$k \equiv 32 \pmod{1350}$	$p_3(1350)$
$k \equiv 17 \pmod{1050}$	$p_4(1050)$
$k \equiv 167 \pmod{1050}$	$p_5(1050)$
$k \equiv 302 \pmod{600}$	$p_2(600)$
$k \equiv 152 \pmod{600}$	$p_3(600)$
$k \equiv 2 \pmod{1200}$	$p_2(1200)$
$k \equiv 234 \pmod{680}$	$p_3(680)$

Congruence	Prime
$k \equiv 64 \pmod{680}$	$p_4(680)$
$k \equiv 117 \pmod{408}$	$p_4(408)$
$k \equiv 66 \pmod{408}$	$p_5(408)$
$k \equiv 1188 \pmod{1224}$	$p_3(1224)$
$k \equiv 1137 \pmod{1224}$	$p_4(1224)$
$k \equiv 321 \pmod{3672}$	$p_6(3672)$
$k \equiv 39 \pmod{255}$	$p_3(255)$
$k \equiv 204 \pmod{255}$	$p_4(255)$
$k \equiv 374 \pmod{765}$	$p_1(765)$
$k \equiv 629 \pmod{765}$	$p_2(765)$
$k \equiv 884 \pmod{1530}$	$p_5(1530)$
$k \equiv 119 \pmod{1530}$	$p_6(1530)$
$k \equiv 29 \pmod{340}$	$p_4(340)$
$k \equiv 279 \pmod{340}$	$p_5(340)$
$k \equiv 534 \pmod{1700}$	$p_4(1700)$
$k \equiv 194 \pmod{1700}$	$p_5(1700)$
$k \equiv 92 \pmod{544}$	$p_2(544)$
$k \equiv 568 \pmod{816}$	$p_3(816)$
$k \equiv 296 \pmod{816}$	$p_4(816)$
$k \equiv 24 \pmod{1632}$	$p_2(1632)$
$k \equiv 104 \pmod{850}$	$p_2(850)$
$k \equiv 184 \pmod{510}$	$p_4(510)$
$k \equiv 99 \pmod{510}$	$p_5(510)$
$k \equiv 434 \pmod{1020}$	$p_6(1020)$
$k \equiv 1114 \pmod{2040}$	$p_3(2040)$
$k \equiv 94 \pmod{2040}$	$p_4(2040)$
$k \equiv 1709 \pmod{3060}$	$p_7(3060)$
$k \equiv 9 \pmod{6120}$	$p_6(6120)$
$k \equiv 21 \pmod{187}$	$p_2(187)$
$k \equiv 123 \pmod{187}$	$p_3(187)$
$k \equiv 344 \pmod{748}$	$p_3(748)$
$k \equiv 157 \pmod{748}$	$p_4(748)$
$k \equiv 174 \pmod{935}$	$p_2(935)$
$k \equiv 89 \pmod{935}$	$p_3(935)$
$k \equiv 939 \pmod{1870}$	$p_1(1870)$
$k \equiv 4 \pmod{1870}$	$p_2(1870)$

Table 12 Covering used in Lemma 2 for $d = 9$, Part 1

Congruence	Prime	Congruence	Prime
$k \equiv 0 \pmod{2}$	11	$k \equiv 29 \pmod{46}$	139
$k \equiv 18 \pmod{23}$	$p_2(23)$	$k \equiv 23 \pmod{46}$	2531
$k \equiv 35 \pmod{46}$	47	$k \equiv 17 \pmod{46}$	$p_8(46)$

Table 12 Covering used in Lemma 2 for $d = 9$, Part 2

Congruence	Prime	Congruence	Prime
$k \equiv 34 \pmod{69}$	277	$k \equiv 16 \pmod{29}$	43037
$k \equiv 28 \pmod{69}$	$p_3(69)$	$k \equiv 10 \pmod{29}$	62003
$k \equiv 22 \pmod{69}$	$p_5(69)$	$k \equiv 4 \pmod{29}$	77843839397
$k \equiv 39 \pmod{92}$	1289	$k \equiv 27 \pmod{58}$	59
$k \equiv 85 \pmod{92}$	18371524594609	$k \equiv 21 \pmod{58}$	$p_2(58)$
$k \equiv 79 \pmod{92}$	$p_5(92)$	$k \equiv 44 \pmod{87}$	4003
$k \equiv 125 \pmod{184}$	2393	$k \equiv 38 \pmod{87}$	72559
$k \equiv 33 \pmod{184}$	$p_4(184)$	$k \equiv 32 \pmod{87}$	$p_3(87)$
$k \equiv 96 \pmod{115}$	31511	$k \equiv 113 \pmod{116}$	349
$k \equiv 73 \pmod{115}$	19707665921	$k \equiv 55 \pmod{116}$	38861
$k \equiv 50 \pmod{115}$	$p_8(115)$	$k \equiv 107 \pmod{116}$	618049
$k \equiv 27 \pmod{115}$	$p_9(115)$	$k \equiv 49 \pmod{116}$	$p_4(116)$
$k \equiv 4 \pmod{115}$	$p_{10}(115)$	$k \equiv 101 \pmod{174}$	638453709757
$k \equiv 67 \pmod{138}$	31051	$k \equiv 95 \pmod{174}$	$p_2(174)$
$k \equiv 61 \pmod{138}$	$p_2(138)$	$k \equiv 89 \pmod{174}$	$p_3(174)$
$k \equiv 55 \pmod{138}$	$p_4(138)$	$k \equiv 257 \pmod{348}$	997369
$k \equiv 3 \pmod{161}$	6763	$k \equiv 83 \pmod{348}$	$p_2(348)$
$k \equiv 49 \pmod{161}$	472341157	$k \equiv 593 \pmod{696}$	13921
$k \equiv 279 \pmod{322}$	967	$k \equiv 419 \pmod{696}$	$p_2(696)$
$k \equiv 233 \pmod{322}$	1569936761	$k \equiv 413 \pmod{522}$	155557
$k \equiv 187 \pmod{322}$	$p_2(322)$	$k \equiv 239 \pmod{522}$	$p_2(522)$
$k \equiv 181 \pmod{207}$	$p_1(207)$	$k \equiv 349 \pmod{406}$	19489
$k \equiv 175 \pmod{276}$	829	$k \equiv 291 \pmod{406}$	243720583
$k \equiv 37 \pmod{276}$	1569889	$k \equiv 233 \pmod{406}$	$p_3(406)$
$k \equiv 169 \pmod{276}$	$p_2(276)$	$k \equiv 175 \pmod{406}$	$p_4(406)$
$k \equiv 31 \pmod{276}$	$p_4(276)$	$k \equiv 117 \pmod{406}$	$p_5(406)$
$k \equiv 157 \pmod{276}$	$p_5(276)$	$k \equiv 59 \pmod{203}$	18620680471
$k \equiv 295 \pmod{552}$	1657	$k \equiv 1 \pmod{203}$	$p_2(203)$
$k \equiv 19 \pmod{552}$	1469409649	$k \equiv 227 \pmod{232}$	233
$k \equiv 301 \pmod{345}$	5521	$k \equiv 169 \pmod{232}$	355193
$k \equiv 232 \pmod{345}$	74521	$k \equiv 111 \pmod{232}$	21591416633
$k \equiv 163 \pmod{345}$	437831	$k \equiv 395 \pmod{464}$	929
$k \equiv 94 \pmod{345}$	19277911	$k \equiv 163 \pmod{464}$	23201
$k \equiv 289 \pmod{414}$	37916801893	$k \equiv 337 \pmod{464}$	182353
$k \equiv 151 \pmod{414}$	$p_2(414)$	$k \equiv 105 \pmod{464}$	4465073
$k \equiv 13 \pmod{414}$	$p_3(414)$	$k \equiv 215 \pmod{261}$	523
$k \equiv 421 \pmod{828}$	4969	$k \equiv 128 \pmod{261}$	670249
$k \equiv 7 \pmod{828}$	111781	$k \equiv 41 \pmod{261}$	44974999
$k \equiv 415 \pmod{483}$	14461987	$k \equiv 209 \pmod{261}$	9113995243
$k \equiv 346 \pmod{483}$	176686231	$k \equiv 122 \pmod{261}$	7299238406959
$k \equiv 277 \pmod{483}$	$p_3(483)$	$k \equiv 899 \pmod{1566}$	1567
$k \equiv 70 \pmod{161}$	$p_3(161)$	$k \equiv 116 \pmod{1566}$	32887
$k \equiv 28 \pmod{29}$	3191	$k \equiv 812 \pmod{1044}$	2089
$k \equiv 22 \pmod{29}$	16763	$k \equiv 551 \pmod{1044}$	31753261

Table 12 Covering used in Lemma 2 for $d = 9$, Part 3

Congruence	Prime
$k \equiv 290 \pmod{1044}$	1452031741
$k \equiv 29 \pmod{1044}$	803137587589
$k \equiv 255 \pmod{290}$	1451
$k \equiv 197 \pmod{290}$	30104611
$k \equiv 139 \pmod{290}$	58765601
$k \equiv 81 \pmod{290}$	2433146345771
$k \equiv 23 \pmod{290}$	$p_5(290)$
$k \equiv 133 \pmod{145}$	9605671
$k \equiv 104 \pmod{145}$	$p_2(145)$
$k \equiv 162 \pmod{290}$	$p_6(290)$
$k \equiv 307 \pmod{580}$	$p_1(580)$
$k \equiv 359 \pmod{435}$	80911
$k \equiv 272 \pmod{435}$	102019681
$k \equiv 185 \pmod{435}$	$p_3(435)$
$k \equiv 98 \pmod{435}$	$p_4(435)$
$k \equiv 446 \pmod{870}$	33931
$k \equiv 11 \pmod{870}$	70350656881
$k \equiv 527 \pmod{609}$	41413
$k \equiv 440 \pmod{609}$	$p_2(609)$
$k \equiv 788 \pmod{1218}$	2437
$k \equiv 179 \pmod{1218}$	4315282651
$k \equiv 92 \pmod{203}$	$p_3(203)$
$k \equiv 5 \pmod{203}$	$p_4(203)$
$k \equiv 22 \pmod{31}$	2791
$k \equiv 10 \pmod{31}$	6943319
$k \equiv 29 \pmod{31}$	$p_3(31)$
$k \equiv 17 \pmod{62}$	$p_1(62)$
$k \equiv 36 \pmod{93}$	$p_1(93)$
$k \equiv 55 \pmod{124}$	2049349
$k \equiv 43 \pmod{124}$	$p_2(124)$
$k \equiv 93 \pmod{186}$	373
$k \equiv 81 \pmod{186}$	44641
$k \equiv 69 \pmod{186}$	3590254957
$k \equiv 57 \pmod{186}$	$p_4(186)$
$k \equiv 231 \pmod{372}$	5209
$k \equiv 195 \pmod{248}$	1489
$k \equiv 71 \pmod{248}$	640543322297
$k \equiv 183 \pmod{248}$	27908132670449
$k \equiv 59 \pmod{248}$	$p_4(248)$
$k \equiv 171 \pmod{248}$	$p_5(248)$
$k \equiv 295 \pmod{496}$	$p_1(496)$
$k \equiv 252 \pmod{279}$	18989357081041
$k \equiv 271 \pmod{310}$	11161
$k \equiv 209 \pmod{310}$	3925963357681

Congruence	Prime
$k \equiv 147 \pmod{310}$	$p_3(310)$
$k \equiv 85 \pmod{310}$	$p_4(310)$
$k \equiv 507 \pmod{744}$	700849
$k \equiv 135 \pmod{744}$	11110153
$k \equiv 495 \pmod{744}$	1220725699657
$k \equiv 123 \pmod{744}$	5419392721
$k \equiv 483 \pmod{744}$	42367299139993
$k \equiv 843 \pmod{1488}$	7573921
$k \equiv 831 \pmod{868}$	8681
$k \equiv 707 \pmod{868}$	106300489
$k \equiv 583 \pmod{868}$	120995729
$k \equiv 459 \pmod{868}$	363981764441
$k \equiv 87 \pmod{434}$	240437
$k \equiv 385 \pmod{434}$	1142669053
$k \equiv 323 \pmod{434}$	$p_3(434)$
$k \equiv 261 \pmod{434}$	$p_4(434)$
$k \equiv 199 \pmod{434}$	$p_5(434)$
$k \equiv 137 \pmod{217}$	3662093
$k \equiv 435 \pmod{465}$	50718862231
$k \equiv 342 \pmod{465}$	$p_2(465)$
$k \equiv 63 \pmod{155}$	311
$k \equiv 1167 \pmod{1674}$	5023
$k \equiv 609 \pmod{1674}$	261315626851
$k \equiv 783 \pmod{1116}$	4967159761
$k \equiv 411 \pmod{1116}$	28975286089
$k \equiv 39 \pmod{1116}$	$p_3(1116)$
$k \equiv 1515 \pmod{1860}$	1861
$k \equiv 1143 \pmod{1860}$	459421
$k \equiv 771 \pmod{1860}$	$p_3(1860)$
$k \equiv 399 \pmod{1860}$	$p_4(1860)$
$k \equiv 27 \pmod{1860}$	$p_5(1860)$
$k \equiv 759 \pmod{930}$	$p_1(930)$
$k \equiv 747 \pmod{1116}$	$p_4(1116)$
$k \equiv 375 \pmod{558}$	1117
$k \equiv 1119 \pmod{2232}$	$p_1(2232)$
$k \equiv 34 \pmod{37}$	2028119
$k \equiv 22 \pmod{37}$	247629013
$k \equiv 10 \pmod{37}$	$p_3(37)$
$k \equiv 35 \pmod{74}$	7253
$k \equiv 23 \pmod{74}$	$p_2(74)$
$k \equiv 11 \pmod{74}$	$p_3(74)$
$k \equiv 36 \pmod{111}$	$p_1(111)$
$k \equiv 24 \pmod{111}$	$p_2(111)$
$k \equiv 12 \pmod{111}$	$p_3(111)$

Table 12 Covering used in Lemma 2 for $d = 9$, Part 4

Congruence	Prime	Congruence	Prime
$k \equiv 37 \pmod{148}$	149	$k \equiv 699 \pmod{777}$	79003664587
$k \equiv 25 \pmod{148}$	3109	$k \equiv 588 \pmod{777}$	531660478441
$k \equiv 13 \pmod{148}$	111149	$k \equiv 477 \pmod{777}$	$p_3(777)$
$k \equiv 1 \pmod{148}$	708840373781	$k \equiv 909 \pmod{1036}$	$p_1(1036)$
$k \equiv 137 \pmod{148}$	$p_5(148)$	$k \equiv 58 \pmod{259}$	2591
$k \equiv 125 \pmod{148}$	$p_6(148)$	$k \equiv 169 \pmod{259}$	64638631
$k \equiv 39 \pmod{222}$	223	$k \equiv 21 \pmod{259}$	$p_3(259)$
$k \equiv 27 \pmod{222}$	4663	$k \equiv 379 \pmod{407}$	46399
$k \equiv 15 \pmod{222}$	$p_3(222)$	$k \equiv 342 \pmod{407}$	390721
$k \equiv 3 \pmod{222}$	$p_4(222)$	$k \equiv 305 \pmod{407}$	2377517312347
$k \equiv 213 \pmod{444}$	$p_1(444)$	$k \equiv 268 \pmod{407}$	$p_4(407)$
$k \equiv 621 \pmod{888}$	54615849443593	$k \equiv 564 \pmod{814}$	409038414731
$k \equiv 177 \pmod{888}$	$p_2(888)$	$k \equiv 934 \pmod{1628}$	24421
$k \equiv 1041 \pmod{1332}$	38629	$k \equiv 120 \pmod{1628}$	115589
$k \equiv 597 \pmod{1332}$	215018101	$k \equiv 1304 \pmod{1628}$	7716573481
$k \equiv 153 \pmod{1332}$	14003835481	$k \equiv 897 \pmod{1628}$	$p_4(1628)$
$k \equiv 1473 \pmod{2664}$	7993	$k \equiv 490 \pmod{1628}$	$p_5(1628)$
$k \equiv 141 \pmod{2664}$	6920827178401	$k \equiv 1711 \pmod{3256}$	3257
$k \equiv 1461 \pmod{1776}$	1777	$k \equiv 83 \pmod{3256}$	$p_2(3256)$
$k \equiv 1017 \pmod{1776}$	7228321	$k \equiv 860 \pmod{1221}$	3306121237
$k \equiv 573 \pmod{1776}$	$p_3(1776)$	$k \equiv 453 \pmod{1221}$	37232500009
$k \equiv 154 \pmod{185}$	$p_1(185)$	$k \equiv 46 \pmod{1221}$	$p_3(1221)$
$k \equiv 117 \pmod{185}$	$p_2(185)$	$k \equiv 1230 \pmod{2442}$	$p_1(2442)$
$k \equiv 191 \pmod{370}$	$p_1(370)$	$k \equiv 141 \pmod{322}$	$p_4(322)$
$k \equiv 6 \pmod{370}$	$p_2(370)$	$k \equiv 95 \pmod{322}$	$p_5(322)$
$k \equiv 253 \pmod{296}$	61544617	$k \equiv 112 \pmod{207}$	$p_2(207)$
$k \equiv 105 \pmod{296}$	$p_2(296)$	$k \equiv 43 \pmod{207}$	$p_3(207)$
$k \equiv 315 \pmod{333}$	96455449	$k \equiv 25 \pmod{345}$	$p_5(345)$
$k \equiv 525 \pmod{555}$	14431	$k \equiv 283 \pmod{414}$	$p_4(414)$
$k \equiv 414 \pmod{555}$	169831	$k \equiv 145 \pmod{414}$	$p_5(414)$
$k \equiv 303 \pmod{555}$	994561	$k \equiv 208 \pmod{483}$	$p_4(483)$
$k \equiv 192 \pmod{555}$	5153778841	$k \equiv 139 \pmod{483}$	$p_5(483)$
$k \equiv 81 \pmod{555}$	$p_5(555)$	$k \equiv 1 \pmod{161}$	$p_4(161)$
$k \equiv 513 \pmod{666}$	902659997773	$k \equiv 251 \pmod{348}$	$p_3(348)$
$k \equiv 291 \pmod{666}$	$p_2(666)$	$k \equiv 77 \pmod{348}$	$p_4(348)$
$k \equiv 69 \pmod{666}$	$p_3(666)$	$k \equiv 245 \pmod{696}$	$p_3(696)$
$k \equiv 649 \pmod{740}$	1481	$k \equiv 71 \pmod{696}$	$p_4(696)$
$k \equiv 501 \pmod{740}$	15541	$k \equiv 65 \pmod{522}$	$p_3(522)$
$k \equiv 353 \pmod{740}$	68821	$k \equiv 53 \pmod{232}$	$p_4(232)$
$k \equiv 205 \pmod{740}$	6336499381	$k \equiv 221 \pmod{232}$	$p_5(232)$
$k \equiv 57 \pmod{740}$	4098464044501	$k \equiv 279 \pmod{464}$	$p_5(464)$
$k \equiv 637 \pmod{740}$	74615611921	$k \equiv 47 \pmod{464}$	$p_6(464)$
$k \equiv 489 \pmod{740}$	$p_7(740)$	$k \equiv 35 \pmod{261}$	$p_6(261)$
$k \equiv 341 \pmod{740}$	$p_8(740)$	$k \equiv 203 \pmod{261}$	$p_7(261)$

Table 12 Covering used in Lemma 2 for $d = 9$, Part 5

Congruence	Prime
$k \equiv 638 \pmod{783}$	$p_1(783)$
$k \equiv 377 \pmod{783}$	$p_2(783)$
$k \equiv 75 \pmod{145}$	$p_3(145)$
$k \equiv 46 \pmod{145}$	$p_4(145)$
$k \equiv 17 \pmod{580}$	$p_2(580)$
$k \equiv 353 \pmod{609}$	$p_3(609)$
$k \equiv 266 \pmod{609}$	$p_4(609)$
$k \equiv 219 \pmod{372}$	$p_2(372)$
$k \equiv 207 \pmod{372}$	$p_3(372)$
$k \equiv 47 \pmod{496}$	$p_2(496)$
$k \equiv 159 \pmod{279}$	$p_2(279)$
$k \equiv 66 \pmod{279}$	$p_3(279)$
$k \equiv 333 \pmod{620}$	$p_1(620)$
$k \equiv 23 \pmod{620}$	$p_2(620)$
$k \equiv 111 \pmod{744}$	$p_6(744)$
$k \equiv 471 \pmod{744}$	$p_7(744)$
$k \equiv 99 \pmod{1488}$	$p_2(1488)$
$k \equiv 335 \pmod{868}$	$p_5(868)$
$k \equiv 211 \pmod{868}$	$p_6(868)$
$k \equiv 75 \pmod{217}$	$p_2(217)$
$k \equiv 13 \pmod{217}$	$p_3(217)$
$k \equiv 249 \pmod{465}$	$p_3(465)$
$k \equiv 156 \pmod{465}$	$p_4(465)$
$k \equiv 423 \pmod{558}$	$p_2(558)$
$k \equiv 237 \pmod{558}$	$p_3(558)$
$k \equiv 51 \pmod{1674}$	$p_3(1674)$
$k \equiv 1503 \pmod{1860}$	$p_6(1860)$
$k \equiv 1131 \pmod{1860}$	$p_7(1860)$
$k \equiv 387 \pmod{930}$	$p_2(930)$
$k \equiv 15 \pmod{930}$	$p_3(930)$

Congruence	Prime
$k \equiv 2235 \pmod{4464}$	$p_1(4464)$
$k \equiv 3 \pmod{4464}$	$p_2(4464)$
$k \equiv 201 \pmod{444}$	$p_2(444)$
$k \equiv 189 \pmod{444}$	$p_3(444)$
$k \equiv 609 \pmod{888}$	$p_3(888)$
$k \equiv 165 \pmod{888}$	$p_4(888)$
$k \equiv 1029 \pmod{1332}$	$p_4(1332)$
$k \equiv 585 \pmod{1332}$	$p_5(1332)$
$k \equiv 129 \pmod{1776}$	$p_4(1776)$
$k \equiv 80 \pmod{185}$	$p_3(185)$
$k \equiv 43 \pmod{185}$	$p_4(185)$
$k \equiv 204 \pmod{333}$	$p_2(333)$
$k \equiv 93 \pmod{333}$	$p_3(333)$
$k \equiv 193 \pmod{740}$	$p_9(740)$
$k \equiv 45 \pmod{740}$	$p_{10}(740)$
$k \equiv 366 \pmod{777}$	$p_4(777)$
$k \equiv 255 \pmod{777}$	$p_5(777)$
$k \equiv 144 \pmod{259}$	$p_4(259)$
$k \equiv 33 \pmod{259}$	$p_5(259)$
$k \equiv 761 \pmod{1036}$	$p_2(1036)$
$k \equiv 613 \pmod{1036}$	$p_3(1036)$
$k \equiv 465 \pmod{518}$	$p_1(518)$
$k \equiv 231 \pmod{407}$	$p_5(407)$
$k \equiv 194 \pmod{407}$	$p_6(407)$
$k \equiv 157 \pmod{814}$	$p_2(814)$
$k \equiv 527 \pmod{814}$	$p_3(814)$
$k \equiv 823 \pmod{1221}$	$p_4(1221)$
$k \equiv 416 \pmod{1221}$	$p_5(1221)$
$k \equiv 9 \pmod{2442}$	$p_2(2442)$

$$p_5(140) = 848654483879497562821.$$

A more complete list of primes can be found in [3, 7]. There are cases however where the number of indicated primes in Table 1 includes 2 primes in a factorization of $\Phi_c(10)$ coming from an unfactored part of $\Phi_c(10)$ which is known to be composite, coprime to the factored part of $\Phi_c(10)$ and the number c, and not of the form u^v for any positive integers u and v with $v \geq 2$.

References

1. J. Brillhart, D. H. Lehmer, J. L. Selfridge, B. Tuckerman, and S. S. Wagstaff, Jr., Factorizations of $b^n \pm 1$, $b = 2, 3, 5, 6, 7, 10, 11, 12$ Up to High Powers, 3rd edition, Contemporary Mathematics, Vol. 22, American Math. Soc., Providence, 2002 (available online).
2. P. Erdős, *Solution to problem 1029: Erdos and the computer*, Mathematics Magazine 52 (1979), 180–181.
3. M. Filaseta, J. Juillerat, and J. Southwick, Data for "Widely Digitally Stable Numbers," http://people.math.sc.edu/filaseta/widelydigitallystable.html.
4. M. Filaseta, M. Kozek, C. Nicol, and J. Selfridge, *Composites that remain composite after changing a digit*, Journal of Combinatorics and Number Theory 2 (2010), 25–36.
5. M. Filaseta and J. Southwick, *Primes that become composite after changing an arbitrary digit*, Math. Comp. 90 (2021), 979–993.
6. J. Hopper and P. Pollack, *Digitally delicate primes*, J. Number Theory 168 (2016), 247–256.
7. J. Juillerat, Widely digitally stable numbers and irreducibility criteria for polynomials with prime values, University of South Carolina, dissertation, 2021.
8. M. S. Klamkin, *Problem 1029*, Mathematics Magazine 51 (1978), p. 69.
9. S. V. Konyagin, *Numbers that become composite after changing one or two digits*, Pioneer Jour. of Algebra, Number Theory and Appl. 6 (2013), 1–7.
10. T. Tao, *A remark on primality testing and decimal expansions*, J. Aust. Math. Soc. 91 (2011), 405–413.

Non-injectivity of Nonzero Discriminant Polynomials and Applications to Packing Polynomials

Kåre S. Gjaldbæk

Abstract We show that an integer-valued quadratic polynomial on $\mathbb{R}^2$ can not be injective on the integer lattice points of any affine convex cone if its discriminant is nonzero. A consequence is the non-existence of quadratic packing polynomials on irrational sectors of $\mathbb{R}^2$.

Keywords Packing polynomials · Cantor polynomials · Fueter-Pólya Theorem

MSC2020 11C08 · 11A67 · 11B34 · 05A15

1 Background

In the seminal 1878 paper *Ein Beitrag zur Mannigfaltigkeitslehre* [2], Cantor introduces the polynomial

$$f(x, y) = x + \frac{(x + y - 1)(x + y + 2)}{2}$$

which bijectively maps $\mathbb{N} \times \mathbb{N}$ onto $\mathbb{N}$, $\mathbb{N}$ denoting the positive integers. Translating, and interchanging the variables, leads to the two *Cantor Polynomials*

$$F(x, y) = \frac{1}{2}(x + y)(x + y + 1) + x \,,$$
$$G(x, y) = \frac{1}{2}(x + y)(x + y + 1) + y \,.$$

K. S. Gjaldbæk (✉)
The Graduate Center of The City University of New York, 365 Fifth Avenue, New York, NY 10016, USA
e-mail: kgjaldbaek@gradcenter.cuny.edu

M. B. Nathanson (ed.), *Combinatorial and Additive Number Theory IV*, Springer Proceedings in Mathematics & Statistics 347,
https://doi.org/10.1007/978-3-030-67996-5_10

These two quadratics are bijections from $\mathbb{N}_0 \times \mathbb{N}_0$ onto $\mathbb{N}_0$, $\mathbb{N}_0$ denoting the non-negative integers.

In 1923, Fueter and Pólya [4] prove that the two Cantor Polynomials are the only two *quadratic* polynomials admitting such a bijection. They further conjecture that a bijection is impossible for a polynomial of degree other than two. Fueter and Pólya's proof relies on a corollary of the Lindemann-Weierstraß theorem which says that the trigonometric and hyperbolic functions evaluated at non-trivial algebraic numbers are transcendental. In 2001, Vsemirnov [10] proves the theorem using elementary methods. Theorem 1 of this paper allows for Fueter and Pólya's strategy to work without the need for the theorem of Lindemann-Weierstraß.

In two papers [6, 7] from 1977, Lew and Rosenberg develop a more general theory, and some of the terminology they introduce has taken hold. They provide a partial result on Fueter and Pólya's conjecture in proving the non-existence of polynomials of degree 3 and 4. The general problem remains open.

For some preliminary results, Lew and Rosenberg consider polynomials on arbitrary sectors, regions that are the convex hull of two half-lines starting at the origin, yet they do not study polynomial bijections from general sectors onto $\mathbb{N}_0$. This is the subject of Nathanson's 2014 paper *Cantor Polynomials for Semigroup Sectors* [8]. Nathanson looks at quadratic polynomial bijections from arbitrary sectors in the first quadrant, in particular the lattice points in the convex cone spanned by the x-axis and the line $y = \alpha x$ for some $\alpha > 0$. Nathanson determines all such polynomials for sectors given by $\alpha = 1/n$, $n \in \mathbb{N}$. Furthermore, he finds two quadratic polynomials for each sector with integer α-value. He raises the question of which rational values for α allow for quadratic polynomial bijections, and whether such is possible for irrational α.

The same year, Stanton and Brandt answer the questions regarding rational sectors. In [9], Stanton determines all quadratic polynomials for $\alpha \in \mathbb{N}$. In addition to the polynomials discovered by Nathanson, she finds two quadratic packing polynomials for the sectors given by $\alpha = 3, 4$ and proves that there are no more. The non-integral rational case is addressed by Brandt in the preprint paper [1].

Regarding the irrational case, to be accurate, Nathanson asks the reader to show that no bijections are possible, although he never names it a conjecture. Corollary 1 of this paper obliges.

2 Quadratic Packing Polynomials on Sectors

Let $\alpha > 0$. Define the sector

$$S(\alpha) = \{(x, y) \in \mathbb{R}^2 \ : \ 0 \leq y \leq \alpha x\} \, .$$

We let $S(\infty)$ denote the first quadrant, i.e. the sector pertaining to the original problem. Adopting the terminology of Lew and Rosenberg, we refer to a bijection from

$S(\alpha) \cap \mathbb{Z}^2$ onto $\mathbb{N}_0$ as a *packing function*, or, in the case of a polynomial, which is our sole focus, *packing polynomial.*

An immediate prerequisite for a packing polynomial is that it is *integer-valued.* That is, it must take integer values on integer lattice points. A consequence of standard results on integer-valued polynomials (see e.g. [5], Chp. X, Lem. 6.4) is that a quadratic packing polynomial on any sector must be of the form

$$P(x, y) = A\frac{x(x-1)}{2} + Bxy + C\frac{y(y-1)}{2} + Dx + Ey + F \quad (1)$$

with $A, B, C, D, E, F \in \mathbb{Z}$. Lew and Rosenberg make the following observation (see [6], Prop. 3.4).

Lemma 1 *Let $P(x, y)$ be a packing polynomial*[1] *on the sector $S(\alpha)$. If $(m, n) \in S(\alpha) \setminus \{(0, 0)\}$ is an integer lattice point, then the homogeneous quadratic part of $P(x, y)$:*

$$P_2(x, y) = \frac{A}{2}x^2 + Bxy + \frac{C}{2}y^2 ,$$

must take only positive values on the ray $\{(xm, xn), x > 0\}$.

An immediate consequence is that we must have $A > 0$.

3 Non-injectivity When Discriminant Is Zero

Let $\omega_1, \omega_2 \in \mathbb{R}^2$ with $\omega_1 \neq \omega_2$ and define the closed convex cone

$$\mathcal{C}(\omega_1, \omega_2) = \{u\omega_1 + v\omega_2 \ : \ u, v \geq 0\}$$

and for $\omega_0 \in \mathbb{R}^2$ the affine convex cone

$$\mathcal{C}_{\omega_0}(\omega_1, \omega_2) = \mathcal{C}(\omega_1, \omega_2) + \omega_0 .$$

Theorem 1 *Let $P : \mathbb{R}^2 \to \mathbb{R}$ be an integer-valued quadratic polynomial. If the discriminant of P is non-zero, then P cannot be injective on the integer lattice points of any affine convex cone.*

Proof Let $P(x, y)$ have the form (1). We will denote its discriminant $\Delta = B^2 - AC$ and use the shorthand notation $D' = D - \frac{A}{2}$, $E' = E - \frac{C}{2}$. Let $\mathcal{C} = \mathcal{C}_{\omega_0}(\omega_1, \omega_2)$ be an arbitrary affine cone. Fix coprime integers r, s with $r \neq 0$. Every lattice point lies on a line $L_{r,s}^{(i)} : y = \frac{s}{r}x + \frac{i}{r}$ for a unique i. For each i, consider the restriction of P to $L_{r,s}^{(i)}$:

[1] In fact, it is only required for the polynomial to be an injection into $\mathbb{N}_0$, a class Lew and Rosenberg call *storing functions.*

$$\begin{aligned}Q_i(x) &= P\left(x, \frac{s}{r}x + \frac{i}{r}\right)\\ &= \frac{1}{2r^2}(Ar^2 + 2Brs + Cs^2)x^2\\ &\quad + \frac{1}{r^2}((Br + Cs)i + r(D'r + E's))x + \text{const.}\end{aligned}$$

If r, s are chosen such that $Ar^2 + 2Brs + Cs^2 \neq 0$, then the values of $Q_i(x)$ are symmetric around

$$x_i = -\frac{(Br + Cs)i + r(D'r + E's)}{Ar^2 + 2Brs + Cs^2}.$$

The corresponding y-coordinate on $L^{(i)}_{r,s}$ is

$$y_i = \frac{(Ar + Bs)i - s(D'r + E's)}{Ar^2 + 2Brs + Cs^2}.$$

This means that, for any choice of r, s with $Ar^2 + 2Brs + Cs^2 \neq 0$, we have

$$P(x_i + r, y_i + s) = P(x_i - r, y_i - s)$$

for all i.

These points of symmetry, (x_i, y_i), fall on the straight line, $L^{\text{sym}}_{r,s}$, with slope $-\frac{Ar+Bs}{Br+Cs}$ which passes through the point

$$(x_0, y_0) = \left(\frac{CD' - BE'}{\Delta}, \frac{AE' - BD'}{\Delta}\right).$$

(To see this, replace i with $\frac{1}{\Delta}((AE' - BD')r - (CD' - BE')s)$ in the formulas for x_i, y_i.) Note that the point (x_0, y_0) does not depend on the choice of r, s. [2] We want to choose r, s such that $(x_i + r, y_i + s)$ and $(x_i - r, y_i - s)$ are both lattice points in $\mathcal{C}$ for some i. This will violate injectivity.

Put $\mathcal{C}_0 = \mathcal{C}_{(x_0,y_0)}(\omega_1, \omega_2)$ and pick an arbitrary lattice point $(m, n) \in \mathcal{C} \cap \mathcal{C}_0$. Choosing

$$\frac{s}{r} = -\frac{A(m - x_0) + B(n - y_0)}{B(m - x_0) + C(n - y_0)}$$

---

[2] When $\Delta \neq 0$, $P(x, y)$ can be rewritten as

$$P(x, y) = \frac{A}{2}(x - x_0)^2 + B(x - x_0)(y - y_0) + \frac{C}{2}(y - y_0)^2 + \frac{D'}{2}x_0 + \frac{E'}{2}y_0 + F.$$

The point (x_0, y_0) is the center of the level curves $P(x, y) = n$ which are ellipses when $\Delta < 0$, or hyperbolas when $\Delta > 0$.

in lowest terms, we have $-\frac{Ar+Bs}{Br+Cs} = \frac{n-y_0}{m-x_0}$. This means that $L_{r,s}^{\text{sym}}$ passes through the lattice point $(m, n) \in \mathcal{C} \cap \mathcal{C}_0$ and therefore infinitely many since its slope is rational. So for infinitely many i, (x_i, y_i) is a lattice point in $\mathcal{C} \cap \mathcal{C}_0$ and eventually we will find an i for which both $(x_i + r, y_i + s)$ and $(x_i - r, y_i - s)$ are lattice points in $\mathcal{C}$. □

4 Applications to Packing Polynomials

Let $P(x, y)$ be a quadratic packing polynomial on the sector $S(\alpha)$ with discriminant Δ. $S(\alpha)$ is a (affine) convex cone, so by the previous section we must have $\Delta = 0$. Employing the strategy of Lew and Rosenberg [6], we consider the regions

$$R_n = \{(x, y) \in S(\alpha) \; : \; 0 \leq P(x, y) \leq n\} \, .$$

For a packing polynomial, each region R_n contains $n + 1$ lattice points. This means that it is a necessary condition that

$$\lim_{n\to\infty} \left(\frac{1}{n} \#(R_n \cap \mathbb{Z}^2) \right) = 1 \, . \tag{2}$$

Furthermore, we have

Lemma 2 *If $P(x, y)$ is a quadratic packing polynomial on $S(\alpha)$, then*

$$\#(R_n \cap \mathbb{Z}^2) = \text{area}(R_n) + O(\sqrt{n}) \, ,$$

as $n \to \infty$.

Proof Since the discriminant is zero, the level curves $P(x, y) = n$ are parabolas and either R_n is bounded for all n or all level curves, including for negative n, fall inside $S(\alpha)$ which is impossible if P is a packing polynomial. We can therefore apply a theorem of Davenport ([3], p. 180) to estimate the number of lattice points in each region R_n. We have

$$|\,\text{area}(R_n) - \#(R_n \cap \mathbb{Z}^2)| \leq h(|\pi_x(R_n)| + |\pi_y(R_n)|) + h^2 \, ,$$

where $|\pi_x(R_n)|$ and $|\pi_y(R_n)|$ denote the lengths of the projections onto the x- and y-axis, respectively, and h is a fixed constant. [3] The length of the y-projection is bounded by the level curve's intersection with the line $y = \alpha x$ or the topmost point on the parabola. The length of the x-projection is bounded by the level curve's

[3] The constant h denotes the maximum number of disjoint intervals one can obtain from intersecting R_n with a line parallel to one of the coordinate axes. Our only concern is that it is bounded by some value.

intersection with the x-axis, the line $y = \alpha x$ or the rightmost point on the parabola. Either is $O(\sqrt{n})$ as $n \to \infty$. □

At this point, we want to note that, since $B^2 = AC$, the homogeneous quadratic part of $P(x, y)$,

$$P_2(x, y) = \frac{1}{2}(\sqrt{A}x \pm \sqrt{C}y)^2 ,$$

is non-negative and vanishes only on the rational line $Ax + By = 0$. By Lem. 1, this means that $P_2(x, y)$ is strictly positive inside $S(\alpha)$ except at the origin.

Lemma 3 *If $P(x, y)$ is a quadratic packing polynomial on $S(\alpha)$, then*

$$\text{area}(R_n) = \frac{n}{2}\int_0^{\arctan\alpha} \frac{d\theta}{\frac{A}{2}\cos^2\theta + B\cos\theta\sin\theta + \frac{C}{2}\sin^2\theta} + O(\sqrt{n})$$

as $n \to \infty$.

Proof Switching to polar coordinates, the equations of the level curves take the form

$$r^2 p_2(\theta) + r p_1(\theta) + F = n ,$$

where $p_2(\theta) = \frac{A}{2}\cos^2\theta + B\cos\theta\sin\theta + \frac{C}{2}\sin^2\theta$ and $p_1(\theta) = D'\cos\theta + E'\sin\theta$. So $r = O(\sqrt{n})$ and $r^2 = \frac{n}{p_2(\theta)} + O(\sqrt{n})$. If A_0 denotes the (possibly empty) area of $S(\alpha)$ bounded by the level curve $P(x, y) = 0$, then the area of R_n is given by

$$\text{area}(R_n) = \frac{1}{2}\int_0^{\arctan\alpha} r^2 d\theta - A_0 = \frac{1}{2}\int_0^{\arctan\alpha} \frac{n}{p_2(\theta)} d\theta + O(\sqrt{n}) .$$

□

Theorem 2 *If $P(x, y)$ is a quadratic packing polynomial on $S(\alpha)$, then*

$$\alpha = \frac{A}{1 - B} .$$

Proof Using Lem. 2 and 3, we can calculate the limit from (2) by computing the integral

$$\begin{aligned}\lim_{n\to\infty}\left(\frac{1}{n}\#(R_n \cap \mathbb{Z}^2)\right) &= \frac{1}{2}\int_0^{\arctan\alpha} \frac{d\theta}{\frac{A}{2}\cos^2\theta + B\cos\theta\sin\theta + \frac{C}{2}\sin^2\theta} \\ &= \int_0^{\alpha} \frac{dt}{A + 2Bt + Ct^2} ,\end{aligned}$$

applying the variable change $t = \tan\theta$. Since, by Thm. 1, $B^2 = AC$, we either have $B = C = 0$, in which case

$$\int_0^\alpha \frac{dt}{A+2Bt+Ct^2} = \int_0^\alpha \frac{dt}{A} = \frac{\alpha}{A},$$

or

$$\int_0^\alpha \frac{dt}{A+2Bt+Ct^2} = \frac{1}{C}\int_0^\alpha \frac{d\theta}{(t+\frac{B}{C})^2} = \frac{1}{B} - \frac{1}{\alpha C + B}.$$

As noted above, this must be equal to 1 if P is a packing polynomial. Solving for α, we get the desired result. □

An immediate consequence of Thm. 2 is the following.

Corollary 1 *There are no irrational sectors allowing for quadratic packing polynomials.*

References

1. Brandt, M.: Quadratic packing polynomials on sectors of $\mathbb{R}^2$. arXiv:1409.0063v1 (2014)
2. Cantor, G.: Ein Beitrag zur Mannigfaltigkeitslehre. Journal fur die reine und angewandte Mathematik **84**, 242–258 (1878)
3. Davenport, H.: On a principle of Lipschitz. Journal of the London Mathematical Society **1**(3), 179–183 (1951)
4. Fueter, R., Pólya, G.: Rationale Abzählung der Gitterpunkte. Vierteljschr. Naturforsch. Ges. Zürich **58**, 380–386 (1923)
5. Lang, S.: Algebra. Springer (2002)
6. Lew, J.S., Rosenberg, A.L.: Polynomial indexing of integer lattice-points I. General concepts and quadratic polynomials. Journal of Number Theory **10**(2), 192–214 (1978)
7. Lew, J.S., Rosenberg, A.L.: Polynomial indexing of integer lattice-points II. Nonexistence results for higher-degree polynomials. Journal of Number Theory **10**(2), 215–243 (1978)
8. Nathanson, M.B.: Cantor polynomials for semigroup sectors. Journal of Algebra and its Applications **13**(5) (2014)
9. Stanton, C.: Packing polynomials on sectors of $\mathbb{R}^2$. Integers **14** (2014)
10. Vsemirnov, M.A.: Two elementary proofs of the Fueter-Pólya theorem on pairing polynomials. Algebra i Analiz **13**(5), 1–15 (2001)

Representing Sequence Subsums as Sumsets of Near Equal Sized Sets

David J. Grynkiewicz

Abstract For a sequence S of terms from an abelian group G of length $|S|$, let $\Sigma_n(S)$ denote the set of all elements that can be represented as the sum of terms in some n-term subsequence of S. When the subsum set is very small, $|\Sigma_n(S)| \leq |S| - n + 1$, it is known that the terms of S can be partitioned into n nonempty sets $A_1, \ldots, A_n \subseteq G$ such that $\Sigma_n(S) = A_1 + \ldots + A_n$. Moreover, if the upper bound is strict, then $|A_i \setminus Z| \leq 1$ for all i, where $Z = \bigcap_{i=1}^{n}(A_i + H)$ and $H = \{g \in G : g + \Sigma_n(S) = \Sigma_n(S)\}$ is the stabilizer of $\Sigma_n(S)$. This allows structural results for sumsets to be used to study the subsum set $\Sigma_n(S)$ and is one of the two main ways to derive the natural subsum analog of Kneser's Theorem for sumsets. In this paper, we show that such a partitioning can be achieved with sets A_i of as near equal a size as possible, so $\lfloor \frac{|S|}{n} \rfloor \leq |A_i| \leq \lceil \frac{|S|}{n} \rceil$ for all i, apart from one highly structured counterexample when $|\Sigma_n(S)| = |S| - n + 1$ with $n = 2$. The added information of knowing the sets A_i are of near equal size can be of use when applying the aforementioned partitioning result, or when applying sumset results to study $\Sigma_n(S)$ (e.g., [20]). We also give an extension increasing the flexibility of the aforementioned partitioning result and prove some stronger results when $n \geq \frac{1}{2}|S|$ is very large.

Keywords Sumset · Subsum · Subsequence sum · Kneser's Theorem · Set partition

2020 Mathematics Subject Classification. 11B75

1 Introduction

Basic Notation

Let G be an abelian group. Following standard conventions in Combinatorial Number Theory (see [21, 22, 37]), by a sequence S of terms from G, we mean a finite,

---

D. J. Grynkiewicz (✉)
University of Memphis, Memphis, TN 38152, USA
e-mail: diambri@hotmail.com

M. B. Nathanson (ed.), *Combinatorial and Additive Number Theory IV*,
Springer Proceedings in Mathematics & Statistics 347,
https://doi.org/10.1007/978-3-030-67996-5_11

unordered string of elements

$$S = g_1 \cdot \ldots \cdot g_\ell$$

with $g_i \in G$ the terms of the sequence S, each term separated via the boldsymbol $\boldsymbol{\cdot}$ (differentiating it from multiplication in circumstances where both operations are in use). Formally, a sequence is considered as an element of the free abelian monoid $\mathcal{F}(G)$ with basis G and operation $\boldsymbol{\cdot}$, giving a standardized system of notation for sequences. Given an element $g \in G$, we let $\mathsf{v}_g(S) \geq 0$ denote the number of occurrences of the term g in S and let $g^{[n]}$ represent the sequence consisting of the element g repeated n times, so that any sequence $S \in \mathcal{F}(G)$ has the form

$$S = \prod_{g\in G}^{\bullet} g^{[\mathsf{v}_g(S)]}.$$

We let $T \mid S$ denote that T is a subsequence of S, so $\mathsf{v}_g(T) \leq \mathsf{v}_g(S)$ for all $g \in G$, and in such case use $T^{[-1]} \boldsymbol{\cdot} S$ or $S \boldsymbol{\cdot} T^{[-1]}$ to denote the sequence obtained by removing from S the terms in T, so $\mathsf{v}_g(T^{[-1]} \boldsymbol{\cdot} S) = \mathsf{v}_g(S) - \mathsf{v}_g(T)$. The support of the sequence S is the set of all elements occurring in S:

$$\mathrm{Supp}(S) = \{g \in G : \ \mathsf{v}_g(S) > 0\}.$$

For a subset $X \subseteq G$, let $S_X \mid S$ denote the subsequence of S consisting of all terms from X, so

$$S_X = \prod_{g\in X}^{\bullet} g^{[\mathsf{v}_g(S)]}.$$

Then $|S| = \ell$ is the length of the sequence,

$$\mathsf{h}(S) = \max\{\mathsf{v}_g(S) : \ g \in G\}$$

is the maximum multiplicity of a term in S, $\sigma(S) = g_1 + \ldots + g_n$ is the sum of S, and

$$\Sigma_n(S) = \{\sigma(T) : \ T \mid S, \ |T| = n\} \subseteq G$$

is the set of n-term subsums of S, for $n \geq 0$.

All intervals are discrete, so $[m, n] = \{x \in \mathbb{Z} : \ m \leq x \leq n\}$. Given subsets $A_1, \ldots, A_n \subseteq G$, their sumset is defined as

$$A_1 + \ldots + A_n = \{a_1 + \ldots + a_n : \ a_i \in A_i\}.$$

The stabilizer of a set $A \subseteq G$ is the subgroup $\mathsf{H}(A) = \{g \in G : \ g + A = A\} \leq G$, which is the largest subgroup H such that A is a union of H-cosets. If $\mathsf{H}(A)$ is trivial, then A is *aperiodic*, and otherwise A is *periodic*. We say that A is *H-periodic* if A is a union of H-cosets, equivalently, if $H \leq \mathsf{H}(A)$. For $x \in G$ and $A, \ B \subseteq G$, we let $\mathsf{r}_{A+B}(x) = |(A - x) \cap B| = |\{(a, b) \in A \times B : \ a + b = x\}|$ denote the number of representations for x as an element of $A + B$, and call $x \in A + B$ a unique expression element when $\mathsf{r}_{A+B}(x) = 1$. For $H \le G$, we let $\phi_H : G \to G/H$ denote the natural homomorphism.

Background

The study of sequence subsums is a classical topic in Combinatorial Number Theory. Often, it is desired that either $0 \in \Sigma_n(A)$, or $|\Sigma_n(S)|$ is large, or $\Sigma_n(S) = G$, and either conditions that guarantee the appropriate outcome, or the structure of sequences failing to satisfy the desired outcome, are sought. The Erdős-Ginzburg-Ziv Theorem [10, 37, 45] and Davenport Constant [22, 37, 43] are two such examples of very well-studied problems along these lines. A selection of other examples may be found here [1, 16, 24, 47–49].

One effective tool for studying $\Sigma_n(S)$, e.g., employed in the original proof of the Erdős-Ginzburg-Ziv Theorem [10], is via setpartitions. Consider a sequence $\mathcal{A} = A_1 \cdot \ldots \cdot A_n$ whose terms A_i are *nonempty* (and finite) subsets of G. We call such a sequence a *setpartition* over G. Note the setpartition $\mathcal{A}$ naturally partitions the terms in its underlying sequence

$$\mathsf{S}(\mathcal{A}) := \prod_{i\in[1,n]}^{\bullet} \prod_{g\in A_i}^{\bullet} g$$

into n nonempty sets. It is then rather immediate that $\sum_{i=1}^n A_i \subseteq \Sigma_n(S)$ when $\mathsf{S}(\mathcal{A}) \mid S$, which allows sumset results to be used for studying $\Sigma_n(S)$. This becomes even more effective if we know there is some setpartition $\mathcal{A} = A_1 \cdot \ldots \cdot A_n$ with $\mathsf{S}(\mathcal{A}) \mid S$ such that equality holds, $\sum_{i=1}^n A_i = \Sigma_n(S)$, for this means the subsums $\Sigma_n(A)$ can be represented as an ordinary sumset, and sumset results directly applied. The more structure that is known for the A_i, the easier and more effective it is to apply the corresponding sumset results. While this cannot hold for a general sequence, we have the striking fact that this is always possible so long as $|\Sigma_n(S)|$ is sufficiently small [26, 37, Theorem 14.1].

Theorem 1 (Partition Theorem) *Let G be an abelian group, let $n \ge 1$, let $S \in \mathcal{F}(G)$ be a sequence of terms from G, and suppose $S' \mid S$ is a subsequence with $\mathsf{h}(S') \le n \le |S'|$. Then there exists a setpartition $\mathcal{A} = A_1 \cdot \ldots \cdot A_n$ with $\mathsf{S}(\mathcal{A}) \mid S$ and $|\mathsf{S}(\mathcal{A})| = |S'|$ such that either*

1. *$|\Sigma_n(S)| \ge |\sum_{i=1}^n A_i| \ge |S'| - n + 1$, or*
2. *$\Sigma_n(S) = \sum_{i=1}^n A_i$, $\mathrm{Supp}(\mathsf{S}(\mathcal{A})^{[-1]} \cdot S) \subseteq Z$ and $|A_i \setminus Z| \le 1$ for all i, where $Z = \bigcap_{i=1}^n (A_i + H)$ and $H = \mathsf{H}(\Sigma_n(S))$.*

□

Theorem 1 ensures that $\Sigma_n(S) = \sum_{i=1}^n A_i$, for some setpartition $\mathcal{A} = A_1 \cdot \ldots \cdot A_n$ with $\mathsf{S}(\mathcal{A}) \mid S$ and $|\mathsf{S}(\mathcal{A})| = |S'|$, provided $|\Sigma_n(S)| \le |S'| - n + 1$, with additional structural information holding when the upper bound is strict. Worth noting,

Theorem 1 can always be applied (so long as $|S| \geq n$) with S' taken to be the maximal subsequence of S with $\mathsf{h}(S') \leq n$. In case Theorem 1.2 holds, this allows us to apply Kneser's Theorem [22, 37, 42, 45] to derive yet more information regarding $\Sigma_n(S)$, which is often incorporated into the statement of Theorem 1 itself (e.g. [37, 38, Theorem 14.1]).

Theorem 2 (Kneser's Theorem) *Let $A_1, \ldots, A_n \subseteq G$ be finite, nonempty subsets with G an abelian group, and let $H = \mathsf{H}(\sum_{i=1}^n A_i)$. Then*

$$|\sum_{i=1}^n A_i| \geq \Big(\sum_{i=1}^n |\phi_H(A_i)| - n + 1\Big)|H| = \sum_{i=1}^n |A_i + H| - (n-1)|H|.$$

Kneser's Theorem is the fundamental lower bound for sumsets in an abelian group. Combining it with Theorem 1 (applied modulo H) yields the analogous result for sequence subsums [38].

Theorem 3 (Subsum Kneser's Theorem) *Let G be an abelian group, let $n \geq 1$, let $S \in \mathcal{F}(G)$ be a sequence of terms from G with $|S| \geq n$, and let $H = \mathsf{H}(\Sigma_n(S))$. Then*

$$|\Sigma_n(S)| \geq \Big(|\phi_H(S')| - n + 1\Big)|H|,$$

where $S' \mid S$ is a maximum length subsequence with $\mathsf{h}(\phi_H(S')) \leq n$. □

Note $|\phi_H(S')| = \sum_{g \in G/H} \max\{n, \mathsf{v}_g(\phi_H(S))\}$. The Subsum Kneser's Theorem can alternatively be derived as a special case of the DeVos-Goddyn-Mohar Theorem [9, 37]. Theorem 3, and the more general Theorem 1, have found numerous use in problems regarding sequence subsums [5, 15, 18, 19, 23, 25, 27–34, 36], extending, complementing or resolving questions of established interest [2–4, 6–8, 10–14, 17, 39–41, 44, 46, 50].

In this paper, we will further strengthen Theorem 1. Theorem 4 applies to the more general object $X + \Sigma_n(S)$ rather than $\Sigma_n(S)$ (which is the case $X = \{0\}$), showing that Theorem 1 holds even if a fixed portion is "frozen" in the set X. For instance, if $X \subseteq \Sigma_m(T)$, then $X + \Sigma_n(S) \subseteq \Sigma_{m+n}(T \cdot S)$, and we obtain the conclusion of Theorem 1 under the restriction of only being able to repartition the terms from S. Theorem 4 also shows that, apart from one highly structured counter-example characterized in Theorem 1.3, the resulting setpartition $\mathcal{A} = A_1 \cdot \ldots \cdot A_n$ can be chosen such that the sizes of the sets A_i are as near equal as possible, i.e., with $||A_i| - |A_j|| \leq 1$ for all $i, j \in [1, n]$ (equivalently, $\lfloor \frac{|\mathsf{S}(\mathcal{A})|}{n} \rfloor \leq |A_i| \leq \lceil \frac{|\mathsf{S}(\mathcal{A})|}{n} \rceil$ for all i). We call such a setpartition *equitable*. While such improvements are not needed for every application of Theorem 1, they can simplify technical issues related to the use of Theorem 1, sometimes in an essential fashion. For example, the results of this paper (Sects. 2 and 3) are needed to prove the main result in the forthcoming paper [20] dealing with refined properties of product-one sequences over a dihedral group.

Theorem 4 *Let G be an abelian group, let $n \geq 1$, let $X \subseteq G$ be a finite, nonempty set, let $L \leq \mathsf{H}(X)$, let $S \in \mathcal{F}(G)$ be a sequence of terms from G, and suppose $S' \mid S$ is a subsequence with $\mathsf{h}(\phi_L(S')) \leq n \leq |S'|$. Then there is a setpartition $\mathcal{A} = A_1 \cdot \ldots \cdot A_n$ with $\mathsf{S}(\mathcal{A}) \mid S$, $|\mathsf{S}(\mathcal{A})| = |S'|$ and $|\phi_L(A_i)| = |A_i|$ for all $i \in [1, n]$ such that one of the following holds.*

1. *$|X + \Sigma_n(S)| \geq |X + \sum_{i=1}^n A_i| \geq (|S'| - n)|L| + |X|$ and $\mathcal{A}$ is equitable.*
2. *$X + \Sigma_n(S) = X + \sum_{i=1}^n A_i$, $\mathcal{A}$ is equitable, $\mathrm{Supp}(\mathsf{S}(\mathcal{A})^{[-1]} \cdot S) \subseteq Z$ and $|A_i \setminus Z| \leq 1$ for all i, where $Z = \bigcap_{i=1}^n (A_i + H)$ and $H = \mathsf{H}(X + \Sigma_n(S))$.*
3. *$n = 2$, $\mathrm{Supp}(\mathsf{S}(\mathcal{A})^{[-1]} \cdot S) \subseteq (A_1 + L) \cap (A_2 + L)$, $X \setminus (\beta + L)$ and $(A_1 + L) \cap (A_2 + L)$ are K-periodic, $\mathsf{H}(X + \Sigma_n(S)) = \mathsf{H}(X + \sum_{i=1}^n A_i) = \mathsf{H}(X) = L$, $\big((A_1 + L) \cup (A_2 + L)\big) \setminus \big((A_1 + L) \cap (A_2 + L)\big)$ is a K-coset, and $|X + \Sigma_n(S)| = (|S'| - n)|L| + |X|$, for some $\beta \in X$ and $K \leq G$ with $L \leq K$ and $K/L \cong (\mathbb{Z}/2\mathbb{Z})^2$.*

In Sect. 3, we will also derive some additional strengthenings of Theorem 4 in the case $n \geq \frac{1}{2}|S'|$ is very large. In particular, we will achieve the same strengthened conclusions recently guaranteed in [38] under a different n is large assumption (Theorem 14). This, in turn, will allow us to derive additional information for S, in particular, when $|S| = 2n$ with $|\Sigma_n(S)| \leq n + 1$ and $\mathsf{h}(S) \leq n$ (Theorem 15).

2 Partitioning Results for General *n*

In this section, we will make heavy use of the arguments used to prove [37, Theorem 14.1] and the following easy consequence of Kneser's Theorem (see [37, Theorem 5.1]).

Theorem 5 *Let G be an abelian group, and let $A, B \subseteq G$ be finite, nonempty subsets. If $|A + B| < |A| + |B| - 1$, then $A + (B \setminus \{x\}) = A + B$ for all $x \in B$.* □

We will also need the following observation, that follows by a routine induction on n.

Lemma 6 *Let G be an abelian group and let $A_1, \ldots, A_n \subseteq G$ be finite, nonempty subsets. Suppose $|\sum_{i=1}^j A_i| \geq |\sum_{i=1}^{j-1} A_i| + |A_j| - 1$ for all $j \in [2, n]$. Then $|\sum_{i=1}^n A_i| \geq \sum_{i=1}^n |A_i| - n + 1$. Moreover, if $|\sum_{i=1}^n A_i| = \sum_{i=1}^n |A_i| - n + 1$, then $|\sum_{i=1}^j A_i| = |\sum_{i=1}^{j-1} A_i| + |A_j| - 1$ for all $j \in [2, n]$.*

Let G be an abelian group, let $X \subseteq G$ be a nonempty subset and let $S \in \mathcal{F}(G)$ be a sequence. A setpartition $\mathcal{A} = A_1 \cdot \ldots \cdot A_n$ with $\mathsf{S}(\mathcal{A}) \mid S$ will be called *maximal relative to* X if any setpartition $\mathcal{B} = B_1 \cdot \ldots \cdot B_n$ with $\mathsf{S}(\mathcal{B}) \mid S$, $|\mathsf{S}(\mathcal{B})| = |\mathsf{S}(\mathcal{A})|$ and $X + \sum_{i=1}^n A_i \subseteq X + \sum_{i=1}^n B_i$ has $X + \sum_{i=1}^n A_i = X + \sum_{i=1}^n B_i$. We simply say $\mathcal{A}$ is maximal relative to X if this is the case with $S = \mathsf{S}(\mathcal{A})$.

Lemma 7 *Let G be an abelian group, let $n \geq 1$, let $X \subseteq G$ be a finite, nonempty subset, let $S \in \mathcal{F}(G)$ be a sequence, let $\mathcal{A} = A_1 \cdot \ldots \cdot A_n$ be a setpartition with $\mathsf{S}(\mathcal{A}) \mid S$ maximal relative to X, and let $H = \mathsf{H}(X + \sum_{i=1}^n A_i)$. Suppose*

$$|X + \sum_{i=1}^{n} A_i| < |X| + \sum_{i=1}^{n} |A_i| - n. \tag{1}$$

Then there exists a setpartition $\mathcal{B} = B_1 \cdot \ldots \cdot B_n$ with $\mathsf{S}(\mathcal{B}) \mid S$, $|\mathsf{S}(\mathcal{B})| = |\mathsf{S}(\mathcal{A})|$ and $X + \sum_{i=1}^n B_i = X + \sum_{i=1}^n A_i$ such that $\mathrm{Supp}(\mathsf{S}(\mathcal{B})^{[-1]} \cdot S) \subseteq Z$ and $|(y + H) \cap B_i| \leq 1$ for all $y \in G \setminus Z$ and $i \in [1, n]$, where $Z = \bigcap_{i=1}^n (B_i + H)$.

Proof In view of (1) and Kneser's Theorem, we conclude that H is nontrivial. Consider an arbitrary setpartition $\mathcal{B} = B_1 \cdot \ldots \cdot B_n$ with $\mathsf{S}(\mathcal{B}) \mid S$, $|\mathsf{S}(\mathcal{B})| = |\mathsf{S}(\mathcal{A})|$ and $X + \sum_{i=1}^n A_i \subseteq X + \sum_{i=1}^n B_i$. Then $X + \sum_{i=1}^n B_i = X + \sum_{i=1}^n A_i$ since $\mathcal{S}(\mathcal{A}) \mid S$ with $\mathcal{A}$ maximal relative to X. Let $Z = \bigcap_{i=1}^n (B_i + H)$. In view of (1) and Lemma 6, there must be some $j \in [1, n]$ such that $|X + \sum_{i=1}^j B_i| < |X + \sum_{i=1}^{j-1} B_i| + |B_j| - 1$, in which case Theorem 5 implies

$$X + \sum_{i=1}^{j-1} B_i + (B_j \setminus \{x\}) = X + \sum_{i=1}^{j} B_i \quad \text{for all } x \in B_j. \tag{2}$$

Since $|X + \sum_{i=1}^j B_i| < |X + \sum_{i=1}^{j-1} B_i| + |B_j| - 1$, Kneser's Theorem implies that $|(x + H') \cap B_j| \geq 2$ for every $x \in B_j$, where $H' = \mathsf{H}(X + \sum_{i=1}^j B_i) \leq H$. In particular,

$$|(x + H) \cap B_j| \geq 2 \quad \text{for all } x \in B_j. \tag{3}$$

Now further restrict $\mathcal{B}$ by assuming $\sum_{i=1}^n |\phi_H(B_i)|$ is maximal (subject to the defining condition for $\mathcal{B}$). Then we must have $B_j \subseteq Z$, where $j \in [1, n]$ is the index defined above. Indeed, if this fails, then there is some $x \in B_j \setminus Z$, and thus also some $k \in [1, n]$ with $\phi_H(x) \notin \phi_H(B_k)$ by definition of Z. We can then remove x from B_j and place it in B_k to yield a new setpartition $\mathcal{B}' = B_1' \cdot \ldots \cdot B_n'$, where $B_j' = B_j \setminus \{x\}$, $B_k' = B_k \cup \{x\}$ and $B_i' = B_i$ for $i \neq j, k$. In view of (2), we have $X + \sum_{i=1}^n A_i = X + \sum_{i=1}^n B_i \subseteq X + \sum_{i=1}^n B_i'$, while in view of (3) and $\phi_H(x) \notin \phi_H(B_k)$, we have $\sum_{i=1}^n |\phi_H(B_i')| = \sum_{i=1}^n |\phi_H(B_i)| + 1$, and now $\mathcal{B}'$ contradicts the maximality of $\sum_{i=1}^n |\phi_H(B_i)|$ for $\mathcal{B}$. Therefore

$$B_j \subseteq Z. \tag{4}$$

Claim A: $|(y + H) \cap B_i| \leq 1$ for all $y \in G \setminus Z$ and $i \in [1, n]$.

Proof Assume by contradiction there is some $k \in [1, n]$ and $y \in B_k \setminus Z$ with $|(y + H) \cap B_k| \geq 2$. Let $\mathcal{C} = C_1 \cdot \ldots \cdot C_n$ be a setpartition with $\mathsf{S}(\mathcal{C}) = \mathsf{S}(\mathcal{B})$, $X + \sum_{i=1}^n C_i = X + \sum_{i=1}^n A_i$, and $C_i \setminus Z = B_i \setminus Z$ and $\phi_H(C_i) = \phi_H(B_i)$ for all i, such that $|C_k|$ is maximal. Since $\phi_H(C_i) = \phi_H(B_i)$ for all i, we still have

$\sum_{i=1}^{n} |\phi_H(C_i)|$ maximal, while $Z = \bigcap_{i=1}^{n}(B_i + H) = \bigcap_{i=1}^{n}(C_i + H)$. Thus, let $j' \in [1, n]$ be an index so that $C_{j'}$ satisfies (2), (3) and (4) for $\mathcal{C}$ (in place of B_j).

Suppose $C_{j'} \subseteq C_k$. Since $y \notin Z$ but $C_{j'} \subseteq Z$ (by (4)), we actually have $C_{j'} \subseteq C_k \setminus \{y\}$, in which case

$$C_{j'} + C_k \subseteq (C_{j'} \cup \{y\}) + (C_k \setminus \{y\}). \tag{5}$$

In such case, we can define a new setpartition $\mathcal{C}' = C'_1 \cdot \ldots \cdot C'_n$ by removing y from C_k and placing it in $C_{j'}$, so $C'_k = C_k \setminus \{y\}$, $C'_{j'} = C_{j'} \cup \{y\}$ and $C'_i = C_i$ for $i \neq k, j'$. Note $y \in B_k \setminus Z = C_k \setminus Z$. In view of (5), we have $X + \sum_{i=1}^{n} A_i = X + \sum_{i=1}^{n} C_i \subseteq X + \sum_{i=1}^{n} C'_i$, while in view of $|(y + H) \cap C_k| = |(y + H) \cap B_k| \geq 2$ (as $y \notin Z$ and $C_k \setminus Z = B_k \setminus Z$) and $\phi_H(y) \notin \phi_H(C_{j'})$ (as $y \notin Z$ and $C_{j'} \subseteq Z$ by (4)), we have $\sum_{i=1}^{n} |\phi_H(C'_i)| = \sum_{i=1}^{n} |\phi_H(C_i)| + 1 = \sum_{i=1}^{n} |\phi_H(B_i)| + 1$, so that $\mathcal{C}'$ contradicts the maximality of $\sum_{i=1}^{n} |\phi_H(B_i)|$ for $\mathcal{B}$. So we instead conclude that $C_{j'} \nsubseteq C_k$. Thus, in view (4), it follows that there is some $x \in C_{j'} \subseteq Z$ with $x \notin C_k$.

In this case, we define a new setpartition $\mathcal{C}' = C'_1 \cdot \ldots \cdot C'_n$ by removing x from $C_{j'}$ and placing it in C_k, so $C'_{j'} = C_{j'} \setminus \{x\}$, $C'_k = C_k \cup \{x\}$ and $C'_i = C_i$ for $i \neq j', k$. In view of (2), we have $X + \sum_{i=1}^{n} A_i = X + \sum_{i=1}^{n} C_i \subseteq X + \sum_{i=1}^{n} C'_i$, while in view of (3) and $x \in Z$, we have $\phi_H(C'_i) = \phi_H(C_i) = \phi_H(B_i)$ and $C'_i \setminus Z = C_i \setminus Z = B_i \setminus Z$ for all i. Thus, since $|C'_k| = |C_k| + 1$, we see that $\mathcal{C}'$ contradicts the maximality of $|C_k|$ for $\mathcal{C}$, completing the claim. □

In view of Claim A, we see that the lemma holds with the setpartition $\mathcal{B}$ unless there is some $y \in \mathrm{Supp}(\mathsf{S}(\mathcal{B})^{[-1]} \cdot S)$ with $y \notin Z$. However, if this were the case, then $\phi_H(y) \notin \phi_H(B_j)$ in view of (4). Define a new setpartition $\mathcal{B}' = B'_1 \cdot \ldots \cdot B'_n$ by removing any term $x \in B_j$ from B_j and placing y into B_j instead, so $B'_j = B_j \setminus \{x\} \cup \{y\}$ and $B'_i = B_i$ for $i \neq j$. In view of (2), we have $X + \sum_{i=1}^{n} A_i = X + \sum_{i=1}^{n} B_i \subseteq X + \sum_{i=1}^{n} B'_i$, while in view of (3) and $\phi_H(y) \notin \phi_H(B_j)$, we have $\sum_{i=1}^{n} |\phi_H(B'_i)| = \sum_{i=1}^{n} |\phi_H(B_i)| + 1$, in which case $\mathcal{B}'$ contradicts the maximality of $\sum_{i=1}^{n} |\phi_H(B_i)|$ for $\mathcal{B}$, completing the proof. □

Lemma 8 *Let G be an abelian group, let $n \geq 1$, let $X \subseteq G$ be a finite, nonempty subset, let $\mathcal{A} = A_1 \cdot \ldots \cdot A_n$ be a setpartition over G maximal relative to X, let $H = \mathsf{H}(X + \sum_{i=1}^{n} A_i)$ and let $Z = \bigcap_{i=1}^{n}(A_i + H)$. Suppose $|(y + H) \cap A_i| \leq 1$ for all $y \in G \setminus Z$ and $i \in [1, n]$, and*

$$|X + \sum_{i=1}^{n} A_i| < |X| + \sum_{i=1}^{n} |A_i| - n + (|H| - 1). \tag{6}$$

Then there exists a setpartition $\mathcal{B} = B_1 \cdot \ldots \cdot B_n$ with $\mathsf{S}(\mathcal{B}) = \mathsf{S}(\mathcal{A})$, $X + \sum_{i=1}^{n} B_i = X + \sum_{i=1}^{n} A_i$ and $Z \subseteq \bigcap_{i=1}^{n}(B_i + H)$ such that $|B_i \setminus Z| \leq 1$ for all i.

Proof In view of (6) and Kneser's Theorem, we conclude that H is nontrivial. Consider a setpartition $\mathcal{B} = B_1 \cdot \ldots \cdot B_n$ with

$$\mathsf{S}(\mathcal{B}) = \mathsf{S}(\mathcal{A}), \quad X + \sum_{i=1}^{n} B_i = X + \sum_{i=1}^{n} A_i, \quad Z = \bigcap_{i=1}^{n}(A_i + H) \subseteq \bigcap_{i=1}^{n}(B_i + H),$$

$$\text{and} \quad |(y+H) \cap B_i| \leq 1 \quad \text{for all } y \in G \setminus Z \text{ and } i \in [1, n]. \tag{7}$$

Since $\mathcal{A}$ satisfies these conditions, it follows that such a setpartition $\mathcal{B}$ exists. Let $e = \sum_{i=1}^{n} |B_i \setminus Z| \geq 0$. Let $I_e \subseteq [1, n]$ be all those indices $i \in [1, n]$ with $B_i \setminus Z$ nonempty, and let $I_Z \subseteq [1, n]$ be all those indices $i \in [1, n]$ with $B_i \subseteq Z$. By re-indexing the B_i, we can w.l.o.g. assume $I_Z = [1, m]$ and $I_e = [m+1, n]$.

Suppose $|X + \sum_{i=1}^{j} B_i| \geq |X + \sum_{i=1}^{j-1} B_i| + |B_j| - 1$ for all $j \in [1, m]$. Then Lemma 6 implies $|X + \sum_{i=1}^{m} B_i| \geq |X| + \sum_{i=1}^{m} |B_i| - m$. Kneser's Theorem implies

$$\begin{aligned} \Big|\Big(X + \sum_{i=1}^{m} B_i\Big) + \sum_{i=m+1}^{n} B_i\Big| &\geq |X + \sum_{i=1}^{m} B_i| + \sum_{i=m+1}^{n} |B_i + H| - (n-m)|H| \quad (8)\\ &\geq |X + \sum_{i=1}^{m} B_i| + \sum_{i=m+1}^{n} |B_i| + e(|H|-1) - (n-m)|H|, \end{aligned}$$

with the second inequality in view of the final condition in (7). Combined with the previous estimate for $|X + \sum_{i=1}^{m} B_i|$, we find

$$|X + \sum_{i=1}^{n} B_i| \geq |X| + \sum_{i=1}^{n} |B_i| - n|H| + (m+e)(|H|-1). \tag{9}$$

Note that $e \geq |I_e| = n - m$. If equality holds, then $|B_i \setminus Z| \leq 1$ follows for all i, completing the proof. Therefore we can instead assume $e \geq n - m + 1$, which combined with (9) yields $|X + \sum_{i=1}^{n} A_i| = |X + \sum_{i=1}^{n} B_i| \geq |X| + \sum_{i=1}^{n} |B_i| - n + (|H| - 1) = |X| + \sum_{i=1}^{n} |A_i| - n + (|H| - 1)$, contrary to hypothesis. So we may instead assume there is some $j \in [1, m]$ such that $|X + \sum_{i=1}^{j} B_i| < |X + \sum_{i=1}^{j-1} B_i| + |B_j| - 1$. In particular, this argument shows that I_Z is nonempty for any setpartition satisfying (7), and Theorem 5 ensures that

$$X + \sum_{i=1}^{j-1} B_i + (B_j \setminus \{x\}) = X + \sum_{i=1}^{j} B_i \quad \text{for all } x \in B_j. \tag{10}$$

In particular, $|B_j| \geq 2$. Let $K = \mathsf{H}(X + \sum_{i=1}^{j} B_i) \leq H$. If $|(y+H) \cap B_j| = 1$ for some $y \in G$, then $|(y+K) \cap B_j| = 1$ as well, whence Kneser's Theorem implies $|(X + \sum_{i=1}^{j-1} B_i) + B_j| \geq |X + \sum_{i=1}^{j-1} B_i| + |B_j + K| - |K| \geq |X + \sum_{i=1}^{j-1} B_i| + |B_j| - 1$, contrary to the definition of j. Therefore we instead conclude that

$$|(y+H) \cap B_j| \geq 2 \quad \text{for all } y \in B_j. \tag{11}$$

Now assume our setpartition $\mathcal{B}$ satisfying (7) is chosen such that

M1. $|I_e|$ is maximal (subject to (7)),
M2. $\sum_{i\in I_e}|B_i|$ is maximal (subject to (7) and M1).

If $|B_i \setminus Z| = 1$ for every $i \in I_e = [m+1, n]$, then the setpartition $\mathcal{B}$ satisfies the conditions of the lemma. Therefore we may assume there is some $k \in I_e = [m+1, n]$ with distinct $y_1, y_2 \in B_k \setminus Z$. Since $j \in [1, m] = I_Z$, we have $B_j \subseteq Z$. Thus $y_1, y_2 \notin B_j$.

Suppose $B_s \subseteq B_k$ for some $s \in I_Z$. Then $B_s \subseteq B_k \setminus \{y_1\}$ and $B_s + B_k \subseteq (B_s \cup \{y_1\}) + (B_k \setminus \{y_1\})$, the former as $y_1 \notin Z$ but $B_s \subseteq Z$ as $s \in I_Z$. In such case, define a new setpartition $\mathcal{B}' = B_1' \cdot \ldots \cdot B_n'$ by setting $B_s' = B_s \cup \{y_1\}$, $B_k' = B_k \setminus \{y_1\}$ and $B_i' = B_i$ for $i \neq s, k$. Then $B_s + B_k \subseteq (B_s \cup \{y_1\}) + (B_k \setminus \{y_1\}) = B_s' + B_k'$ ensures that $X + \sum_{i=1}^n A_i = X + \sum_{i=1}^n B_i \subseteq X + \sum_{i=1}^n B_i'$, and equality must hold as $\mathcal{A}$ is maximal relative to X. By definition, $\mathsf{S}(\mathcal{B}') = \mathsf{S}(\mathcal{B}) = \mathsf{S}(\mathcal{A})$. Since $B_s \subseteq Z$, we have still have $|(y+H) \cap B_i'| \leq 1$ for all i and $y \in G \setminus Z$, while $Z \subseteq \bigcap_{i=1}^n B_i'$ follows since $y_1 \notin Z$. Thus $\mathcal{B}'$ satisfies (7). Since $y_1 \notin Z$ and $y_2 \in B_k' \setminus Z$, we see $I_e \cup \{j\} \subseteq [1, n]$ is the subset of indices $i \in [1, n]$ for which $B_i' \setminus Z$ is nonempty, meaning $\mathcal{B}'$ contradicts the maximality condition M1 for $\mathcal{B}$. So we instead assume $B_s \not\subseteq B_k$ for all $s \in I_Z$. In particular, $B_j \not\subseteq B_k$, meaning there is some $x \in B_j \setminus B_k$.

In this case, define a new setpartition $\mathcal{B}' = B_1' \cdot \ldots \cdot B_n'$ by setting $B_j' = B_j \setminus \{x\}$, $B_k' = B_k \cup \{x\}$ and $B_i' = B_i$ for $i \neq j, k$. In view of (10), we have $X + \sum_{i=1}^n A_i = X + \sum_{i=1}^n B_i \subseteq X + \sum_{i=1}^n B_i'$, and equality must hold as $\mathcal{A}$ is maximal relative to X. By definition, $\mathsf{S}(\mathcal{B}') = \mathsf{S}(\mathcal{B}) = \mathsf{S}(\mathcal{A})$. Since $x \in B_j \subseteq Z$, we have still have $|(y+H) \cap B_i'| \leq 1$ for all i and $y \in G \setminus Z$, while $Z \subseteq \bigcap_{i=1}^n B_i'$ follows in view of (11). Thus $\mathcal{B}'$ satisfies (7). Since $x \in B_j \subseteq Z$, we see $I_e \subseteq [1, n]$ is still the subset of indices $i \in [1, n]$ for which $B_i' \setminus Z$ is nonempty, meaning $\mathcal{B}'$ satisfies M1. However, since $|B_k' = |B_k| + 1$, $k \in I_e$ and $j \notin I_e$, the maximality of $\sum_{i\in I_e}|B_i|$ for $\mathcal{B}$ is contradicted by $\mathcal{B}'$. □

Lemma 9 *Let G be an abelian group, let $n \geq 0$, let $X \subseteq G$ be a finite, nonempty subset, let $S \in \mathcal{F}(G)$ be a sequence, and let $\mathcal{A} = A_1 \cdot \ldots \cdot A_n$ be a setpartition with $\mathsf{S}(\mathcal{A}) \mid S$, $\mathrm{Supp}(\mathsf{S}(\mathcal{A})^{[-1]} \cdot S) \subseteq Z$, and $|A_i \setminus Z| \leq 1$ for all i, where $Z = Z + H \subseteq \bigcap_{i=1}^n (A_i + H)$ and $H \leq \mathsf{H}(X + \sum_{i=1}^n A_i)$. Then the following hold.*

1. $X + \Sigma_n(S) = X + \sum_{i=1}^n A_i$.
2. *If $Z = g + H$ for some $g \in G$, then $X + \Sigma_\ell(S) = X + \sum_{i=1}^n A_i + (\ell - n)g$ for any $\ell \in [n, n + |\mathsf{S}(\mathcal{A})^{[-1]} \cdot S|]$.*

Proof 1. Note $X + \sum_{i=1}^n A_i \subseteq X + \Sigma_n(S)$ holds trivially. Let $T \mid S$ with $|T| = n$ be arbitrary. Since $X + \sum_{i=1}^n A_i$ is H-periodic by hypothesis, to establish the reverse inclusion, it suffices to show $\sigma(\phi_H(T)) \in \sum_{i=1}^n \phi_H(A_i)$. Write $T = x_1 \cdot \ldots \cdot x_s \cdot y_{s+1} \cdot \ldots \cdot y_n$, with the x_i the terms of T with $x_i \in G \setminus Z$, and the y_i the terms of T with $y_i \in Z$. In view of $\mathrm{Supp}(\mathsf{S}(\mathcal{A})^{[-1]} \cdot S) \subseteq Z$ and $|A_i \setminus Z| \leq 1$ for all i, we can re-index the A_i so that $x_i \in A_i$ for $i \in [1, s]$. But now, since $y_j \in Z = Z + H \subseteq \bigcap_{i=1}^n (A_i + H) \subseteq A_j + H$ for all $j \geq s+1$, it follows that $\sigma(\phi_H(T)) = \phi_H(x_1) + \ldots + \phi_H(x_s) + \phi_H(y_{s+1}) + \ldots + \phi_H(y_n) \in \sum_{i=1}^n \phi_H(A_i)$, completing Item 1.

2. By translating all terms of S appropriately by $-g$, we can w.l.o.g. assume $g = 0$, whence $H \subseteq \bigcap_{i=1}^n (A_i + H)$. In particular, $H \cap A_i \neq \emptyset$ and $|A_i \setminus H| \leq 1$ for all i. Since $X + \sum_{i=1}^n A_i$ is H-periodic with $\mathsf{S}(\mathcal{A}) \mid S$, $\text{Supp}(\mathsf{S}(\mathcal{A})^{[-1]} \cdot S) \subseteq g + H = H$ and $n \leq \ell \leq n + |\mathsf{S}(\mathcal{A})^{[-1]} \cdot S|$, we trivially have $X + \sum_{i=1}^n A_i = X + \sum_{i=1}^n A_i + (\ell - n)g \subseteq X + \Sigma_\ell(S)$. To show the reverse inclusion, let $T = g_1 \cdot \ldots \cdot g_\ell$ be an arbitrary ℓ-term subsequence of S. Since $\text{Supp}(\mathsf{S}(\mathcal{A})^{[-1]} \cdot S) \subseteq g + H = H$ and $|A_i \setminus H| \leq 1$ and for all i, there are at most n non-zero terms in $\phi_H(T)$, and by re-indexing, we can w.l.o.g. assume $\phi_H(g_i) = 0$ for $i > n$. Then, since $H \cap A_i \neq \emptyset$, $\text{Supp}(\mathsf{S}(\mathcal{A})^{[-1]} \cdot S) \subseteq H$ and $|A_i \setminus H| \leq 1$ for all i, it follows that $\phi_H(g_1) + \ldots + \phi_H(g_\ell) = \phi_H(g_1) + \ldots + \phi_H(g_n) \in \sum_{i=1}^n \phi_H(A_i)$. Hence, since $X + \sum_{i=1}^n A_i$ is H-periodic, we conclude that $X + \sigma(T) = X + g_1 + \ldots + g_\ell \subseteq X + \sum_{i=1}^n A_i$. Since T was an arbitrary ℓ-term subsequence of S, this establishes the reverse inclusion $X + \Sigma_\ell(S) \subseteq X + \sum_{i=1}^n A_i$. □

Let G be an abelian group and $A \subseteq G$ a subset. We say A is *quasi-periodic* if there is a subset $A_\emptyset \subseteq A$ such that $A \setminus A_\emptyset$ is nonempty and periodic with $A_\emptyset$ contained in a $\mathsf{H}(A \setminus A_\emptyset)$-coset. If $H \leq G$ is a nontrivial subgroup, then an *H-quasi-periodic decomposition* is a partition $A = (A \setminus A_\emptyset) \cup A_\emptyset$ with $A_\emptyset$ a subset of an H-coset and $A \setminus A_\emptyset$ H-periodic (or empty). It is *reduced* if $A_\emptyset$ is not quasi-periodic. As is easily derived,

$$\mathsf{H}(A) = \mathsf{H}(A_\emptyset) \leq H \quad \text{when } \emptyset \neq A_\emptyset \subset A_\emptyset + H. \tag{12}$$

If $A_\emptyset$ is quasi-periodic, as exhibited by $A'_\emptyset \subseteq A_\emptyset$, then $A = (A \setminus A'_\emptyset) \cup A'_\emptyset$ is a quasi-periodic decomposition with $A'_\emptyset \subset A_\emptyset$. Every finite set $A \subseteq G$ has a reduced quasi-periodic decomposition, and this decomposition is unique unless $A \cup \{\alpha\}$ is periodic for some $\alpha \notin A$ (see [35, Proposition 2.1]). The Kemperman Structure Theorem [37, Theorem 9.1] implies that, if $A,\ B \subseteq G$ are finite, nonempty subsets with $|A + B| = |A| + |B| - 1$ and either $A + B$ aperiodic or containing a unique expression element, then there are H-quasi-periodic decompositions $A = (A \setminus A_\emptyset) \cup A_\emptyset$, $B = (B \setminus B_\emptyset) \cup B_\emptyset$ and $A + B = \Big((A + B) \setminus (A_\emptyset + B_\emptyset)\Big) \cup (A_\emptyset + B_\emptyset)$ with the pair $(A_\emptyset, B_\emptyset)$ satisfying one of four possible structural types (I)–(IV), each with explicitly defined restrictions on where, and how many, unique expression elements there are. We will make use of this theory, referencing the details regarding Kemperman's Critical Pair Theory rather than repeating the rather lengthy statements and details here.

Lemma 10 *Let G be an abelian group, let $H \leq G$ be a subgroup with $|H| \geq 3$, and let $Y \subseteq G$ be a finite subset such that $Y \setminus \{y_0\}$ is H-periodic (or empty) for some $y_0 \in Y$.*

1. *Y is aperiodic with $Y = (Y \setminus \{y_0\}) \cup \{y_0\}$ its unique reduced quasi-periodic decomposition.*
2. *If there is a K-quasi-periodic decomposition $Y = Y_1 \cup Y_0$, then $y_0 \in Y_0$ and $Y_0 \setminus \{y_0\}$ is H-periodic. Moreover, if $|Y_0| \geq 2$, then $H \leq K$.*
3. *If $A,\ B \subseteq G$ with $A + B = Y$ and $|A + B| = |A| + |B| - 1$, then there are $a_0 \in A$ and $b_0 \in B$ such that $A \setminus \{a_0\}$ and $B \setminus \{b_0\}$ are H-periodic with $a_0 + b_0 = y_0$.*

Proof We may w.l.o.g. assume $H = \mathsf{H}(Y \setminus \{y_0\})$. We may assume $Y \setminus \{y_0\}$ is nonempty, else Items 1–3 all hold trivially. Item 1 follows from [35, Proposition 2.1] and [35, Comment c.6].

2. Suppose $Y = Y_1 \cup Y_0$ is a K-quasi-periodic decomposition and let $Y_0 = (Y_0 \setminus Y_\emptyset) \cup Y_\emptyset$ be a reduced quasi-periodic decomposition of Y_0. Then $(Y \setminus Y_\emptyset) \cup Y_\emptyset$ is a reduced quasi-periodic decomposition of Y with either $\mathsf{H}(Y \setminus Y_\emptyset) = \mathsf{H}(Y_0 \setminus Y_\emptyset) \leq \langle Y_0 - Y_0 \rangle \leq K$ or $Y_0 = Y_\emptyset$ (by (12)). Hence Item 1 ensures that $Y_\emptyset = \{y_0\}$ with $H = \mathsf{H}(Y \setminus \{y_0\}) = \mathsf{H}(Y \setminus Y_\emptyset) \leq K$ if $|Y_0| \geq 2$. If $Y_0 = \{y_0\} = Y_\emptyset$, then $Y_1 = Y \setminus \{y_0\} = Y \setminus Y_\emptyset$ is H-periodic. Otherwise, $H \leq K$ ensures Y_1 is H-periodic, while $Y \setminus Y_\emptyset = Y \setminus \{y_0\}$ is H-periodic by hypothesis. In either case, Y_1 and $Y \setminus Y_\emptyset$ are both H-periodic, and it follows that $(Y \setminus Y_\emptyset) \setminus Y_1 = Y_0 \setminus Y_\emptyset = Y_0 \setminus \{y_0\}$ is also H-periodic. Item 2 now follows.

3. Suppose $A + B = Y$ and $|A + B| = |A| + |B| - 1$. Since we know $Y = A + B$ is aperiodic by Item 1, we can directly apply the Kemperman Structure Theorem [35, Proposition 2.1] to $A + B$ yielding associated K-quasi-periodic decompositions $A = (A \setminus A_\emptyset) \cup A_\emptyset$, $B = (B \setminus B_\emptyset) \cup B_\emptyset$ and $Y = (Y \setminus (A_\emptyset + B_\emptyset)) \cup (A_\emptyset + B_\emptyset)$. If the pair $(A_\emptyset, B_\emptyset)$ has type (IV), then $Y \cup \{\beta\}$ is periodic for some $\beta \in G \setminus Y$. In such case, any reduced quasi-periodic decomposition $Y = Y_1 \cup Y_0$ must have $|Y_0| \geq 2$ or $|\mathsf{H}(Y_1)| = 2$ (cf. [35, Proposition 2.1]), contrary to Item 1. If the pair $(A_\emptyset, B_\emptyset)$ has type (III), then Y is periodic, contrary to Item 1. If the pair $(A_\emptyset, B_\emptyset)$ has type (II), then either $Y \cup \{\beta\}$ is periodic for some $\beta \in G$, yielding the same contradiction as before, or else $Y = (Y \setminus (A_\emptyset + B_\emptyset)) \cup (A_\emptyset + B_\emptyset)$ is a reduced quasi-periodic decomposition of Y (by [35, Comment c.3]) with $|A_\emptyset + B_\emptyset| \geq 3$, again contrary to Item 1. We are left to conclude that $(A_\emptyset, B_\emptyset)$ has type (I), so w.l.o.g. $|A_\emptyset| = 1$, say with $A_\emptyset = \{a_0\}$. Hence $A + B = Y$ is a union of $a_0 + B$ with a K-periodic set. In particular, $(Y \setminus (a_0 + B_\emptyset)) \cup (a_0 + B_\emptyset)$ is a K-quasi-periodic decomposition of Y. If $|B_\emptyset| \geq 2$, then Item 2 yields $H \leq K$ with $y_0 \in a_0 + B_0$ and $(a_0 + B_\emptyset) \setminus \{y_0\}$ H-periodic. Letting $b_0 = y_0 - a_0 \in B_0$, Item 3 follows. Therefore instead assume $|B_\emptyset| = 1$, say $B_\emptyset = \{b_0\}$. Moreover, in view of [35, Proposition 2.2], we can assume $a_0 + b_0 \in A + B$ is the only unique expression element.

In this case $\Big((A + B) \setminus \{a_0 + b_0\}\Big) \cup \{a_0 + b_0\}$ is a reduced K-quasi-periodic decomposition, in which case Item 1 implies that $a_0 + b_0 = y_0$ and $(A + B) \setminus \{a_0 + b_0\} = Y \setminus \{y_0\}$. Since $(y_0 + H) \cap (A + B) = (y_0 + H) \cap Y = \{y_0\}$ with $y_0 = a_0 + b_0$, we must have $(a_0 + H) \cap A = \{a_0\}$ and $(b_0 + H) \cap B = \{b_0\}$. If $|A_0| = |B_0| = 1$, then Item 3 follows trivially. Therefore we can instead w.l.o.g. assume $|A_0| \geq 2$. Since $a_0 + b_0 = y_0 \in A + B = Y$ is the only unique expression, we have $(A \setminus \{a_0\}) + B = Y \setminus \{y_0\}$ with $\mathsf{H}((A \setminus \{a_0\}) + B) = \mathsf{H}(Y \setminus \{y_0\}) = H$. Kneser's Theorem now implies

$$\begin{aligned} |A| + |B| - 2 &= |(A \setminus \{a_0\}) + B| \geq |(A \setminus \{a_0\}) + H| + |B + H| - |H| \\ &\geq |A \setminus \{a_0\}| + |B| - 1, \end{aligned} \tag{13}$$

with the latter inequality in view of $(b_0 + H) \cap B = \{b_0\}$. Thus we must have equality in (13). In particular, equality holding in the second inequality in (13) forces $(A \setminus \{a_0\}) + H = A \setminus \{a_0\}$ and $(B \setminus \{b_0\}) + H = B \setminus \{b_0\}$, i.e., $A \setminus \{a_0\}$ and $B \setminus \{b_0\}$ are H-periodic, completing Item 3. □

Lemma 11 *Let G be an abelian group and let $A,\ B \subseteq G$ be finite, nonempty subsets with $H = \mathsf{H}(A + B) = \mathsf{H}(B)$. Then it follows that $|(A \cup \{x\}) + B| \geq |(A \cup \{x\}) + H| + |B + H| - |H|$ for any $x \in G$.*

Proof By hypothesis, $\phi_H(A) + \phi_H(B)$ is aperiodic. Thus $(\phi_H(A) \cup \{\phi_H(x)\}) + \phi_H(B)$ is either still aperiodic, in which case Kneser's Theorem implies $|\phi_H(A \cup \{x\}) + \phi_H(B)| \geq |\phi_H(A \cup \{x\})| + |\phi_H(B)| - 1$, or else it is periodic, and thus strictly contains the aperiodic subset $\phi_H(A) + \phi_H(B)$. In such case, applying Kneser' Theorem to $\phi_H(A) + \phi_H(B)$ yields $|\phi_H(A \cup \{x\}) + \phi_H(B)| \geq |\phi_H(A)| + |\phi_H(B)| \geq |\phi_H(A \cup \{x\})| + |\phi_H(B)| - 1$. In either case, since $(A \cup \{x\}) + B$ is H-periodic in view of $H = \mathsf{H}(B)$, the desired conclusion follows by multiplying the inequality by $|H|$.

Lemma 12 *Let G be an abelian group, let $n \geq 1$, let $X \subseteq G$ be a finite, nonempty subset, let $S \in \mathcal{F}(G)$ be a sequence, and let $\mathcal{A} = A_1 \cdot \ldots \cdot A_n$ be a setpartition with $\mathsf{S}(\mathcal{A}) = S$ and*

$$|X + \sum_{i=1}^{n} A_i| \geq \min\{|S| - n + |X|,\ |X + \Sigma_n(S)|\}.$$

Then one of the following holds.

1. *$n = 2$, $X \setminus \{\beta\}$ and $A_1 \cap A_2$ are K-periodic, $(A_1 \cup A_2) \setminus (A_1 \cap A_2)$ is a K-coset, $X + \Sigma_n(S) = X + A_1 + A_2$ is aperiodic and $|X + \Sigma_n(S)| = |S| - n + |X|$, for some $\beta \in X$ and $K \leq G$ with $K \cong (\mathbb{Z}/2\mathbb{Z})^2$.*
2. *There exists an* equitable *setpartition $\mathcal{B} = B_1 \cdot \ldots \cdot B_n$ with $\mathsf{S}(\mathcal{B}) = S$ and $|X + \sum_{i=1}^{n} B_i| \geq \min\{|S| - n + |X|,\ |X + \Sigma_n(S)|\}$. Moreover, if $|X + \Sigma_n(S)| \leq |S| - n + |X|$ and $|A_i \setminus Z| \leq 1$ for all i, where $H = \mathsf{H}(X + \sum_{i=1}^{n} A_i)$ and $Z = \bigcap_{i=1}^{n}(A_i + H)$, then $Z = \bigcap_{i=1}^{n}(B_i + H)$, $X + \sum_{i=1}^{n} B_i = X + \sum_{i=1}^{n} A_i = X + \Sigma_n(S)$ and $|B_i \setminus Z| \leq 1$ for all i.*

Proof If $|S| = \sum_{i=1}^{n} |A_i| \leq n + 1$ or $n = 1$, then $\mathcal{A}$ is trivially equitable, and Item 2 follows taking $\mathcal{B}$ to be $\mathcal{A}$. Therefore we may assume $|S| \geq n + 2$ and $n \geq 2$. In particular, G is nontrivial. Let $\mathcal{A} = A_1 \cdot \ldots \cdot A_n$ be an arbitrary setpartition with $\mathsf{S}(\mathcal{A}) = S$ and

$$|X + \sum_{i=1}^{n} A_i| \geq \min\{|X| + \sum_{i=1}^{n} |A_i| - n,\ |X + \Sigma_n(S)|\}. \tag{14}$$

Note $\mathcal{A}$ exists by hypothesis. If, for the original setpartition $\mathcal{A}$, we have that $|X + \Sigma_n(S)| \leq |S| - n + |X|$ and $|A_i \setminus Z| \leq 1$ for all i, where $H = \mathsf{H}(X + \sum_{i=1}^{n} A_i)$ and $Z = \bigcap_{i=1}^{n}(A_i + H)$, then fix the set $Z \subseteq G$ and only consider setpartitions $\mathcal{A}$ also satisfying

$$Z \subseteq \bigcap_{i=1}^{n}(A_i + H) \quad \text{and} \quad |A_i \setminus Z| \leq 1 \quad \text{for all } i, \tag{15}$$

for the fixed set Z. Otherwise, simply let $Z = H = G$. In either case, let $e = \sum_{i=1}^{n} |A_i \setminus Z|$.

Note $|X + \Sigma_n(S)| \leq |S| - n + |X| = |X| + \sum_{i=1}^{n} |A_i| - n$ and $|A_i \setminus Z| \leq 1$ for all i combined with (14) imply $X + \sum_{i=1}^{n} A_i = X + \Sigma_n(S)$, combined with $\sum_{i=1}^{n} |A_i| = |S| \geq n + 2$ imply Z is nonempty, and combined with the definition of e imply $e \leq n$. Moreover, if $e = n$, then the n terms from the sets $A_i \setminus Z$ for $i = 1, \ldots, n$ cannot all be equal modulo H, else they would be included in the set Z by definition. What this means is that an arbitrary setpartition $\mathcal{A}$ satisfying (14) and (15) must have $X + \sum_{i=1}^{n} A_i = X + \Sigma_n(S)$ and $Z = \bigcap_{i=1}^{n}(A_i + H)$, so that the quantities $H = \mathsf{H}(X + \sum_{i=1}^{n} A_i) = \mathsf{H}(X + \Sigma_n(S))$ and $Z = \bigcap_{i=1}^{n}(A_i + H)$ remain invariant as we range over all setpartitions satisfying (14) and (15). This is also trivially the case when $Z = H = G$. Now choose a setpartition $\mathcal{A}$ with $\mathsf{S}(\mathcal{A}) = S$ satisfying (14) and (15) that is as equitable as possible, meaning one such that

$$\sum_{i=1}^{n} |A_i|^2 \quad \text{is minimal.} \tag{16}$$

Assume by contradiction that $\mathcal{A}$ is not equitable, so

$$m := \min_{i \in [1,n]} \{|A_i|\} \leq \max_{i \in [1,n]} \{|A_i|\} - 2.$$

Note this ensures that H is nontrivial, lest $|A_i| \in \{|Z|, |Z| + 1\}$ for all i. Let $I_m \subseteq [1, n]$ be the subset of indices $i \in [1, n]$ with $|A_i| = m$, let $I_{m+1} \subseteq [1, n]$ be the subset of indices $i \in [1, n]$ with $|A_i| = m + 1$, let $I_{m+2} \subseteq [1, n]$ be the subset of indices $i \in [1, n]$ with $|A_i| \geq m + 2$, let $I_Z \subseteq [1, n]$ be all indices $i \in [1, n]$ with $A_i \subseteq Z$, let $I_e \subseteq [1, n]$ be all indices $i \in [1, n]$ with $A_i \setminus Z$ nonempty, and let

$$J_Z = (I_m \cup I_{m+1}) \cap I_Z \quad \text{and} \quad J_e = (I_m \cup I_{m+1}) \cap I_e.$$

Since $\mathcal{A}$ is not equitable, I_{m+2} and I_m are nonempty. Consider $k \in I_{m+2}$ and $s \in I_m$. Since $k \in I_{m+2}$ and $|A_k \setminus Z| \leq 1$, we have $|A_k \cap Z| \geq |A_k| - 1 \geq m + 1$. Let $\alpha_1, \ldots, \alpha_r \in A_k \cap Z$ be those elements contained in Z with $(\alpha_i + H) \cap A_k = \{\alpha_i\}$, and let $Z_1 = \{\alpha_1, \ldots, \alpha_r\} + H \subseteq Z$. Then we have $|A_k \cap (Z \setminus Z_1)| = |A_k \cap Z| - r \geq |A_k| - 1 - r \geq m + 1 - r$. Since $s \in I_m$ and $Z \subseteq A_s + H$, it follows that $|(Z \setminus Z_1) \cap A_s| \leq |A_s \cap Z| - r \leq |A_s| - r = m - r$, in which case the pigeonhole principle guarantees there is some $y \in (A_k \setminus A_s) \cap Z$ with $|(y + H) \cap A_k| \geq 2$. Consequently, the indices k and s and element y in the hypothesis of the following claim always exist. Moreover, if $k \in I_{m+2} \cap I_Z$, then we get the improved estimate $|A_k \cap Z| = |A_k| \geq m + 2$, in which case the above argument yields at least *two* elements $y,\ y' \in A_k$ satisfying the hypotheses of Claim A.

Claim A: $A_s \nsubseteq A_k$ and $X + \sum_{i=1, i\neq k}^{n} A_i + (A_k \setminus \{y\}) \neq X + \sum_{i=1}^{n} A_i$ for any $k \in I_{m+2}$, $s \in I_m$ and $y \in (A_k \setminus A_s) \cap Z$ with $|(y+H) \cap A_k| \geq 2$.

Proof If $X + \sum_{i=1, i\neq k}^{n} A_i + (A_k \setminus \{y\}) = X + \sum_{i=1}^{n} A_i$, then we can remove y from A_k and place it in A_s to yield a new setpartition $\mathcal{B} = B_1 \cdot \ldots \cdot B_n$, where $B_k = A_k \setminus \{y\}$, $B_s = A_s \cup \{y\}$, and $B_i = A_i$ for all $i \neq k, s$, such that $\mathsf{S}(\mathcal{B}) = S$, $X + \sum_{i=1}^{n} A_i \subseteq X + \sum_{i=1}^{n} B_i$, $Z \subseteq \bigcap_{i=1}^{n}(B_i + H)$ (since $|(y+H) \cap A_k| \geq 2$), and $|B_i \setminus Z| = |A_i \setminus Z| \leq 1$ for all i (since $y \in Z$). Thus $\mathcal{B}$ satisfies (14) and (15). However, since $|A_k| \geq |A_s| + 2$, it follows that $|A_k|^2 + |A_s|^2 > |B_k|^2 + |B_s|^2$, so that $\mathcal{B}$ contradicts the minimality of (16) for $\mathcal{A}$. Therefore we instead conclude that $X + \sum_{i=1, i\neq k}^{n} A_i + (A_k \setminus \{y\}) \neq X + \sum_{i=1}^{n} A_i$. If $A_s \subseteq A_k$, then let $y \in (A_k \setminus A_s) \cap Z$ be any element with $|(y+H) \cap A_k| \geq 2$, which exists as argued above Claim A. Then $A_s \subseteq A_k \setminus \{y\}$, whence $A_s + A_k \subseteq (A_s \cup \{y\}) + (A_k \setminus \{y\})$, and removing y from A_k and placing it in A_s yields a new setpartition $\mathcal{B}$ that contradicts the minimality of (16) for $\mathcal{A}$ as before. Therefore $A_s \nsubseteq A_k$, completing the claim. □

Let $L = \mathsf{H}(X + \sum_{i \in I_m \cup I_{m+1}} A_i)$ and re-index the A_i such that $J_Z = (I_m \cup I_{m+1}) \cap I_Z = [1, n_Z]$, $J_e = (I_m \cup I_{m+1}) \cap I_e = [n_Z + 1, n_e]$ and $I_{m+2} = [n_e + 1, n]$.

Claim B: $|A_k + L| \geq |A_k| + 3(|L| - 1)$ for any $k \in I_{m+2}$.

Proof Let $k \in I_{m+2}$ and $s \in I_m$. Since $X + \sum_{i=1, i\neq k}^{n} A_i$ is L-periodic (as $k \notin I_m \cup I_{m+1}$), it follows from Claim A that any element $y \in A_k \cap Z$ satisfying the hypotheses of Claim A must be the unique element from its L-coset in A_k. Since $L \leq H$, the same is true of any element $y \in A_k$ which is the unique element from its H-coset in A_k. Since there is always at least one element satisfying the hypotheses of Claim A, at least two when $k \in I_Z$, and also an element from $A_k \setminus Z$ which is the unique element from its H-coset in A_k when $k \in I_e$, the claim follows if there is any $y \in A_k \cap Z$ that is the unique element from its H-coset in A_k, so we instead assume $|(y+H) \cap A_k| \geq 2$ for all $y \in A_k \cap Z$. Thus

$$|(A_k \cap Z) \setminus A_s| \geq |A_k \cap Z| - |A_s| + 1 = |A_k \cap Z| - m + 1,$$

is the number of elements $y \in A_k \cap Z$ satisfying the hypothesis of Claim A, with the inequality following since Claim A ensures that $A_s \nsubseteq A_k$. If $k \in I_Z$, we obtain at least $|(A_k \cap Z) \setminus A_s| \geq |A_k \cap Z| - m + 1 = |A_k| - m + 1 \geq 3$ elements satisfying the hypotheses of Claim A. If $j \in I_e$, we obtain at least $|(A_k \cap Z) \setminus A_s| \geq |A_k \cap Z| - m + 1 \geq |A_k| - m \geq 2$ elements satisfying the hypotheses of Claim A, as well as the element from $A_k \setminus Z$, which is the unique element from its H-coset in A_k. In either case, the claim follows. □

Claim C: Either $|X + \sum_{i \in J_Z} A_i| \leq |X| + \sum_{i \in J_Z} |A_i| - |J_Z| - \max\{|L| - 1, 1\}$, or else L is trivial and $|X + \sum_{i=1}^{j} A_i| = |X + \sum_{i=1}^{j-1} A_i| + |A_j| - 1$ for all $j \in [1, n]$.

Proof Suppose

$$|X+\sum_{i\in J_Z}A_i|\geq|X|+\sum_{i\in J_Z}|A_i|-|J_Z|-\max\{|L|-1,\ 1\}+1. \tag{17}$$

Note $J_Z\cup J_e=I_m\cup I_{m+1}$ and $L=\mathsf{H}(X+\sum_{i\in J_Z}A_i+\sum_{i\in J_e}A_i)$. Kneser' Theorem implies

$$|X+\sum_{i\in I_m\cup I_{m+1}}A_i|\geq|X+\sum_{i\in J_Z}A_i|+\sum_{i\in J_e}|A_i+L|-|J_e||L|. \tag{18}$$

By definition, each A_i with $i\in J_e\subseteq I_e$ has some element $z\in A_i\setminus Z$ which is the unique element from its H-coset in A_i, meaning $(z+H)\cap A_i=\{z\}$. Since $L\leq H$, it is also the unique element from its L-coset in A_i, ensuring that $|A_i+L|\geq|A_i|+|L|-1$ for $i\in J_e$. Combining this observation with (17) and (18), we conclude that

$$|X+\sum_{i\in I_m\cup I_{m+1}}A_i|\geq|X|+\sum_{i\in I_m\cup I_{m+1}}|A_i|-|I_m\cup I_{m+1}|-\max\{|L|-1,\ 1\}+1. \tag{19}$$

Let $k\in I_{m+2}=[n_e+1,n]$. Since $X+\sum_{i=1}^{n_e}A_i$ is L-periodic, it follows that $X+\sum_{i=1}^{k-1}A_i$ is also L-periodic. Consequently, in view of Claim A, it follows that there is a unique expression element in the sumset $\phi_L\big(X+\sum_{i=1}^{k-1}A_i\big)+\phi_L(A_k)$. Thus Theorem 5 implies that $|\phi_L\big(X+\sum_{i=1}^{k-1}A_i\big)+\phi_L(A_k)|\geq|\phi_L\big(X+\sum_{i=1}^{k-1}A_i\big)|+|\phi_L(A_k)|-1$ for $k\in I_{m+2}$. Lemma 6 now yields

$$|X+\sum_{i=1}^{n}A_i|\geq|X+\sum_{i\in I_m\cup I_{m+1}}A_i|+\sum_{i\in I_{m+2}}|A_i+L|-|I_{m+2}|-(|L|-1)|I_{m+2}|. \tag{20}$$

By Claim B, we have $|A_k+L|\geq|A_k|+3(|L|-1)$ for all $k\in I_{m+2}$. Combining this observation with (19) and (20), we deduce that

$$|X+\sum_{i=1}^{n}A_i|\geq|X|+\sum_{i=1}^{n}|A_i|-n+2|I_{m+2}|(|L|-1)-\max\{|L|-1,1\}+1. \tag{21}$$

Suppose $|X+\sum_{i=1}^{n}A_i|\leq|X|+\sum_{i=1}^{n}|A_i|-n$. Then, since $|I_{m+2}|\geq1$, we must have equality in (21) with L trivial, and equality must hold in all estimates used to derive (21). Since L is trivial, Kneser's Theorem implies that $|X+\sum_{i=1}^{j}A_i|\geq|X|+\sum_{i=1}^{j}|A_i|-j$ for all $j\in I_m\cup I_{m+1}=[1,n_e]$. But now Lemma 6 ensures that equality must hold in all these estimates, as otherwise (19) holds strictly, implying (21) also holds strictly. We also have equality holding in the modulo L estimates used to derive (20), meaning $|X+\sum_{i=1}^{k}A_i|=|X|+\sum_{i=1}^{k}|A_i|-k$ for all $k\geq n_e+1$, for otherwise (20) holds strictly, and thus (21) as well. The claim now follows. So it remains to consider the case when

$$|X+\sum_{i=1}^{n}A_i| > |X|+\sum_{i=1}^{n}|A_i|-n=|S|-n+|X|,$$

in which case $Z=H=G$.

In this case, since $n\in I_{m+2}$ by our choice of indexing, Claim A ensures that every element $y\in A_n\setminus A_s$ is part of a unique expression element in $\left(X+\sum_{i=1}^{n-1}A_i\right)+A_n$, where $s\in I_m$. If some $y\in A_n\setminus A_s$ is part of exactly one unique expression element, then we have $|X+\sum_{i=1}^{n-1}A_i+(A_n\setminus\{y\})|=|X+\sum_{i=1}^{n}A_i|-1\geq|X|+\sum_{i=1}^{n}|A_i|-n$. Thus removing y from A_n and placing it in A_s yields a new setpartition contradicting the minimality of (16) for $\mathcal{A}$. Therefore we can assume each $y\in A_n\setminus A_s$ is part of at least two unique expression elements. Since $A_s\not\subseteq A_n$ by Claim A, we have $|A_n\setminus A_s|\geq 3$, so there are at least three such elements in A_n, and thus at least two elements in $A_n\setminus\{y\}$ that are each part of at least two unique expression elements in the sumset $\left(X+\sum_{i=1}^{n-1}A_i\right)+(A_n\setminus\{y\})$, for any fixed $y\in A_n\setminus A_s$. In consequence, the Kemperman Structure Theorem [37, Theorem 9.1] [35, Proposition 2.2] implies

$$|\left(X+\sum_{i=1}^{n-1}A_i\right)+(A_n\setminus\{y\})|\geq|X+\sum_{i=1}^{n-1}A_i|+|A_n\setminus\{y\}|. \tag{22}$$

If L is trivial or $|I_{m+2}|\geq 2$, then repeating the arguments used to derive (20) and (21) for $A_1\cdot\ldots\cdot A_{n-1}$ rather than $A_1\cdot\ldots\cdot A_n$ yields $|X+\sum_{i=1}^{n-1}A_i|\geq|X|+\sum_{i=1}^{n-1}|A_i|-(n-1)$. Combined with (22), we obtain the inequality $|X+\sum_{i=1}^{n-1}A_i+(A_n\setminus\{y\})|\geq|X|+\sum_{i=1}^{n}|A_i|-n$, and now removing y from A_n and placing it in A_s yields a new setpartition contradicting the minimality of (16) for $\mathcal{A}$. Therefore we now assume L is nontrivial and $|I_{m+2}|=1$, meaning $I_m\cup I_{m+1}=[1,n-1]$. In this case, since $H=G$ ensures $n\in I_Z\cap I_{m+2}$ so that there is at least one element from $A_n\setminus\{y\}$ satisfying the hypothesis of Claim A, we can repeat the arguments used to derive (20) for $A_1\cdot\ldots\cdot A_{n-1}\cdot(A_n\setminus\{y\})$ rather than $A_1\cdot\ldots\cdot A_n$ to find

$$\begin{aligned}|X+\sum_{i=1}^{n-1}A_i+(A_n\setminus\{y\})| &\geq|X+\sum_{i=1}^{n-1}A_i|+|(A_n\setminus\{y\})+L|-|L| && (23)\\ &\geq|X+\sum_{i\in I_m\cup I_{m+1}}A_i|+|A_n|+|L|-3, && (24)\end{aligned}$$

with the latter inequality as Claim B ensures $|(A_n\setminus\{y\})+L|-|A_n\setminus\{y\}|\geq 2(|L|-1)$. Combining (23) with (19) and using that L is nontrivial, we obtain $|X+\sum_{i=1}^{n-1}A_i+(A_n\setminus\{y\})|\geq|X|+\sum_{i=1}^{n}|A_i|-n$. Once again, removing y from A_n and placing it in A_s yields a new setpartition contradicting the minimality of (16) for $\mathcal{A}$, completing Claim C. □

We now split the proof into two cases depending on which outcome holds in Claim C.

CASE 1: $|X + \sum_{i \in J_Z} A_i| \le |X| + \sum_{i \in J_Z} |A_i| - |J_Z| - \max\{|L| - 1,\ 1\}$ holds for every setpartition $\mathcal{A} = A_1 \cdot \ldots \cdot A_n$ with $\mathsf{S}(\mathcal{A}) = S$ satisfying (14), (15) and (16).

For $j \in [0, n_Z]$, let $K_j = \mathsf{H}(X + \sum_{i=1}^{j} A_i) \le L$. In view of the case hypothesis, let $j \in J_Z = [1, n_Z]$ be the minimal index such that $|X + \sum_{i=1}^{j} A_i| \le |X| + \sum_{i=1}^{j} |A_i| - j - \max\{|K_j| - 1,\ 1\}$.

Claim D: $K_j = \mathsf{H}(X + \sum_{i=1}^{j} A_i) = \mathsf{H}(X + \sum_{i=1}^{j-1} A_i)$ and $|X + \sum_{i=1}^{j} A_i| < |X + \sum_{i=1}^{j-1} A_i| + |A_j| - 1$.

Proof If $j = 1$, then $|X + A_1| \le |X| + |A_1| - 2$, so that Kneser's Theorem implies K_1 is nontrivial with $|X| + |A_1| - |K_1| \ge |X + A_1| \ge |X + K_1| + |A_1| - |K_1|$ (both upper bounds for $|X + A_1|$ follow from the definition of j). Thus $X + K_1 = X$, ensuring that $K_0 = K_1$, and the claim follows. Therefore we now assume $j \in [2, n_Z]$. Hence the minimality of j ensures

$$|X + \sum_{i=1}^{j-1} A_i| \ge |X| + \sum_{i=1}^{j-1} |A_i| - (j-1) - \max\{|K_{j-1}| - 1,\ 1\} + 1. \tag{25}$$

Applying Kneser's Theorem again, we find

$$|X + \sum_{i=1}^{j} A_i| \ge |X + \sum_{i=1}^{j-1} A_i + K_j| + |A_j| - |K_j|. \tag{26}$$

If $K_{j-1} \neq K_j$, then K_j is nontrivial and $|X + \sum_{i=1}^{j-1} A_i + K_j| \ge |X + \sum_{i=1}^{j-1} A_i| + |K_{j-1}|$, which combined with (25) and (26) yields $|X + \sum_{i=1}^{j} A_i| \ge |X| + \sum_{i=1}^{j} |A_i| - j + 1 - (|K_j| - 1) = |X| + \sum_{i=1}^{j} |A_i| - j + 1 - \max\{|K_j| - 1,\ 1\}$, contrary to the definition of j. Therefore $K_{j-1} = K_j$. If $|X + \sum_{i=1}^{j} A_i| \ge |X + \sum_{i=1}^{j-1} A_i| + |A_j| - 1$, then (25) yields

$$\begin{aligned} |X + \sum_{i=1}^{j} A_i| &\ge |X| + \sum_{i=1}^{j} |A_i| - j + 1 - \max\{|K_{j-1}| - 1,\ 1\} \\ &= |X| + \sum_{i=1}^{j} |A_i| - j + 1 - \max\{|K_j| - 1,\ 1\}, \end{aligned}$$

again contrary to the definition of j. Therefore we are left to conclude that $|X + \sum_{i=1}^{j} A_i| < |X + \sum_{i=1}^{j-1} A_i| + |A_j| - 1$, and the claim follows. □

All the above is valid for any $\mathcal{A}$ with $\mathsf{S}(\mathcal{A}) = S$ which satisfies (14), (15) and (16). We now impose additional extremal conditions on $\mathcal{A}$:

(a) $|J_Z|$ is minimal (subject to satisfying (14), (15) and (16)), say $|J_Z| = n_Z$, with the A_i indexed so that $J_Z = [1, n_Z]$.
(b) j is maximal (subject to satisfying (14), (15), (16) and (a)).
(c) $\sum_{i=1}^{j} |\phi_K(A_i)|$ is maximal, where $K = \mathsf{H}(X + \sum_{i=1}^{j-1} A_i)$ (subject to satisfying (14), (15), (16), (a) and (b)).

Let $k \in I_{m+2}$ be fixed and set $K := K_j \leq L \leq H$, so

$$K = \mathsf{H}(X + \sum_{i=1}^{j-1} A_i) = \mathsf{H}(X + \sum_{i=1}^{j} A_i)$$

by Claim D. Since Claim D ensures $|(X + \sum_{i=1}^{j-1} A_i) + A_j| < |X + \sum_{i=1}^{j-1} A_i| + |A_j| - 1$, it follows from Kneser's Theorem applied to $(X + \sum_{i=1}^{j-1} A_i) + A_j$ that K is nontrivial and

$$|A_j + K| - |A_j| \leq |K| - 2. \tag{27}$$

In particular,

$$|(x + K) \cap A_j| \geq 2 \quad \text{for all } x \in A_j. \tag{28}$$

Since $j \in J_Z \subseteq I_Z$, we have $A_j \subseteq Z$. Let $Z_0 = (A_j + K) \setminus (A_k + K) \subseteq Z$. Let $X_0 \subseteq Z_0 \cap A_j$ be a subset consisting of one element from A_j for every K-coset contained in Z_0. Thus

$$|Z_0| = |K|\,|X_0|$$

with Z_0 the union of all K-cosets that intersect A_j but not A_k.
Claim E: There is a subset $Y \subseteq (A_k \cap Z) \setminus (A_j + K)$ with $|Y| \geq \max\{1, |X_0|\}$ and $(A_k \setminus Y) + H = A_k + H$.

Proof Let us first show there is some $y \in (A_k \cap Z) \setminus (A_j + K)$ such that $|(y + H) \cap A_k| \geq 2$. Let $y_0 \in A_k \cap Z$ be an element satisfying the hypotheses of Claim A. Then $|(y_0 + H) \cap A_k| \geq 2$. Moreover, since $X + \sum_{i \in I_m \cup I_{m+1}} A_i$ is K-periodic with $k \in I_{m+2}$, the conclusion of Claim A ensures that $(y_0 + K) \cap A_k = \{y_0\}$. If $y_0 \notin A_j + K$, then taking $y = y_0$ yields the desired element y. Therefore we may assume $y_0 \in A_j + K$. Then $Z_0 \cup (y_0 + K) \subseteq A_j + K$ with $\big(Z_0 \cup (y_0 + K)\big) \cap A_k = \{y_0\}$. Thus $|A_k \cap (A_j + K)| \leq |A_j + K| - |Z_0| - |K| + 1 \leq |A_j| - |Z_0| - 1$, with the second inequality in view of (27). It follows that

$$\begin{aligned} |(A_k \cap Z) \setminus (A_j + K)| &= |A_k \cap Z| - |A_k \cap (A_j + K)| \\ &\geq (|A_k| - 1) - (|A_j| - |Z_0| - 1) = |A_k| - |A_j| + |Z_0|. \end{aligned} \tag{29}$$

In particular, since $|A_k| - |A_j| \geq (m + 2) - (m + 1) = 1$ (as $k \in I_{m+2}$ and $j \in I_m \cup I_{m+1}$), we conclude that $(A_k \cap Z) \setminus (A_j + K)$ is nonempty. If there is some $y \in (A_k \cap Z) \setminus (A_j + K)$ with $|(y + H) \cap A_k| \geq 2$, then the desired element y is found. Otherwise, we conclude that each $y \in (A_k \cap Z) \setminus (A_j + K)$ is the unique element from its H-coset in A_k. However, since $y \in Z \subseteq A_j + H$, it follows that $(y + H) \cap A_j$ is also nonempty, say with $y' \in (y + H) \cap A_j$. Since $y \notin A_j + K$, we have $y \notin y' + K$, ensuring that $y' + K \neq y + K$. Thus, as $(y + H) \cap A_k = \{y\}$, we conclude that $(y' + K) \cap A_k$ is empty, meaning $y' + K \subseteq Z_0$. As this is true for each $y \in (A_k \cap Z) \setminus (A_j + K)$, with the corresponding sets $y' + K$ each lying in the distinct cosets $y + H$ for $y \in (A_k \cap Z) \setminus (A_j + K)$ (as each such y is the unique element from its H-coset in A_k), it follows that

$$|Z_0| \geq |(A_k \cap Z) \setminus (A_j + K)|\,|K| \geq |(A_k \cap Z) \setminus (A_j + K)|.$$

However, applying this estimate in (29) yields the contradiction $m + 2 \leq |A_k| \leq |A_j| \leq m + 1$. Thus the existence of the desired element $y \in (A_k \cap Z) \setminus (A_j + K)$ with $|(y + H) \cap A_k| \geq 2$ is established, and we assume $|X_0| \geq 2$ as the claim is now complete taking $Y = \{y\}$ when $|X_0| \leq 1$.

By definition, $Z_0 \cap A_k = \emptyset$ and $Z_0 \subseteq A_j + K$. Thus $|A_k \cap (A_j + K)| \leq |A_j + K| - |Z_0| \leq |A_j| - |Z_0| + |K| - 2 = |A_j| - |K|(|X_0| - 1) - 2$, with the second inequality in view of (27). It follows that

$$\begin{aligned} |(A_k \cap Z) \setminus (A_j + K)| &= |A_k \cap Z| - |A_k \cap (A_j + K)| \\ &\geq |A_k| - |A_j| + |K|(|X_0| - 1) + 1. \end{aligned} \tag{30}$$

Let $Z_0' = \Big((A_k \cap Z) \setminus (A_j + K) + H\Big) \cap (Z_0 + H)$ and partition

$$\Big((A_k \cap Z) \setminus (A_j + K) + H\Big) = Z_0' \cup Z_1.$$

Since each H-coset in Z_0' contains a K-coset from Z_0, we have

$$|X_0| \geq |Z_0'|/|H|. \tag{31}$$

Let $Y \subseteq (A_k \cap Z) \setminus (A_j + K)$ be obtained by taking the set $(A_k \cap Z) \setminus (A_j + K)$ and removing one element from $(A_k \cap Z) \setminus (A_j + K)$ from each of the $|Z_0'|/|H|$ H-cosets contained in Z_0'. Then

$$|Y| = |(A_k \cap Z) \setminus (A_j + K)| - |Z_0'|/|H| \geq |(A_k \cap Z) \setminus (A_j + K)| - |X_0|, \tag{32}$$

with the inequality in view of (31), and $(\alpha + H) \cap (A_k \setminus Y)$ is nonempty for every $\alpha + H \subseteq Z_0'$ (as one element for each of these H-cosets was left out of Y, and thus remains in $A_k \setminus Y$). For each $\alpha + H \subseteq Z_1 \subseteq Z$, we have $\alpha + H \subseteq Z \subseteq A_j + H$, and thus there is some $\alpha' \in \alpha + H$ with $(\alpha' + K) \cap A_j$ nonempty. Since $\alpha + H \nsubseteq Z_0'$, it follows by definition of Z_0' and Z_0 that $(\alpha' + K) \cap A_k$ is nonempty, and necessarily disjoint from $(A_k \cap Z) \setminus (A_j + K)$, and thus also from $Y \subseteq (A_k \cap Z) \setminus (A_j + K)$. In consequence, we have $(A_k \setminus Y) + H = A_k + H$. If $|Y| \geq |X_0|$, the claim is complete, so we instead assume $|Y| \leq |X_0| - 1$, in which case (32) implies

$|(A_k \cap Z) \setminus (A_j + K)| \leq 2|X_0| - 1$. However, using this estimate in (30) along with $|A_k| - |A_j| \geq (m+2) - (m+1) = 1$ yields $(|K| - 2)(|X_0| - 1) + 1 \leq 0$, which is not possible since K is nontrivial (as noted above (27)) and $|X_0| \geq 1$, completing the claim. □

In view of Claim E, there is a nonempty subset $Y \subseteq (A_k \cap Z) \setminus (A_j + K)$ with $(A_k \setminus Y) + H = A_k + H$ and $|Y| = \max\{1, |X_0|\}$. Define a new setpartition $\mathcal{B} = B_1 \cdot \ldots \cdot B_n$ by setting $B_j = (A_j \setminus X_0) \cup Y$, $B_k = (A_k \setminus Y) \cup X_0$ and $B_i = A_i$ for all $i \neq k, j$. Since $K = \mathsf{H}(X + \sum_{i=1}^{j-1} A_i)$, (28) ensures that $\phi_K(A_j) = \phi_K(A_j \setminus X_0)$ and $X + \sum_{i=1}^{j} A_i = X + \sum_{i=1}^{j-1} A_i + (A_j \setminus X_0)$. As a result,

$$X + \sum_{i=1}^{n} A_i \subseteq X + \sum_{\substack{i=1\\ i\neq j,k}}^{n} A_i + (A_j \setminus X_0) + (A_k \cup X_0). \tag{33}$$

By definition of Y, we have $\phi_K(Y)$ disjoint from $\phi_K(A_j)$, and thus also from $\phi_K(X_0)$, while the definition of X_0 ensures that $\phi_K(X_0)$ is disjoint from $\phi_K(A_k)$ with $\phi_K(A_j \setminus X_0) = \phi_K(A_j) \subseteq \phi_K(A_k \cup X_0) \setminus \phi_K(Y)$. It follows that

$$\begin{aligned}&\phi_K(A_j \setminus X_0) + \phi_K(A_k \cup X_0) \subseteq \\ &\Big(\phi_K(A_j \setminus X_0) \cup \phi_K(Y)\Big) + \Big(\phi_K(A_k \cup X_0) \setminus \phi_K(Y)\Big) \subseteq \phi_K(B_j) + \phi_K(B_k).\end{aligned}$$

As a result, since $X + \sum_{i=1}^{j-1} A_i = X + \sum_{i=1}^{j-1} B_i$ is K-periodic, we conclude from (33) that

$$X + \sum_{i=1}^{n} A_i \subseteq X + \sum_{\substack{i=1\\ i\neq j,k}}^{n} A_i + (A_j \setminus X_0) + (A_k \cup X_0) \subseteq X + \sum_{i=1}^{n} B_i,$$

so (14) holds for $\mathcal{B}$. Since $X_0 \subseteq A_j \subseteq Z$ (as $j \in J_Z$), Claim E ensures that $Z \subseteq A_k + H = B_k + H$ and $|B_k \setminus Z| = |A_k \setminus Z| \leq 1$. Since $\phi_K(A_j) \subseteq \phi_K(B_j)$ with $Y \subseteq Z$ by definition, we have $Z \subseteq B_j + H = A_j + H$ and $|B_j \setminus Z| = |A_j \setminus Z| = 0$. Consequently, $Z \subseteq \bigcap_{i=1}^{n}(B_i + H)$ and $|B_i \setminus Z| \leq 1$ for all i, in which case $\mathcal{B}$ satisfies (15). We have $|B_j| = |A_j| - |X_0| + |Y|$, $|B_k| = |A_j| - |Y| + |X_0|$, and $|B_i| = |A_i|$ for $i \neq j, k$. Let I'_m, I'_{m+1}, I'_{m+2}, I'_e, I'_Z and J'_Z be the associated quantities I_m, I_{m+1}, I_{m+2}, I_e, I_Z and J_Z for $\mathcal{B}$ rather than $\mathcal{A}$.

Suppose $|X_0| = 0$. Then $|Y| = 1$, $|B_j| = |A_j| + 1$ and $|B_k| = |A_k| - 1$. If $|A_k| \geq |A_j| + 2$, then $\mathcal{B}$ contradicts the minimality of (16) for $\mathcal{A}$. Otherwise, we have $|A_k| = |A_j| + 1 = m + 2$, $|B_j| = |A_k| = m + 2$ and $|B_k| = |A_j| = m + 1$ so that $\sum_{i=1}^{n} |B_i|^2 = \sum_{i=1}^{n} |A_i|^2$, meaning $\mathcal{B}$ satisfies the extremal condition (16). Now $I'_{m+1} = I_{m+1} \setminus \{j\} \cup \{k\}$, $I'_m = I_m$ and $I'_{m+2} = I_{m+2} \setminus \{k\} \cup \{j\}$. If $k \in I_e$, then $k \in I'_e$ (as $Y \subseteq Z$), in which case $J'_Z = J_Z \setminus \{j\}$, contradicting the minimality condition (a) for $\mathcal{A}$. On the other hand, if $k \in I_Z$, then $J'_Z = J_Z \setminus \{j\} \cup \{k\}$, so condition

(a) holds for $\mathcal{B}$. Swapping the indices on B_k and B_j, so now $B_k = (A_j \setminus X_0) \cup Y$ and $B_j = (A_k \setminus Y) \cup X_0$, we obtain $J'_Z = J_Z$ and $I'_{m+1} \cup I'_m = I_{m+1} \cup I_m$. Since $A_i = B_i$ for $i < j$, the definition of j ensures $j' \geq j$, where j' is the associated quantity for $\mathcal{B}$ corresponding to the index j for $\mathcal{A}$, while the extremal condition given in (b) forces $j' \leq j$. Thus $j' = j$. However, since $k \in I_Z$, there are at least two elements y satisfying the hypotheses of Claim A for $\mathcal{A}$, which in view of the conclusion of Claim A and $K = \mathsf{H}(X + \sum_{i=1}^{j-1} A_i) = \mathsf{H}(X + \sum_{i=1}^{j'-1} B_i)$, means both these elements are the unique element from their K-coset in A_k. As at most one of them can be contained in the singleton set Y, we conclude that $B_j = A_k \setminus Y$ contains some $y \in B_j$ with $|(y + K) \cap B_j| = 1$. However, in such case, (28) could not hold for the index j in $\mathcal{B}$, contradicting that it must hold for $j' = j$ by the arguments above. So we instead conclude that $|X_0| \geq 1$,

Since $|X_0| \geq 1$, we have $|X_0| = |Y|$, $|B_j| = |A_j|$ and $|B_k| = |A_k|$, so (16) holds for $\mathcal{B}$ with $I'_m = I_m$, $I'_{m+1} = I_{m+1}$ and $I'_{m+2} = I_{m+2}$. Since $Y \subseteq Z$ and $A_j \subseteq Z$ (as $j \in J_Z$), we conclude that $J'_Z = J_Z$, meaning condition (a) holds for $\mathcal{B}$. As argued in the previous case, we must have $j' = j$, so that condition (b) holds. In particular, we must have $K = \mathsf{H}(X + \sum_{i=1}^{j-1} A_i) = \mathsf{H}(X + \sum_{i=1}^{j'-1} B_i)$. By definition of Y, each $y \in Y$ is disjoint from $A_j + K$. Thus, since $|Y| = |X_0| \geq 1$ and $\phi_K(A_j \setminus X_0) = \phi_K(A_j)$, we conclude that $|\phi_K(B_j)| > |\phi_K(A_j)|$, in which case the maximality condition (c) for $\mathcal{A}$ is contradicted by $\mathcal{B}$, completing CASE 1.

CASE 2: L is trivial and $|X + \sum_{i=1}^{j} A_i| = |X + \sum_{i=1}^{j-1} A_i| + |A_j| - 1$ for all $j \in [1, n]$, for some setpartition $\mathcal{A} = A_1 \cdot \ldots \cdot A_n$ with $\mathsf{S}(\mathcal{A}) = S$ satisfying (14), (15) and (16), where $J_Z = [1, n_Z]$, $J_e = [n_Z + 1, n_e]$ and $I_{m+2} = [n_e + 1, n]$.

Let $Y = X + \sum_{i=1}^{n-1} A_i$ and $V = X + \sum_{i=1}^{n} A_i = Y + A_n$. In view of the case hypothesis and Lemma 6, we have $|X + \sum_{i=1}^{j} A_i| = |X| + \sum_{i=1}^{j} |A_i| - j$ for all $j \in [1, n]$. In particular, $|X + \sum_{i \in J_Z} A_i| = |X| + \sum_{i \in J_Z} |A_i| - |J_Z|$ and

$$|V| = |X + \sum_{i=1}^{n} A_i| = |X| + \sum_{i=1}^{n} |A_i| - n = |S| - n + |X|. \tag{34}$$

The former equality combined with Claim C ensures that the hypotheses of CASE 2 hold for $\mathcal{A}$ under any re-indexing of the A_i with $i \in I_{m+2} = [n_e + 1, n]$, allowing us to freely assume an arbitrary set A_k with $k \in I_{m+2}$ occurs with $k = n$. We aim to either contradict the extremal condition (16) or show that Item 1 holds.

Claim F: If $H = \mathsf{H}(X + \sum_{i=1}^{n} A_i)$, then $A_k \subseteq Z$ with $|(y + H) \cap A_k| \geq 2$ for all $k \in I_{m+2}$ and $y \in A_k$.

Proof Suppose $H = \mathsf{H}(X + \sum_{i=1}^{n} A_i)$. Note H is nontrivial as remarked after (16). By our choice of indexing, $n \in I_{m+2}$. Let $s \in I_m$. Each $y \in A_n$ satisfying the hypothesis of Claim A is a unique expression element in $Y + A_n$, of which there is at least one. Thus, since $Y + A_n$ is H-periodic with $|Y + A_n| = |Y| + |A_n| - 1$ by case hypothesis, we can apply the Kemperman Structure Theorem directly to

$Y + A_n$ to conclude that there are H-quasi-periodic decompositions (cf. [35, Comment c.14]) $Y = Y_1 \cup Y_0$ and $A_n = X_1 \cup X_0$ with $|Y_0| + |X_0| = |H| + 1$. Moreover, in view of [37, Theorem 5.1] and H nontrivial, either all unique expression elements are contained in $Y_0 + X_0$, or $|X_0| = 1$ with all unique expression elements involving the unique element in X_0, or $|Y_0| = 1$ with all unique expression elements involving the unique element in Y_0. If $|X_0| = 1$, then $|Y_0| + |X_0| = |H| + 1$ ensures $|Y_0| = |H| \geq 2$, in which case all unique expression elements in $Y + A_n$ must involve the unique element from X_0. However, this contradicts that there is an element $y \in A_n$ satisfying Claim A, which is part of a unique expression element in $Y + A_n$ but not the unique element from its H-coset. Therefore $|X_0| \geq 2$, ensuring that $|(y + H) \cap A_n| \geq 2$ for all $y \in A_n$. Since any element from $A_n \setminus Z$ is the unique element from its H-coset in A_n, it follows that $A_n \subseteq Z$. Repeating the above argument for an arbitrary A_k with $k \in I_{m+2}$ (using an appropriate re-indexing), we conclude that $|(y + H) \cap A_k| \geq 2$ for all $y \in A_k$, and that $A_k \subseteq Z$, which completes the claim. □

Let $s \in I_m$ be arbitrary. Recall $n \in I_{m+2}$ by our choice of indexing. In view of Claim F, any element $y \in A_n \setminus A_s$ satisfies the hypotheses of Claim A (this is trivially true if $H = G$). Thus, since Claim A ensures that $A_s \not\subseteq A_n$, we conclude that there are $|A_n \setminus A_s| \geq 3$ elements satisfying the hypotheses of Claim A. Each such $y \in A_n \setminus A_s$ is part of a unique expression element in $Y + A_n$ by Claim A. As there are at least three such y, and since $|Y + A_n| = |Y| + |A_n| - 1$ by case hypothesis, the Kemperman Structure Theorem [37, Theorem 9.1] [35, Proposition 2.2] ensures this is only possible if there are K-quasi-periodic decompositions $Y = (Y \setminus \{y\}) \cup \{y\}$ and $A_n = (A_n \setminus A_\emptyset) \cup A_\emptyset$ with $y + A_\emptyset \subseteq Y + A_n$ the subset of all unique expression elements in $Y + A_n$, and

$$K = \langle A_\emptyset - A_\emptyset \rangle.$$

In particular, $A_n \setminus A_s \subseteq A_\emptyset$ with each $x \in A_\emptyset$ being part of exactly one unique expression element $y + x \in Y + A_n$. Moreover, since $A_\emptyset$ is a subset of a K-coset, we have $|K| \geq |A_\emptyset| \geq |A_n \setminus A_s| \geq 3$. Consequently, in view of the case hypothesis and Lemma 10.3, it follows by a short inductive argument that there are $a_i \in A_i$ for $i \in [1, n-1]$ and $\beta \in X$ such that $X \setminus \{\beta\}$ and $A_i \setminus \{a_i\}$ for $i \in [1, n-1]$ are K-periodic with $\beta + a_1 + \ldots + a_{n-1} = y$.

Let $x_1, \ldots, x_r \in A_n \setminus A_s \subseteq A_\emptyset$ be the $r \geq 3$ distinct elements in $A_n \setminus A_s$. By translating all terms of S appropriately, we can w.l.o.g. assume $a_s = 0$. Consider an arbitrary $x_t \in A_n \setminus A_s \subseteq A_\emptyset$. By the above work, we have $Y + (A_n \setminus \{x_t\}) = V \setminus \{y + x_t\}$. If $V \setminus \{y + x_t\}$ is aperiodic, then Lemma 11 together with Kneser's Theorem implies that $|X + (A_s \cup \{x_t\}) + \sum_{\substack{i=1\\i\neq s}}^{n-1} A_i + (A_n \setminus \{x_t\})| \geq |A_s \cup \{x_t\}| + |X + \sum_{\substack{i=1\\i\neq s}}^{n-1} A_i + (A_n \setminus \{x_t\})| - 1 \geq |X| + \sum_{i=1}^{n} |A_i| - n$, in which case moving x_t from A_n to A_s yields a new setpartition satisfying (14) and (15) (in view of Claim F), thus contradicting the minimality of (16) for $\mathcal{A}$. Therefore we instead conclude $H_t := \mathsf{H}(V \setminus \{y + x_t\})$ is nontrivial for every $x_t \in A_n \setminus A_s \subseteq A_\emptyset$. Since there are at least two such elements, [35, Proposition 2.1] implies the H_t are distinct cardinality two subgroups for $t \in A_n \setminus A_s$. Moreover, $V \cup \{\alpha\}$ is $(H_1 + \ldots + H_r)$-periodic for the unique element

$$\alpha \in (y + x_t + H_t) \setminus \{y + x_t\}. \tag{35}$$

Since $V \setminus \{y + x_t\}$ is periodic, [35, Comment c.6] implies

$$V = X + \sum_{i=1}^{n} A_i \quad \text{is aperiodic.} \tag{36}$$

Thus, since H is nontrivial (as noted after (16)), we conclude that $H \neq \mathsf{H}(X + \sum_{i=1}^{n} A_i)$, leaving us in the situation where $H = Z = G$ with I_e empty.

We must have $\phi_K(\beta) + \sum_{i=1}^{n} \phi_K(a_i) \in \phi_K(X) + \sum_{i=1}^{n} \phi_K(A_i)$ a unique expression element, where $a_n \in A_\emptyset$, for otherwise $V = X + \sum_{i=1}^{n} A_i$ will be K-periodic (as $X \setminus \{\beta\}$, $A_n \setminus A_\emptyset$ and $A_i \setminus \{a_i\}$ for $i \in [1, n-1]$ are all K-periodic), contrary to (36). In particular,

$$V = X + \sum_{i=1}^{n} A_i = Z_{[1,n]} \cup (\beta + \sum_{i=1}^{n-1} a_i + A_\emptyset) = Z_{[1,n]} \cup (y + A_\emptyset)$$

for some K-periodic set $Z_{[1,n]} \subseteq G$. Since $X + \sum_{i=1}^{n-1} A_i + (A_n \setminus \{x_t\}) = Z_{[1,n]} \cup (y + A_\emptyset \setminus \{x_t\})$ is a K-quasi-periodic decomposition having $H_t = \mathsf{H}\Big(Z_{[1,n]} \cup (y + A_\emptyset \setminus \{x_t\})\Big)$ and $y + A_\emptyset \setminus \{x_t\}$ a nonempty, proper subset of a K-coset, we must also have (by (12))

$$H_t = \mathsf{H}(A_\emptyset \setminus \{x_t\}) \leq K \quad \text{for any } x_t \in A_n \setminus A_s. \tag{37}$$

As a result, since $X \setminus \{\beta\}$ and $A_i \setminus \{a_i\}$ are K-periodic, it follows that $|X + H_t| = |X| + 1$ and $|A_i + H_t| = |A_i| + 1$ for all $i \in [1, n-1]$. Moreover, $A_n \setminus \{x_t\}$ is H_t-periodic. Now $0 = a_s \in A_s$ with $A_s \setminus \{0\}$ H_t-periodic. Hence, if $\{x_t, a_s\} = \{x_t, 0\} \neq H_t$, then $|(A_s \cup \{x_t\}) + H_t| = |A_s| + 3$. In such case, Lemma 11 together with Kneser's Theorem implies $|X + (A_s \cup \{x_t\}) + \sum_{i=1, i\neq s}^{n-1} A_i + (A_n \setminus \{x_t\})| \geq |X + H_t| + |(A_s \cup \{x_t\}) + H_t| + \sum_{i=1, i\neq s}^{n-1} |A_i + H_t| + |(A_n \setminus \{x_t\}) + H_t| - n|H_t| = |X| + \sum_{i=1}^{n} |A_i| - n + 1$, in which case moving x_t from A_n to A_s yields a new setpartition contradicting the minimality of (16) for $\mathcal{A}$. Therefore we instead conclude that

$$H_t = \{0, x_t\} \leq K, \qquad \text{for each } x_t \in A_n \setminus A_s,$$

is a cardinality two subgroup. Repeating the above arguments using any $i \in I_m$ in place of s, we find that $a_i = 0$ for all $i \in I_m$ (as $\{a_i, x_t\}$ must equal a single H_t-coset with $x_t \in H_t$ the unique nonzero element of H_t). Since $H_t = \{0, x_t\}$, it follows from (35) that

$$\alpha = y + 2x_t = y.$$

Thus $V \cup \{y\} = Z_{[1,n]} \cup \Big(y + (A_\emptyset \cup \{0\})\Big)$ is $(H_1 + \ldots + H_r)$-periodic with $H_1 + \ldots + H_r \leq K$, ensuring that $A_\emptyset \cup \{0\}$ is also $(H_1 + \ldots + H_r)$-periodic.

Suppose $|A_n| \geq m + 3$ and let $K' = H_1 + H_2 = \{0, x_1, x_2, x_1 + x_2\} \leq K$. As just noted, $A_\emptyset \cup \{0\}$ is $(H_1 + \ldots + H_r)$-periodic, and thus also K'-periodic with $K' \leq K$. Consequently, $A_n \setminus \{x_1, x_2\} = Z' \cup \{x_1 + x_2\}$ with $Z' := A_n \setminus K'$ a K'-periodic set and $x_1 + x_2$ the unique element from its K'-coset in $A_n \setminus \{x_1, x_2\}$. The sets $X \setminus \{\beta\}$ and $A_i \setminus \{a_i\}$ for $i \in [1, n-1]$ are all K-periodic, and thus also K'-periodic. The set $\Big(A_n \setminus \{x_1, x_2\}\Big) \setminus \{x_1 + x_2\} = Z' = A_n \setminus K'$ is also K'-periodic with $\phi_{K'}(x_1 + x_2) = 0$. It follows that $\phi_{K'}(y) = \phi_{K'}(\beta) + \sum_{i=1}^{n-1} \phi_{K'}(a_i) \in \phi_{K'}(X) + \sum_{i=1}^{n} \phi_{K'}(A_i)$ must be a unique expression element, as otherwise $X + \sum_{i=1}^{n} A_i$ would be K'-periodic, contradicting (36), and thus

$$X + (A_s \cup \{x_1, x_2\}) + \sum_{\substack{i=1\\ i\neq s}}^{n-1} A_i + (A_n \setminus \{x_1, x_2\})$$
$$= V \setminus (y + K') \cup (y + \{0, x_1, x_2\} + \{x_1 + x_2\}) = V$$

Removing x_1 and x_2 from A_n and placing them in A_s now yields a new setpartition with the same cardinality sumset as $\mathcal{A}$, contradicting the minimality of (16) for $\mathcal{A}$ (as $|A_n| \geq m + 3$). So we conclude that $|A_n| = m + 2$. Re-indexing the A_k with $k \in I_{m+2}$ and repeating these arguments for any A_k with $k \in I_{m+2}$, we conclude that $|A_k| = m + 2$ for all $k \in I_{m+2}$.

Since $A_n \setminus A_s \subseteq A_\emptyset$, we have $A_n \setminus A_\emptyset \subseteq A_s$. Hence, since $A_n \setminus A_\emptyset$ is K-periodic and $0 \in A_s$ is the unique element from it K-coset in A_s, it follows that $A_n \setminus A_\emptyset \subseteq A_s \setminus \{0\}$. If $A_n \setminus A_\emptyset \neq A_s \setminus \{0\}$, then $A_s \setminus \{0\}$ and $A_n \setminus A_\emptyset$ being K-periodic ensures $|A_s| \geq |A_n \setminus A_\emptyset| + |K| + 1 \geq |A_n \setminus A_\emptyset| + |A_\emptyset| + 1 \geq |A_n| + 1$, which is not possible. We are left to conclude $A_n \setminus A_\emptyset = A_s \setminus \{0\}$. Thus $m - 1 = |A_s \setminus \{0\}| = |A_n \setminus A_\emptyset|$, implying $m + 2 = |A_n| = (m - 1) + |A_\emptyset|$ and $|A_\emptyset| = 3$, and since $A_n \setminus A_s \subseteq A_\emptyset$ is a set of size at least three, we conclude that $A_\emptyset = A_n \setminus A_s$ and $A_n \cap A_s = A_n \setminus A_\emptyset = A_s \setminus \{0\}$. In particular, $K = \langle A_\emptyset - A_\emptyset \rangle = H_1 + H_2 + H_3 = \langle x_1, x_2, x_3 \rangle$. Since $V \cup \{y\} = Z_{[1,n]} \cup \Big(y + (A_\emptyset \cup \{0\})\Big)$ is K-periodic, with $Z_{[1,n]}$ a K-periodic set, it follows that $\{0\} \cup A_\emptyset = \{0, x_1, x_2, x_3\} = K$ is an elementary 2-group of order 4, whence $K \cong (\mathbb{Z}/2\mathbb{Z})^2$. Repeating these arguments for any $s \in I_m$ and $k \in I_{m+2}$, it follows that there exists a K-periodic subset $W \subseteq G \setminus K$ such that

$$A_s = W \cup \{0\} \quad \text{and} \quad A_k = W \cup (K \setminus \{0\}) \quad \text{for every } s \in I_m \text{ and } k \in I_{m+2}. \quad (38)$$

Since $A_s \setminus \{0\}$ is K-periodic, we have $|A_s \setminus \{0\}| = m - 1$ divisible by $|K| = 4$. Any $j \in I_{m+1}$ also has $A_j \setminus \{a_j\}$ K-periodic, whence $m = |A_j| - 1$ is divisible by 4. Since $m - 1$ and m cannot both be divisible by 4, it follows that I_{m+1} is empty.

Suppose $|I_{m+2}| \geq 2$. Then (38) implies $A_{n-1} = A_n$. Since $A_{n-1} \setminus \{a_{n-1}\}$ is K-periodic, we have $|A_{n-1}| \equiv 1 \mod |K|$. Since $A_n \setminus A_\emptyset$ is K-periodic with $|A_\emptyset| = 3$, we have $|A_n| \equiv 3 \mod |K|$. However, since $|K| = 4$, this contradicts that $A_{n-1} = A_n$. So we conclude that $|I_{m+2}| = 1$.

We now know $A_1 = A_2 = \ldots = A_{n-1} = W \cup \{0\}$ and $A_n = W \cup \{x_1, x_2, x_3\}$ with $W \subseteq G \setminus K$ and $X \setminus \{\beta\}$ K-periodic sets and $K = \{0, x_1, x_2, x_3\}$. If $n \geq 3$, then consider the setpartition $\mathcal{A} = A'_1 \cdot \ldots \cdot A'_n$ with $A'_i = W \cup \{0\}$ for $i \in [1, n-3]$, $A'_{n-2} = W \cup \{x_1\}$, $A'_{n-1} = W \cup \{0, x_2\}$ and $A'_n = W \cup \{0, x_3\}$. Then $X + \sum_{i=1}^n A'_i = V \cup \{\beta\} = V + K$, so that $\mathcal{A}'$ contradicts the minimality of (16) for $\mathcal{A}$ in view of (34). Therefore $n = 2$ (as we assumed $n \geq 2$ at the very start of the proof). It is now readily checked that $\mathcal{A} = A_1 \cdot A_2$ with $A_1 = W \cup \{x\}$ and $A_2 = W \cup (K \setminus \{x\})$, for $x \in K$, are the only setpartitions partitioning the terms of S with

$$|X + \sum_{i=1}^{2} A_i| \geq \min\{|X| + \sum_{i=1}^{2} |A_i| - 2,\ |X + \Sigma_2(S)|\}$$
$$= |X| + \sum_{i=1}^{2} |A_i| - 2 = |V|,$$

so the original setpartition $\mathcal{A}$ from the hypotheses must have this form. As the above works shows Item 1 holds for such $\mathcal{A}$, the case and proof is complete. □

We can now proceed with the proof of Theorem 4.

Proof (*Proof Theorem* 4) The case when L is nontrivial follows by applying the case L trivial to $\phi_L(S') \mid \phi_L(S)$. So it suffices to handle the case when L is trivial, which we now assume. Let $\mathcal{A} = A_1 \cdot \ldots \cdot A_n$ be a setpartition with $\mathsf{S}(\mathcal{A}) \mid S$ and $|\mathsf{S}(\mathcal{A})| = |S'|$ with $|X + \sum_{i=1}^n A_i|$ maximal. In view of [37, Proposition 10.1], the hypotheses $S' \mid S$ and $n \leq |S'| \leq \mathsf{h}(S')$ are equivalent to such a setpartition existing. Then $\mathcal{A}$ is a setpartition with $\mathsf{S}(\mathcal{A}) \mid S$ maximal relative to X.

Suppose $|X + \sum_{i=1}^n A_i| \geq |X| + \sum_{i=1}^n |A_i| - n = |S'| - n + |X|$. Note we trivially have $|X + \Sigma_n(S)| \geq |X + \Sigma_n(\mathsf{S}(\mathcal{A}))| \geq |X + \sum_{i=1}^n A_i|$. Applying Lemma 12 to $\mathcal{A}$ allows us to assume $\mathcal{A}$ is equitable (by replacing $\mathcal{A}$ by a modified setpartition as need be, potentially losing that $|X + \sum_{i=1}^n A_i|$ is maximal), yielding Item 1, unless Lemma 7.1 holds. Assume this is the case. By translating all terms of S appropriately, we can w.l.o.g. assume we have $(A_1 \cup A_2) \setminus (A_1 \cap A_2) = K$ with $0 \in A_1$ and $K \setminus \{0\} \subseteq A_2$. If there is some $x \in \mathrm{Supp}(S)$ with $x \in K$, then the setpartition $\mathcal{A}' = A'_1 \cdot A'_2$ defined by $A'_1 = (A_1 \setminus K) \cup \{x, x_1\}$ and $A'_2 = (A_2 \setminus K) \cup \{x, x_2\}$, where $x_1, x_2 \in K \setminus \{x\}$ are distinct, is an equitable setpartition with $X + A'_1 + A'_2 = X + A_1 + A_2 + K$ and $|X + A'_1 + A'_2| = |X + A_1 + A_2| + 1$. Item 1 holds in this case. If there is some $x \in \mathrm{Supp}(S)$ with $x \notin K$ and $x \notin A_1 \cap A_2$, then (12) ensures that $H = \mathsf{H}(X + A_1 + A_2 \setminus \{y\}) = \mathsf{H}(A_2 \setminus \{y\})$ with $K \setminus \{0, y\}$ an H-coset, for any $y \in K \setminus \{0\}$. In such case, the setpartition $\mathcal{A}' = A'_1 \cdot A'_2$ defined by $A'_1 = A_1 \cup \{x\}$ and $A'_2 = A_2 \setminus \{y\}$, where $y \in K \setminus \{0\}$, is an equitable setpartition, while Lemma 11 and Kneser's Theorem imply $|X + A'_1 + A'_2| \geq |(A_1 \cup \{x\}) + H| + |X + (A_2 \setminus \{y\}) +$

$H| - |H| \geq |X+H| + |(A_1 \cup \{x\}) + H| + |(A_2 \setminus \{y\}) + H| - 2|H| = |X| + |A_1| + |A_2| - 1$. Item 1 follows in this case as well. Otherwise, we have $\mathrm{Supp}(\mathsf{S}(\mathcal{A})^{[-1]} \cdot S) \subseteq A_1 \cap A_2$, and the remaining conclusions needed for Item 3 to hold follow from Lemma 7.1.

Next instead suppose $|X + \sum_{i=1}^n A_i| < |X| + \sum_{i=1}^n |A_i| - n = |S'| - n + |X|$. Let $H = \mathsf{H}(X + \sum_{i=1}^n A_i)$ and $Z = \bigcap_{i=1}^n (A_i + H)$. In view of Lemma 7, it follows that we can assume $\mathrm{Supp}(\mathsf{S}(\mathcal{A})^{[-1]} \cdot S) \subseteq Z$ and $|(y+H) \cap A_i| \leq 1$ for all $y \in G \setminus Z$ and $i \in [1, n]$ (by replacing $\mathcal{A}$ by a modified setpartition as need be). This allows us to apply Lemma 8 to add the stronger assumption that $|A_i \setminus Z| \leq 1$ for all i (by replacing $\mathcal{A}$ by a modified setpartition as need be). But now Lemma 9 ensures that $X + \sum_{i=1}^n A_i = X + \Sigma_n(\mathsf{S}(\mathcal{A})) = X + \Sigma_n(S)$. Thus $H = \mathsf{H}(X + \sum_{i=1}^n A_i) = \mathsf{H}(X + \Sigma_n(S))$, and we can apply Lemma 12. Since $|\Sigma_n(S)| = |\Sigma_n(\mathsf{S}(\mathcal{A}))| = |X + \sum_{i=1}^n A_i| < |X| + \sum_{i=1}^n |A_i| - n = |S'| - n + |X|$, Lemma 8.1 cannot hold. Thus Lemma 8.2 allows us to further assume $\mathcal{A}$ is equitable (again, by replacing $\mathcal{A}$ by a modified setpartition as need be), and Item 2 follows, completing the proof. □

3 Partitioning Results for Large *n*

In this section, we derive stronger results in the case our setpartition $\mathcal{A} = A_1 \cdot \ldots \cdot A_n$ satisfies $\sum_{i=1}^n |A_i| \leq 2n$.

Lemma 13 *Let G be an abelian group, let $n \geq 1$, let $X \subseteq G$ be a finite, nonempty subset, let $\mathcal{A} = A_1 \cdot \ldots \cdot A_n$ be a setpartition over G, and let $H = \mathsf{H}(X + \sum_{i=1}^n A_i)$ and $Z = \bigcap_{i=1}^n (A_i + H)$. Suppose $\sum_{i=1}^n |A_i| \leq 2n$, $|A_i \setminus Z| \leq 1$ for all $i \in [1, n]$, and*

$$|X + \sum_{i=1}^n A_i| < |X + H| + \Big(\sum_{i=1}^n |A_i| - n\Big)|H|.$$

Then H is nontrivial and $Z = \alpha + H$ for some $\alpha \in G$.

Proof Kneser's Theorem implies

$$|X + \sum_{i=1}^n A_i| \geq |X + H| + \sum_{i=1}^n |A_i + H| - n|H|. \tag{39}$$

If $Z = \emptyset$, then the hypothesis $|A_i \setminus Z| \leq 1$ implies $|A_i + H| = |H||A_i| = |H|$ for all i. Thus (39) implies $|X + \sum_{i=1}^n A_i| \geq |X + H| + \Big(\sum_{i=1}^n |A_i| - n\Big)|H|$, contrary to hypothesis. Therefore Z is nonempty. If H is trivial, then (39) implies $|X + \sum_{i=1}^n A_i| \geq |X| + \sum_{i=1}^n |A_i| - n = |X + H| + \Big(\sum_{i=1}^n |A_i| - n\Big)|H|$, contrary to hypothesis. Therefore H is nontrivial. Note that Z is H-periodic by its definition. If $|Z| \geq 2|H|$, then $|A_i + H| \geq 2|H|$ for all i, so that (39) and the hypothesis $|S| \leq 2n$ imply

$$|X+\sum_{i=1}^{n}A_i|\geq|X+H|+n|H|\geq|X+H|+\Big(\sum_{i=1}^{n}|A_i|-n\Big)|H|,$$

contrary to hypothesis. Therefore $|Z|=|H|$, completing the proof. □

We now derive our strengthening of Theorem 4 for large n, mirroring the main result from [38] (which obtained the same conclusion assuming n is large with respect to the exponent).

Theorem 14 *Let G be an abelian group, let $n\geq 1$, let $X\subseteq G$ be a finite, nonempty subset, let $L\leq \mathsf{H}(X)$, let $S\in\mathcal{F}(G)$ be a sequence, and let $S'\mid S$ be a subsequence with $\mathsf{h}(\phi_L(S'))\leq n\leq|S'|$. Suppose $|S'|\leq 2n$. Then one of the following holds:*

1. *$n=2$, $|S'|=|S|=|\operatorname{Supp}(\phi_L(S))|$, $\operatorname{Supp}(\phi_L(S))=\alpha+K/L$ for some $K\leq G$ and $\alpha\in G$ with $L\leq K$ and $K/L\cong(\mathbb{Z}/2\mathbb{Z})^2$, $X\setminus(\beta+L)$ is K-periodic (or empty) for some $\beta\in X$, and $X+\Sigma_n(S)=X+(K\setminus L)+2\alpha$ with $|X+\Sigma_n(S)|=|X|+2|L|=(|S|-n)|L|+|X|$.*
2. *There exists an* equitable *setpartition $\mathcal{A}=A_1\cdot\ldots\cdot A_n$ with $\mathsf{S}(\mathcal{A})\mid S$, $|\mathsf{S}(\mathcal{A})|=|S'|$, $|\phi_L(A_i)|=|A_i|$ for all $i\in[1,n]$, and $|X+\Sigma_n(S)|\geq|X+\sum_{i=1}^{n}A_i|\geq(|S'|-n)|L|+|X|$.*
3. *There exists an* equitable *setpartition $\mathcal{A}=A_1\cdot\ldots\cdot A_n$ with $\mathsf{S}(\mathcal{A})\mid S$, $|\mathsf{S}(\mathcal{A})|=|S'|$ and $|\phi_L(A_i)|=|A_i|$ for all $i\in[1,n]$, a subgroup $K\leq H=\mathsf{H}(X+\Sigma_n(S))$ with $L<K$ proper, and $\alpha\in G$ such that*
   - *(a) $X+\Sigma_n(S)=X+\sum_{i=1}^{n}A_i$,*
   - *(b) $\operatorname{Supp}(\mathsf{S}(\mathcal{A})^{[-1]}\cdot S)\subseteq\alpha+K=\bigcap_{i=1}^{n}(A_i+K)$ and $|A_i\setminus(\alpha+K)|\leq 1$ for all i,*
   - *(c) $|X+\Sigma_n(S)|\geq|X+H|+|S_{G\setminus(\alpha+H)}|\cdot|H|$ and also $|X+\Sigma_n(S)|\geq|X+K|+|S_{G\setminus(\alpha+K)}|\cdot|K|$,*
   - *(d) $L+\sum_{i\in I_K}A_i=\alpha|I_K|+K$, where $I_K\subseteq[1,n]$ is the nonempty subset of all $i\in[1,n]$ with $A_i\subseteq\alpha+K$.*

Proof As with the proof of Theorem 4, it suffices to prove the case when L is trivial, as we can then apply this case to $\phi_L(S')\mid\phi_L(S)$. We divide the proof into two mains cases.

CASE 1: $|X+\Sigma_n(S)|<|S'|-n+|X|$.

We will show Item 3 holds. In this case, let $\mathcal{A}=A_1\cdot\ldots\cdot A_n$ be an arbitrary setpartition resulting from the application of Theorem 1.2 to $S'\mid S$. By Theorem 1.2, $\mathcal{A}$ is equitable, so $|A_i|\leq 2$ for all i (as $|S'|\leq 2n$), $\mathsf{S}(\mathcal{A})\mid S$, $|\mathsf{S}(\mathcal{A})|=|S'|$, (a) holds, $\operatorname{Supp}(\mathsf{S}(\mathcal{A})^{[-1]}\cdot S)\subseteq Z$, and $|A_i\setminus Z|\leq 1$ for all i, where $Z=\bigcap_{i=1}^{n}(A_i+H)$ and $H=\mathsf{H}(X+\Sigma_n(S))$. By case hypothesis, $|X+\sum_{i=1}^{n}A_i|=|X+\Sigma_n(S)|<|X|+\sum_{i=1}^{n}|A_i|-n\leq|X|+\Big(\sum_{i=1}^{n}|A_i|-n\Big)|H|$, so that Lemma 13 implies H is nontrivial and

$$Z=\alpha+H\quad\text{for some }\alpha\in G.$$

But now Theorem 1.2 implies (b) holds with $K = H$, in which case $n + |S_{G\setminus(\alpha+H)}| = \sum_{i=1}^{n} |\phi_H(A_i)|$, and now (c) holds with $K = H$ by Kneser's Theorem. Since H is nontrivial, the case when G is trivial is complete, allowing us to proceed by induction on $|G| \in \mathbb{N} \cup \{\infty\}$.

Let $I_H \subseteq [1, n]$ be all those indices $i \in [1, n]$ with $A_i \subseteq \alpha + H$, and let $I'_H \subseteq I_H$ all those indices $i \in I_H$ with $|A_i| = 1$. If $I_H = [1, n]$, then $\sum_{i=1}^{n} A_i = \sum_{i \in I_H} A_i = \alpha|I_H| + H$ and (d) holds with $K = H$, yielding Item 3 with $K = H$, as desired. Therefore, we may assume $I_H \subset [1, n]$ is a proper subset. Since (b) and (c) hold with $K = H$, it follows that

$$\begin{aligned} |X + \Sigma_n(S)| &\geq |X + H| + |S_{G\setminus(\alpha+H)}| \cdot |H| &(40)\\ &= |X + H| + (n - |I_H|)|H| \geq |X| + (n - |I_H|)|H|. &(41)\end{aligned}$$

Since $\mathcal{A}$ is equitable with $|S'| \leq 2n$, we have $|A_i| \leq 2$ of all i, and thus $|S'| \leq 2n - |I'_H|$. Hence the case hypothesis yields $|X + \Sigma_n(S)| \leq |S'| - n - 1 + |X| \leq n - 1 + |X| - |I'_H|$, which combines with (40) and $I_H \subset [1, n]$ proper to yield

$$|I_H \setminus I'_H| \geq (n - |I_H|)(|H| - 1) + 1 \geq |H|. \qquad (42)$$

Consequently,

$$\sum_{i \in I_H} |A_i| - |I_H| + 1 = \sum_{i \in I_H \setminus I'_H} |A_i| - |I_H \setminus I'_H| + 1 \geq |H| + 1, \qquad (43)$$

with the final inequality following by combining (42) with the fact that $|A_i| = 2$ for $i \in I_H \setminus I'_H$. As a result, if $|\sum_{i \in I_H} A_i| \geq \min\{|H|, \sum_{i \in I_H} |A_i| - |I_H| + 1\} = |H|$, then $|\sum_{i \in I_H} A_i| = |H|$ follows, in which case (d) holds with $K = H$, completing the proof as before. By translating all terms appropriately, we can w.l.o.g. assume $\alpha = 0$.

Let $T = S_H$, let $T' = \mathsf{S}(\prod^{\bullet}_{i \in I_H} A_i)$, let $n' = |I_H| \geq |H| > 0$, and let $H' = \mathsf{H}(\{0\} + \Sigma_{n'}(T)) \leq H$. By re-indexing the A_i, we can w.l.o.g. assume that $I_H = [1, n']$. Since T' is the sequence partitioned by the setpartition $A_1 \cdot \ldots \cdot A_{n'}$, it follows that $\mathsf{h}(T') \leq n' \leq |T'|$ (see [37, Proposition 10.1]). Since the setpartition $A_1 \cdot \ldots \cdot A_n$ is equitable with $|S| \leq 2n$, we have $|A_i| \in \{1, 2\}$ for all i, so $|T'| \leq 2n'$. Since $T \in \mathcal{F}(H)$, we trivially have

$$|\Sigma_{n'}(T)| \leq |H| < \sum_{i \in I_H} |A_i| - |I_H| + 1 = |T'| - n' + 1, \qquad (44)$$

with the second inequality following from (43). If $H = G$, then (a) becomes $X + \Sigma_n(S) = X + \sum_{i=1}^{n} A_i = G$ with $|S_{G\setminus(\alpha+G)}| = 0$, in which case (b)–(d) all follow trivially with $K = H = G$. Therefore we may assume $H < G$ is a proper, nontrivial subgroup, and since the stabilizer $H = \mathsf{H}(X + \Sigma_n(S))$ of a finite set must be finite, it follows that we can apply the induction hypothesis to $\{0\} + \Sigma_{n'}(T)$ using $T' \mid T$.

Then Item 3 must hold for $\Sigma_{n'}(T)$ in view of (43). Let $\mathcal{B} = B_1 \cdot \ldots \cdot B_{n'}$ be the resulting setpartition and let $\beta + K$, where $K \leq H' \leq H$, be the resulting coset. Let $I_K \subseteq [1, n'] = I_H$ be the subset of indices $i \in [1, n']$ with $B_i \subseteq \beta + K$. By re-indexing the B_i, we can w.l.o.g. assume $I_K = [1, n'']$, where $n'' = |I_K|$.

Define a new setpartition $\mathcal{A}' = A'_1 \cdot \ldots \cdot A'_n$ as follows. Set $A'_i = B_i$ for $i \in [1, n']$. Since each A_i with $i \in [1, n] \setminus I_H = [n' + 1, n]$ contains a term from outside $H = \alpha + H$, it follows that $|A_i| = 2$ and $|H \cap A_i| = |A_i| - 1 = 1$ for all $i \notin I_H$. Since $\sum_{i=1}^{n'} |B_i| = \sum_{i=1}^{n'} |A_i| = \sum_{i=1}^{n} |A_i \cap H|$, we have $|\mathsf{S}(\mathcal{B})^{[-1]} \cdot S_H| \geq \sum_{i=n'+1}^{n} |A_i \cap H|$. This means we can take each set A_i with $i \notin I_H$ and replace the element from $A_i \cap H$ with a separate term from $\mathsf{S}(\mathcal{B})^{[-1]} \cdot S_H$ to yield the set A'_i. As $|A_i \cap H| = 1$ for all $i \notin I_H$, we are guaranteed that $|A'_i| = |A_i|$ for all i.

By translating all terms of S by $-\beta \in H$, we can w.l.o.g. assume $\beta = 0$. Since all terms of $\mathsf{S}(\mathcal{B})^{[-1]} \cdot S_H$ are from $K = \beta + K$ by (b) (holding for $\mathcal{B}$), it follows that each A'_i, with $i > n'$, has $|A'_i \setminus K| = 1$. Since (b) holds for $\mathcal{B}$, we also have $|A'_i \setminus K| = |B_i \setminus K| \leq 1$ for all $i \leq n'$ with $K = \bigcap_{i=1}^{n}(A_i + K)$. Thus, since $\sum_{i=1}^{n'} B_i = \sum_{i=1}^{n'} A'_i$ is K-periodic, Lemma 4.1 implies $\sum_{i=1}^{n} A'_i = \Sigma_n(S) = \sum_{i=1}^{n} A_i$. Hence (a) holds for $\mathcal{A}' = A'_1 \cdot \ldots \cdot A'_n$. If $\sum_{i=1}^{n'} B_i = \sum_{i=1}^{n'} A'_i = H$, then (a)–(d) all hold for $\mathcal{A}'$ with $K = H$, completing the proof. Therefore we may assume $|\sum_{i=1}^{n'} B_i| \leq |H| - |K|$ (as $\sum_{i=1}^{n'} B_i \subseteq H$ is K-periodic). Thus (c) for $\mathcal{B}$ ensures that $|T_{H \setminus K}| \leq |H/K| - 2$. The first part of (c) was already established. If the second fails for $\mathcal{A}'$, then it follows that

$$\begin{aligned}|X + H| + |S_{G \setminus H}|\,|H| \leq |X + \Sigma_n(S)| &< |X + K| + |S_{G \setminus K}|\,|K| \\ &\leq |X + H| + (|S_{G \setminus H}| + |H/K| - 2)|K|,\end{aligned}$$

implying $|S_{G \setminus H}|(|H/K| - 1) \leq |H/H| - 2$, which forces $|S_{G \setminus H}| = 0$. However, in such case (a)–(d) all hold for $\mathcal{A}$ with $K = H$. Thus we can assume both parts of (c) hold for $\mathcal{A}'$ using K. In view of the construction of the A'_i and (b) for $\mathcal{B}$, it follows that (b) holds for $\mathcal{A}'$ with K, while (d) holds for $\mathcal{A}'$ with K as it holds for $\mathcal{B}$. But now (a)–(d) all hold for $\mathcal{A}'$ with subgroup $K \leq H$, which completes CASE 1.

CASE 2: $|X + \Sigma_n(S)| \geq |S'| - n + |X|$

Apply Theorem 4 to $S' \mid S$. If either Theorem 1.1 or Theorem 1.2 holds, then the case hypothesis ensures there exists an equitable setpartition $\mathcal{A} = A_1 \cdot \ldots \cdot A_n$ with $\mathsf{S}(\mathcal{A}) \mid S$ and $|\mathsf{S}(\mathcal{A})| = |S'|$ such that

$$|X + \Sigma_n(S)| \geq |X + \sum_{i=1}^{n} A_i| \geq |S'| - n + |X|. \tag{45}$$

It follows that Item 2 holds in this case. Therefore we may instead assume Theorem 1.3 holds, and let $\mathcal{A} = A_1 \cdot A_2$ be the resulting setpartition. Then $|A_1| \equiv 1 \mod 4$ and $|A_2| \equiv 3 \mod 4$ with $|A_1| + |A_2| = |S'| \leq 2n = 4$. It follows that $|A_1| = 1$,

$|A_2| = 3$ and $A_1 \cap A_2 = \emptyset$. Item 1 now follows from Theorem 1.3, completing the case and proof. □

Finally, we conclude with the following application of Theorem 14, deriving some structural information regarding S when, in particular, $|S| = 2n$ with $|\Sigma_n(S)| \leq n + 1$ and $\mathsf{h}(S) \leq n$.

Theorem 15 *Let G be an abelian group, let $n \geq 1$, and let $S \in \mathcal{F}(G)$ be a sequence with $|S| > n$. Suppose $|\Sigma_n(S)| \leq m + 1$, where*

$$m = \min\{n, |S| - n, |S| - \mathsf{h}(S)\}.$$

Then one of the following holds, with Items 1–4 only possible if $|\Sigma_n(S)| = m + 1$ or $|\operatorname{Supp}(S)| = 1$.

1. $n = 2$, $|S| = |\operatorname{Supp}(S)|$, *and* $\operatorname{Supp}(S) = x + K$ *for some* $K \leq G$ *and* $x \in G$ *with* $K \cong (\mathbb{Z}/2\mathbb{Z})^2$.
2. $m = 2$ *and* $\operatorname{Supp}(S) = x + K$ *for some* $K \leq G$ *and* $x \in G$ *with* $K \cong \mathbb{Z}/3\mathbb{Z}$.
3. $|\operatorname{Supp}(S)| \leq 2$.
4. $\operatorname{Supp}(S) \subseteq \{x - d, x, x + d\}$ *for some* $x, d \in G$ *with* $\mathsf{v}_x(S) = \mathsf{h}(S) \geq |S| - m$.
5. *There exists* $x \in G$ *and a setpartition* $\mathcal{A} = A_1 \cdot \ldots \cdot A_n$ *with* $\mathsf{S}(\mathcal{A}) \mid S$, $|\mathsf{S}(\mathcal{A})| = n + m$, $\sum_{i=1}^n A_i = \Sigma_n(S)$, $\operatorname{Supp}(\mathsf{S}(\mathcal{A})^{[-1]} \cdot S) \subseteq x + H$, $|A_i| \leq 2$ *and* $(x + H) \cap A_i \neq \emptyset$ *for all* $i \in [1, n]$, *and* $|\sum_{i=1}^n A_i| = |\sum_{i=1, i \neq j}^n A_i|$ *for some* $j \in [1, n]$,

   where $H = \mathsf{H}(\Sigma_n(S))$ *is nontrivial.*

Proof If $\mathsf{h}(S) = |S|$, then $|\operatorname{Supp}(S)| = 1$, and Item 3 holds. Therefore we may assume $\mathsf{h}(S) < |S|$, and so, since $|S| > n$, may let m be the maximal integer in $[1, n]$ such that there is a subsequence $S' \mid S$ with $|S'| = n + m$ and $\mathsf{h}(S') \leq n$. If $m = n$, then $2n \leq |S|$ and $\mathsf{h}(S) \leq |S| - n$, whence $m = n = \min\{n, |S| - n, |S| - \mathsf{h}(S)\} \geq 1$. If $S' = S$, then $|S| = |S'| = n + m \leq 2n$ and $\mathsf{h}(S) \leq |S| - m = n$, whence $m = |S| - n = \min\{n, |S| - n, |S| - \mathsf{h}(S)\} \geq 1$. If $m < n$ and S' is a proper subsequence, then the maximality of m ensures that $(S')^{[-1]} \cdot S$ has only one distinct term, say x. Now $\mathsf{v}_x(S') \leq n$. If $\mathsf{v}_x(S') < n$, then $S'' = S' \cdot x$ is a subsequence with $|S''| = |S'| + 1 = n + m + 1$, $\mathsf{h}(S'') \leq n$ and $m + 1 \leq n$, so $m + 1$ contradicts the maximality of m. Therefore $\mathsf{h}(S) = \mathsf{v}_x(S) = |S| - |S'| + n = |S| - m \geq |S| - n$ in this case, implying $\mathsf{h}(S) = |S| - |S'| + n \geq n$ and $m = |S| - \mathsf{h}(S) = \min\{n, |S| - n, |S| - \mathsf{h}(S)\} \geq 1$. In consequence, in all possible cases, we deduce that

$$m = \min\{n, |S| - n, |S| - \mathsf{h}(S)\} \geq 1. \tag{46}$$

Let $H = \mathsf{H}(\Sigma_n(S))$.

By hypothesis, $|\Sigma_n(S)| \leq m + 1 = |S'| - n + 1$. If $|\Sigma_n(S)| \leq m$, then Theorem 1.2 applied to $S' \mid S$ (with $X = \{0\}$) yields a setpartition $\mathcal{A} = A_1 \cdot \ldots \cdot A_n$ with $\mathsf{S}(\mathcal{A}) \mid S$, $|\mathsf{S}(\mathcal{A})| = |S'| = n + m$, $\sum_{i=1}^n A_i = \Sigma_n(S)$, $\operatorname{Supp}(\mathsf{S}(\mathcal{A})^{[-1]} \cdot S) \subseteq Z$, and $|A_i| \leq 2$ and $|A_i \setminus Z| \leq 1$ for all i, where $Z = \bigcap_{i=1}^n (A_i + H)$. By indexing the A_i appropriately, we can assume $|A_i| = 2$ for $i \in [1, m]$. By Lemma 13, H is nontrivial and $Z = x + H$ for some $x \in G$. If $|\sum_{i=1}^{j} A_i| > |\sum_{i=1}^{j-1} A_i|$ for all $j \in [2, m]$, then it follows that $|\Sigma_n(S)| = |\sum_{i=1}^{n} A_i| \geq m + 1$, contrary to assumption. Thus there is some $j \in [2, m]$ with $|\sum_{i=1}^{j} A_i| = |\sum_{i=1}^{j-1} A_i|$, meaning Item 5 holds. It remains to consider the case when $|\Sigma_n(S)| = m + 1 = |S'| - n + 1 = \sum_{i=1}^{n} |A_i| - n + 1$.

Suppose $m = 1$. If $1 = m = |S| - \mathsf{h}(S)$, then $\mathsf{h}(S) = |S| - 1$, implying $|\operatorname{Supp}(S)| \leq 2$, so Item 3 follows. If $1 = m = n$, then $|\operatorname{Supp}(S)| = |\Sigma_1(S)| = |\Sigma_n(S)| \leq m + 1 = 2$, and Item 3 follows. If $1 = m = |S| - n$, then $n = |S| - 1$ and $|\operatorname{Supp}(S)| = |\Sigma_1(S)| = |\sigma(S) - \Sigma_{|S|-1}(S)| = |\Sigma_n(S)| \leq m + 1 = 2$, and Item 3 again follows. So we may now assume $m \geq 2$.

Apply Theorem 14 (with $X = \{0\}$) to $\Sigma_n(S)$ with $S' \mid S$. If Theorem 14.1 holds, then Item 1 follows. Otherwise, in view of $|\Sigma_n(S)| \leq m + 1 = |S'| - n + 1$, let $\mathcal{A} = A_1 \cdot \ldots \cdot A_n$ be the resulting *equitable* setpartition with $Z = \bigcap_{i=1}^{n}(A_i + H)$,

$$\mathsf{S}(\mathcal{A}) \mid S, \quad |\mathsf{S}(\mathcal{A})| = |S'| = n + m, \quad \sum_{i=1}^{n} A_i = \Sigma_n(S) \qquad (47)$$
$$\text{and} \quad |A_i| = 2 \text{ for all } i \in [1, m].$$

CASE 1: For any setpartition $\mathcal{A}$ satisfying (47), we have $|\sum_{i=1}^{j} A_i| \geq |\sum_{i=1}^{j-1} A_i| + 1$ for all $j \in [2, m]$.

In this case, Lemma 6 implies $|\sum_{i=1}^{n} A_i| \geq m + 1$, with equality only possible if equality holds in each estimate $|\sum_{i=1}^{j} A_i| \geq |\sum_{i=1}^{j-1} A_i| + 1$ for $j \in [2, m]$. As this is the case, $|\sum_{i=1}^{j} A_i| = |\sum_{i=1}^{j-1} A_i| + 1$ for all $j \in [2, m]$. Moreover, this must be true under any re-indexing of the A_i with $i \in [1, m]$, whence each A_i is an arithmetic progression with a common difference $d \in G$, and each $\sum_{i=1}^{j} A_i$ is also an arithmetic progression with difference d and length $j + 1$ for $j \in [1, m]$. In particular,

$$3 \leq m + 1 \leq \operatorname{ord}(d),$$

and $\Sigma_n(S) = \sum_{i=1}^{n} A_i$ is an arithmetic progression with difference d, whence either H is trivial or $H = \langle d \rangle$. Thus $\sum_{\substack{i=1 \\ i \neq j}}^{n} A_i$ is aperiodic for any $j \in [1, m]$. Moreover, if $H = \langle d \rangle$, then $m = \operatorname{ord}(d) - 1$ and $\Sigma_n(S) = \sum_{i=1}^{n} A_i$ is a single H-coset, which in view of $|S| > n$ is only possible if $\operatorname{Supp}(S)$ is contained in a single H-coset.

Suppose some pair A_i and A_j are disjoint with $i, j \in [1, m]$, say $A_m = \{x, x + d\}$ and $A_{m-1} = \{y, y + d\}$. Then $y \notin \{x + d, x, x - d\}$ and

$$\Sigma_2(x \cdot (x + d) \cdot y \cdot (y + d)) = \{x + y, x + y + d, x + y + 2d, 2y + d, 2x + d\}$$

is a set of cardinality at least 4. Thus, since

$$A_1 + \ldots + A_{m-2} + \Sigma_2(x \cdot (x + d) \cdot y \cdot (y + d)) + A_{m+1} + \ldots + A_n \subseteq \Sigma_n(S)$$

with $|\Sigma_n(S)| = m + 1$, we must have $m \geq 3$. Now $\sum_{i=1}^{m-2} A_i$ is an arithmetic progression with difference d and length $2 \leq m - 1 \leq \mathrm{ord}(d) - 2$, but $\{x, y\}$ is *not* an arithmetic progression with difference d since $y \notin \{x + d, x, x - d\}$. It follows that $|\sum_{i=1}^{m-2} A_i + \{x, y\}| \geq m + 1$. Thus, since $|\Sigma_n(S)| = m + 1$, we conclude that

$$|\Sigma_n(S)| = |\sum_{i=1}^{m-2} A_i + \{x, y\}| = |\sum_{i=1}^{m-2} A_i + \{x, y\} + \{x + d, y + d\}|, \tag{48}$$

implying that $\sum_{i=1}^{m-2} A_i + \{x, y\}$ is a translate of $\Sigma_n(S) = \sum_{i=1}^{m-2} A_i + \{x, y\} + \{x + d, y + d\} + A_{m+1} + \ldots + A_n$ with $x - y \in H = \mathsf{H}(\sum_{i=1}^{m-2} A_i + \{x, y\})$. In such case, H is nontrivial as $x \neq y$, so we have $m = \mathrm{ord}(d) - 1$ and $\mathrm{Supp}(S) \subseteq x + H = x + \langle d \rangle$ by the observation at the end of the previous paragraph. Letting $\mathcal{A}' = A_1' \cdot \ldots \cdot A_n'$, where $A_{m-1}' = \{x, y\}$, $A_m' = \{x + d, y + d\}$ and $A_i' = A_i$ for $i \neq m - 1, m$, it follows in view of (48) that Item 5 holds. So we can now assume $A_i \cap A_j \neq \emptyset$ for all $i, j \in [1, m]$. Thus, since each A_i is an arithmetic progression with difference d, it follows that there must be some $x \in \bigcap_{i=1}^{m} A_i$ (this is trivially true if $\mathrm{ord}(d) = 3$, as then $m \leq \mathrm{ord}(d) - 1 = 2$). Thus $A_1 \cup \ldots \cup A_m \subseteq \{x - d, x, x + d\}$ with $x \in A_i$ for all $i \in [1, m]$.

If there is some $y \in \mathrm{Supp}(S) \setminus \{x - d, x, x + d\}$, then we can exchange the term equal to $x \pm d$ in A_m with y, resulting in a set $A_m' = \{x, y\}$ that is not an arithmetic progression with difference d, while A_1 remains an arithmetic progression with difference d as $m \geq 2$. Since $\sum_{i=1, i\neq m}^{n} A_i$ is aperiodic, Knseser's Theorem ensures the resulting setpartition (replacing A_m by A_m') satisfies (47), and so repeating the above arguments using the setpartition $A_1 \cdot \ldots A_{m-1} \cdot A_m' \cdot A_{m+1} \cdot \ldots \cdot A_n$ completes the proof. Therefore we may instead assume $\mathrm{Supp}(S) \subseteq \{x - d, x, x + d\}$. Indeed, we may assume $\mathrm{Supp}(S) = \{x - d, x, x + d\}$, else Item 3 holds. If $\mathrm{ord}(d) = 3$, then $2 \leq m \leq \mathrm{ord}(d) - 1$ forces $m = 2$. In this case, $\mathrm{Supp}(S) = x + K$ with $K = \{0, d, -d\}$ a subgroup of size 3, and Item 2 follows. Therefore we can assume $\mathrm{ord}(d) \geq 4$.

Suppose there is a term $y \in \mathrm{Supp}((A_1 \cdot \ldots \cdot A_m)^{[-1]} \cdot S)$ with $y \neq x$. Since $\mathrm{Supp}(S) = \{x - d, x, x + d\}$, we have $y = x \pm d$, say w.l.o.g. $y = x + d$. If $A_i = \{x, x + d\}$ for all $i \in [1, m]$, then either $|\mathrm{Supp}(S)| = 2$, yielding Item 3, or else we can exchange y for some $y' = x - d \in \mathrm{Supp}((A_1 \cdot \ldots \cdot A_m)^{[-1]} \cdot S)$. Thus, swapping y as need be, we obtain that there is some A_i with $i \in [1, m]$, say A_m, with $y \notin A_m$. Then w.l.o.g. $y = x + d$ and $A_m = \{0, x - d\}$. Note we either have $y \in \mathrm{Supp}(\mathsf{S}(\mathcal{A})^{[-1]} \cdot S)$ or $A_k = \{y\} = \{x + d\}$ for some $k > m$. Define a new setpartition $\mathcal{A}' = A_1' \cdot \ldots \cdot A_n'$ with $A_i' = A_i$ for $i \leq m$, $A_m' = \{x - d, x + d\}$, and either $A_i' = A_i$ for all $i > m$ (if $y \in \mathrm{Supp}(\mathsf{S}(\mathcal{A})^{[-1]} \cdot S)$) or else $A_k' = \{x\}$ and $A_i' = A_i$ for all $i \in [m + 1, n] \setminus \{k\}$ (if $y \notin \mathrm{Supp}(\mathsf{S}(\mathcal{A})^{[-1]} \cdot S)$). Since $\mathrm{ord}(d) \geq 4$, it follows that A_m' is not an arithmetic progression with difference d, while each A_i' with $i \in [1, m - 1]$ is. Since $m \geq 2$, repeating the above arguments using the setpartition $\mathcal{A}'$ completes the proof (as $\sum_{i=1, i\neq m}^{n} A_i$ is aperiodic, Kneser's Theorem ensures (47) holds for $\mathcal{A}'$). So we instead assume $\mathrm{Supp}((A_1 \cdot \ldots \cdot A_m)^{[-1]} \cdot S) \subseteq \{x\}$. Combined with $x \in A_i$ for all $i \in [1, m]$, we find $\mathsf{v}_x(S) = \mathsf{h}(S) \geq |S| - m$, and now Item 4 holds, completing CASE 1.

CASE 2: There is some setpartition $\mathcal{A}$ satisfying (47) with $|\sum_{i=1}^{n} A_i| = |\sum_{\substack{i=1\\ i\neq j}}^{n} A_i|$ for some $j \in [2, m]$.

By case hypothesis and Kneser's Theorem, $H = \mathsf{H}(\Sigma_n(S))$ is nontrivial. If $m + 1 = |\Sigma_n(S)| = |H|$, then $\Sigma_n(S)$ is an H-coset, which in view of $|S| > n$ is only possible if $\mathrm{Supp}(S) \subseteq x + H$ for some $x \in G$. Hence Item 5 holds in view of the case hypothesis. So we now assume $|\sum_{i=1}^{n} A_i| = |\Sigma_n(S)| \geq 2|H|$. Thus $\sum_{i=1}^{n} |\phi_H(A_i)| \geq n + 1$. Let $\mathcal{B} = B_1 \cdot \ldots \cdot B_n$ be a setpartition with $\mathsf{S}(\mathcal{B}) \mid S$, $|\mathsf{S}(\mathcal{B})| = n + m$, and $\sum_{i=1}^{n} B_i = \Sigma_n(S)$ such that, letting $I_2 \subseteq [1, n]$ be the subset of all $i \in [1, n]$ with $|\phi_H(B_i)| \geq 2$, the following hold

M1. For each $i \in I_2$, there is some $b_i \in B_i$ such that $|\phi_H\big(B_i \setminus (b_i + H)\big)| = |B_i \setminus (b_i + H)|$, and
M2. either $\sum_{i\in I_2} |\phi_H(B_i)| > 2|I_2|$ or $|\phi_H(B_{i'})| = |B_{i'}|$ for some $i' \in I_2$.

Since $|A_i| \leq 2$ for all i and $\sum_{i=1}^{n} |\phi_H(A_i)| \geq n + 1$, $\mathcal{A}$ satisfies all these hypotheses. Let $I_1 = [1, n] \setminus I_2$ be the subset of all $i \in [1, n]$ with $|\phi_H(B_i)| = 1$, and re-index the B_i so that $I_1 = [1, |I_1|]$. Kneser's Theorem implies

$$\begin{aligned}
|\sum_{i=1}^{n} B_i| &\geq \sum_{i\in I_2} |B_i + H| - (|I_2| - 1)|H| \\
&\geq \sum_{i\in I_2} |B_i| + (|I_2| + 1)(|H| - 1) - (|I_2| - 1)|H| \\
&= \sum_{i\in I_2} |B_i| - |I_2| + (2\,H| - 1),
\end{aligned}$$

with the latter inequality in view of conditions M1 and M2 (note that $\sum_{i\in I_2} |\phi_H(B_i)| \geq 2|I_2|$ holds trivially in view of $|\phi_H(B_i)| \geq 2$ for $i \in I_2$). Combined with the inequality $|\sum_{i=1}^{n} B_i| = |\Sigma_n(S)| \leq |S'| - n + 1 = \sum_{i=1}^{n} |B_i| - n + 1 = \sum_{i=1}^{n} |B_i| - |I_1| - |I_2| + 1$, we find

$$\sum_{i\in I_1} |B_i| \geq |I_1| + 2\,H| - 2. \tag{49}$$

Consequently, I_1 is nonempty, and since we trivially have $|\sum_{i\in I_1} B_i| \leq |H|$ (as each B_i with $i \in I_1$ is contained in an H-coset), it follows that $|\sum_{i\in I_1} B_i| \leq |H| \leq \sum_{i\in I_1} |B_i| - |I_1| - (|H| - 2) < \sum_{i\in I_1} |B_i| - |I_1| + 1$, with the later inequality holding since H is nontrivial (as noted at the start of the case). Lemma 6 now implies there is some $j \in [2, |I_1|]$ with $|\sum_{i=1}^{j} B_i| < |\sum_{i=1}^{j-1} B_i| + |B_j| - 1$, in which case Theorem 5 implies

$$\sum_{i=1}^{j-1} B_i + (B_j \setminus \{y\}) = \sum_{i=1}^{j} B_i \quad \text{for all } y \in B_j. \tag{50}$$

In particular, $|B_j| \geq 2$. Also, since $|B_i| \geq 2$ for all $i \in I_2$, and since $\sum_{i=1}^{n} |B_i| = |S'| \leq 2n$, it follows that

$$\sum_{i \in I_1} |B_i| \leq 2|I_1| \quad \text{and} \quad |I_1| \geq 2\,H| - 2 \geq |H|, \tag{51}$$

with the latter inequality above following from the former combined with (49).

Now additionally assume that our setpartition $\mathcal{B}$ is chosen, subject to $\mathsf{S}(\mathcal{B}) \mid S$, $|\mathsf{S}(\mathcal{B})| = n + m$, $\sum_{i=1}^{n} B_i = \Sigma_n(S)$, M1 and M2, so that

M3. $\sum_{i=1}^{n} |\phi_H(B_i)|$ is maximal.

Since $\mathcal{A}$ satisfies the defining conditions for $\mathcal{B}$, we have $\sum_{i=1}^{n} |\phi_H(B_i)| \geq \sum_{i=1}^{n} |\phi_H(A_i)| \geq n+1$, ensuring that I_2 is nonempty. We claim that this ensures $B_j + H \subseteq B_i$ for all $i \in [1, n]$, where $j \in I_1$ is the index defined above. Indeed, if this fails, then there is some $x \in B_j$ and $k \in [1, n]$ with $\phi_H(x) \notin \phi_H(B_k)$. In this case, remove x from B_j and place it in B_k to yield a new setpartition $\mathcal{B} = B_1' \cdot \ldots \cdot B_n'$, where $B_j' = B_j \setminus \{x\}$, $B_k' = B_k \cup \{x\}$ and $B_i' = B_i$ for $i \neq j, k$. In view of (50), we have $\mathsf{S}(\mathcal{B}') = \mathsf{S}(\mathcal{B})$, $|\mathsf{S}(\mathcal{B}')| = |\mathsf{S}(\mathcal{B})| = n + m$ and $\sum_{i=1}^{n} B_i' = \Sigma_n(S)$. Since $\phi_H(x) \notin \phi_H(B_k)$, it follows that x is the unique element from its H-coset in B_k', so M1 and M2 also hold for $\mathcal{B}'$. However, since $|\phi_H(B_j)| = |\phi_H(B_j')| = 1$ and $|\phi_H(B_k')| = |\phi_H(B_k)| + 1$, we see that $\mathcal{B}'$ contradicts the maximality of $\sum_{i=1}^{n} |\phi_H(B_i)|$ for $\mathcal{B}$ given in M3. Therefore, $B_j + H \subseteq B_i$ for all i, as claimed. Letting $x \in B_j$ and recalling that B_j is contained in an H-coset (as $j \in I_1$), it follows that $x + H = \bigcap_{i=1}^{n} (B_i + H)$. Likewise, if there were some $y \in \mathrm{Supp}(\mathsf{S}(\mathcal{B})^{[-1]} \cdot S)$ with $\phi_H(y) \neq \phi_H(x)$, then we could remove x from B_j and place y in B_j to yield a new setpartition $\mathcal{B} = B_1' \cdot \ldots \cdot B_n'$, where $B_j' = B_j \setminus \{x\} \cup \{y\}$ and $B_i' = B_i$ for $i \neq j$, which would again contradict the maximality of $\mathcal{B}$ given in M3. Therefore we may assume otherwise. In summary,

$$\mathrm{Supp}(\mathsf{S}(\mathcal{B})^{[-1]} \cdot S) \subseteq x + H = \bigcap_{i=1}^{n} (B_i + H). \tag{52}$$

Claim A: $(y + H) \cap B_i = \{y\}$ for any $i \in [1, n]$ and $y \in B_i \setminus (x + H)$.

Proof Assume by contradiction there is some $k \in [1, n]$ and $y \in B_k \setminus (x + H)$ with $|(y + H) \cap B_k| = r \geq 2$. Since $B_i \subseteq x + H$ for each $i \in I_1$ by (52), we must have $k \in I_2$. Let $\mathcal{C} = C_1 \cdot \ldots \cdot C_n$ be a setpartition with $\mathsf{S}(\mathcal{C}) = \mathsf{S}(\mathcal{B})$ and $\sum_{i=1}^{n} B_i = \Sigma_n(S)$ such that $C_i = B_i$ for all $i \in I_2 \setminus \{k\}$, $C_k \setminus B_k \subseteq x + H$, $C_k \cap B_k = B_k \setminus \{y_1, \ldots, y_t\}$, $C_{|I_1|+1-i} \setminus (x + H) = \{y_i\} \subset C_{|I_1|+1-i}$ for $i \in [1, t]$, where $y_1, \ldots, y_t \in (y + H) \cap B_k$ are $t \in [0, r - 1]$ distinct elements, and (subject to these conditions) $|(x + H) \cap C_k|$ is maximal, and then (subject to prior conditions) $t \geq 0$ is maximal. Note $\mathcal{B}$ satisfies these conditions with $t = 0$, so $\mathcal{C}$ exists. The defining conditions for

$\mathcal{C}$ ensure

$$I_2' = I_2 \cup [|I_1| + 1 - t, |I_1|]$$

is the subset of indices $i \in [1, n]$ with $|\phi_H(C_i)| \geq 2$ and that M1 holds for all C_i with $i \in I_2' \setminus \{k\}$.

Suppose $t = r - 1$. The defining conditions for $\mathcal{C}$ along with M1 for $\mathcal{B}$ ensure all elements from $C_k \setminus (\{x, y\} + H)$ are the unique element from their H-coset in C_k with $|(y + H) \cap C_k| = r - t$. Thus, since $t = r - 1$, we see that M1 holds for $\mathcal{C}$. The defining conditions for $\mathcal{C}$ ensure $\phi_H(C_i) = \phi_H(B_i)$ for $i \in I_2$ and $i \in I_1 \setminus [|I_1| + 1 - t, |I_1|]$, while $\phi_H(B_{|I_1|+1-i}) \subset \phi_H(C_{|I_1|+1-i})$ for $i \in [1, t]$; moreover, $C_i = B_i$ for $i \in I_2 \setminus \{k\}$. Thus, since M2 holds for $\mathcal{B}$, it also holds for $\mathcal{C}$ (note $i' \neq k$ in M2 as $|(y + H) \cap B_k| \geq 2$), and $\sum_{i=1}^n |\phi_H(C_i)| = \sum_{i=1}^n |\phi_H(B_i)| + t = \sum_{i=1}^n |\phi_H(B_i)| + r - 1 > \sum_{i=1}^n |\phi_H(B_i)|$. Hence $\mathcal{C}$ contradicts the maximality condition M3 for $\mathcal{B}$. So we instead assume $t < r - 1$, meaning $|(y + H) \cap C_k| = r - t \geq 2$.

Suppose $(x + H) \cap C_k = x + H$. Then $C_i \subseteq x + H \subseteq C_k$ for all $i \in [1, |I_1| - t]$ (since $\phi_H(C_i) = \phi_H(B_i)$ for all $i \notin [|I_1| + 1 - t, |I_1|]$). Let $y_{t+1} \in (y + H) \cap C_k$. In view of (51), we have $|I_1| \geq |H| \geq r \geq t + 2$, so we can define a new setpartition $\mathcal{C}' = C_1' \cdot \ldots \cdot C_n'$, where $C_k' = C_k \setminus \{y_{t+1}\}$, $C_{|I_1|-t}' = C_{|I_1|-t} \cup \{y_{t+1}\}$, and $C_i' = C_i$ for all $i \neq k, |I_1| - t$. Then $\mathsf{S}(\mathcal{C}') = \mathsf{S}(\mathcal{C})$. We have $C_{|I_1|-t} \subseteq x + H = (x + H) \cap C_k$ and $y_{t+1} \in y + H \neq x + H$. Thus $C_{|I_1|-t} \subseteq C_k \setminus \{y_{t+1}\}$ and $C_{|I_1|-t} + C_k \subseteq (C_{|I_1|-t} \cup \{y_{t-1}\}) + (C_k \setminus \{y_{t+1}\})$, ensuring $\Sigma_n(S) = \sum_{i=1}^n C_i \subseteq \sum_{i=1}^n C_i' \subseteq \Sigma_n(S)$, forcing equality to hold. But now, since $t + 1 \leq r - 1$, we see that $\mathcal{C}'$ contradicts the maximality of t for $\mathcal{C}$. So we instead conclude that

$$(x + H) \cap C_k \subset x + H. \tag{53}$$

Note $\rho := |H| - |(x + H) \cap C_k| \geq 1$ by (53). Since $C_k \setminus B_k \subseteq x + H$ and $C_k \cap B_k = B_k \setminus \{y_1, \ldots, y_t\}$ with $t < r$, it follows from M1 for $\mathcal{B}$ that

$$|(C_k + H) \setminus C_k| \geq (|\phi_H(C_k)| - 2)(|H| - 1) + t + \rho \geq t + 1. \tag{54}$$

We also have $|(C_i + H) \setminus C_i| \geq (|\phi_H(C_i)| - 1)(|H| - 1) \geq |H| - 1$ for $i \in I_2 \setminus \{k\}$ (by M1 for $\mathcal{B}$), either $|\phi_H(C_k)| \geq 3$ (improving the final estimate in (54) by $|H| - 1$) or $|(C_i + H) \setminus C_i| \geq 2(|H| - 1)$ for some $i \in I_2 \setminus \{k\}$ (by M2 for $\mathcal{B}$, noting that $i' \neq k$ in view of $|(y + H) \cap B_k| \geq 2$), and $|(C_i + H) \setminus C_i| = |H| - 1$ for $i \in [|I_1| + 1 - t, |I_1|]$. As a result,

$$\sum_{i \in I_2'} |(C_i + H) \setminus C_i| \geq \sum_{i \in I_2'} |C_i| + |I_2'|(|H| - 1) + t + 1.$$

Combining this estimate with Kneser's Theorem, we obtain $|\sum_{i=1}^n C_i| \geq \sum_{i \in I_2'} |C_i + H| - (|I_2'| - 1)|H| \geq \sum_{i \in I_2'} |C_i| - |I_2'| + t + 1 + |H|$. Combined with the inequality $|\sum_{i=1}^n C_i| = |\Sigma_n(S)| \leq |S'| - n + 1 = \sum_{i=1}^n |C_i| - n + 1 = \sum_{i=1}^n |C_i| - |I_1'| - |I_2'| + 1$, where $I_1' := [1, n] \setminus I_2' = [1, |I_1| - t]$, we find

$$\sum_{i\in I_1'}|C_i| \geq |I_1'| + |H| + t \geq |I_1'| + |H|. \tag{55}$$

Consequently, since we trivially have $|\sum_{i\in I_1'} C_i| \leq |H|$ (as each C_i with $i \in I_1'$ is contained in an H-coset), it follows that we have $|\sum_{i\in I_1'} C_i| \leq |H| < \sum_{i\in I_1'}|C_i| - |I_1'| + 1$. As before, this ensures via Lemma 6 and Theorem 5 that there is some $j' \in [2, |I_1'|]$ with

$$\sum_{i=1}^{j'-1} C_i + (C_{j'} \setminus \{z\}) = \sum_{i=1}^{j} C_i \quad \text{for all } z \in C_{j'}. \tag{56}$$

Suppose $C_{j'} \subseteq C_k$. Let $y_{t+1} \in (y+H) \cap C_k$. In view of (51), we have $|I_1| \geq |H| \geq r \geq t+2$, so we can define a new setpartition $\mathcal{C}' = C_1' \cdot \ldots \cdot C_n'$, where $C_k' = C_k \setminus \{y_{t+1}\}$, $C_{j'}' = C_{j'} \cup \{y_{t+1}\}$, and $C_i' = C_i$ for all $i \neq k, j'$. Then $\mathsf{S}(\mathcal{C}') = \mathsf{S}(\mathcal{C})$. We have $C_{j'} \subseteq C_k$, and thus $C_{j'} \subseteq C_k \setminus \{y_{t+1}\}$ (since $j' \in I_1'$ ensures $C_{j'} \subseteq x + H \neq y + H$ and $y_{t+1} \in y + H$). Hence $C_{j'} + C_k \subseteq (C_{j'} \cup \{y_{t+1}\}) + (C_k \setminus \{y_{t+1}\})$, ensuring $\Sigma_n(S) = \sum_{i=1}^n C_i \subseteq \sum_{i=1}^n C_i' \subseteq \Sigma_n(S)$, forcing equality to hold. But now, since $t+1 \leq r-1$, we see that $\mathcal{C}'$ contradicts the maximality of t for $\mathcal{C}$ (re-indexing the C_i' with $i \in I_1'$ so that $j' = |I_1'|$). So we instead conclude that $C_{j'} \nsubseteq C_k$.

Since $C_{j'} \nsubseteq C_k$ and $C_{j'} \subseteq x + H$ (as $j' \in I_1'$), there is some $z \in C_{j'} \setminus C_k$ with $z \in x + H$. Define a new setpartition $\mathcal{C}' = C_1' \cdot \ldots \cdot C_n'$, where $C_{j'}' = C_{j'} \setminus \{z\}$, $C_k' = C_k \cup \{z\}$, and $C_i' = C_i$ for all $i \neq j', k$. Then $\mathsf{S}(\mathcal{C}') = \mathsf{S}(\mathcal{C})$, and (56) ensures $\Sigma_n(S) = \sum_{i=1}^n C_i \subseteq \sum_{i=1}^n C_i' \subseteq \Sigma_n(S)$, in which case equality holds. But now $|(x+H) \cap C_k'| = |(x+H) \cap C_k| + 1$, so that $\mathcal{C}'$ contradicts the maximality of $|(x+H) \cap C_k|$ for $\mathcal{C}$, completing Claim A □

Since H is nontrivial (as noted at the start of CASE 2) and $m+1 = |\Sigma_n(S)| \leq |S'| - n + 1 = \sum_{i=1}^n |B_i| - n + 1$, Claim A allows us to apply Lemma 8 to $\mathcal{B}$ (with $X = \{0\}$), giving the existence of a setpartition $\mathcal{C} = C_1 \cdot \ldots \cdot C_\ell$ with $\mathsf{S}(\mathcal{C}) = \mathsf{S}(\mathcal{B})$, $\sum_{i=1}^n C_i = \sum_{i=1}^n B_i = \Sigma_n(S)$, $(x+H) \subseteq Z = \bigcap_{i=1}^n (C_i + H)$, and $|C_i \setminus Z| \leq 1$ for all i. If $Z \neq x + H$, then $m = n$ and $|\phi_H(C_i)| = 2$ for all i (recall $|S'| = n + m \leq 2n$), whence Kneser's Theorem implies $|S'| - n + 1 = |\Sigma_n(S)| = |\sum_{i=1}^n C_i| \geq (n+1)|H| \geq (|S'| - n + 1)|H|$, contradicting that H is nontrivial. Therefore $Z = x + H$. It necessarily follows that $\sum_{i=1}^n |\phi_H(C_i)| = \sum_{i=1}^n |\phi_H(B_i)|$ since $x + H = \bigcap_{i=1}^n (B_i + H) = \bigcap_{i=1}^n (C_i + H)$ with $|(y+H) \cap C_i| \leq 1$ and $|(y+H) \cap B_i| \leq 1$ for all $i \in [1, n]$ and $y + H \neq x + H$ (cf. Claim A and Lemma 8). Applying Lemma 12 (with $X = \{0\}$) allows us to replace $\mathcal{C}$ with a setpartition having all the defining properties for $\mathcal{C}$ and which is equitable (Lemma 12.1 cannot hold since $H = \mathsf{H}(\Sigma_n(S))$ is nontrivial), so we gain that $|C_i| \leq 2$ for all i. In doing so, we find that $\mathcal{C}$ now satisfies the defining conditions for $\mathcal{B}$. Thus we can w.l.o.g. assume the setpartition $\mathcal{B}$ defined above has $|B_i| \leq 2$ for all i. In view of (49), there are at least $2|H| - 2 \geq |H|$ sets B_i with $|B_i| = 2$ and $i \in I_1$. Since $|\sum_{i \in I_1} B_i| \leq |H|$ in view of each B_i being contained in an H-coset for $i \in I_1$, it now follows by a simple greedy algorithm [28, Proposition 2.2] that there is a subset $J_1 \subset I_1$ with $|J_1| \leq |H| - 1$ and $|\sum_{i \in J_1} B_i| = |\sum_{i \in I_1} B_i|$. Recalling (52), we find Item 5 holds using the setpartition $\mathcal{B}$, completing the case and proof. □

References

1. S. Adhikari, Plus-minus weighted zero-sum constants: a survey, in Analytic number theory, modular forms and q-hypergeometric series, 1-9, *Springer Proc. Math. Stat.* 221, Springer (2017).
2. A. Bialostocki and P. Dierker, On the Erdős-Ginzburg-Ziv theorem and the Ramsey numbers for stars and matchings, *Discrete Math.*, **110** (1992), no. 1-3, 1–8.
3. A. Bialostocki and M. Lotspeich, Developments of the Erdős-Ginzburg-Ziv Theorem I, in Sets, graphs and numbers, *Colloq. Math. Soc. János Bolyai* **60** (1992), North-Holland, Amsterdam, 1992, 97–117
4. A. Bialostocki, P. Erdős and H. Lefmann, Monochromatic and zero-sum sets of nondecreasing diameter, *Discrete Math.* **137** (1995), no. 1-3, 19–34.
5. A. Bialostocki, P. Dierker, D. J. Grynkiewicz and M. Lotspeich, On some developments of the Erdős-Ginzburg-Ziv theorem II, *Acta Arith.* **110** (2003), no. 2, 173–184.
6. B. Bollobás and I. Leader, The number of k-sums modulo k, *J. Number Theory* **78** (1999), 27–35.
7. H. Q. Cao, An Addition Theorem on the Cyclic Group $\mathbb{Z}/p^{\alpha}q^{\beta}\mathbb{Z}$, *Electronic J. Combin.* **13** (2006) (electronic).
8. Y. Caro, Zero-sum problems–a survey, *Discrete Math.* **152** (1996), no. 1-3, 93–113.
9. M. DeVos, L. Goddyn and B. Mohar, A generalization of Knesers addition theorem, *Adv. Math.* **220** (2009), no. 5, 1531–1548.
10. P. Erdős, A. Ginzburg, A. Ziv, Theorem in Additive Number Theory, *Bull. Res. Council Israel* **10F** (1961), 41–43.
11. P. Erdős and R. L. Graham, Old and new problems and results in combinatorial number theory, Monographies de L'Enseignement Mathmatique [Monographs of L'Enseignement Mathmatique] **28** (1980), Université de Genéve, L'Enseignement Mathématique, Geneva, pp. 95.
12. P. Erdős and E. Szemerédi, On a problem of Graham, *Publ. Math. Debrecen*, **23** (1976), no. 1–2, 123–127.
13. Z. Füredi and D. J. Kleitman, The minimal number of zero sums, in Combinatorics, Paul Erdős is eighty **1**, Bolyai Soc. Math. Stud., János Bolyai Math. Soc. (1993), Budapest, 159–172.
14. W. Gao, Addition theorems for finite abelian groups, *J. Number Theory* **53** (1995), no.2, 241–246.
15. W. Gao, A. Geroldinger and D. J. Grynkiewicz, Inverse Zero-Sum Problems III, *Acta Arithmetica* **141** (2010), 103–152.
16. W. Gao, D. Han, H. Zhang, The EGZ-constant and short zero-sum sequences over finite abelian groups,*J. Number Theory* 162 (2016), 601–613.
17. W. Gao, R. Thangadurai and J. Zhuang, Addition theorems on the cyclic groups of order p^l, *Discrete Mathematics* **308** (2008), no. 10, 2030–2033.
18. W. Gao, D. J. Grynkiewicz and X. Xia, On n-Sums in an Abelian Group, *Combin. Prob. and Comput.* **25** (2016), no. 3, 419–435.
19. W. Gao, Y. Li, P. Yuan and J. Zhuang, On the structure of long zero-sum free sequences and n-zero-sum free sequences over finite cyclic groups, *Arch. Math. (Basel)* **105** (2015), no. 4, 361–370.
20. A. Geroldinger, D. J. Grynkiewicz and J. Oh, *title*, submitted.

21. A. Geroldinger and I. Ruzsa, *Combinatorial Number Theory and Additive Group Theory*, Birkhäuser (2009), Basel.
22. A. Geroldinger and F. Halter-Koch, *Non-Unique Factorizations: Algebraic, Combinatorial and Analytic Theory*, Pure and Applied Mathematics: A series of Monographs and Textbooks **278** (2006), Chapman & Hall, an imprint of Taylor & Francis Group, Boca Raton, FL.
23. A. Geroldinger and D. J. Grynkiewicz, The Large Davenport Constant I: Groups with a Cyclic, Index 2 Subgroup, *J. Pure and Applied Alg.* **217** (2013), no. 5, 863–885.
24. E. Balandraud and B. Girard, A Nullstellensatz for sequences over $\mathbf{F}_p$, *Combinatorica* 34 (2014), no. 6, 657–688.
25. D. J. Grynkiewicz, On four colored sets with nondecreasing diameter and the Erdős-Ginzburg-Ziv Theorem, *J. Combin. Theory, Ser. A* **100** (2002), no. 1, 44–60.
26. D. J. Grynkiewicz, On a partition analog of the Cauchy-Davenport theorem, *Acta Math. Hungar.* **107** (2005), no. 1–2, 161–174.
27. D. J. Grynkiewicz, On a conjecture of Hamidoune for subsequence sums, *Integers* **5** (2005), no. 2, A7 (electronic).
28. D. J. Grynkiewicz, An extension of the Erdős-Ginzburg-Ziv theorem to hypergraphs, *European J. Combin.* **26** (2005), no. 8, 1154–1176.
29. D. J. Grynkiewicz, A weighted Erdős-Ginzburg-Ziv Theorem, *Combinatorica* **26** (2006), no. 4, 445–453.
30. D. J. Grynkiewicz, On the number of m-term zero-sum subsequences, *Acta Arith.* **121** (2006), no. 3, 275–298.
31. D. J. Grynkiewicz, Note on a Conjecture of Graham, *European J. Combin.*. **32** (2011), no. 8, 1336–1344.
32. D. J. Grynkiewicz and R. Sabar, Monochromatic and zero-sum sets of nondecreasing modified diameter, *Electron. J. Combin.* **13** (2006), no. 1, Research Paper 28 (electronic).
33. D. J. Grynkiewicz and A. Schultz, A Five Color Zero-Sum Generalization, in *Graphs and Combin.* **22** (2006), 335–339.
34. D. J. Grynkiewicz, O. Ordaz, M. T. Varela and F. Villarroel, On Erdős-Ginzburg-Ziv inverse theorems, *Acta Arith.* **129** (2007), no. 4, 307–318.
35. D. J. Grynkiewicz, Quasi-Periodic Decompositions and the Kemperman Structure Theorem, *European J. Combin.* 26 (2005), 559–575.
36. D. J. Grynkiewicz, E. Marchan and O. Ordaz, A Weighted Generalization of Two Theorems of Gao, *Ramanujan J.* **28** (2012), no. 3, 323–340.
37. D. J. Grynkiewicz, *Structural Additive Theory*, Developments in Mathematics **30**, Springer (2013), Switzerland.
38. D. J. Grynkiewicz, Iterated Sumsets and Setpartitions, to appear in *Ramanujan J.*
39. Y. O. Hamidoune, O. Ordaz and A. Ortuño, On a combinatorial theorem of Erdős, Ginzburg and Ziv, *Combin. Probab. Comput.* **7** (1998), no. 4, 403–412.
40. Y. O. Hamidoune, Subsequence Sums, *Combin. Propab. Comput.* **12** (2003), 413–42.
41. M. Kisin, The number of zero sums modulo m in a sequence of length n, *Mathematika* **41** (1994), no. 1, 149–163.
42. M. Kneser, Ein Satz über abelsche Gruppen mit Anwendungen auf die Geometrie der Zahlen, *Math. Z.* **61** (1955), 429–434.
43. B. Landman and A. Robertson, *Ramsey Theory over the Integers*, 2nd Edition, Student Mathematical Library 73, American Mathematical Society, Providence (2014).
44. H. B. Mann, Two addition theorems, *J. Combinatorial Theory* **3** (1967), 233–235.
45. M. B. Nathanson, *Additive Number Theory: Inverse Problems and the Geometry of Sumsets*, Springer (1996), Harrisonburg, VA.
46. J. E. Olson, An addition theorem for finite abelian groups, *J. Number Theory* **9** (1977), no. 1, 63–70.
47. O. Ordaz, A. Philipp, I. Santos, and W. A. Schmid, On the Olson and the strong Davenport constants, *J. Théor. Nombres Bordeaux* 23 (2011), no. 3, 715–750.
48. G. Wang, W. Gao, Y. Li, and J. Peng, A unifying look at zero-sum invariants, *Int. J. Number Theory* 14 (2018), no. 3, 705–711.

49. P. Yuan and X. Zeng, On the structure of long unsplittable minimal zero-sum sequences, *Acta Arith.* 176 (2016), no. 2, 131–159.
50. P. Yuan and X. Zeng, Two conjectures on an addition theorem, *Acta Arith.* **148** (2011), no. 4, 395–411.

Bounds on Point Configurations Determined by Distances and Dot Products

Slade Gunter, Eyvi Palsson, Ben Rhodes, and Steven Senger

Abstract We study a family of variants of Erdő's unit distance problem, concerning distances and dot products between pairs of points chosen from a large finite point set. Specifically, given a large finite set of n points E, we look for bounds on how many subsets of k points satisfy a set of relationships between point pairs based on distances or dot products. We survey some of the recent work in the area and present several new, more general families of bounds.

Keywords Point Configurations

1 Introduction

Paul Erdős introduced the single distance problem in [8]. It asks how often a single distance can occur between pairs of points chosen from a large finite point set in the plane. While there have been many generalizations and variants of this original problem involving distances, there has also been work on pursuing similar questions for dot products, as in [12, 24]. Moreover, configurations involving more than two points have also seen attention. See [1, 3, 15] for examples and applications. The book by Brass et al. [5], details many related problems on point configurations.

We offer some upper and lower bounds on general families of point configurations determined by distances or dot products between pairs of points. In what follows, if two quantities, $X(n)$ and $Y(n)$, vary with respect to some natural number parameter, n, then we write $X(n) \lesssim Y(n)$ if there exist constants, C and N, both independent of n, such that for all $n > N$, we have $X(n) \leq CY(n)$. If $X(n) \lesssim Y(n)$ and $Y(n) \lesssim X(n)$, we write $X(n) \approx Y(n)$. By convention, we will always assume that the parameters associated to the configurations, h and k, are like constants compared to the size parameter n.

S. Gunter · E. Palsson · B. Rhodes · S. Senger (✉)
Missouri State University, Springfield, U.S.
e-mail: stevensenger@gmail.com

M. B. Nathanson (ed.), *Combinatorial and Additive Number Theory IV*,
Springer Proceedings in Mathematics & Statistics 347,
https://doi.org/10.1007/978-3-030-67996-5_12

1.1 Background

We begin by recording some of the best estimates for some often studied point configurations. In 1984, Spencer, Szemerédi, and Trotter gave the least known upper bound on the number of times that a distance can occur in a large finite point set, in [22].

Theorem 1 ([22]) *Given a large finite set of n points in the plane, the maximum number of times that any distance can occur is bounded above by* $\lesssim n^{\frac{4}{3}}$.

There is still a wide gap between this bound and Erdős' conjectured upper bound of $n^{1+\epsilon}$, for any $\epsilon > 0$. However, an easy consequence of the celebrated Szemerédi–Trotter point-line incidence theorem from [23] yields the same bound for dot products, which happens to be sharp. Though they did not explicitly state this bound, it is a direct corollary of their main result. We make this connection explicit below.

Theorem 2 ([23]) *Given a large finite set of n points in the plane, the maximum number of times that any nonzero dot product can occur is bounded above by* $\lesssim n^{\frac{4}{3}}$. *Moreover, this bound is sharp.*

1.1.1 *k*-Chains

The first type of generalization that we introduce here is a *k-chain*, which is a sequence of distinct points restricted by the values of the distances or dot products determined by successive pairs. We borrow notation from related problems on distances in [4, 19, 20]. Specifically, if we fix a k-tuple of real numbers, $(\alpha_1, \alpha_2, \ldots, \alpha_k)$, then a distance *k-chain* of that type is a $(k+1)$-tuple of points, $(x_1, x_2, \ldots, x_{k+1})$, such that for all $j = 1, \ldots, k$, we have $|x_j - x_{j+1}| = \alpha_j$. A dot product k-chain is defined similarly, except with $\alpha_j = x_j \cdot x_{j+1}$ in place of the distance. For example, if we fix a triple of real numbers, (α, β, γ), a distance 3-chain of that type will be a set of four points, where the distance between the first two points is α, the distance between the middle two points is β, and the distance between last two points is γ. We follow convention and refer to 2-chains as *hinges*. In [2, 21], Dan Barker and the fourth listed author gave the following tight bounds on the number of hinges (2-chains) in a large finite point set in the plane.

Theorem 3 ([2, 21]) *Given a large, finite set of n points in* $\mathbb{R}^2$, *and a pair of nonzero real numbers* (α_1, α_2), *the maximum number distance or dot product hinges of type* (α_1, α_2) *is no more than* $\lesssim n^2$. *Moreover, this bound is tight.*

While distances and dot products have had similar behavior in these first few estimates, their paths diverge for longer chains. In [20], Adam Sheffer, and the second and fourth listed authors gave upper and lower bounds on the maximum possible number of occurrences of k-chains of a given type in large finite point sets in the plane. Many of these bounds were improved by Frankl and Kupavskii, in [11], where they give the best known upper bounds on longer distance k-chains, which we record below.

Theorem 4 (Theorem 2 in [11]) *Given a large, finite set of n points in* $\mathbb{R}^2$, *any* $\epsilon > 0$, *and a natural number k, the maximum number of distance k-chains of the type* $(\alpha_1, \ldots, \alpha_k)$ *that can exist in the set is no more than*

$$\lesssim \begin{cases} n^{\frac{k+3}{3}} & \text{if } k \equiv 0 \pmod 3, \\ n^{\frac{k+3}{3}+\epsilon} & \text{if } k \equiv 1 \pmod 3, \\ n^{\frac{k+4}{3}} & \text{if } k \equiv 2 \pmod 3. \end{cases}$$

In the case of dot products, the best known upper bounds are from Shelby Kilmer, Caleb Marshall, and the fourth listed author, in [16]. Here and in what follows, we will bound results on dot products that are nonzero, as there are degeneracies that can occur for these types of estimates when we consider zero dot products. See [16] for more details.

Theorem 5 ([16]) *Given a large, finite set of n points in* $\mathbb{R}^2$ *and a natural number k, the maximum number of dot product k-chains of the type* $(\alpha_1, \ldots, \alpha_k)$ *that can exist in the set is* $\lesssim n^{\frac{2(k+1)}{3}}$.

Although the upper bounds are clearly larger in the case of dot products, it is unclear if they are closer to or further from the truth than what is known about distances, as they exhibit fundamentally distinct behaviors for $k \geq 3$, as the following results will show. We now turn our attention to lower bounds on the maximum number of k-chains of a given type we can construct.

Theorem 6 (Theorem 2 in [11]) *Given a large, finite natural number n, and a small natural number k, there exists a set of n points in* $\mathbb{R}^2$, *and a k-tuple* $(\alpha_1, \ldots, \alpha_k)$, *such that the set has a number k-chains of the type* $(\alpha_1, \ldots, \alpha_k)$ *that is at least*

$$\gtrsim \begin{cases} n^{\frac{k+3}{3}} & \text{if } k \equiv 0 \pmod 3, \\ n^{\frac{k+2}{3}+\epsilon} & \text{if } k \equiv 1 \pmod 3, \\ n^{\frac{k+4}{3}} & \text{if } k \equiv 2 \pmod 3. \end{cases}$$

In the case of dot products, there is a family of constructions demonstrating a lower bound for the maximum number of dot product k-chains in a large finite set of points in the plane.

Proposition 1 ([16]) *Given a large, finite natural number n, and a small natural number k, there exists a set of n points in* $\mathbb{R}^2$ *and a k-tuple of nonzero real numbers* $(\alpha_1, \ldots, \alpha_k)$, *for which there are at least* $n^{\lceil (k+1)/2 \rceil}$ *instances of k-chains of the type* $(\alpha_1, \ldots, \alpha_k)$ *in the set.*

Comparing these results shows that the behaviors of distances and dot products are distinct for $k \geq 4$. Indeed, if we fix $k = 4$, Theorem 1 gives that there exists a set of n points in the plane with $\gtrsim n^3$ occurrences of some dot product 4-chain, while Theorem 4 shows that the number of occurrences of any type of distance 4-chain can never exceed $\lesssim n^{\frac{7}{3}} + \epsilon$, for any $\epsilon > 0$. See also [6, 13, 14] for some related results in vector spaces over finite fields and modules over finite rings of integers.

So far, we have only considered sequences of points. However, to deal with more general relationships between points we will need a way to organize our information.

1.1.2 Graphs

Let a *graph G* be defined by a pair of sets, $V(G)$ and $E(G)$, where the first set has elements called *vertices*, and the second set consists of two-element subsets of $V(G)$ called *edges*. The *degree* of a vertex is the number of edges it is an element of. In a given graph, if two vertices are listed in the same edge, we will call them *adjacent*. Finally, we will define an *edge-weighted graph* as a graph G paired with a *weight function*, $w : E(G) \to \mathbb{R}$. Using edge-weighted graphs, we can specify a k-chain as a graph on $k+1$ vertices with edges connecting successive vertices. Each edge will have a weight corresponding to the α_j. A *subgraph* of G is a graph G' with $V(G') \subseteq V(G)$ and $E(G') \subseteq E(G)$. Further, G' is an *induced subgraph* if it is a subgraph with all possible edges from $E(G)$ for the vertices present in $V(G')$.

1.1.3 Three Dimensions

Thus far, we have only considered point sets in the plane. The reason for this is that in dimensions four and higher, there is a simple construction due to Lenz that renders the unit distance problem trivial as stated. There is a similar such construction in three dimensions for dot products. These constructions can then be modified to quickly make many related questions trivial unless there are further restrictions on the point sets. See [5, 16] for more details. That said, we include here a few distance results in three dimensions that do not appear to follow from such simple adaptations of the Lenz example.

Let $u_3(n)$ denote the maximum number of times the unit distance can occur in any large finite set of n points in $\mathbb{R}^3$. In [9], Erdős proved that for any large n, $u_3(n) \gtrsim n^{\frac{4}{3}} \log\log n$ pairs of points separated by the unit distance. In [25], Josh Zahl showed $u_3(n) \lesssim n^{\frac{295}{197}+\epsilon}$, for any $\epsilon > 0$.

Currently, the best known upper bounds on the number of occurrences of a given type of k-chain in a large finite set of n points in $\mathbb{R}^3$ are due to Frankl and Kupavskii, in [11].

Theorem 7 ([11]) *Given a large, finite set of n points in* $\mathbb{R}^3$ *and a natural number k, the maximum number of distance k-chains of the type* $(\alpha_1, \ldots, \alpha_k)$ *that can exist in the set, is* $\lesssim n^{\frac{k}{2}+1+\epsilon}$*, for any* $\epsilon > 0$.

For even k, this essentially settles the problem, as it nearly matches the lower bound $\gtrsim n^{\frac{k}{2}+1}$, from [20]. In the case that k is odd, there is a wider gap between the upper and lower bounds, but Frankl and Kupavskii have the best results. We refer the interested reader to Theorem 4 and Proposition 2 in [11] for full details, as the lower bound is somewhat difficult to express concisely.

1.2 Main Results

We now introduce the definitions we use to quantify many of the results in this paper. Given an edge-weighted graph G, we say that a k-tuple of points $X = (x_1, \ldots, x_k)$ forms a *distance G-configuration* if there is a bijection from $\varphi : V(G) \to X$ so that for every edge $\{u, v\} \in E(G)$, we have that $w(\{u, v\})$ is the distance from $\varphi(u)$ to $\varphi(v)$. Similarly define a *dot product G-configuration* where we replace distance by dot product.

Definition 1 Given any large, finite set of n points in $\mathbb{R}^2$, and a graph G, define the maximum number of distance G-configurations that can exist in E to be $f(G; n)$. Similarly, define the maximum number of dot product G-configurations that can exist in the set to be $g(G; n)$.

With this definition in tow, we now focus on giving estimates on the number of G-configurations for various families of edge-weighted graphs. Also note that, in many cases, the particular edge weights will not affect the results, only whether or not an edge is specified in the graph. In light of this, we will not explicitly mention the weights unless they are pertinent. We now name several special families of graphs, and translate known bounds or give new bounds on the configurations they define.

A *path* P_k is a graph consisting of $k-1$ distinct edges where each edge shares a vertex with the next for a total of k distinct vertices. Theorems 4, 5, 6, and 1 can be restated using this notation by expressing upper or lower bounds on $f(P_{k+1}; n)$ or $g(P_{k+1}; n)$.

1.2.1 Specific Families in Two Dimensions

A *cycle* C_k is formed by adding an edge to P_k so that the first and last vertices in the path are adjacent. Because any C_k contains a P_k, the upper bounds on $f(P_{k+1}; n)$ and $g(P_{k+1}; n)$ imply the same upper bounds on $f(C_{k+1}; n)$ and $g(C_{k+1}; n)$, respectively. We pause for a moment to focus on the triangle, C_3. In the case of distances, the best bounds that we have for $f(C_3; n)$ follow from the fact any pair of distinct triangles can share no more than one pair of vertices. So for each pair of points (x, y) separated by a given distance α, we have at most 4 choices for z such that (x, y, z) or (y, x, z) form a given triangle (distance C_3-configuration). Combining this with the trivial lower bound gives is

$$n \leq f(C_3; n) \lesssim f(P_1; n) \lesssim n^{\frac{4}{3}}. \tag{1}$$

See [5, 7], and the references contained therein for more details. In the case of dot products, we could potentially fix a pair of points (x_1, x_3) and have as many as n choices for x_2 to make a given dot product C_3-configuration. However, we can recover the same upper bound with more careful analysis.

Theorem 8 *Given any large finite natural number* n*, we have that*

$$n \lesssim g(C_3; n) \lesssim n^{\frac{4}{3}}.$$

A *k-star* is a graph on $k+1$ vertices where one vertex is adjacent to the other k, but none of the other vertices are adjacent to one another.

Theorem 9 *Given any large finite natural number* n*, and an edge-weighted* k*-star* G*, we have that*

$$f(G; n) \approx g(G; n) \approx n^k.$$

A *tree* is a graph with no cycle as a subgraph. In a tree, a vertex with degree one is called a *leaf*. A *binary tree* is a tree where every vertex has degree one, two, or three. Often times, it is useful to designate one vertex the *root* of a tree. Such a tree is then called *rooted*. The neighbors of the root are called its *children*, and the root is called the *parent* of its children. Any neighbors that these vertices have, excluding the root, will be called their children, and so on. The number of edges the path from the root to any vertex is called the *height* of that vertex. A *perfect binary tree of height* h has exactly 2^h leaves, each of height h. This definition forces every vertex, except for the leaves, to have exactly two children. A c-ary tree is a generalization of a binary tree where each parent can have up to c children. So when $c = 2$, this is a binary tree. Let $T_{c,h}$ denote the perfect c-ary tree of height h.

Theorem 10 *Given any large finite natural number* n*, and* $c \geq 2$*, we have that*

$$f(T_{c,h}; n) \approx n^{c^h}.$$

We can get a similar result for dot products, but there are technical obstructions to it holding in full generality. So we record here a special case that is relatively easy to state, and explore the more general bound in another estimate that we present later. Let $T_{c,h,\alpha}$ denote an edge-weighted perfect c-ary tree of height h whose weights are all the same $\alpha \neq 0$.

Theorem 11 *Given any large finite natural number* n*, and* $\alpha \neq 0$*, we have that*

$$g(T_{c,h,\alpha}; n) \approx n^{c^h}.$$

1.2.2 General Families in Two Dimensions

Where the previous results have been based on rather specific families of graphs, we now move on to some more general families of graphs. We can always get an

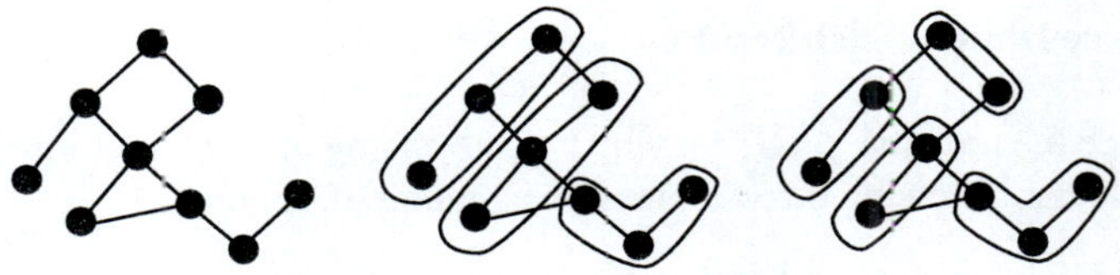

Fig. 1 This is a graph that can be (0, 3)-covered and (3, 1)-covered

upper bound on the number of G-configurations by decomposing G into subgraphs H_j such that $V(G) = \cup_j V(H_j)$, and taking the product of the upper bounds on the various components. This is in effect ignoring any restrictions imposed on the original configuration by edges between the H_j. That is, we are losing the information provided by the weights of edges in the set $(E(G) \setminus \cup_j E(H_j))$. In the most extreme case, we would just set each of the H_j to be a single vertex from G, and we would get the trivial upper bound of $n^{|V(G)|}$.

To this end, we introduce a definition and illustrate its use to help bridge the gap between the special cases in the previous subsection and more general graphs that could arise. Given any graph G, we say that G can be (s, t)-*covered* if every vertex can be included in a set of s disjoint copies of P_2 and t disjoint copies of P_3. Note that if a graph can be (s, t)-covered for a pair of parameters s and t, then $|V(G)| = 2s + 3t$, as each P_2 has two vertices, each P_3 has three, and these subgraphs are disjoint (Fig. 1).

Theorem 12 *Given any large finite natural number n, and an edge-weighted graph G, that can be (s, t)-covered, we have that*

$$f(G; n) \lesssim n^{\frac{4}{3}s+2t} \text{ and } g(G; n) \lesssim n^{\frac{4}{3}s+2t}.$$

As a simple corollary of Theorem 12, we get many nontrivial upper bounds for arbitrary G-configurations by applying the previous result to a subgraph of G that can be (s, t)-covered, and then crudely bounding the number of choices for the remaining vertices by n.

Corollary 1 *Given any large finite natural number n, and an edge-weighted graph G, with a subgraph H that can be (s, t)-covered, we have that*

$$f(G; n) \lesssim n^{\frac{4}{3}s+2t+|V(G)\setminus V(H)|} \text{ and } g(G; n) \lesssim n^{\frac{4}{3}s+2t+|V(G)\setminus V(H)|}.$$

This corollary allows us to prove a dot product version of Theorem 10. Here we only state and prove the version for binary trees.

Corollary 2 *Given any large finite natural number n,*

$$g(T_{2,h}; n) \lesssim \begin{cases} n^{\frac{2}{3}(2^{h+1}-1)} & \text{if } h \text{ odd},\\ n^{1+\frac{4}{3}(2^{h}-1)} & \text{if } h \text{ even}. \end{cases}$$

1.2.3 Three Dimensional Results

We introduce the notation $f_3(G; n)$, which is the analog of $f(G; n)$ for point sets in $\mathbb{R}^3$. We have the following three-dimensional analog of Theorem 9.

Theorem 13 *Given any large finite natural number n, and an edge-weighted k-star G, we have that*

$$f_3(G; n) \approx n^k.$$

The situation is a bit different for the three-dimensional version of Theorem 10. When $c \geq 3$, we get similar result, again by restricting that all of the edge weights are the same $\alpha \neq 0$.

Theorem 14 *Given any large finite natural number n, and a $c \geq 3$, we have that*

$$f_3(T_{c,h,\alpha}; n) \approx n^{c^h}.$$

However, in the special case that $c = 2$, we can no longer be sure the the upper bound matches the lower bound, so we have the following more precise yet technical estimate.

Theorem 15 *Given a large natural number n, an $\alpha \neq 0$, and any $\epsilon > 0$,*

$$f(T_{c,h,\alpha}; n) \lesssim n^{c^h\left(1+\frac{1}{c^2-1}\right)+\frac{(-1)^h}{2(c+1)}+\frac{1}{2(1-c)}+\epsilon}$$

Applying this in the case that $c = 2$ implies the following.

Corollary 3 *Given any large finite natural number n, we have that*

$$n^{2^h} \lesssim f_3(T_{2,h,\alpha}; n) \lesssim n^{\frac{1}{6}(2^{h+3}-3+(-1)^h)}.$$

Finally, in the absence of any other discernible graph structures that we can identify, we have a three-dimensional companion to Theorem 1. A *matching* in a graph is a set of edges that share no vertices. For a graph G, let m_G denote the maximum number of edges in a matching of G, and let r_G be the number of remaining vertices. Specifically,

$$m_G := \max\{E(H) : H \text{ is a matching in } G\} \text{ and } r_G := |V(G) - 2m_G|.$$

Theorem 16 *Given any large finite natural number n, any $\epsilon > 0$, and an edge-weighted graph G, we have*

$$f_3(G; n) \lesssim n^{\frac{295}{197}m_G+r_G+\epsilon}.$$

We conclude the statement of our main results with a concrete comparison. Let us fix our attention on a perfect binary tree of height three whose edge weights are all the same $\alpha \neq 0$. Theorem 3 gives an upper bound of $f_3(T_{2,3,\alpha}) \lesssim n^{\frac{14}{3}+\epsilon}$, for any $\epsilon > 0$. If we apply Theorem 16, we first look for the largest matching we can find in $T_{2,3,\alpha}$. The best we can do is $m_{T_{2,3,\alpha}} = 2$, which gives us three vertices left over, or $r_{T_{2,3,\alpha}} = 3$. So we get an upper bound of $f_3(T_{2,3,\alpha}) \lesssim n^{\frac{590}{197}+3+\epsilon}$, for any $\epsilon > 0$, which is considerably worse than the estimate given by Theorem 3.

1.3 *Organization of This Paper*

In Sect. 2, we flesh out some of the tools that we will use to prove the main results. To this end, we outline proofs of some known results mentioned above. Section 3 has the proofs of the new results.

2 Preliminaries

Distance configurations can be explored by looking at the incidences of points and families of circles centered at those points. The analogous geometric objects for dot product configurations are special families of lines.

2.1 *The α-Line for a Point p*

Given a point $p \in \mathbb{R}^2 \setminus \{(0,0)\}$, the set of points in the plane that have dot product α with p is a line we call the *α-line* of p, $\ell_\alpha(p)$. We call any line through the origin a *radial* line. A quick calculation using the definition of the dot product gives us that for any $p \in \mathbb{R}^2 \setminus \{(0,0)\}$ and any real number α, we have that $\ell_\alpha(p)$ will be perpendicular to the unique radial line through p. We include the following lemmata from [2].

Lemma 1 *If p and r are two points in $\mathbb{R}^2 \setminus \{(0,0)\}$ that do not lie on the same radial line, and $\alpha, \beta \in \mathbb{R}$, then there exists exactly one point $q \in \mathbb{R}^2 \setminus \{(0,0\}$ such that $p \cdot q = \alpha$ and $q \cdot r = \beta$.*

Given a point $p \in \mathbb{R}^2$, let $|p|$ denote the distance from p to the origin Fig. 2.

Lemma 2 *If p and r are two points in $\mathbb{R}^2 \setminus \{(0,0)\}$ such that $\ell_\alpha(p)$ coincides with $\ell_\beta(r)$, then p and r must lie on the same radial line through the origin, and $|p|/|r| = \beta/\alpha$.*

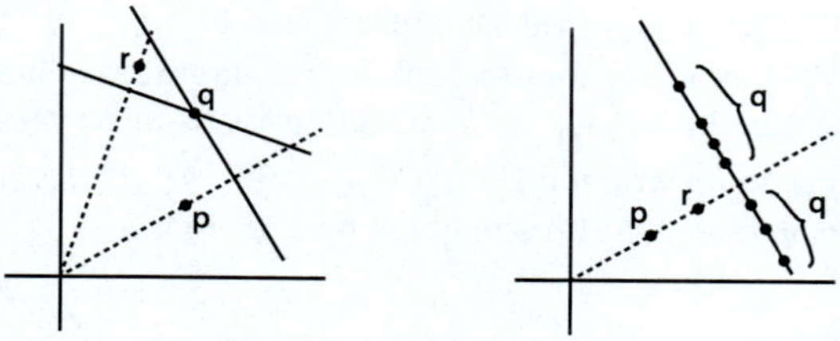

Fig. 2 Here, we are looking for triples of points (p, q, r) such that $p \cdot q = \alpha$ and $q \cdot r = \beta$. On the left, we have points p and r on distinct radial lines, so q is the unique intersection of $\ell_\alpha(p)$ and $\ell_\beta(r)$ for positive numbers α and β. On the right, we see that p and r are on the same radial line, so for some specific choices of positive real numbers α and β, we can have $\ell_\alpha(p)$ coincide with $\ell_\beta(r)$, giving potentially many choices for points q

2.2 *Sketch of Proof of Theorem 3*

We take a moment to give an outline of the proof of Theorem 3 to illustrate the basic ideas used in the main results below. This highlights some of the fundamental similarities and differences between distances and dot products.

In the case of distances, Theorem 3 follows from the observation that there are about n^2 pairs of points (x_1, x_3) and for each such pair, there at most two intersections of the circles of radii α_1 and α_2 centered at x_1 and x_3, respectively. These two intersections are the only possible locations for x_2 if (x_1, x_2, x_3) is to be a distance hinge of type (α_1, α_2).

The companion result for dot product hinges is similar, but a bit more involved. We follow the same program as with distances, but for each pair of points, (x_1, x_3), we apply either Lemmas 1 or 2, depending on whether or not $\ell_{\alpha_1}(x_1)$ coincides with $\ell_{\alpha_2}(x_3)$. In the first case, everything follows as with distances, but in the second case, we have to count a bit more carefully. See [2] for more details.

2.3 *The Szemerédi–Trotter Theorem*

The celebrated Szemerédi–Trotter Theorem from [23] is a key component of many of the dot product proofs. We also note that this result provided the foundation for Theorem 1, which is heavily leaned on in the study of distance configurations.

Theorem 17 *Given n points and m lines in the plane, the number of point-line pairs, such that the point lies on the line is*

$$\lesssim \left(n^{\frac{2}{3}}m^{\frac{2}{3}} + n + m\right).$$

Theorem 2 follows from this by considering the set of n points in the plane, and the set of their α-lines. Lemma 2 then guarantees that there will be n unique such lines, and the result follows.

3 Proofs

3.1 Proof of Theorem 8

The lower bound follows using a geometric progression along any line through the origin. We give an explicit construction for points along the x-axis. Let the set of points be $\{(ag^j, 0) : j = 1 \dots n\}$, for $a, g > 0$. We then have $\gtrsim n$ triples of points of the form $(x_1, x_2, x_3) = ((ag^{j-1}, 0), (ag^j, 0), (ag^{j+1}, 0))$, with the property that $x_1 \cdot x_2 = a^2 g^{2j-1}$, $x_2 \cdot x_3 = a^2 g^{2j+1}$, and $x_3 \cdot x_1 = a^2 g^{2j}$, and the lower bound is achieved.

For the upper bound, we want to get a bound on the number of triples of points of the form (x_1, x_2, x_3) such that $x_1 \cdot x_2 = \alpha_1$, $x_2 \cdot x_3 = \alpha_2$, and $x_3 \cdot x_1 = \alpha_3$. We start by considering every pair of points from the set that could possibly be x_1 and x_3 from such a triple. Theorem 2, we know that there can be no more than $\lesssim n^{\frac{4}{3}}$ point pairs that have dot product α_3, so we know that there can be no more than $\lesssim n^{\frac{4}{3}}$ choices for the pair (x_1, x_3).

For each such pair of points, either they do not lie on the same radial line or they do lie on the same radial line. If they do not lie on the same radial line, then Lemma 1 guarantees that $\ell_{\alpha_1}(x_1)$ intersects $\ell_{\alpha_2}(x_3)$ in exactly one point, and this point is the only potential candidate for an x_2 that would make (x_1, x_2, x_3) a dot product C_3-configuration. Therefore, there are at most $n^{\frac{4}{3}}$ dot product C_3-configurations with x_1 and x_3 on distinct radial lines.

We now turn our attention to the pairs of points (x_1, x_3) that satisfy $x_3 \cdot x_1 = \alpha_3$ and have both points on the same radial line. Because x_1 and x_3 share the same radial line, we know that $\ell_{\alpha_1}(x_1)$ and $\ell_{\alpha_2}(x_3)$ will have the same slope, as both of these lines must be perpendicular to the radial line containing x_1 and x_3. If $\ell_{\alpha_1}(x_1)$ does not coincide with $\ell_{\alpha_2}(x_3)$, then they must not intersect, as they have the same slope. Therefore, there is no possible location for a point x_2 that would make (x_1, x_2, x_3) a dot product C_3-configuration.

3.1.1 Coincident α-Lines

So the last (and most involved) possibility to check is when x_1 and x_3 satisfy $x_3 \cdot x_1 = \alpha_3$ and we have that $\ell_{\alpha_1}(x_1)$ coincides with $\ell_{\alpha_2}(x_3)$. In this case, we appeal to Lemma 2, which tells us that

$$\frac{|x_1|}{|x_3|} = \frac{\alpha_1}{\alpha_2}.$$

$$|x_1| = |x_3|\frac{\alpha_1}{\alpha_2}. \tag{2}$$

But recall that $x_3 \cdot x_1 = |x_3||x_1| \cos\theta$, where θ is the angle made by x_1, the origin, and x_3. But x_1 and x_3 are on the same radial line, so $\cos\theta = \pm 1$. Combining this with the fact that $x_3 \cdot x_1 = \alpha_3$, gives us

$$\pm\alpha_3 = |x_3||x_1|.$$

Combining this with (2) yields

$$\pm\alpha_3 = |x_3|^2\frac{\alpha_1}{\alpha_2},$$

which means that x_3 must live on a circle of radius $\sqrt{\alpha_3\alpha_2/\alpha_1}$ centered at the origin. Call this circle $\mathcal{C}$. Recall that since $\ell_{\alpha_1}(x_1)$ and $\ell_{\alpha_2}(x_3)$ coincide, any point on this line will be a suitable choice for x_2 to make (x_1, x_2, x_3) a dot product C_3-configuration. We now pause to state a small fact about antipodal points on a circle, whose technical proof we delay.

Claim 1 If p and p' are antipodal points on a circle centered at the origin, then for any nonzero α, we have that $\ell_\alpha(p)$ is distinct from $\ell_\alpha(p')$.

So we will finish this case by estimating the number of incidences of points from our set (putative choices for x_2) and lines of form $\ell_{\alpha_2}(x_3)$ for choices of x_3 on the circle $\mathcal{C}$. Notice that each distinct choice of x_3 must give rise to a different line of the form $\ell_{\alpha_2}(x_3)$, because each point $x_3 \in \mathcal{C}$ will have lie on a different radial line, except for its antipodal point x_3', but in that case, we can apply Claim 1 to guarantee that $\ell_{\alpha_2}(x_3)$ will not coincide with $\ell_{\alpha_2}(x_3')$ for nonzero choices of α_2. Therefore, our final count amounts to estimating the incidences between $\lesssim n$ points and $\lesssim n$ distinct lines in the plane. By appealing to Theorem 17, we see that there can be no more than $n^{\frac{4}{3}}$ such incidences, completing the proof.

3.1.2 Proof of Claim 1

Proof By way of contradiction, suppose $\ell_\alpha(p)$ coincides with $\ell_\alpha(p')$. Because p and p' are antipodal points on a circle centered at the origin, they must lie on the same radial line. Let q be the intersection of $\ell_\alpha(p)$ and the radial line through p and p'. Let θ be the angle determined by p, the origin, and q. Define ϕ to be the angle between p', the origin, and q. Because p, p', q, and the origin all lie on a line, with p and p' on opposite sides of the origin, we must have that $\{\theta, \phi\} = \{0, \pi\}$. However, by our assumptions and the definition of dot product, we have that

$$|p||q| \cos\theta = p \cdot q = \alpha = p' \cdot q = |p'||q| \cos\phi.$$

Because p and p' lie on the same circle, we know that $|p| = |p'|$. So comparing the left and right hand sides of the above equation yields $\cos\theta = \cos\phi$, which implies that $1 = -1$, which is a contradiction. □

3.2 Proof of Theorem 9

Start by labeling the vertices of G as $v_1, \ldots, v_{k+1}$, where each of the vertices v_1 through v_k has degree 1, and the vertex v_{k+1} has degree k.

Now we show that $f(G; n) \lesssim n^k$. To see this, we apply Theorem 4 for 2-chains to see that there are no more than n^2 triples of points of the form (x_1, x_{k+1}, x_2) such that $|x_1 - x_{k+1}| = w(v_1, v_{k+1})$ and $|x_2 - x_{k+1}| = w(v_2, v_{k+1})$. For each such triple, we then have no more than n choices for each of the other x_j, for a total of n^k.

We then prove that $f(G; n) \gtrsim n^k$. Start by assuming the origin is in our set. Then arrange about n/k points on k circles centered at the origin, whose radii are the edge weights, $w(v_j, v_{k+1})$. If any weights repeat, just put more points on the circle of the appropriate radius until we have placed about n points.

The same results follow for the upper and lower bounds of $g(G; n)$, except by appealing to Theorem 3 for the upper bound, and with appropriate α-lines in place of circles for the lower bound.

3.3 Proofs of Theorems 10 and 11

We start by proving Theorem 10, and later describe how to modify the proof to prove Theorem 11. Call the root of G v_1. Let the children of v_1 be called $v_2, v_3, \ldots v_{c^2-1}$. Continuing in this way, we get that the vertices at height k will have indices i satisfying $c^k \leq i < c^{k+1}$. Because G is perfect and of height h, we will have a total of $c^{h+1} - 1$ vertices. Now, we have c^h leaves, and n choices for each of them, giving us a total of n^{2^h} choices for set of leaves. Every pair of leaves sharing a parent in the graph determine exactly two possibilities for points from the set that could serve as parents, just as in the proof of Theorem 3. Since there will be c^{h-1} such relationships, the total number of distance G-configurations is no more than $c^{h-1}n^{c^h}$, as claimed.

To see the lower bound, we will construct a point set, E, with $\gtrsim n^{c^k}$ instances of a given G-configuration. Consider H, the induced subgraph of G formed by removing the leaves from G. Explicitly construct a single instance of an H-configuration by picking a point to correspond to the root, then picking points at the prescribed distances from the root corresponding to the first generation of children, and so on, until we have a single instance of the prescribed H-configuration. We will finish by processing each leaf of H in the following way. Suppose v is a leaf in H. Then in G, v has c children. Let w_j be the weight of the edge between v and its jth child. Now, for each child, arrange $c^{-h}n$ points on the sphere of radius w_j centered at the point of E corresponding to the vertex v. Do this for every leaf of H. Notice that by construction, we have $c^{-h}n \approx n$ choices of points to correspond to every leaf of G, and moreover, these choices are independent.

To prove Theorem 11, we follow the same logic, except with α-lines in place of circles. The restriction that all of the edge weights in G are the same nonzero value α prevents the existence of potentially pathological sub-configurations (such as in the final case of the proof of Theorem 8) where the points corresponding to two leaf vertices might not uniquely determine the point corresponding to their parent vertex.

3.4 *Proof of Theorem 12*

Because G is (s, t)-covered, we know that for each occurrence of the given distance G-configuration, we must have at least s pairs of points separated by fixed distances determined by the weights on their corresponding edges, and at least t triples of points that form hinges with distance relations also defined by the appropriate edge weights. Theorem 1 guarantees that there can be no more than $n^{\frac{4}{3}}$ pairs of points separated by any given distance, and Theorem 3 tells us that there can be no more than n^2 occurrences of any hinge. Putting these together, we get

$$f(G; n) \lesssim \left(n^{\frac{4}{3}}\right)^s (n^2)^t,$$

as claimed. To prove the analogous result for dot products, we follow the same logic, but appeal to Theorem 2 in place of Theorem 1 above.

3.5 *Proof of Corollary 2*

We begin with the case that h is odd. If $h = 1$, our graph is just P_2, and this amounts to the hinge bound, Theorem 3. If $h = 3$, we cover the root and its children by a P_2, and the other two generations are covered by four copies of P_2. In general, we see that the vertices corresponding to every other generation of the tree can be decomposed into copies of P_2, by greedily separating out hinges from every other generation. Because h is odd, we know that there will be an even number of generations, so no points will be left out. As there are $2^{h+1} - 1$ vertices in G, and they are all broken up into triples, the total number of triples is $\frac{1}{3}\left(2^{h+1} - 1\right)$. Theorem 3 has guaranteed that no such triple can occur more than n^2 times, for a total upper bound of

$$g(G; n) \lesssim n^{\frac{2}{3}(2^{h+1}-1)}.$$

In the case that h is even, note that G can be decomposed into a root and two perfect binary trees of height $h - 1$ (Fig. 3). So we apply the odd height estimate to each of these trees, and have no more than n choices for the root, yielding

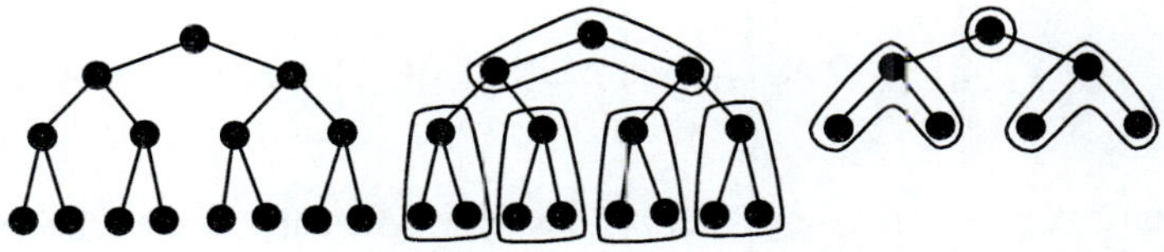

Fig. 3 This is a perfect binary tree of height 3, the same tree with a (0, 5)-covering shown, and a perfect binary tree of height 2 with a (0, 2)-covering on all but its root, which must be considered separately

$$g(G; n) \lesssim n^{\frac{2}{3}(2^h-1)} n^{\frac{2}{3}(2^h-1)} n.$$

3.6 Proof of Theorem 14

The lower bounds follow from a similar construction to the one in the proof of Theorem 10, except with spheres in place of circles. The proof of the upper bounds also follow by similar reasoning, except that instead of the intersections of two circles determining constantly many points, we use that three spheres of the same radius will intersect in constantly many points.

3.7 Proof of Theorem 15

The basic idea is to decompose the tree into subgraphs that can be estimated separately. We begin by estimating the longest chain, which is a path from one leaf, through the root, and to another leaf. Therefore, this path will be P_{2h+1}, which is a distance $2h$-chain. We will call this path the *outer chain*. Next, there will be $c - 2$ perfect c-ary trees of height $h - 1$ whose roots were children of the original root. This is because the root of the big tree has c children, but two of them were accounted for in the outer chain.

Notice that each of these two children of the root will have one child in the outer chain, and their other $c - 1$ children (grand-children of the root) as yet unaccounted for. This gives us a total of $2c - 2$ children that are not yet counted in this generation. Each of these, being at height two from the root, will then be the root of their own perfect c-ary trees of height $h - 2$. This pattern continues with the next children of vertices in the outer chain that are as yet unaccounted for. There will again be $c - 1$ children from each of the outer chain vertices, and these children will again be the roots of their own perfect c-ary trees of successively lower heights. Putting all of these pieces together, we get the following recursive relationship.

$$f(T_{c,h,\alpha}; n) \lesssim f(P_{2h+1}; n) f(T_{c,h-1,\alpha}; n)^{c-2} \prod_{j=0}^{h-2} f(T_{c,j,\alpha})^{2c-2}. \quad (3)$$

To simplify the analysis, we will let $a_{c,h}$ be the exponent of n from the best estimate we have for $f(T_{c,h,\alpha}; n)$. That is,

$$a_{c,h} = \min\{x : \text{ we can show } f(T_{c,h,\alpha}; n) \lesssim n^x\}.$$

We will work with $a_{c,h}$, acknowledging that it might not correspond to the best possible estimate for $f(T_{c,h,\alpha}; n)$, but it will give us an upper bound. Continuing, we translate what we know about each of the components of (3) into expressions related to $a_{c,h}$.

Recall that a P_{2h+1}-configuration is just a $2h$-chain, so by Theorem 7, we know that $f(P_{2h+1}; n) \lesssim n^{\frac{2h}{2}+1+\epsilon}$, for any $\epsilon > 0$. So this term will give us a factor of $n^{h+\epsilon}$. We also notice that $T_{c,0,\alpha}$ is just a single point, so $f(T_{c,0,\alpha}; n) = n$. This gives us that $a_{c,0} = 1$. Similarly, we can see that $T_{c,1,\alpha}$ is a c-star. So Theorem 13 tells us that $f(T_{c,1,\alpha}; n) \approx n^c$. This gives us that $a_{c,1} = c$. Now we can rewrite (3) in terms of $a_{c,h}$. We get the following recurrence relation, with two initial conditions.

$$a_{c,h} = (h + \epsilon) + (c - 2)a_{c,h-1} + \sum_{j=0}^{h-2} (2c - 2)a_{c,j}; \quad a_{c,0} = 1; \quad a_{c,1} = c. \quad (4)$$

This recurrence relation has the solution

$$a_{c,h} = c^h \left(1 + \frac{1}{c^2 - 1}\right) + \frac{(-1)^h}{2(c + 1)} + \frac{1}{2(1 - c)} + \epsilon.$$

This implies the desired result (Fig. 4).

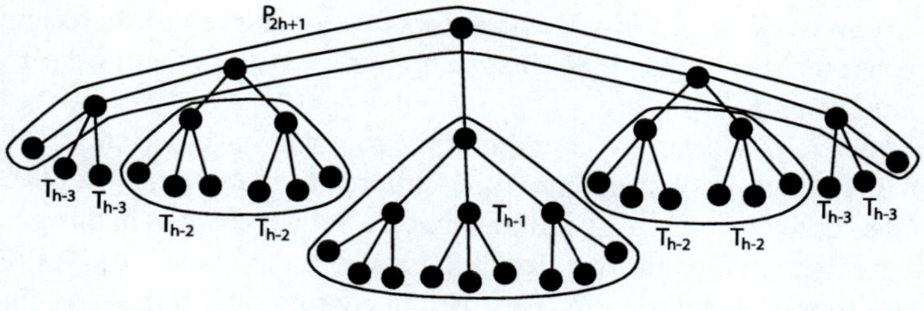

Fig. 4 This is the root and next three generations of a $T_{3,3}$, with the various components labeled as in the proof of Theorem 3

3.8 Proof of Theorem 16

This proof is similar to that of Theorem 12. We find a maximal matching of G consisting of m_G edges. Each of these edges can have no more than $n^{\frac{295}{197}+\epsilon}$ pairs of points that can represent them. There are no more than n choices for any of the r_G vertices of G that have yet to be accounted for. Putting this all together yields the desired result.

References

1. P. Bahls, *Channel assignment on Cayley graphs*. J. Graph Theory, 67: 169–177 (2011). https://doi.org/10.1002/jgt.20523
2. D. Barker and S. Senger, *Upper bounds on pairs of dot products*. Journal of Combinatorial Mathematics and Combinatorial Computing, Volume 103, November, 2017, pp. 211–224.
3. J. J. Benedetto and M. Fickus, *Finite normalized tight frames*. Adv. Comput. Math. 18, pp. 357–385 (2003).
4. M. Bennett, A. Iosevich, and K. Taylor, *Finite chains inside thin subsets of* $\mathbb{R}^d$, Analysis and PDE, volume 9, no. 3 (2016).
5. P. Brass, W. Moser, and J. Pach, *Research Problems in Discrete Geometry*. Springer (2000), 499 pp.
6. D. Covert and S. Senger, *Pairs of dot products in finite fields and rings*, Nathanson M. (eds) Combinatorial and Additive Number Theory II. CANT 2015, CANT 2016. Springer Proceedings in Mathematics & Statistics, vol 220. Springer, Cham.
7. J. DeWitt, K. Ford, E. Goldstein, S. Miller, G. Moreland, E. Palsson, and S. Senger, *Dimensional lower bounds for Falconer type incidence and point configuration theorems*, with Journal d'Analyse Mathématique **139**, 143–154 (2019).
8. P. Erdős, *On sets of distances of n points*. Amer. Math. Monthly **53** (1946) 248–250.
9. P. Erdős, *On sets of distances of n points in Euclidean space,* Magyar Tudományos Akadémia Matemakai Kutató Intézet Közleményi **5** (1960) 165–169.
10. J. Fox, J. Pach, A. Suk, A. Sheffer, and J. Zahl, *A semi-algebraic version of Zarankiewicz's problem*. Journal of the European Mathematical Society, Volume 19, Issue 6, 2017, pp. 1785–1810. https://doi.org/10.4171/JEMS/705.
11. N. Frankl and A. Kupavskii, *Almost sharp bounds on the number of discrete chains in the plane*, arXiv:1912.00224.
12. J. Garibaldi, A. Iosevich, and S. Senger, *Erdős distance problem,* AMS Student Library Series, 56, (2011).
13. D. Hart, A. Iosevich, D. Koh, and M. Rudnev, *Averages over hyperplanes, sum-product theory in finite fields, and the Erdős-Falconer distance conjecture*. Trans. Amer. Math. Soc., 363 (2011) pp. 3255–3275.
14. A. Iosevich, M. Rudnev, *Erdös distance problem in vector spaces over finite fields*. Trans. Amer. Math. Soc. 359 (2007), no. 12, 6127–6142.
15. A. Iosevich and S. Senger, *Orthogonal systems in vector spaces over finite fields*. Electronic J. of Combinatorics, Volume 15, December (2008).
16. S. Kilmer, C. Marshall, and S. Senger, *Dot product chains*, arXiv:2006.11467.
17. B. Lund, *Incidences and pairs of dot products*, arXiv:1509.01072.
18. B. Lund, A. Sheffer, and F. de Zeeuw, *Bisector energy and few distinct distances*, F. Discrete Comput Geom (2016) 56: 337. https://doi.org/10.1007/s00454-016-9783-5.
19. Y. Ou and K. Taylor, *Finite point configurations ad the regular value theorem in a fractal setting*, arXiv:2005.12233 (2020).

20. E. Palsson, A. Scheffer, and S. Senger, *On the number of discrete chains*, arXiv:1902.08259, (2019) (submitted).
21. S. Senger, *Explorations of the Erdős-Falconer distance problem and related applications*, Dissertation, Univ. of Missouri (2011).
22. J. Spencer, E. Szemerédi, W. T. Trotter. *Unit distances in the Euclidean plane*, Graph theory and combinatorics (1984): 293–303.
23. E. Szemerédi and W. T. Trotter, Jr., *Extremal problems in discrete geometry*, Combinatorica 3 (1983), no. 3–4, pp. 381–392.
24. S. Steinerberger, *A note on the number of different inner products generated by a finite set of vectors*. Discrete Mathematics, 310, (2010), pp. 1112–1117.
25. J. Zahl, *Breaking the 3/2 Barrier for Unit Distances in Three Dimensions*, Int. Math. Res. Notices, rnx336, (2018).

Distribution of Missing Differences in Diffsets

Scott Harvey-Arnold, Steven J. Miller, and Fei Peng

Abstract Lazarev, Miller and O'Bryant [11] investigated the distribution of $|S+S|$ for S chosen uniformly at random from $\{0, 1, \ldots, n-1\}$, and proved the existence of a divot at missing 7 sums (the probability of missing exactly 7 sums is less than missing 6 or missing 8 sums). We study related questions for $|S-S|$, and show some divots from one end of the probability distribution, $P(|S-S|=k)$, as well as a peak at $k=4$ from the other end, $P(2n-1-|S-S|=k)$. A corollary of our results is an asymptotic bound for the number of complete rulers of length n.

Keywords Difference sets · Diffsets · Missing differences

1 Introduction

1.1 Background

Let S be a typical subset of $[n] := \{0, 1, \ldots, n-1\}$; in other words, we choose S uniformly at random, or equivalently each integer in $[n]$ is independently chosen to be in S with probability $1/2$. Define

$$S+S := \{x+y : x, y \in S\} \text{and} S-S := \{x-y : x, y \in S\}. \tag{1.1}$$

This work was partially supported by NSF grant DMS1561945, Carnegie Mellon University, and Williams College. We thank Joshua Siktar for the constructive comments, and Carnegie Mellon University for the research opportunity and the AFS computing platform.

S. Harvey-Arnold · F. Peng
Carnegie Mellon University, Pittsburgh, PA, USA

S. J. Miller (✉)
Williams College, Williamstown, MA, USA
e-mail: sjm1@williams.edu

M. B. Nathanson (ed.), *Combinatorial and Additive Number Theory IV*,
Springer Proceedings in Mathematics & Statistics 347,
https://doi.org/10.1007/978-3-030-67996-5_13

We refer to these as the *sumset* and the *diffset* of S, and we denote the cardinality of a set A by $|A|$.

The sizes of the sumset and the diffset have been compared extensively. As addition is commutative and subtraction is not, it was conjectured that as $n \to \infty$ almost all sets S should be difference dominated: $|S - S| > |S + S|$. Thus while sum-dominant sets were known to exist, and constructions for infinite families were given, they were thought to be rare. This conjecture turns out to be false; Martin and O'Bryant [13] proved that for a small but positive proportion of all subsets of $[n]$, the sumset has a larger cardinality than the diffset. This result holds if instead of choosing each element with probability 1/2 we instead choose with a fixed probability $p > 0$; however, if p is allowed to decay to zero with n then Hegarty and Miller [8] proved almost all sets are difference dominated. For these and related results see [1–7, 9, 10, 12, 14–25].

The distribution of $|S + S|$ has also been studied.

When S is chosen uniformly at randomly from $[n]$, Lazarev, Miller and O'Bryant [11] proved an unusual "divot" occurs in the limiting probability distribution of $|S + S|$ (the existence of the limiting distribution was shown by Zhao [25]). In particular, the limiting probability of missing 7 sums is less than that of missing 6 (or 8):

$$\lim_{n\to\infty} P(2n-1-|S+S|=7) < \lim_{n\to\infty} P(2n-1-|S+S|=6) < \lim_{n\to\infty} P(2n-1-|S+S|=8). \tag{1.2}$$

Further, [11] gave rigorous bounds for $\lim_{n\to\infty} P(2n-1-|S+S|=k)$ for $0 \le k < 32$, which imply that there are no more divots until $k = 27$. It is unknown whether there could be more divots later. Figure 1 of their paper is reproduced here with permission as Fig. 1.

However, the probability distribution of $|S - S|$, the size of the diffset, has not been extensively investigated. One reason for the success in $|S + S|$ and the lack of progress for $|S - S|$ is: it was believed that the diffset is significantly harder to exhaustively investigate. For many sets, their properties can be determined by decomposing S as $L \cup M \cup R$, where L and R are respectively the left and right fringe elements and M is the middle; typically L and R are of bounded size independent of n, so most elements in S are in M. As there are many ways to write a number as a sum or difference of elements, most elements in $[n] + [n]$ or $[n] - [n]$ are realized, especially since a typical S has on the order of $n/2$ elements and thus generates on the order of $n^2/4$ pairs. The difference is for the fringe elements, where there are fewer representations and thus a greater chance of an element not being obtained.[1] For sumsets the left and right fringes do not interact, with the left fringe $L + L$ and the right $R + R$; this is not the case for the diffset, where the fringes are $L - R$ and its negative $R - L$. As a result, to determine whether an extremal element is in

[1] An integer $m \le n$ can be written as $m + 1$ sums of pairs of elements from $[n]$, and if m is modest it is thus unlikely that none of these pairs have both elements in S; however, if m is small then an element can have a significant probability of not occurring. For example, if $0 \in S$ but $1 \notin S$ then $1 \notin S + S$.

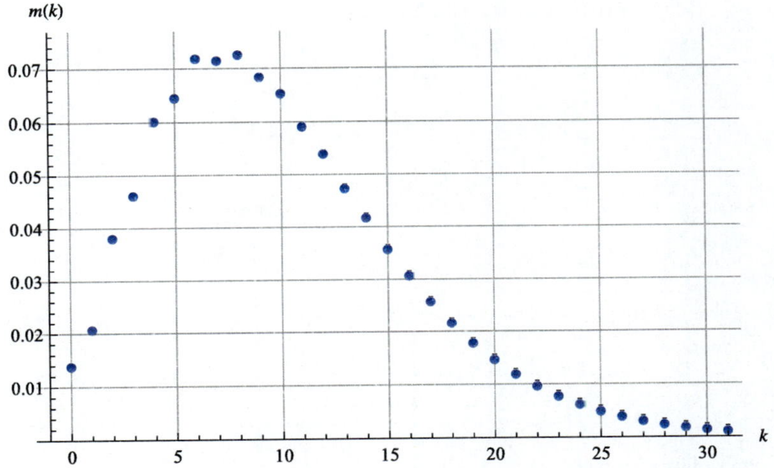

Fig. 1 Experimental values of $m(k)$, the probability $S + S$ is missing exactly k sum, when each $m \in [n]$ is in S with probability $1/2$. The vertical bars depict the values allowed by the most rigorous bounds in [11]. In most cases, the allowed interval is smaller than the dot indicating the experimental value. The data comes from generating 2^{28} sets uniformly forced to contain 0 from $[0, 256)$

$S + S$, only one fringe matters while for $S - S$, both ends must be considered. The computational complexity is hence (believed to be) *squared*, which makes the diffset distribution significantly harder to exhaustively investigate.

We note that this is not exactly the case. In fact, this paper will focus on the probability distribution of $|S - S|$, and derive similar results to those of [11].

1.2 *Distribution of* $|S - S|$ *when* $n = 35$

We display the probability distribution when $n = 35$ in Fig. 2. We exhaustively listed every subset of $[n]$ and recorded the corresponding $|S - S|$. The probability distribution is exactly the frequencies divided by 2^{35}.

One would immediately note that $|S - S|$ is either 0 or odd, as x and $-x$ are either both in or both out of $S - S$, and 0 is in iff $S \neq \emptyset$. Apart from that, there are two observations from Fig. 2.

1. The left part (i.e., the distribution of small-sized diffsets) is "messy", with noticeable divots. For combinatorial reasons, there are divots at 5, 9 and 15 that seem to preserve when n grows; by "a divot at 5" we mean

$$P(|S - S| = 3) > P(|S - S| = 5) < P(|S - S| = 7). \tag{1.3}$$

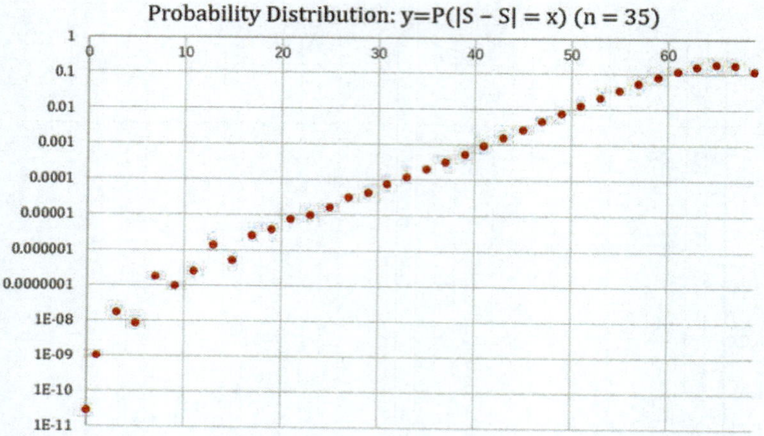

Fig. 2 Probability distribution of $P(|S - S| = x)$ when $n = 35$

2. The right part (the distribution of large-sized diffsets) is "smooth" (i.e. divot-less). Also, there is a peak at *missing 4 differences*. That is,

$$\forall k \neq 4, P(2n - 1 - |S - S| = 4) > P(2n - 1 - |S - S| = k). \tag{1.4}$$

(When $n = 35$, this is saying $|S - S| = 65$, represented by the third rightmost red dot in Fig. 2, is the most likely cardinality of the diffset).

These observations seem to continue to hold for larger n, though our investigations are no longer exhaustive but instead are random samples from the space.

For conciseness, let

$$P_n^{\mathrm{H}}(k) := P(|S - S| = k), \quad P_n^{\mathrm{M}}(k) := P(2n - 1 - |S - S| = k). \tag{1.5}$$

Here, H means *having* differences whereas M means *missing*. They are two complementary perspectives.

1.3 Main Results

We show that Observations 1 and 2 hold for sufficiently large n.

Proposition 1.1 *Observation 1 is true for $n \geq 12$. That is, $\forall n \geq 12$,*

$$P_n^H(3) > P_n^H(5) < P_n^H(7),$$
$$P_n^H(7) > P_n^H(9) < P_n^H(11),$$

$$\text{and}\, P_n^H(13) \;>\; P_n^H(15) \;<\; P_n^H(17). \tag{1.6}$$

(Note that when $n = 11$, Observation 1 fails because $P_n^H(13) = 269 \times 2^{-11} < 275 \times 2^{-11} = P_n^H(15)$.) We will only exhibit the proof for the divot at 5 to provide some insights, since those for the further claims would be equally elementary but much more verbose.

Theorem 1.2 *Observation 2 is true for sufficiently large n. That is,*

$$\exists N : \forall n \geq N, \forall k \neq 4 : \; P_n^M(4) \;>\; P_n^M(k). \tag{1.7}$$

(Note that when $n = 14$, Observation 2 fails because $P_n^M(4) = 2975 \times 2^{-14} = P_n^M(2)$. We think (but cannot prove) that 15 is a valid candidate for N.)

Similar to Theorem 1.9 in [11], we have the following result about limiting probabilities, which can be used to prove Theorem 1.2.

Theorem 1.3 *The limiting probability distribution of missing differences, $\ell(k) := \lim_{n\to\infty} P_n^M(k)$, is well-defined, positive on (and only on) even k's, adds up to 1, and satisfies*

$$\ell(4) \;>\; \ell(2) \;>\; \ell(6) \;>\; \ell(0) \;>\; \ell(8) \;>\; \ell(10) \;>\; \ell(12) \;>\cdots>\; \ell(20). \tag{1.8}$$

Rigorous bounds for $\ell(k)$ are given in Theorem 3.18. As a corollary, we provide an asymptotic bound for the OEIS sequence A103295, which counts the number of complete rulers.[2]

Theorem 1.4 *The OEIS sequence A103295 satisfies $a_n \sim c \cdot 2^n$, where $0.2433 < c < 0.2451$.*[3]

2 Results about Having (Few) Differences

We give a few straightforward results on having few differences.

Definition 2.1 A sequence Q has a divot at i if Q_i is smaller than the nearest non-zero neighbor on each side of the sequence.

Note in the above definition we require the neighbors to be non-zero; this is important as otherwise there would be divots at every even number in $[2n-1]$.

Proposition 2.2 *For all $n \geq 4$, P_n^H has a divot at 5: $P_n^H(3) > P_n^H(5) < P_n^H(7)$.*

---

[2] See Definition 3.22.

[3] By the "$\sim$" symbol, we mean that $\lim_{n\to\infty} a_n/(c \cdot 2^n) = 1$.

Proof We have the following characterizations, where we abbreviate a set S being an arithmetic progression[4] by writing that S is an AP.

- $|S - S| = 3 \iff |S| = 2$.
- $|S - S| = 5 \iff |S| = 3 \text{and} S$ is an AP (e.g., $\{3, 8, 13\}$).
- $|S - S| = 7 \iff |S| = 3 \text{and} S$ is not an AP,or$|S| = 4 \text{and} S$ is an AP.

Thus, by counting arithmetic progressions, the following equations hold:

$$\begin{aligned} 2^n P_n^{\mathrm{H}}(3) &= \binom{n}{2} \\ 2^n P_n^{\mathrm{H}}(5) &= \binom{\lfloor \frac{n}{2} \rfloor}{2} + \binom{\lfloor \frac{n+1}{2} \rfloor}{2} \\ 2^n P_n^{\mathrm{H}}(7) &= \binom{n}{3} - 2^n P_n^{\mathrm{H}}(5) + \sum_{i=0}^{2} \binom{\lfloor \frac{n+i}{3} \rfloor}{2}. \end{aligned} \tag{2.1}$$

When $n \geq 4$, we have

$$P_n^{\mathrm{H}}(3) > P_n^{\mathrm{H}}(5) \leq \frac{\binom{n}{3}}{2^n} - P_n^{\mathrm{H}}(5) < P_n^{\mathrm{H}}(7). \tag{2.2}$$

In view of the proof, for any fixed k we see that $P_n^{\mathrm{H}}(k)$ can be written in a closed form in terms of n. Thus, the proof for divots at 9 and 15 can be done similarly and is left out. (The reason for the numbers to be specifically 5, 9 and 15 is that their immediately-smaller sizes (3, 7 and 13) are the sizes of $|S - S|$ when 2/3/4 elements are chosen in general positions to form S (and hence very likely), and their immediately-larger sizes (7, 11 and 17) need less "coincidences" (collisions of differences) to happen.)

We will now conclude Theorem 1.1.

3 Results about Missing (Few) Differences

3.1 *Intuitively Measuring the Limiting Probabilities*

We show that the limiting probability of having k differences, and that of missing k differences, exist. The latter (Claim 3.2) is a special case of Theorem 1.3 in [25], but as some parts of this argument will be used later, we provide details.

Claim 3.1 For all $k \geq 0$, $\lim_{n\to\infty} P_n^{\mathrm{H}}(k) = 0$.

[4] This means there are integers a, d and m such that $S = \{a, a + d, a + 2d, \ldots, a + md\}$.

Proof The claim follows immediately by noting $P(|S-S|=k) \le P(|S| \le k) \to 0$. We include it to compare it with the next claim.

Claim 3.2 For all $k \ge 0$, $\lim_{n\to\infty} P_n^{\mathrm{M}}(k)$ exists and $\sum_{i=0}^{\infty} \lim_{n\to\infty} P_n^{\mathrm{M}}(i) = 1$.

Proof Recall that when k is odd, we have $P_n^{\mathrm{M}}(k) = 0$ for all $n \neq \frac{k+1}{2}$. We are interested in evens.

$\forall k \ge 0, \forall m > k, \forall \varepsilon > 0, \forall n > 2m, \forall S \subseteq [n]$, if $\{0, \ldots, n-m-1\} \subseteq S-S$, then

$$\begin{aligned} |(S-S) \cap \{n-m, \ldots, n-1\}| = m-k &\iff |(S-S) \cap \{0, \ldots, n-1\}| = n-k \\ &\iff |S-S| = 2n-1-2k. \end{aligned} \tag{3.1}$$

Thus

$$\begin{aligned} &\left|P_n^{\mathrm{M}}(2k) - P\left(|(S-S) \cap \{n-m, \ldots, n-1\}| = m-k\right)\right| \\ &\le P(\{0, \ldots, n-m-1\} \nsubseteq S-S). \end{aligned} \tag{3.2}$$

The main term is constant with respect to n:

$$\begin{aligned} &P\left(|(S-S) \cap \{n-m, \ldots, n-1\}| = m-k\right) \\ =\ &P\left(|((S \cap \{n-m, \ldots, n-1\}) - (S \cap \{0, \ldots, m-1\})) \cap \{n-m, \ldots, n-1\}| = m-k\right) \\ =\ &P_{S_1 \subseteq [n]\setminus[n-m], S_2 \subseteq [m]}\left(|(S_1 - S_2) \cap \{n-m, \ldots, n-1\}| = m-k\right) \\ =\ &P_{S \subseteq [2m]}\left(|(S-S) \cap \{m, \ldots, 2m-1\}| = m-k\right) \\ =:\ &f_k(m). \end{aligned} \tag{3.3}$$

By Lemma 11 in [13],[5]

$$\begin{aligned} P(\{0, \ldots, n-m-1\} \nsubseteq S-S) &\le \sum_{i=0}^{n-m-1} P(i \notin S-S) \\ &\le \sum_{i=0}^{\lfloor \frac{n}{2} \rfloor - 1} \left(\frac{3}{4}\right)^{\frac{n}{3}} + \sum_{i=\lfloor \frac{n}{2} \rfloor}^{n-m-1} \left(\frac{3}{4}\right)^{n-i} \\ &< \left(\frac{3}{4}\right)^{\frac{n}{3}} \cdot \frac{n}{2} + \left(\frac{3}{4}\right)^{m+1} \cdot 4 \end{aligned}$$

[5] It states that if A is a uniformly randomly chosen subset of $[n]$, then

$$P(k \notin S-S) \begin{cases} \le \left(\frac{3}{4}\right)^{n/3} & 1 \le k \le \frac{n}{2} \\ \le \left(\frac{3}{4}\right)^{n-k} & \frac{n}{2} \le k \le n-1. \end{cases}$$

$$< \varepsilon + 4\left(\frac{3}{4}\right)^{m+1} \quad \text{for sufficiently large } n. \tag{3.4}$$

For sufficiently large n,

$$\left|P_n^{\mathrm{M}}(2k) - f_k(m)\right| < \varepsilon + 4\left(\frac{3}{4}\right)^{m+1}. \tag{3.5}$$

By the arbitrariness of m and ε, $\{P_n^{\mathrm{M}}(2k)\}_n$ is Cauchy and so converges. The rest of the claim follows from non-negativity of the limits and the fact that $\sum_{i=0}^{2n-1} P_n^{\mathrm{M}}(i) = 1$.

Remark 3.3 The bounded error, $\varepsilon + 4\left(\frac{3}{4}\right)^{m+1}$, is irrelevant to k, so the convergence is uniform.

Definition 3.4 Let $\ell(k) := \lim_{n\to\infty} P_n^{\mathrm{M}}(k)$.

Lemma 3.5 *For all $k \geq 0$, we have $\ell(2k+2) \geq \ell(2k)/2$.*

Proof We have

$$\begin{aligned}
&P_n^{\mathrm{M}}(2k+2)\\
=&P(|S-S| = 2n-1-2(k+1))\\
\geq&P(n-1 \notin S \wedge |(S-S) \cap \{-n+2,\ldots,n-2\}| = 2n-1-2(k+1))\\
=&P(n-1 \notin S)\cdot P_{S\subseteq[n-1]}(|S-S| = 2(n-1)-1-2k)\\
=&\frac{1}{2}P_{n-1}^{\mathrm{M}}(2k).
\end{aligned} \tag{3.6}$$

Note the left and right hand sides converge to $\ell(2k+2)$ and $\ell(2k)/2$ respectively.

Corollary 3.6 *For all $k \geq 0$, $\lim_{n\to\infty} P_n^M(2k) > 0$.*

Compared with the distribution of having-differences (Claim 3.1), this shows that the direction we view matters. We see non-zero limits at this end.

Remark 3.7 By Remark 5.1 (in Appendix),

$$\begin{aligned}
P_{36}^{\mathrm{M}}(0) &= \frac{8342197304}{2^{36}} \approx 0.1214,\\
P_{36}^{\mathrm{M}}(2) &= \frac{12668987317}{2^{36}} \approx 0.1843,\\
P_{36}^{\mathrm{M}}(4) &= \frac{12894355828}{2^{36}} \approx 0.1876,\\
P_{36}^{\mathrm{M}}(6) &= \frac{10879185718}{2^{36}} \approx 0.1583,
\end{aligned}$$

$$P_{36}^{\mathrm{M}}(8) = \frac{8208838614}{2^{36}} \approx 0.1195. \tag{3.7}$$

This gives us a sensible (but not rigorous) estimate of $\ell(k)$.

We do have a rigorous bound of $\ell(k)$, in view of the proof for Claim 3.2.

Proposition 3.8 *For all* $m > k$, $|\ell(2k) - f_k(m)| \leq 4(\frac{3}{4})^{m+1}$.

Proof Replace $P_n^{\mathrm{M}}(2k)$ by $\ell(2k)$ in Eq. (3.5).

One would like to use this fact to prove Theorem 1.3, since $f_k(m)$ is finitely computable. Unfortunately this quickly becomes unrealistic because it takes $4^m m^2$ computations to exhaustively determine $f_k(m)$, and to reduce the uncertainty to the level of $(\ell(0) - \ell(8))/2 \approx (0.1214 - 0.1195)/2$ (which is the bottleneck difference) we need $m \geq 28$. In 2020, it took our laptop[6] around 40 seconds to run $m = 17$ with this method, and thus it would need around 14.4 years to computationally verify the theorem. We need something better.

3.2 Using Conditional Probabilities

Lemma 3.9 *The conditional probability of* $k \notin S - S$, *given that* $0, n-1 \in S$, *is bounded by the following:*

$$P\left(k \notin S - S \middle| 0, n-1 \in S\right) \begin{cases} = 0 & k = n-1 \\ = \frac{4}{9} \cdot \left(\frac{3}{4}\right)^{n-k} & \frac{n}{2} \leq k < n-1 \\ \leq \frac{4}{9} \cdot \left(\frac{3}{4}\right)^{\frac{n}{3}} & 0 \leq k < \frac{n}{2}. \end{cases} \tag{3.8}$$

Proof For all $k < n$ let $D := \{\{a, b\} : a, b \in [n], |a - b| = k\}$. We say $D' \subseteq D$ is *mutually disjoint* if $\forall p_1, p_2 \in D'$, $p_1 \cap p_2 = \emptyset$. If $D' \subseteq D$ is mutually disjoint and $0, n-1 \in \bigcup D'$ (that is, D' contains the pairs $\{0, k\}$ and $\{n-1-k, n-1\}$), then

$$\begin{aligned} P\left(k \notin S - S \middle| 0, n-1 \in S\right) &= P\left(D \cap \mathscr{P}(S) = \emptyset \middle| 0, n-1 \in S\right) \\ &\leq P\left(D' \cap \mathscr{P}(S) = \emptyset \middle| 0, n-1 \in S\right) \\ &= \prod_{p \in D'} \left(1 - 2^{-|p \setminus \{0, n-1\}|}\right) \\ &= \begin{cases} 0 & k = n-1 \\ \frac{4}{9} \cdot \left(\frac{3}{4}\right)^{|D'|} & 0 \leq k < n-1. \end{cases} \end{aligned} \tag{3.9}$$

When $2k > n-1$, D is already mutually disjoint and has size $n - k$; otherwise, we can find a mutually disjoint D' with $|D'| \geq n/3$, and let $0, n-1 \in \bigcup D'$ without loss of generality. We hence conclude the lemma.

[6] CPU: i7-6500U @ 2.5GHz, RAM: 8GB.

The conditional probability distribution requiring $0, n-1 \in S$ is compared with the usual probability distribution without such restriction. We define similar notions to P_n^{M}, f_k.

Definition 3.10 Let

$$\begin{aligned} Q_n^{\mathrm{M}}(k) &:= P\left(|S-S| = 2n-1-k \middle| 0, n-1 \in S\right); \\ g_k(m) &:= P_{S\subseteq[2m]}\left(|(S-S)\cap\{m,\ldots,2m-1\}| \;=\; m-k \middle| 0, 2m-1 \in S\right). \end{aligned}$$

Proposition 3.11 *$\forall k \geq 0$, $\forall m > k$, $\forall \varepsilon > 0$ and for sufficiently large n,*

$$\left|Q_n^M(2k) - g_k(m)\right| \;<\; \varepsilon + \frac{16}{9}\cdot\left(\frac{3}{4}\right)^{m+1}.$$

Proof This follows from an analagous argument as in Claim 3.2. By Lemma 3.9, the uncertainty is 4/9 the original one.

Definition 3.12 We have $j(k) := \lim_{n\to\infty} Q_n^{\mathrm{M}}(k)$.

Proposition 3.13 *Note $j(k)$ is well-defined; in addition, for all $m > k$ we have*

$$|j(2k) - g_k(m)| \;<\; \frac{16}{9}\left(\frac{3}{4}\right)^{m+1}.$$

Proof The proof is similar to that of Proposition 3.8.

Lemma 3.14 *For $k \in 2\mathbb{N}$,*

$$\ell(k) \;=\; \frac{j(k)}{4} + \ell(k-2) - \frac{\ell(k-4)}{4}.$$

Proof

$$\begin{aligned} P_n^{\mathrm{M}}(k) &= P(|S-S| = 2n-1-k) \\ &= \frac{1}{4}P\left(|S-S| = 2n-1-k \middle| 0, n-1 \in S\right) \\ &\quad + \frac{1}{2}P\left(|S-S| = 2n-1-k \middle| 0 \notin S\right) \\ &\quad + \frac{1}{2}P\left(|S-S| = 2n-1-k \middle| n-1 \notin S\right) \\ &\quad - \frac{1}{4}P\left(|S-S| = 2n-1-k \middle| 0, n-1 \notin S\right) \\ &= \frac{1}{4}Q_n^{\mathrm{M}}(k) + \frac{1}{2}P_{n-1}^{\mathrm{M}}(k-2) + \frac{1}{2}P_{n-1}^{\mathrm{M}}(k-2) - \frac{1}{4}P_{n-2}^{\mathrm{M}}(k-4). \quad (3.10) \end{aligned}$$

The left and right hand sides converge to $\ell(k)$ and $\frac{j(k)}{4} + \ell(k-2) - \frac{\ell(k-4)}{4}$ respectively.

Corollary 3.15 *For* $k \in 2\mathbb{N}$,

$$j(k) = 4\ell(k) - 4\ell(k-2) + \ell(k-4), \text{ and} \ell(k) = \sum_{i=0}^{\infty} \frac{i+1}{2^{i+2}} j(k-2i).$$

Remark 3.16 It's better to focus on and compute the j sequence than the ℓ sequence, for the following reasons.

- Using the same value of m, estimating the j sequence will produce less uncertainty than estimating the ℓ sequence. In view of Propositions 3.8 and 3.13, given $f_k(m)$ and $g_k(m)$, which are finitely computable, $\ell(2k)$ is within $4\left(\frac{3}{4}\right)^{m+1}$ from $f_k(m)$, while $j(2k)$ is within only $\frac{16}{9}\left(\frac{3}{4}\right)^{m+1}$ from $g_k(m)$, reducing by a factor of 4/9.
- When estimating $\ell(0) - \ell(8)$, which is the bottleneck difference regarding Theorem 1.2, the uncertainty coming from the j sequence would be further compressed while that from ℓ would be amplified. Say each term in the j sequence has an uncertainty of e, then by Corollary 3.15, the uncertainty of $\ell(0) - \ell(8)$ is $63e/64$, whereas if we estimated the ℓ sequence honestly the uncertainty would be $2e$.
- What's more, it is 4x faster to compute $g_k(m)$ than $f_k(m)$ because the condition $0, n-1 \in S$ reduces two degrees of freedom.

Approximately,[7] the j method is $4^{\log_{3/4}(\frac{4}{9}\cdot\frac{63}{128})} \times 4 \approx 6064$ times faster than the ℓ method to verify Theorem 1.3. Divide the 14.4 years mentioned earlier by this modifier, and things start to become feasible.

Armed with these results, we are ready now to prove Theorem 1.3.

3.3 Calculations and Results

Calculation 3.17 *The code in Appendix 6 calculates the data in Table 1.*

This provides the numerical bounds we need.

Theorem 3.18 *The following inequalities hold:*

$$0.12165 < \ell(0) < 0.12255$$
$$0.18434 < \ell(2) < 0.18614$$

[7] This is a rough estimate: the computational complexities of $f_k(m)$ and $g_k(m)$ are both asymptotically $4^m \cdot m^2$, but when m is decreased we only counted the boost coming from the 4^m factor, neglecting that from the quadratic term; also, m is always an integer, so there are floor-and-ceiling errors.

Table 1 Values of $g_k(m)$ when $m = 23$

k	$g_k(23)$
0	$8592305829704/2^{44}$
1	$4442759682300/2^{44}$
2	$2367846591103/2^{44}$
3	$1174068145740/2^{44}$
4	$559669653171/2^{44}$
5	$256031157923/2^{44}$
6	$114186380080/2^{44}$
7	$49736070308/2^{44}$
8	$21123843993/2^{44}$
9	$8778930083/2^{44}$
10	$3543398884/2^{44}$
11	$1378772067/2^{44}$
12	$508048560/2^{44}$
13	$174732658/2^{44}$
14	$54900922/2^{44}$
15	$15344643/2^{44}$
16	$3692910/2^{44}$
17	$737437/2^{44}$
18	$116855/2^{44}$
19	$13885/2^{44}$
20	$1134/2^{44}$
21	$55/2^{44}$
22	$1/2^{44}$

$$
\begin{aligned}
0.18713 &< \ell(4) < 0.18959 \\
0.15728 &< \ell(6) < 0.16019 \\
0.11801 &< \ell(8) < 0.12119 \\
0.08188 &< \ell(10) < 0.08523 \\
0.05355 &< \ell(12) < 0.05700 \\
0.03334 &< \ell(14) < 0.03685 \\
0.01981 &< \ell(16) < 0.02335 \\
0.01115 &< \ell(18) < 0.01471 \\
0.00580 &< \ell(20) < 0.00937.
\end{aligned}
\tag{3.11}
$$

Proof Follows from Proposition 3.13, Corollary 3.15 and Calculation 3.17. (Calculation 3.17 calculates $g_k(23)$, which Proposition 3.13 uses to bound $j(2k)$, and then Corollary 3.15 reveals the relationship between ℓ and j.)

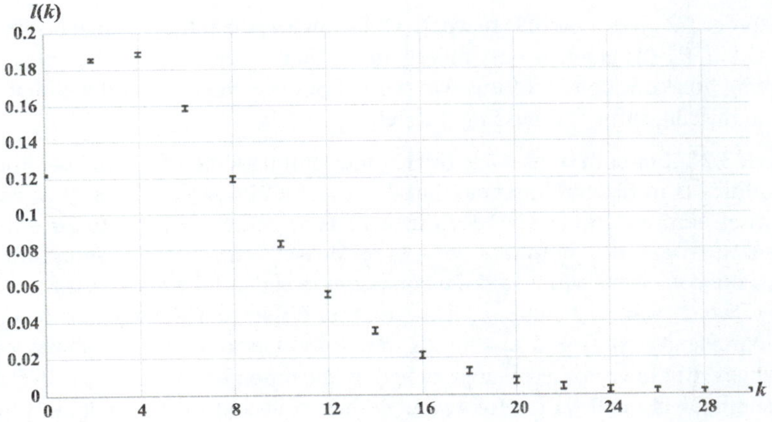

Fig. 3 Bounds on $\ell(k)$ for $0 \le k \le 30$. (Odd k's are omitted.)

Proof (Theorem 1.3) Combine Claim 3.2, Corollary 3.6 and Theorem 3.18.

The rigorous bounds are illustrated in Fig. 3.

After exhibiting the following auxiliary result, we will prove Theorem 1.2.

Lemma 3.19 ([13], Theorem 3) $\sum_{i=0}^{\infty} i \cdot \ell(i) = 6$.

Theorem 3.20 *For all* $k \neq 4$, $\ell(k) < \ell(4)$.

Proof Theorem 1.3 includes the case for $k \le 20$. When $k > 20$, by Lemmas 3.5 and 3.19,

$$\begin{aligned} 2(k-6) \cdot \ell(k) &\le \sum_{i=k}^{\infty} (k-6) \cdot \ell(i) \\ &< \sum_{i=6}^{\infty} (i-6) \cdot \ell(i) \\ &= \sum_{i=0}^{\infty} (i-6) \cdot \ell(i) + 6\ell(0) + 4\ell(2) + 2\ell(4) \\ &< \sum_{i=0}^{\infty} i \cdot \ell(i) - 6\sum_{i=0}^{\infty} \ell(i) + (6+4+2)\ell(4) \\ &= 12\ell(4). \end{aligned} \tag{3.12}$$

Thus for all $k \neq 4$, $\ell(k) < \ell(4)$.

Proof (Theorem 1.2) Follows from Theorem 3.20 and Remark 3.3.

Theorem 1.2 gives a partial answer to Question 5.2; the rather strange occurrence of $P_n^{\mathrm{M}}(2) = P_n^{\mathrm{M}}(4)$ happens only finitely many times.

Having gone through all of this, we present our comment about the comparison between missing sums and missing differences:

Remark 3.21 For both sumset and diffset, the natural method to bound the limiting probabilities is to blatantly assume that the set of all possible values (0 to $2n-2$ for sumset, $-(n-1)$ to $(n-1)$ for diffset) has its middle part (m to $2n-m-2$ for sumset, $-(n-m-1)$ to $(n-m-1)$ for diffset) *entirely* in the sumset/diffset. Using a finite m, there will be a finite computation and a finite error term.

Now, when using the same m, it is true that diffsets have a squared computational complexity compared to sumsets, precisely due to the independence of the two fringes. But the error term, approached by the union bound, is also different.

Assuming k is small w.r.t n, for a number that's k away from the endpoints (e.g. k for sumset and $n-k-1$ for diffset), the probability that it is *not* in $S+S$ is about $0.75^{k/2+1}$, since $k \notin S+S$ means the $\lfloor k/2 \rfloor + 1$ disjoint pairs

$$\{0, k\}, \{1, k-1\}, \ldots, \{\lfloor k/2 \rfloor, \lceil k/2 \rceil\}$$

are all missing from S. The probability that such a k-away number is not in $S-S$ is, however, 0.75^{k+1}, since $n-k-1$ is missing from $S-S$ if and only if none of the pairs

$$\{0, n-k-1\}, \{1, n-k\}, \ldots, \{k, n-1\}$$

are in S. As a result, when k ranges from m to $n-1$, the accumulative error terms for sumset and diffset are respectively constant multiples of $0.75^{m/2}$ and 0.75^m. To reach a similar error term, the sumset calculation would need to choose an m twice as large as the diffset. This would make the two computation complexities asymptotically the same, as naturally computing the theoretical limiting probability is $\Omega(4^m m^2)$ for diffset, $\Omega(2^m m^2)$ for sumset.

3.4 About Rulers

Our numerical bounds can show the asymptotics of a ruler-counting sequence.

Definition 3.22 A *ruler* of length L is any subset $R \subseteq \{0, \ldots, L\}$. It is *complete* if it can measure every distance shorter or equal to its length; that is, $\{0, \ldots, L\} \subseteq R - R$.

Lemma 3.23 *Let a_n be the number of complete rulers of length n; then* $a_{n-1} \sim \ell(0) \cdot 2^n$.

Proof $S \subseteq [n]$ is a complete ruler of length $n-1$ iff $|S-S| = 2n-1$, so the number of complete rulers of length $n-1$ is equal to $P_n^{\mathrm{M}}(0) \cdot 2^n$, which goes to $\ell(0) \cdot 2^n$.

Proof (Theorem 1.4) Plug the bounds from Theorem 3.18 into Lemma 3.23. Here $c = 2\ell(0)$.

4 Conjectures

Here are some conjectures that would be nice to be true.

Recall Proposition 1.1; intuitively, when $k \ll n$, k elements from $[n]$ in general positions usually gives $|S - S| = k(k-1) + 1$. On the other hand, to have $|S - S| = k(k-1) + 3$ requires a maximal appearance of coincidences (repeated differences). Hence we have the following conjecture about the divots in P_n^{H}.

Conjecture 4.1 For every $k > 1, k(k-1) + 3$ is a divot of P_n^{H} for sufficiently large n. Furthermore, they are the only divots.

We also noticed that once a divot appears in P_n^{H}, it seems to never move again:

Conjecture 4.2 If k is a divot of P_n^{H}, then it is also a divot of $P_{n'}^{\mathrm{H}}$ for all $n' > n$.

About missing differences, we proved Theorem 1.2 by limits, hence not giving an explicit threshold N such that every $n \geq N$ satisfies Observation 2. Experimental data suggest that 15 might be enough already, so we guess:

Conjecture 4.3 For all $n \geq 15, \forall k \neq 4, P_n^{\mathrm{M}}(4) > P_n^{\mathrm{M}}(k)$.

Recall that in Theorem 1.3, we compared the limiting probabilities of missing 0, 2, 4, 6, ..., 20 differences, and found no divot. What about missing 22, or more? In fact, any two limiting probabilities can be approximated to arbitrary precision using our method (given sufficient computation), but we couldn't bound infinitely many of them at the same time. Nevertheless, both intuition and experimental data seem to suggest that the decay after $\ell(4)$ should go on forever. Thus, we leave the following conjecture.

Conjecture 4.4 In fact, $\ell(4) > \ell(6) > \ell(8) > \ell(10) > \cdots$. Equivalently, the sequence ℓ has no divot.

5 Distribution of $|S - S|$ when $n \leq 36$

Remark 5.1 Denoting the table by T, $T_{n,k}/2^n = P_n^{\mathrm{H}}(k) = P_n^{\mathrm{M}}(2n-1-k)$ (Tables 2 and 3).

Question 5.2 *Observe that when* $n = 3, 11, 12, 14$, $P_n^M(2) = P_n^M(4)$. *Such frequent repetition of large numbers doesn't look so random. Is there any reason behind it? Will it happen again?*

Table 2 Number of $S \subseteq [n]$ with $|S - S| = k$. ($n \leq 24$)

n n	0	1	2	3	4	5	6	7	8	9	10	11	12	13	14	15	16	17	18	19	20	21	22	23	24
k																									
0	1	1	1	1	1	1	1	1	1	1	1	1	1	1	1	1	1	1	1	1	1	1	1	1	1
1	0	1	2	3	4	5	6	7	8	9	10	11	12	13	14	15	16	17	18	19	20	21	22	23	24
3	0	0	1	3	6	10	15	21	28	36	45	55	66	78	91	105	120	136	153	171	190	210	231	253	276
5	0	0	0	1	2	4	6	9	12	16	20	25	30	36	42	49	56	64	72	81	90	100	110	121	132
7	0	0	0	0	3	8	17	31	51	77	112	155	208	272	348	436	539	656	789	939	1107	1293	1500	1727	1976
9	0	0	0	0	0	4	10	17	27	43	62	85	113	148	189	236	289	352	423	501	588	687	795	913	1042
11	0	0	0	0	0	0	9	25	47	77	113	170	237	319	413	531	666	825	1000	1206	1430	1691	1970	2289	2630
13	0	0	0	0	0	0	0	17	49	97	169	269	409	606	863	1195	1607	2115	2735	3492	4393	5450	6690	8130	9790
15	0	0	0	0	0	0	0	0	33	93	177	275	402	549	730	967	1238	1562	1932	2355	2829	3345	3946	4613	5343
17	0	0	0	0	0	0	0	0	0	63	187	377	629	973	1417	1978	2688	3628	4765	6151	7794	9781	12089	14774	17861
19	0	0	0	0	0	0	0	0	0	0	128	377	747	1228	1850	2642	3633	4849	6340	8278	10580	13381	16603	20474	24909
21	0	0	0	0	0	0	0	0	0	0	0	248	747	1509	2507	3770	5338	7271	9641	12469	15909	20315	25533	31893	39392
23	0	0	0	0	0	0	0	0	0	0	0	0	495	1472	2975	4999	7519	10654	14499	19129	24681	31221	38903	48354	59263
25	0	0	0	0	0	0	0	0	0	0	0	0	0	988	2975	6022	10104	15278	21596	29249	38430	49408	62377	77572	95318
27	0	0	0	0	0	0	0	0	0	0	0	0	0	0	1969	5911	11985	20192	30501	43062	58148	76121	97667	123155	153424
29	0	0	0	0	0	0	0	0	0	0	0	0	0	0	0	3911	11880	24103	40524	61350	86236	115893	150319	190510	236824
31	0	0	0	0	0	0	0	0	0	0	0	0	0	0	0	0	7857	23734	48377	81542	123470	174352	234160	304245	385858
33	0	0	0	0	0	0	0	0	0	0	0	0	0	0	0	0	0	15635	47474	96676	162994	246765	347050	465537	602109
35	0	0	0	0	0	0	0	0	0	0	0	0	0	0	0	0	0	0	31304	94885	193562	326913	494449	696108	931109
37	0	0	0	0	0	0	0	0	0	0	0	0	0	0	0	0	0	0	0	62732	190623	388606	656644	993569	1396647
39	0	0	0	0	0	0	0	0	0	0	0	0	0	0	0	0	0	0	0	0	125501	380805	776640	1312446	1985532
41	0	0	0	0	0	0	0	0	0	0	0	0	0	0	0	0	0	0	0	0	0	250793	763402	1557467	2633237
43	0	0	0	0	0	0	0	0	0	0	0	0	0	0	0	0	0	0	0	0	0	0	503203	1528095	3117611
45	0	0	0	0	0	0	0	0	0	0	0	0	0	0	0	0	0	0	0	0	0	0	0	1006339	3061916
47	0	0	0	0	0	0	0	0	0	0	0	0	0	0	0	0	0	0	0	0	0	0	0	0	2014992
49	0	0	0	0	0	0	0	0	0	0	0	0	0	0	0	0	0	0	0	0	0	0	0	0	0

Table 3 Number of $S \subseteq [n]$ with $|S - S| = k$. $(24 \leq n \leq 36)$

n	24	25	26	27	28	29	30	31	32	33	34	35	36
k													
0	1	1	1	1	1	1	1	1	1	1	1	1	1
1	24	25	26	27	28	29	30	31	32	33	34	35	36
3	276	300	325	351	378	406	435	465	496	528	561	595	630
5	132	144	156	169	182	196	210	225	240	256	272	289	306
7	1976	2248	2544	2864	3211	3584	3985	4415	4875	5365	5888	6443	7032
9	1042	1184	1338	1504	1682	1876	2084	2305	2541	2795	3064	3349	3651
11	2630	3010	3419	3876	4357	4886	5443	6060	6707	7410	8143	8940	9776
13	9790	11699	13868	16325	19094	22202	25674	29543	33832	38569	43786	49515	55787
15	5343	6158	7029	7980	9024	10164	11384	12696	14093	15597	17216	18941	20767
17	17861	21464	25554	30192	35439	41365	47972	55334	63485	72583	82597	93598	105615
19	24909	30034	35835	42560	50164	58778	68336	79218	91199	104572	119214	135569	153328
21	39392	48297	58729	70921	85023	101393	120236	141992	166842	195124	227418	263837	304894
23	59263	72166	86779	103803	122773	144495	168711	195948	226062	259777	297046	338522	383708
25	95318	116803	141545	170669	203518	241453	283954	332047	385486	445578	511668	585268	666132
27	153424	188936	230785	281634	340918	411385	492735	587687	696368	821738	964188	1126614	1309990
29	236824	290286	351743	422400	502848	598252	705828	831558	972438	1134483	1314383	1519559	1747229
31	385858	480260	589088	713474	855957	1018020	1202962	1419676	1664732	1947773	2265195	2627654	3032028
33	602109	759570	939048	1145157	1379205	1646202	1948206	2289594	2673659	3121284	3619723	4191609	4824889
35	931109	1202343	1512270	1865592	2266137	2720935	3236533	3821295	4483176	5231412	6075752	7058965	8161491
37	1396647	1867806	2404100	3013664	3697776	4468556	5330593	6293553	7368022	8567388	9903780	11391366	13047575
39	1985532	2792117	3726584	4795360	5994044	7342144	8845276	10520512	12382684	14456863	16757210	19313503	22151419
41	2633237	3984017	5596451	7469425	9586795	11966365	14608625	17543417	20782662	24369445	28318130	32680465	37482058
43	3117611	5270104	7970998	11195574	14913983	19131301	23822819	29022146	34739876	41039669	47936336	55509344	63800433
45	3061916	6244117	10557091	15968677	22417023	29862931	38239392	47566626	57804101	69047026	81288502	94666428	109216351
47	2014992	6125358	12494664	21122722	31935586	44822674	59651353	76346946	94783970	115036473	137031262	160950680	186816887
49	0	4035985	12278446	25038586	42321005	63983506	89749444	119386846	152607226	189351319	229343035	272803379	319629353
51	0	0	8080448	24564954	50090752	84658919	127967673	179465499	238552257	304816636	377630128	456991110	542473471
53	0	0	0	16169267	49200792	100303312	169496641	256144840	359073831	477185749	609113912	754212597	911317415

(continued)

Table 3 (continued)

n	24	25	26	27	28	29	30	31	32	33	34	35	36
k													
55	0	0	0	0	32397761	98478615	200765677	339187677	512453496	718291220	953949620	1217261287	1505590283
57	0	0	0	0	0	64826967	197164774	401837351	678805584	1025433250	1436715877	1907636501	2432498687
59	0	0	0	0	0	0	129774838	394536002	804070333	1358091161	2051059855	2873264810	3813305230
61	0	0	0	0	0	0	0	259822143	789993459	1609586119	2717986051	4104228068	5747795503
63	0	0	0	0	0	0	0	0	520063531	1580640910	3220331421	5437313809	8208838614
65	0	0	0	0	0	0	0	0	0	1040616486	3163602123	6444236200	10879185718
67	0	0	0	0	0	0	0	0	0	0	2083345793	6330608624	12894355828
69	0	0	0	0	0	0	0	0	0	0	0	4168640894	12668987317
71	0	0	0	0	0	0	0	0	0	0	0	0	8342197304
73	0	0	0	0	0	0	0	0	0	0	0	0	0

6 Code for Estimating $j(2k)$

```
#include <stdio.h>
#include <time.h>
#include <math.h>
long long cnts[100];
int main() {
    int m = 23, d; // Measure prob of missing n-1, ...\
    n-m diffs
    clock_t begin = clock();
    double j[100], l[100], jerr = pow(0.75, m+1) * 16 \
   / 9, lerr = 0;
    long long cnt, r1 = (1LL << 2*m), r2 = 2*m-1;
    for(long long S = (1LL << (2*m-1)) + 1; S < r1; S\
   +=2) {
        cnt = 0;
        for(d = m; d < r2; d++) if(!(S & (S >> d))) \
   cnt++;
        cnts[cnt]++;
    }
    for(int i=0; i<m; i++) {
        j[2*i] = 1.0 * cnts[i] / pow(2, 2*m-2);
        printf("j(%d) = %f+-%e\t(G%d = %lld/2^%d)\n", \
   2*i,
        j[2*i], jerr, i, cnts[i], 2*m-2);
    }
    printf("\n");
    for(int i=0; i<m; i++) {
        l[2*i] = 0;
        for(int k=0; k<=i; k++)
            l[2*i] += (k+1) / pow(2, k+2) * j[2*(i-k)\
   ];
        lerr += (i+1) / pow(2, i+2) * jerr;
        printf("l(%d) in (%f,%f)\n", 2*i, l[2*i]-lerr,\
    l[2*i]+lerr);
    }
    clock_t end = clock();
    double time_spent = (double)(end - begin) / \
   CLOCKS_PER_SEC;
    printf("\nIn %f sec.\n", time_spent);
    return 0;
}
```

Remark 6.1 The algorithm is $\Omega(4^m m^2)$. When $m = 23$, it runs for 92.73 hours on our laptop (with -O3 optimization level). The reader is welcome to confirm our calculations or achieve better bounds.

References

1. M. Asada, S. Manski, S. J. Miller and H. Suh, *Fringe pairs in generalized MSTD sets*, International Journal of Number Theory **13** (2017), no. 10, 2653–2675.
2. A. Bower, R. Evans, V. Luo and S. J. Miller, *Coordinate sum and difference sets of d-dimensional modular hyperbolas*, INTEGERS #A31, 2013, 16 pages.
3. H. Chu, N. Luntzlara, S. J. Miller and L. Shao, *Generalizations of a Curious Family of MSTD Sets Hidden By Interior Blocks*, to appear in Integers.
4. H. Chu, N. McNew, S. J. Miller, V. Xu and S. Zhang, *When Sets Can and Cannot Have MSTD Subsets*, Journal of Integer Sequences **21** (2018), Article 18.8.2.
5. T. Do, A. Kulkarni, S. J. Miller, D. Moon and J. Wellens, *Sums and Differences of Correlated Random Sets*, Journal of Number Theory **147** (2015), 44–68.
6. T. Do, A. Kulkarni, S. J. Miller, D. Moon, J. Wellens and J. Wilcox, *Sets Characterized by Missing Sums and Differences in Dilating Polytopes*, Journal of Number Theory **157** (2015), 123–153.
7. P. V. Hegarty, *Some explicit constructions of sets with more sums than differences*, Acta Arith. **130** (2007), 61–77.
8. P. V. Hegarty and S. J. Miller, *When almost all sets are difference dominated*, Random Structures Algorithms **35** (2009), 118–136.
9. A. Hemmady, A. Lott and S. J. Miller, *When almost all sets are difference dominated in* $\mathbb{Z}/\mathbb{Z}$, Integers **17** (2017), Paper No. A54, 15 pp.
10. G. Iyer, O. Lazarev, S. J. Miller and L. Zhang, *Generalized More Sums Than Differences Sets*, Journal of Number Theory **132** (2012), no. 5, 1054–1073.
11. O. Lazarev, S. J. Miller and K. O'Bryant, *Distribution of Missing Sums in Sumsets*, Experimental Mathematics **22** (2013), no. 2, 132–156.
12. J. Marica, *On a conjecture of Conway*, Canad. Math. Bull. **12** (1969), 233–234.
13. G. Martin and K. O'Bryant, *Many sets have more sums than differences*, Additive Combinatorics, Providence, RI, 2007, 287–305.
14. S. J. Miller, B. Orosz and D. Scheinerman, *Explicit constructions of infinite families of MSTD sets*, Journal of Number Theory **130** (2010), 1221–1233.
15. S. J. Miller and C. Peterson, *A geometric perspective on the MSTD question*, Discrete and Computational Geometry **62** (2019), no. 4, 832–855.
16. S. J. Miller and D. Scheinerman, *Explicit constructions of infinite families of MSTD sets*, Additive Number Theory: Festschrift In Honor of the Sixtieth Birthday of Melvyn B. Nathanson (David Chudnovsky and Gregory Chudnovsky, editors), Springer-Verlag, 2010.
17. S. J. Miller and K. Vissuet, *Most Subsets are Balanced in Finite Groups*, Combinatorial and Additive Number Theory, CANT 2011 and 2012 (Melvyn B. Nathanson, editor), Springer Proceedings in Mathematics & Statistics (2014), 147–157.
18. S. Miller, V. Xu and X. Zhang, *MSTD Subsets and Properties of Divots in the Distribution of Missing Sums*, Combinatorial and Additive Number Theory, 05/26/16.
19. M. B. Nathanson, *Problems in additive number theory I*, Additive combinatorics, Providence, RI, 2007, 263–270.
20. M. B. Nathanson, *Sets with more sums than differences*, Integers **7** (2007), #A5.
21. I. Z. Ruzsa, *On the cardinality of* $A + A$ *and* $A - A$, Combinatorics Year, North-Holland-Bolyai Tàrsulat, Keszthely, 1978, 933–938.

22. I. Z. Ruzsa, *Sets of sums and differences*, Séminaire de Théorie des Nombres de Paris, Birkhäuser, Boston, 1984, 267–273.
23. I. Z. Ruzsa, *On the number of sums and differences*, Acta Math. Sci. Hungar. **59** (1992), 439–447.
24. Y. Zhao, *Constructing MSTD sets using bidirectional ballot sequences*, J. Number Theory **130** (2010), 1212–1220.
25. Y. Zhao, *Sets characterized by missing sums and differences*, J. Number Theory **131** (2011), 2107–2134.

Recent Progress in Hilbert Cubes Theory

Norbert Hegyvári

Abstract In 1978 Nathanson obtained several results on sumsets contained in infinite sets of integers. Later the author investigated how big a Hilbert cube avoiding a given infinite sequence of integers can be. In the present paper we concentrate on *Densities and Hilbert cubes, Hilbert cubes which avoid given sets, and* Δ*-degenerate Hilbert cubes*. The aim is to collect some results in the past and some related recent problems.

Keywords Subset sums · Hilbert cubes · Additive combinatorics

AMS 2010 Primary 11B30 · 11L03 · Secondary 11B75

1 Introduction

In 1892 Hilbert defined an affine d-dimensional cube which is nowadays called Hilbert cube as follows: Let $x_0, a_1 < a_2 < \cdots < a_d$ sequence of integers. Hilbert cube is the set

$$H(x_0, a_1, a_2, \ldots, a_d) = \Big\{x_0 + \sum_{1\le i\le d} \{0, a_i\}\Big\} = x_0 + \{0, a_1\} + \cdots + \{0, a_d\}.$$

Theorem 1.1 (Hilbert 1892) *For every* r *and* d*, there exists (a least) number* $h(d; r)$ *so that for every colouring* $\chi : [1; h(d; r)] \mapsto [1; r]$ *there exists an affine d-dimensional monochromatic cube.*

N. Hegyvári (✉)
Institute of Mathematics, ELTE TTK, Eötvös University, Pázmány st. 1/c, Budapest 1117, Hungary
e-mail: hegyvari@renyi.hu

Hungary and Alfréd Rényi Institute of Mathematics, Hungarian Academy of Science, 127, Budapest 1364, Hungary

M. B. Nathanson (ed.), *Combinatorial and Additive Number Theory IV*,
Springer Proceedings in Mathematics & Statistics 347,
https://doi.org/10.1007/978-3-030-67996-5_14

This is one of the first Ramsey-type result in the number theory. Hilbert cubes have many applications and these days this topic can be called *Hilbert cubes theory*. Starting our discussion, from many important results we would only like to mention one, the well-known density version which was an important tool in the theorem of Szemerédi:

Theorem 1.2 (Szemerédi 1969) *If the lower density* ($\liminf_{N\to\infty} |S \cap [1, N]|/N > 0$) *of S is positive then S contains a Hilbert cube with dimension at least*

$$d \gg \log\log N.$$

In the present paper we concentrate on *Densities and Hilbert cubes*, *Hilbert cubes which avoid given sets*, and Δ*-degenerate Hilbert cubes*. The aim is to collect some results in the past and some related recent problems.

2 Notation

Throughout the paper, $[n] := \{1, 2 \ldots, n\}$, and for $H = H(x_0, a_1 < a_2 < \cdots < a_d)$ $\dim(H) := d$ denotes the dimension of the Hilbert cube H, and $|H|$ its size. When $x_0 = 0$, $A = \{a_1 < a_2 < \cdots < a_d\}$ then instead of H, the usual notation is $FS(A)$ (and said to be *subset sum of* A), so $H = x_0 + FS(A))$, where $A = \{a_1 < a_2 < \cdots, < a_d\}$.

Infinite Hilbert cubes are defined in a natural way: if $\{a_1 < a_2 < \cdots < a_d < \cdots\}$ is a sequence of integers and $H_d = H(x_0, a_1, a_2, \ldots, a_d)$, then $H := \cup_d H_d$ is an infinite Hilbert-cube and $FS(A) = H(0, a_1 < a_2 < \cdots < a_d < \cdots)$.

For a real number $\Delta \in (0, 1]$ the cube $H := H(x_0, a_1, a_2, \ldots, a_d)$ is said to be $\Delta-$degenerate, if $\frac{\log_2 |H|}{d} = \Delta$. ($\log_2 x$ means $\ln x/\ln(2)$.) The case $\Delta = 1$ is equivalent to $|H| = 2^d$, i.e., all sums are distinct and in this case we say that H is *non-degenerate*. Note that understanding the behaviour of Hilbert cubes with degeneracy $\Delta < 1$ is more difficult compared to the case of $\Delta = 1$ (that is, when all the sums are distinct).

The additive energy of a set A is defined by

$$E(A) = |\{(x_1, x_2, x_3, x_4) \in A \times A \times A \times A : x_1 + x_2 = x_3 + x_4\}|.$$

Let G be a countable torsion group and let $H_1 \subseteq H_2 \subseteq \cdots \subseteq H_n \subseteq \cdots$ be a sequence of finite subgroups of G. Then G is said to be σ-finite with respect to $\{H_n\}$ if $G = \cup_{n=1}^{\infty} H_n$. Let $A \subseteq G$. The asymptotic upper density of A is defined by

$$\overline{d}(A) := \limsup_{n\to\infty} \frac{|A \cap H_n|}{|H_n|}.$$

3 Some Previous Results

Densities and Hilbert cubes.
As we have seen already, a set with positive upper density contains a *finite* cube with dimension $c \log \log n$. A question would naturally arise: Does it contain an *infinite* Hilbert cube? In the end of 70s E.G. Strauss proved the following surprising result (see [6]):

Theorem 3.1 (E.G. Strauss) *For every $\varepsilon > 0$, there exists a sequence of integers with density $> 1 - \varepsilon$, which does not contain an infinite Hilbert cube.*

Nevertheless Nathanson showed that this does not occur with sets having total density [7]:

Theorem 3.2 (M.B. Nathanson) *Every sequence of integers with density 1 contains an infinite Hilbert cube.*

In the next section we give a generalization of Theorem 3.2 to σ-finite groups (see more on σ-finite groups in [9]). Namely we prove

Theorem 3.3 *Let $G = \{G_n\}_{n=1}^{\infty}$ be a σ-finite group respect to $\{G_n\}$. Let $A \subseteq G$ and assume $\overline{d}(A) = 1$. Then A contains an infinite Hilbert cube.*

It would be interesting to find a similar example of Strauss in these type of groups.

It is also an interesting problem to look at which condition implies a Hilbert cube (or equivalently a subset sum of an infinite sequence of integers) that has asymptotic density. U. Zannier raised the question that the condition $a_{n+1} \leq 2a_n$ implies that there exists asymptotic density of $FS(A)$. Indeed Ruzsa proved [10]:

Theorem 3.4 (Ruzsa) *Let $A = \{a_1 < a_2 < \cdots\} \subseteq \mathbb{N}$ with $a_{n+1} \leq 2a_n$. Then the density $d(FS(A))$ exists.*

Later Erdős raised a stronger version of Zannier's question [10]:

Question 1 (Erdős) *Is it true that the condition $a_n \leq a_1 + a_2 + \cdots + a_{n-1} + c$ ($c > 0$ integers) also implies that $\mathbf{d}(FS(A))$ exists?*

Write $s_{n-1} = a_1 + a_2 + \cdots + a_{n-1}$. The main difficulty of this problem is when the quantity $a_n - s_{n-1}$ changes its sign infinitely many times.

Let $F(n) = |a_n - s_{n-1}|$. We could prove the following partial answer [3]:

Theorem 3.5 (Hegyvári-Hennecart-Pach) *Assume that for the function $F(n) = \theta(s_{n-1})$ with $\theta(k) \ll \frac{k}{(\log k)^2}$. Then $\mathbf{d}(FS(A))$ exists.*

Hilbert cubes avoid given sets.
Bergelson and Ruzsa investigated the existence of Hilbert cubes which avoid classical additive sets. For example they proved that there is an infinite Hilbert-cube which avoids the sequence of squares (see [1]).

In [5] I investigated what the dimension of the biggest Hilbert cube is that avoids a given sequence of integers. Let us remark that the finite case is almost trivial. Let $A \subseteq$

$[N]$, and we are looking for the dimension of a Hilbert cube $H \subseteq [N]$ for which $A \cap H = \emptyset$. Clearly there is an interval $[x, y]$ with length at least $(N - |A|)/(|A| + 1)$ containing no element from A. Hence, the Hilbert cube

$$H(x, 1, 2, \cdots, \lfloor\sqrt{(N - |A|)/(|A| + 1)}\rfloor)$$

avoids the set A. Clearly this argument does not work in the infinite case. Nevertheless I proved

Theorem 3.6 *Let A be a sequence of integers and let $\omega : \mathbb{N} \to \mathbb{R}^+$ be any function and assume that $\omega(x) \to \infty$ as $x \to \infty$. Then there exists an infinite cube $H = H(x_0, a_1 < a_2 < \cdots <)$ which avoids A and for which*

$$\limsup_{n\to\infty} \frac{dim H(n)}{\sqrt{n/A(n) \cdot \omega(n) \cdot \log^2 n}} > 0,$$

where $dim H(n) = |\{a_1 < a_2 < \cdots <\} \cap [n]|$.

In [4] we investigate some modular analogous of the above mentioned questions. We prove that if a cube H avoids some "arithmetic" set in $\mathbb{F}_p$, then the dimension of the cube cannot be big. We use the expression "arithmetic sets" for *classical* subsets of $\mathbb{F}_p$; for instance, for sets of the form $\{2^n : n \in \mathbb{Z}\}$, $\{2^n 3^k : n, k \in \mathbb{Z}\}$, or for sets satisfying some special property. I just quote one result:

Theorem 3.7 (Hegyvári-Pach) *Let $A, B \subseteq \mathbb{F}_p$ be subsets such that $0 \notin A + B$, and H be a Δ-degenerate Hilbert cube. Denote the reciprocal set of the sumset $A + B$ by $\mathcal{P}(A, B) = \left\{\frac{1}{a+b} : a \in A, b \in B\right\}$. If $\mathcal{P}(A, B) \cap H = \emptyset$, then*

$$\dim H \leq \frac{2}{\Delta} \log_2 \left(\frac{p^{3/2}}{\sqrt{|A||B|}}\right) + \frac{4}{\Delta}.$$

In the proof we combined combinatorial arguments with Fourier analytic methods (see [4]).

On Δ-degenerate Hilbert cubes.
As we noted understanding the behaviour of Hilbert cubes with degeneracy $\Delta < 1$ is more difficult compared to the case of $\Delta = 1$.

Indeed, we prove the following proposition (this may be a folklore with different notations, however I did not find any literature).

Proposition 3.8 *Assume that the set $X \subseteq \mathbb{N}$ contains a Δ–degenerate Hilbert cube $H := H(x_0, a_1, a_2, \ldots, a_d)$ with $\Delta < 1$. Then X contains at least $2^{d(1-\Delta)-1}$ many non-trivial three-term arithmetic progressions.*

Proof Since $\Delta < 1$, there is an element $z \in H$ which is covered at least $R := 2^{d(1-\Delta)}$ many times, i.e. there are sets $B_i = \{a_{i,1}, a_{i,2}, \ldots a_{i,s_i}\}$, $1 \leq i \leq R$ such that

$$z = a_{i,1} + a_{i,2} + \cdots + a_{i,s_i}.$$

Now form a system of 'sunflowers' from $\{B_j\}_{j=1}^{R}$, i.e. a collection of sets whose pairwise intersection is constant.

So we have a system of sets $S_1, S_2, \ldots, S_T$; $1 \le T \le \lfloor R/2 \rfloor$, where each $S_j := \{B_{j_1}, B_{j_2}, \ldots, B_{j_{u_j}}\}$ we have

$$B_{j_k} \cap B_{j_m} = K_j := \bigcap_i B_{j_i}$$

for every $k \neq m$, and each S_j is maximal in the sense that we can not extend a further set B from the rest. Clearly $\sum_{j=1}^{T} u_j = R$. This system is not necessarily unique, (say form it by greedy algorithm). Finally define a system as follows: for $1 \le j \le T$ let $S'_j := \{B_{j_1} \setminus K_j, B_{j_2} \setminus K_j, \ldots, B_{j_{u_j}} \setminus K_j\}$ (S'_j contains 'petals' of the sunflowers).

Observe that for every $k \neq m$, $(B_{j_k} \setminus K_j) \cap (B_{j_m} \setminus K_j) = \emptyset$, and $\sum_{a \in (B_{j_k} \setminus K_j)} a = \sum_{a' \in (B_{j_m} \setminus K_j)} a' \in FS(A)$. Write this common sum by z_j.

By the maximality of the systems we have $z_j \neq z_{j'}$ whenever $j \neq j'$. Furthermore for every $1 \le j \le T$

$$x_0 + i x_j \in H \subseteq X; \quad 1 \le i \le u_j,$$

i.e. S_j determines $u_j - 1$ many three-term arithmetic progressions. Summing up we have at least

$$\sum_{j=1}^{T} (u_j - 1) = \sum_{j=1}^{T} u_j - T = R - T \ge R/2 = 2^{d(1-\Delta)-1}$$

many three-term arithmetic progressions as we stated. □

Recall, the additive energy of a given set A counts the number of solutions $x_1 + x_4 = x_3 + x_2$, $(x_1, x_2, x_3, x_4) \in A^4$. Let us observe that one of the solution forms a Hilbert cube with dimension 2. Indeed, $H(x_1, (x_3 - x_1), (x_2 - x_1)) = \{x_1, x_1 + x_3 - x_1, x_1 + x_2 - x_1, x_1 + x_3 - x_1 + x_2 - x_1\} = \{x_1, x_3, x_2, x_4\}$ using that $x_1 + x_3 - x_1 + x_2 - x_1 = x_4$. So the additive energy $E(A)$ counts all 2-dimensional Hilbert cubes containing A.

The study of the number of additive sets in $[n]$ goes back almost 30 years; we merely mention here the famous Erdős–Cameron conjecture (see e.g. [2]).

We say that H is $\Delta_{\le}$-degenerate Hilbert cube if $\frac{\log_2 |H|}{d} \le \Delta$.

In this section we are going to investigate the number of $\Delta_{\le}$-degenerate Hilbert cubes in $[n]$. We prove

Theorem 3.9 *Let* $\mathcal{H} = \{\{a_i\}_{i=1}^{d} \subseteq [n] : H(x_0, a_1, \ldots, a_d)$ *is* $\Delta_{\le}$ *– degenerate Hilbert cube*$\}$. *We have*

$$n^{(1+o_d(1))\Delta d} \le |\mathcal{H}| \le n^{(\Delta d)} 3^{d^2}.$$

4 Proofs

First we are going to prove Theorem 3.3.

Proof of Theorem 3.3 Using an iteration process we will show an infinite Hilbert cube in the form $H := H(0, a_1, a_2, \dots, a_d, \dots)$. Write $H_n := H(0, a_1, a_2, \dots, a_n)$. Let a_1 be any element of G, so H_1 is defined, and assume that for $n \geq 1$ the sequence of $H_1 \subseteq H_2 \subseteq \dots \subseteq H_n \subseteq \dots$ has been defined with $H_n \subseteq G_{k_n}$. Let $0 < \varepsilon < 1/|G_{k_n}|$ be any real number. Since $\overline{d}(A) = 1$ we have a sequence G_{i_j} such that $|A \cap G_{i_j}|/|G_{i_j}| > 1 - \varepsilon$. It implies that $|G_{i_j} \setminus A| < \varepsilon|G_{i_j}|$. Choose an i_j for which $G_{k_n} \subseteq G_{i_j}$. We have

$$|G_{i_j} \setminus A| < \varepsilon|G_{i_j}| < \frac{|G_{i_j}|}{|G_{k_n}|} = ind(G_{i_j}/G_{k_n}).$$

It implies that there is a coset $a_{n+1} + G_{k_n}$ for which $A \cap (a_{n+1} + G_{k_n}) = a_{n+1} + G_{k_n}$. Hence

$$a_{n+1} + H_n \subseteq a_{n+1} + G_{k_n} \subset A,$$

i.e. $\{0, a_{n+1}\} + H_n = H_{n+1} \subset A$ as we wanted. □

Now we turn to the proof of Theorem 3.9.

Proof of Theorem 3.9 First we prove the upper bound. Since for every $H \in \mathcal{H}$, $|H(x_0, a_1, a_2, \dots, a_d)| = FS(A)$, our task is to count the number of $A \subseteq [n]$ for which $|A| \leq 2^{(\Delta d)}$.

Let $1 \leq j \leq d$. We say that the element a_j is *doubler element* if

$$FS(a_1 < a_2 < \dots < a_{j-1}) \cap \{a_j + FS(a_1 < a_2 < \dots < a_{j-1})\} = \emptyset \quad (1)$$

Since $FS(a_1 < a_2 < \dots < a_j) = \{0, a_j\} + FS(a_1 < a_2 < \dots < a_{j-1})$, thus if a_j is a doubler element then $|FS(a_1 < a_2 < \dots < a_j)| = 2|FS(a_1 < a_2 < \dots < a_{j-1})|$. Now let us denote the number of double elements in A by D. Clearly

$$2^{(\Delta d)} \geq |\mathcal{H}| = |FS(a_1 < a_2 < \dots < a_d)| \geq 2^D.$$

and so D is at most Δd.

Assume now that a_j is not a doubler element. It implies that $a_j \in \{x - x' : x, x' \in FS(a_1 < a_2 < \dots < a_{j-1})\}$, hence we have that a_j can be written in the form $a_j = \sum_{i \neq j} \delta_i a_i;\ \delta_i \in \{1, +1, -1\}$, and thus the number of non-doubler elements is at most 3^d.

Now we are in the position to give an upper estimation for $\mathcal{H}$. Write $r := \lfloor \Delta d \rfloor$. We can select $\binom{d}{r}$ subscripts j where a_j is a doubler element, the number of possible values of the doublers being at most n^r. Finally, the number of non-doublers is at most $(3^d)^{d-r}$. Thus we have

$$|\mathcal{H}| \le \binom{d}{r} \cdot n^r \cdot (3^d)^{d-r} \le d^r \cdot n^r \cdot 3^{d^2-kr} \le n^r \cdot 3^{d^2} \le n^{(\Delta d)} 3^{d^2}.$$

using the inequalities $d^r < 3^{dr}$ and $r \le \Delta d$.

Now we turn to the lower estimation.

Let $0 < \Delta < 1$ and d be an integer large enough. Define the integer d_1 by

$$d_1 := \max\{x : 2\log_2(d-x) + x - 1 \le \lfloor \Delta d \rfloor\}$$

and $k = d - d_1$. If Δ is fixed and d is large enough then d_1 exists, $d_1 < \Delta d$ and $d_1 = (1 + o_d(1))\Delta d$. Consider the following Hilbert cube $H = (0, 1, 2, \dots k, u_1, u_2, \dots, u_{d_1})$, where $u_1, u_2, \dots, u_{d_1} \in [n]$. We claim that $H \in \mathcal{H}$.

Decompose H into the sum of two disjoint cubes $H = H_1 + H_2$ where $H_1 := H_1(0, 1, 2, \dots, k)$ and $H_2 := H_2(0, u_1, u_2, \dots, u_{d_1})$. Now $H_1 = \{0, 1, 2, \dots, \binom{k+1}{2}\}$ hence $|H_1| \sim \frac{k^2}{2}$, and $|H_2| \le 2^{d_1}$ and so

$$|H| = |H_1 + H_2| \le |H_1||H_2| \le$$

$$\frac{(d-d_1)^2}{2} 2^{d_1} = 2^{2\log_2(d-d_1)+d_1-1} \le 2^{\Delta d}.$$

The d_1-tuple $u_1, u_2, \dots, u_{d_1}$ can be chosen from $\{k+1, \dots, n\}$ at least $\binom{n-k}{d_1} = n^{(1+o_d(1))\Delta d}$ in many ways, hence

$$|\mathcal{H}| \ge n^{(1+o_d(1))\Delta d},$$

as we wanted. □

Acknowledgements H.N. was supported by National Research, Development and Innovation Office of Hungary (Grant Nr. K129335).

References

1. V. Bergelson and I. Z. Ruzsa, Squarefree numbers, IP sets and ergodictheory, Paul Erd?os and his Mathematics I, Bolyai Society MathematicalStudies, 11, Budapest, 2002, 147–160
2. B. Green: "The CameronErdös conjecture". Bulletin of the London Mathematical Society. 36 (6): 769778
3. N. Hegyvári, F. Hennecart, P.P. Pach, On the density of sumsets and product sets Australasian Journal of Combinatorics 74 : 1 pp. 1–16. , 16 p. (2019)
4. N. Hegyvári, P.P. Pach, Hilbert Cubes meet arithmetic sets, to appear in J. of Numb.Th.
5. N. Hegyvári, On additive and multiplicative Hilbert cubes, Journal of Combinatorial Theory, Series A **115** (2008) 354–360
6. N. Hindman: Ultrafilters and combinatorial number theory. Number theory, Carbondale 1979 (Proc. Southern Illinois Conf., Southern Illinois Univ., Carbondale, Ill., 1979), pp. 119–184, Lecture Notes in Math., 751, Springer, Berlin, 1979

7. M.B. Nathanson, Sumsets contained in infinite sets of integers, Journal of Combinatorial Theory, Series A, **28 (2)** (1980) 150–155
8. M.B. Nathanson, Inverse Theorems for Subset Sums Transactions of the American Mathematical Society Vol. 347, No. 4 (1995), pp. 1409–1418
9. M.B. Nathanson Additive Number Theory: Inverse Problems and the Geometry of Sumsets, Graduate Texts in Mathematics (165), Springer; 1996 edition
10. I.Z. Ruzsa, The density of the set of sums, Acta Arith. 58 (1991), 169172

Intrinsic Characterization of Representation Functions and Generalizations

Charles Helou

Abstract Given a set A of natural numbers, i.e., nonnegative integers, we establish an intrinsic characterization of the representation function of A, which to every natural number n associates the number $r_A(n)$ of ordered pairs (a, b) of elements $a, b \in A$ such that $a + b = n$, thus answering an open problem stated many years ago by Mel Nathanson. We also establish similar characterizations of the characteristic function $\chi_A(n)$, which is equal to 1 or 0 according as the natural number n lies or does not lie in A, and the counting function $A(n)$, which gives the number of elements a of A satisfying $a \leq n$. We then generalize to representation functions as sums of more than two elements of A.

Keywords Representation functions · Counting and characteristic functions · m-ary sums

AMS Classification Numbers 11B34 · 11B13 · 11B75

1 Introduction

For a subset A of the set $\mathbb{N} = \{0, 1, 2, \ldots\}$ of natural numbers, the (binary) representation function $r_A(n)$ of A is defined, for $n \in \mathbb{N}$, by

$$r_A(n) = |\{(a, b) \in A \times A : a + b = n\}|.$$

In a 2008 paper, Nathanson [4] states emphatically
"The simplest *inverse problem for representation functions* is:
What functions are representation functions?"
(See also [3]).

C. Helou (✉)
Penn State Brandywine, 25 Yearsley Mill Rd, Media, PA 19063, USA
e-mail: cxh22@psu.edu

M. B. Nathanson (ed.), *Combinatorial and Additive Number Theory IV*,
Springer Proceedings in Mathematics & Statistics 347,
https://doi.org/10.1007/978-3-030-67996-5_15

In this paper, we present an answer to this question by providing an intrinsic characterization of representation functions in the form of a functional equation. We then do the same for the counting function of A, defined by

$$A(n) = |A \cap [0, n]| = |\{a \in A : a \leq n\}|,$$

and for the characteristic function of A, defined by

$$\chi_A(n) = \begin{cases} 1 & \text{if } n \in A, \\ 0 & \text{if } n \notin A \end{cases}.$$

It is to be noted that each of the three functions $r_A(n)$, $\chi_A(n)$ and $A(n)$ completely determines the set A, and they can be extended to the field $\mathbb{R}$ of real numbers, by simply replacing $n \in \mathbb{N}$ by $x \in \mathbb{R}$ in the definitions.

We then turn to representation functions by sums of more than two elements of A, namely the m-ary representation function, for $m \geq 2$, defined by

$$r_{A,m}(n) = |\{(a_1, a_2, \dots, a_m) \in A^m : a_1 + a_2 + \cdots + a_m = n\}|,$$

and we intrinsically characterize $r_{A,p}(n)$, for any prime number p, by a functional equation. The results here extend and generalize those in [2].

Another binary representation function encountered in the literature [1] is defined by $r'_A(n) = |\{(a, b) \in A \times A : a \leq b \text{ and } a + b = n\}|$. It is related to $r_A(n)$ by the relations

$$r_A(n) = \begin{cases} 2r'_A(n) - 1, & \text{if } 2 \mid n \text{ and } \frac{n}{2} \in A \\ 2r'_A(n), & \text{otherwise} \end{cases},$$

$$r'_A(n) = \begin{cases} \frac{1}{2}(r_A(n) + 1), & \text{if } 2 \mid n \text{ and } \frac{n}{2} \in A \\ \frac{1}{2}r_A(n), & \text{otherwise} \end{cases}.$$

For $r'_A(n)$, the problem of its intrinsic characterization is still open.

2 Characterization of Binary Representation Functions

Theorem 1 *A function $u : \mathbb{N} \longrightarrow \mathbb{N}$ is the representation function of a subset A of $\mathbb{N}$ if and only if it satisfies, for all $n \in \mathbb{N}$, the relation*

$$u(n) = \frac{1}{2}\left(n + 1 - \sum_{k=0}^{n}(-1)^{u(2k)u(2(n-k))}\right). \tag{1}$$

And, in this case, the set A is given by

$$A = \{n \in \mathbb{N} : u(2n) \equiv 1 \pmod 2\}. \tag{2}$$

Proof Assume first that u is the representation function of a subset A of $\mathbb{N}$, i.e. $u(n) = r_A(n) = |R_A(n)|$ for all $n \in \mathbb{N}$, where

$$R_A(n) = \{(a, b) \in A \times A : a + b = n\}.$$

Then, for $0 \leq k \leq n$,

$$(-1)^{r_A(2k)r_A(2(n-k))} = \begin{cases} 1, & \text{if } r_A(2k) \text{ or } r_A(2(n-k)) \equiv 0 \pmod 2 \\ -1, & \text{if } r_A(2k) \equiv r_A(2(n-k)) \equiv 1 \pmod 2 \end{cases}.$$

Now, $r_A(2k) = |R_A(2k)| \equiv 1 \pmod 2$ if and only if $k \in A$, since the elements of $R_A(2k)$ come in double pairs (a, b) and (b, a), which are different unless $a = b = k$. So

$$(-1)^{r_A(2k)r_A(2(n-k))} = \begin{cases} 1, & \text{if } k \notin A \text{ or } (n-k) \notin A \\ -1, & \text{if } k \in A \text{ and } (n-k) \in A \end{cases}.$$

Hence

$$\sum_{k=0}^{n} (-1)^{r_A(2k)r_A(2(n-k))} = \sum_{\substack{k \notin A \text{ or } (n-k) \notin A \\ 0 \leq k \leq n}} 1 + \sum_{\substack{k \in A \text{ and } (n-k) \in A \\ 0 \leq k \leq n}} (-1) = |I| - |J|,$$

where

$$I = \{0 \leq k \leq n : k \notin A \text{ or } (n-k) \notin A\}, \quad J = \{0 \leq k \leq n : k \in A \text{ and } (n-k) \in A\}.$$

Clearly, $I \cap J = \emptyset$ and $I \cup J = \{0, 1, \ldots, n\}$, so that $|I| = n + 1 - |J|$, and therefore

$$\sum_{k=0}^{n} (-1)^{r_A(2k)r_A(2(n-k))} = |I| - |J| = n + 1 - 2\,J|.$$

Moreover, setting $a = k$ and $b = n - k$ in the definition of J, we get

$$|J| = |\{0 \leq k \leq n : k \in A,\ (n-k) \in A\}| = |\{(a, b) \in A \times A : a + b = n\}| = r_A(n).$$

Hence

$$\sum_{k=0}^{n} (-1)^{r_A(2k)r_A(2(n-k))} = n + 1 - 2\,J| = n + 1 - 2r_A(n),$$

i.e.,

$$r_A(n) = \frac{1}{2}\left(n + 1 - \sum_{k=0}^{n}(-1)^{r_A(2k)r_A(2(n-k))}\right).$$

Thus, the representation function of any subset A of $\mathbb{N}$ satisfies condition (1).

Conversely, assume that a function $u : \mathbb{N} \longrightarrow \mathbb{N}$ satisfies condition (1), and let $A = \{n \in \mathbb{N} : u(2n) \equiv 1 \pmod 2\}$. Then

$$\sum_{k=0}^{n}(-1)^{u(2k)u(2(n-k))} = \sum_{k\in I} 1 + \sum_{k\in J}(-1) = |I| - |J|,$$

where

$$I = \{0 \leq k \leq n : u(2k) \equiv 0\ (mod\ 2) \text{ or } u(2(n-k)) \equiv 0\ (mod\ 2)\} = \\ = \{0 \leq k \leq n : k \notin A \text{ or } (n-k) \notin A\},$$

and

$$J = \{0 \leq k \leq n : u(2k) \equiv u(2(n-k)) \equiv 1 \pmod 2\} = \\ = \{0 \leq k \leq n : k \in A \text{ and } (n-k) \in A\}.$$

As in the first part of the proof, I and J partition the set of integers $\{0, 1, 2, \ldots, n\}$, so that $|I| + |J| = n + 1$ and, as above,

$$|J| = |\{(a, b) \in A \times A : a + b = n\}| = r_A(n),$$

so that

$$|I| - |J| = n + 1 - 2J| = n + 1 - 2r_A(n).$$

It follows that

$$\sum_{k=0}^{n}(-1)^{u(2k)u(2(n-k))} = |I| - |J| = n + 1 - 2r_A(n),$$

i.e.,

$$r_A(n) = \frac{1}{2}\left(n + 1 - \sum_{k=0}^{n}(-1)^{u(2k)u(2(n-k))}\right) = u(n),$$

for all $n \in \mathbb{N}$. Here, the last equality follows from the assumption that u satisfies (1). Thus u is the representation function of the set A, defined by $A = \{n \in \mathbb{N} : u(2n) \equiv 1 \pmod 2\}$.

3 Counting and Characteristic Functions for Binary Sums

Proposition 1 *If a function* $v : \mathbb{N} \longrightarrow \mathbb{N}$ *is the counting function of a subset* A *of* $\mathbb{N}$*, then it satisfies the relation*

$$\sum_{k=0}^{n} v(k)v(n-k) = \sum_{k=0}^{n}(n-k+1)\sum_{j=0}^{k}(v(j)-2v(j-1)+v(j-2))v(k-j), \tag{3}$$

for all $n \in \mathbb{N}$*, where by definition* $v(x) = 0$ *if* $x \notin \mathbb{N}$.

And the set A *is given by*

$$A = \{n \in \mathbb{N} : v(n) \equiv v(n-1)+1 \pmod 2\}. \tag{4}$$

Proof As v is the counting function of $A \subset \mathbb{N}$, we have

$$v(n) = A(n) = |\{a \in A : a \leq n\} = \sum_{k=0}^{n} \chi_A(k),$$

for all $n \in \mathbb{N}$. Then the generating series of $A(n)$ is

$$\begin{aligned} h_A(X) &= \sum_{n=0}^{\infty} A(n)X^n = \sum_{n=0}^{\infty}\left(\sum_{k=0}^{n}\chi_A(k)\right)X^n = \\ &= \sum_{n=0}^{\infty}\sum_{k=0}^{n}\chi_A(k)X^kX^{n-k} = \sum_{k=0}^{\infty}\sum_{n=k}^{\infty}\chi_A(k)X^kX^{n-k} = \\ &= \sum_{k=0}^{\infty}\chi_A(k)X^k\sum_{n=k}^{\infty}X^{n-k} = \sum_{k=0}^{\infty}\chi_A(k)X^k\sum_{m=0}^{\infty}X^m = \\ &= \frac{f_A(X)}{1-X}, \end{aligned} \tag{5}$$

where

$$f_A(X) = \sum_{k=0}^{\infty}\chi_A(k)X^k = \sum_{a\in A}X^a$$

is the generating series of the characteristic function $\chi_A(n)$ of A, and whose square

$$g_A(X) = f_A(X)^2 = \sum_{n=0}^{\infty}\left(\sum_{k=0}^{n}\chi_A(k)\chi_A(n-k)\right)X^n = \sum_{n=0}^{\infty}r_A(n)X^n \tag{6}$$

is the generating series of the representation function $r_A(n)$ of A.

Squaring in (5) gives

$$\begin{aligned}h_A(X)^2 &= \left(\sum_{n=0}^{\infty} A(n)X^n\right)^2 = \sum_{n=0}^{\infty}\left(\sum_{k=0}^{n} A(k)A(n-k)\right)X^n\\ &= \frac{f_A(X)^2}{(1-X)^2} = \frac{d}{dX}\left(\frac{1}{1-X}\right)g_A(X) = \left(\sum_{j=0}^{\infty}(j+1)X^j\right)\left(\sum_{k=0}^{\infty} r_A(k)X^k\right) =\\ &= \sum_{n=0}^{\infty}\left(\sum_{j+k=n}(j+1)r_A(k)\right)X^n = \sum_{n=0}^{\infty}\left(\sum_{k=0}^{n}(n-k+1)r_A(k)\right)X^n. \end{aligned} \tag{7}$$

Thus, for all $n \in \mathbb{N}$,

$$\sum_{k=0}^{n} A(k)A(n-k) = \sum_{k=0}^{n}(n-k+1)r_A(k). \tag{8}$$

Moreover, by (6), we have

$$\begin{aligned}\sum_{n=0}^{\infty} r_A(n)X^n &= f_A(X)^2 = h_A(X)^2(1-X)^2 = \sum_{n=0}^{\infty}\left(\sum_{k=0}^{n} A(k)A(n-k)\right)X^n(1-2X+X^2) =\\ &= \sum_{n=0}^{\infty}\left(\sum_{k=0}^{n} A(k)A(n-k) - 2\sum_{k=0}^{n} A(k)A(n-k-1) + \sum_{k=0}^{n} A(k)A(n-k-2)\right)X^n =\\ &= \sum_{n=0}^{\infty}\left(\sum_{k=0}^{n} A(k)(A(n-k) - 2A(n-k-1) + A(n-k-2))\right)X^n,\end{aligned}$$

so that, for all $n \in \mathbb{N}$,

$$\begin{aligned}r_A(n) &= \sum_{k=0}^{n} A(k)(A(n-k) - 2A(n-k-1) + A(n-k-2)) =\\ &= \sum_{j=0}^{n}(A(j) - 2A(j-1) + A(j-2))A(n-j).\end{aligned} \tag{9}$$

Substituting (9) into (8) yields, for all $n \in \mathbb{N}$,

$$\sum_{k=0}^{n} A(k)A(n-k) = \sum_{k=0}^{n}(n-k+1)\sum_{j=0}^{k}(A(j) - 2A(j-1) + A(j-2))A(k-j), \tag{10}$$

which shows that the counting function of any subset A of $\mathbb{N}$ satisfies (3). As to the set A, by Theorem 1, it is given by $A = \{n \in \mathbb{N} : r_A(2n) \equiv 1 \pmod 2\}$, i.e., in view of (9),

$$A = \left\{ n \in \mathbb{N} : \sum_{j=0}^{2n} (A(j) - 2A(j-1) + A(j-2))A(2n-j) \equiv 1 \pmod 2 \right\}. \tag{11}$$

Moreover,

$$\sum_{j=0}^{2n} (A(j) - 2A(j-1) + A(j-2))A(2n-j) \equiv \sum_{j=0}^{2n} (A(j) + A(j-2))A(2n-j) \equiv \tag{12}$$

$$\equiv \sum_{j=0}^{2n} A(j)A(2n-j) + \sum_{j=0}^{2n} A(j-2)A(2n-j) \pmod 2,$$

and

$$\sum_{j=0}^{2n} A(j)A(2n-j) = \sum_{j=0}^{n-1} A(j)A(2n-j) + \sum_{j=n+1}^{2n} A(j)A(2n-j) + A(n)^2 = \tag{13}$$

$$= 2\sum_{j=0}^{n-1} A(j)A(2n-j) + A(n)^2 \equiv A(n)^2 \pmod 2,$$

and

$$\sum_{j=0}^{2n} A(j-2)A(2n-j) = \sum_{j=0}^{n} A(j-2)A(2n-j) + \sum_{j=n+2}^{2n} A(j-2)A(2n-j) + \tag{14}$$

$$+ A(n-1)^2 = 2\sum_{j=0}^{n} A(j-2)A(2n-j) + A(n-1)^2 \equiv A(n-1)^2 \pmod 2.$$

So

$$\sum_{j=0}^{2n} (A(j) - 2A(j-1) + A(j-2))A(2n-j) \equiv A(n)^2 + A(n-1)^2 \equiv \tag{15}$$

$$\equiv (A(n) - A(n-1))^2 \equiv A(n) - A(n-1) \pmod 2,$$

and therefore

$$A = \{n \in \mathbb{N} : A(n) - A(n-1) \equiv 1 \pmod 2\}.$$

Proposition 2 *A function* $v : \mathbb{N} \longrightarrow \mathbb{N}$ *is the counting function of a subset* A *of* $\mathbb{N}$ *if and only if it is an increasing function in the broad sense, i.e.,*

$$v(n+1) \geq v(n), \quad \forall n \in \mathbb{N}. \tag{16}$$

And the set A *is given by*

$$A = \{n \in \mathbb{N} : v(n) > v(n-1)\}, \tag{17}$$

where by definition $v(x) = 0$ *if* $x \notin \mathbb{N}$.

Proof The proof is a straightforward verification. □

Proposition 3 *A function* $w : \mathbb{N} \longrightarrow \{0, 1\}$ *is the characteristic function of a subset* A *of* $\mathbb{N}$ *if and only if it satisfies the condition*

$$w(n) = \frac{1}{2}\left(1 - (-1)^{w(n)}\right), \quad \forall n \in \mathbb{N}. \tag{18}$$

And the set A *is given by*

$$A = \{n \in \mathbb{N} : w(n) = 1\}, \tag{19}$$

Proof Indeed,

$$\frac{1}{2}\left(1 - (-1)^{w(n)}\right) = \begin{cases} 1, & \text{if } w(n) = 1, \\ 0, & \text{if } w(n) = 0. \end{cases}$$

4 Representation Functions for m-ary Sums

Let A be a subset of $\mathbb{N}$, and $m \geq 2$ be an integer. The m-ary representation function of A is defined, for $n \in \mathbb{N}$, by

$$r_{A,m}(n) = |R_{A,m}(n)|, \tag{20}$$

where

$$R_{A,m}(n) = \{(a_1, a_2, \ldots, a_m) \in A^m : a_1 + a_2 + \cdots + a_m = n\}.$$

Lemma 1 *Let* A *be a subset of* $\mathbb{N}$, *and* $m, n \in \mathbb{N}$, *with* $m \geq 2$. *Then*

$$r_{A,m}(n) = \sum_{(m_1, m_2, \ldots, m_h)} \binom{m}{m_1}\binom{m - m_1}{m_2} \cdots \binom{m - m_1 - m_2 - \cdots - m_{h-1}}{m_h}, \tag{21}$$

where the summation is over all tuples $(m_1, m_2, \ldots, m_h)$, *with* $1 \leq h \leq m$, *such that* $m_1 \geq m_2 \geq \cdots \geq m_h \geq 1$, $m_1 + m_2 + \cdots + m_h = m$, *and there exists* $(a_1, a_2,$

$\ldots, a_m) \in R_{A,m}(n)$, with exactly m_1 of $a_1, a_2, \ldots, a_m$ equal to some $a_{i_1} \in A$, exactly m_2 of $a_1, a_2, \ldots, a_m$ equal to some $a_{i_2} \in A, \ldots$, exactly m_h of $a_1, a_2, \ldots, a_m$ equal to some $a_{i_h} \in A$, where $a_{i_1}, a_{i_2}, \ldots, a_{i_h}$ are distinct elements of A, thus exhausting all $(a_1, a_2, \ldots, a_m)$ in $R_{A,m}(n)$.

Proof For any representation $n = a_1 + a_2 + \cdots + a_m$, where $a_1, \ldots, a_m \in A$, i.e., $(a_1, \ldots, a_m) \in R_{A,m}(n)$, with exactly m_1 of $a_1, a_2, \ldots, a_m$ equal to some $a_{i_1} \in A$, exactly m_2 of $a_1, a_2, \ldots, a_m$ equal to some $a_{i_2} \in A, \ldots$, exactly m_h of $a_1, a_2, \ldots, a_m$ equal to some $a_{i_h} \in A$, where $a_{i_1}, a_{i_2}, \ldots, a_{i_h}$ are distinct, $m_1 \geq m_2 \geq \cdots \geq m_h \geq 1$ and $m_1 + m_2 + \cdots + m_h = m$, there are exactly

$$\nu(m_1, m_2, \ldots, m_h) = \binom{m}{m_1}\binom{m-m_1}{m_2}\cdots\binom{m-m_1-m_2-\cdots-m_{h-1}}{m_h}$$

m-tuples in $R_{A,m}(n)$ whose coordinates are $a_1, a_2, \ldots, a_m$ in some rearrangement. Here the binomial coefficient factor $\binom{m}{m_1}$ gives the number of choices of the m_1 positions of a_{i_1} among the m coordinates of $(a_1, a_2, \ldots, a_m)$, the binomial coefficient factor $\binom{m-m_1}{m_2}$ gives the number of choices of the m_2 positions of a_{i_2} among the remaining $m - m_1$ coordinates of $(a_1, a_2, \ldots, a_m), \ldots, \binom{m-m_1-\cdots-m_{h-1}}{m_h}$ gives the number of choices of the m_h positions of a_{i_h} among the remaining $m - m_1 - \cdots - m_{h-1}$ coordinates of $(a_1, a_2, \ldots, a_m)$.

Lemma 2 *Let A be a subset of $\mathbb{N}$, and p a prime number. Then, for any $k \in \mathbb{N}$,*

$$k \in A \iff r_{A,p}(kp) \equiv 1 \pmod{p}. \tag{22}$$

More precisely,

$$r_{A,p}(kp) \equiv \begin{cases} 1 \pmod{p}, & \text{if } k \in A, \\ 0 \pmod{p}, & \text{if } k \notin A. \end{cases} \tag{23}$$

Proof If $k \in A$, then the representation $kp = k + k + \cdots + k$ (p summands) corresponds to the unique element $(k, k, \ldots, k) \in R_{A,p}(kp)$.

For any other representation $kp = a_1 + a_2 + \cdots + a_p$, where $a_1, \ldots, a_p \in A$ are not all equal and not necessarily distinct, with exactly m_1 of $a_1, a_2, \ldots, a_p$ equal to some $a_{i_1} \in A$, exactly m_2 of $a_1, a_2, \ldots, a_p$ equal to some $a_{i_2} \in A, \ldots$, exactly m_h of $a_1, a_2, \ldots, a_p$ equal to some $a_{i_h} \in A$, where $a_{i_1}, a_{i_2}, \ldots, a_{i_h}$ are distinct, $m_1 \geq m_2 \geq \cdots \geq m_h \geq 1$ and $m_1 + m_2 + \cdots + m_h = p$, by Lemma 1, there are exactly

$$\nu(m_1, m_2, \ldots, m_h) = \binom{p}{m_1}\binom{p-m_1}{m_2}\cdots\binom{p-m_1-m_2-\cdots-m_{h-1}}{m_h}$$

p-tuples in $R_{A,p}(kp)$ whose coordinates are $a_1, a_2, \ldots, a_p$ in some rearrangement. Since p is a prime number, p divides $\binom{p}{m_1}$ for any integer $1 \leq m_1 \leq p-1$, which is the case for any representation of kp other than the trivial one $kp = k + k + \cdots + k$ (if $k \in A$). So p divides the number $\nu(m_1, m_2, \ldots, m_h)$ of rearrangements of any representation $kp = a_1 + a_2 + \cdots + a_p$, with m_1 (resp. m_2, ..., resp. m_h) equal summands among $a_1, a_2, \ldots, a_p$, other than the trivial one for which $h = 1$ and $m_1 = p$. Thus $r_{A,p}(kp) = |R_{A,p}(kp)|$, which is equal to the sum of all those numbers $\nu(m_1, m_2, \ldots, m_h)$, is divisible by p if $k \notin A$, and $r_{A,p}(kp) \equiv 1 \pmod{p}$ if $k \in A$.

To establish an intrinsic characterization of p-ary representation functions, we need the following result about the evaluation of some multiple sums.

Lemma 3 *For any positive integers m, n, we have*

$$\sum_{k_1=0}^{n} \sum_{k_2=0}^{n-k_1} \sum_{k_3=0}^{n-k_1-k_2} \cdots \sum_{k_m=0}^{n-k_1-k_2-\cdots-k_{m-1}} 1 = \binom{m+n}{m}. \tag{24}$$

Proof We proceed by induction on m, with an arbitrary $n \geq 1$. The equality trivially holds for $m = 1$, since $\sum_{k_1=0}^{n} 1 = n + 1 = \binom{1+n}{n}$. Assume, inductively, that the equality holds for some $m \geq 1$ and for all $n \geq 1$. Then

$$\sum_{k_1=0}^{n} \sum_{k_2=0}^{n-k_1} \sum_{k_3=0}^{n-k_1-k_2} \cdots \sum_{k_m=0}^{n-k_1-k_2-\cdots-k_{m-1}} \sum_{k_{m+1}=0}^{n-k_1-\cdots-k_{m-1}-k_m} 1 = \tag{25}$$

$$= \sum_{k_1=0}^{n} \binom{m+n-k_1}{m} = \sum_{k=m}^{m+n} \binom{k}{m},$$

where the first equality results from the induction assumption for m, applied to the inner sum starting at the sum over k_2, with n replaced by $n - k_1$, and the second equality results from the change of summation index in the second sum from k_1 to $k = m + n - k_1$. Now, by a straightforward induction on n, we have

$$\sum_{k=m}^{m+n} \binom{k}{m} = \binom{m+n+1}{m+1}. \tag{26}$$

Hence, combining the last two identities, (25) and (26), we see that the equality, (24), in the statement, holds for $m + 1$ and all $n \in \mathbb{N}$. This completes the proof by induction of the Lemma.

Theorem 2 *Let p be a prime number. A function $u : \mathbb{N} \longrightarrow \mathbb{N}$ is the p-ary representation function of a subset A of $\mathbb{N}$, i.e., $u(n) = r_{A,p}(n)$ for all $n \in \mathbb{N}$, if and only if it satisfies the following relation, for all $n \in \mathbb{N}$,*

$$u(n) = \frac{1}{1-\zeta_p}\left(\binom{n+p-1}{n} - \sum_{k_1+k_2+\cdots+k_p=n} \zeta_p^{u(k_1p)u(k_2p)\ldots u(k_pp)}\right), \tag{27}$$

where ζ_p is a primitive pth root of unity in $\mathbb{C}$, e.g., $\zeta_p = e^{2i\pi/p}$, and the summation is over all p-tuples $(k_1, k_2, \ldots, k_p) \in \mathbb{N}^p$ such that $k_1 + k_2 + \cdots + k_p = n$.

Moreover, in this case, the set A is given by

$$A = \{n \in \mathbb{N} : u(np) \equiv 1 \pmod{p}\}. \tag{28}$$

Proof Assume first that $u = r_{A,p}$ is the p-ary representation function of a subset A of $\mathbb{N}$. Then, we have, in view of Lemma 2,

$$\sum_{k_1+k_2+\cdots+k_p=n} \zeta_p^{r_{A,p}(k_1 p) r_{A,p}(k_2 p) \ldots r_{A,p}(k_p p)} = \sum_{\substack{k_1,\ldots,k_p \in A \\ k_1+\cdots+k_p=n}} \zeta_p + \sum_{\substack{k_1 \notin A \text{ or } \ldots \text{ or } k_p \notin A \\ k_1+\cdots+k_p=n}} 1 =$$

$$= \zeta_p r_{A,p}(n) + \sum_{\substack{0 \le k_1,\ldots,k_p \le n \\ k_1+\cdots+k_p=n}} 1 - \sum_{\substack{k_1,\ldots,k_p \in A \\ k_1+\cdots+k_p=n}} 1 =$$

$$= (\zeta_p - 1) r_{A,p}(n) + \sum_{k_1=0}^{n} \sum_{k_2=0}^{n-k_1} \sum_{k_3=0}^{n-k_1-k_2} \cdots \sum_{k_{p-1}=0}^{n-k_1-k_2\cdots-k_{p-2}} 1. \tag{29}$$

Moreover, by Lemma 3,

$$\sum_{k_1=0}^{n} \sum_{k_2=0}^{n-k_1} \cdots \sum_{k_{p-1}=0}^{n-k_1-k_2\cdots-k_{p-2}} 1 = \binom{p-1+n}{p-1} = \tag{30}$$

$$= \binom{n+p-1}{n}.$$

Substituting the last relation into the previous one and solving for $r_{A,p}$ yields, for every $n \in \mathbb{N}$,

$$r_{A,p}(n) = \frac{1}{1-\zeta_p} \left(\binom{n+p-1}{n} - \sum_{k_1+k_2+\cdots+k_p=n} \zeta_p^{r_{A,p}(k_1 p) r_{A,p}(k_2 p) \ldots r_{A,p}(k_p p)} \right), \tag{31}$$

so that $r_{A,p}$ satisfies the relation (27). Also, by Lemma 2, we have

$$A = \{n \in \mathbb{N} : r_{A,p}(pn) \equiv 1 \pmod{p}\}, \tag{32}$$

so that $r_{A,p}$ satisfies the condition (28).

Conversely, if u satisfies the relation (27) and we define A by (28), we have, by definition of A,

$$\sum_{k_1+k_2+\cdots+k_p=n} \zeta_p^{u(k_1p)u(k_2p)\ldots u(k_pp)} = \sum_{\substack{k_1,\ldots,k_p\in A\\ k_1+\cdots+k_p=n}} \zeta_p + \sum_{\substack{k_1\notin A \text{ or } \ldots \text{ or } k_p\notin A\\ k_1+\cdots+k_p=n}} 1 =$$

$$= \zeta_p r_{A,p}(n) + \sum_{\substack{0\le k_1,\ldots,k_p\le n\\ k_1+\cdots+k_p=n}} 1 - \sum_{\substack{k_1,\ldots,k_p\in A\\ k_1+\cdots+k_p=n}} 1 =$$

$$= (\zeta_p - 1) r_{A,p}(n) + \sum_{k_1=0}^{n}\sum_{k_2=0}^{n-k_1}\sum_{k_3=0}^{n-k_1-k_2} \cdots \sum_{k_{p-1}=0}^{n-k_1-k_2\cdots-k_{p-2}} 1, \tag{33}$$

and since, by Lemma 3,

$$\sum_{k_1=0}^{n}\sum_{k_2=0}^{n-k_1} \cdots \sum_{k_{p-1}=0}^{n-k_1-k_2\cdots-k_{p-2}} 1 = \binom{n+p-1}{n}, \tag{34}$$

then

$$\sum_{k_1+k_2+\cdots+k_p=n} \zeta_p^{u(k_1p)u(k_2p)\ldots u(k_pp)} = (\zeta_p - 1) r_{A,p}(n) + \binom{n+p-1}{n}, \tag{35}$$

so that

$$r_{A,p}(n) = \frac{1}{1-\zeta_p}\left(\binom{n+p-1}{n} - \sum_{k_1+k_2+\cdots+k_p=n} \zeta_p^{u(k_1p)u(k_2p)\ldots u(k_pp)}\right) = u(n), \tag{36}$$

for all $n \in \mathbb{N}$. Thus $u = r_{A,p}$ is the p-ary representation function of the subset A of $\mathbb{N}$ defined by (28).

Remark 1 The result in Lemma 2 is not valid in general, for an arbitrary integer $m \geq 2$ and a subset A of $\mathbb{N}$.

For example, if $m = 4$ and $A = \{0, 1, 2, 4, 6, 8, \ldots\}$, then the value of $r_{A,4}(n)$ is the coefficient of x^n in

$$(1 + x + x^2 + x^4 + x^6 + x^8 + \cdots)^4 = 1 + 4x + 10x^2 + 16x^3 + 23x^4 + 28x^5 + 38x^6 + 44x^7 + 59x^8 + \cdots,$$

so that $r_{A,4}(4) = 23 \not\equiv 1 \pmod 4$ and $r_{A,4}(8) = 59 \not\equiv 1 \pmod 4$, even though $1, 2 \in A$.

Another example, with $m \geq 2$ and $A = \mathbb{N}$, gives

$$r_{\mathbb{N},m}(n) = \binom{m+n-1}{n} = \frac{m(m+1)\ldots(m+n-1)}{n!}, \quad \forall n \in \mathbb{N}.$$

Indeed,

$$r_{\mathbb{N},m}(n) = |\{(k_1, k_2, \ldots, k_m) \in \mathbb{N}^m : k_1 + k_2 + \cdots k_m = n\}| = \sum_{\substack{0 \le k_1,\ldots,k_m \le n: \\ k_1+\cdots+k_m=n}} 1 =$$

$$= \sum_{k_1=0}^{n} \sum_{k_2=0}^{n-k_1} \cdots \sum_{k_{m-1}=0}^{n-k_1-k_2-\cdots-k_{m-2}} 1 = \binom{m+n-1}{m-1} = \binom{m+n-1}{n},$$

by Lemma 3.

In general,

$$r_{\mathbb{N},m}(km) = \binom{(k+1)m-1}{m-1} = \frac{(km+1)(km+2)\ldots(km+m-1)}{(m-1)!} \not\equiv 1 \pmod{m},$$

even though $k \in \mathbb{N}$. Thus, for instance, if $m = 4$, then

$$r_{\mathbb{N},4}(4k) = \frac{(4k+1)(4k+2)(4k+3)}{6} = \frac{(4k+1)(4k+3)(2k+1)}{3}$$
$$\equiv 2k+1 \pmod{4},$$

so that $r_{\mathbb{N},4}(4k) \equiv 1 \pmod{4}$ if and only if $k \equiv 0 \pmod{2}$, whereas all integers $k \ge 0$ lie in $\mathbb{N}$, including the odd integers.

If $m = 6$, then

$$r_{\mathbb{N},6}(6k) = \frac{(6k+1)(6k+2)(6k+3)(6k+4)(6k+5)}{120} =$$
$$= \frac{(6k+1)(3k+1)(2k+1)(3k+2)(6k+5)}{10}$$
$$\equiv 1 - \frac{k(3k-1)}{2} \pmod{6}.$$

Indeed,
$(6k+1)(6k+5) \equiv 5 \pmod{12}$, and
$(3k+1)(3k+2) \equiv 3k(k+1)+2 \pmod{12}$, and
$(3k+1)(3k+2)(2k+1) \equiv (3k(k+1)+2)(2k+1) \equiv 2+k-3k^2 \pmod{12}$.

So $\frac{(6k+1)(3k+1)(2k+1)(3k+2)(6k+5)}{10} \equiv \frac{5(2+k-3k^2)}{10} \equiv 1 + \frac{k(1-3k)}{2} \pmod{6}$.

Therefore $r_{\mathbb{N},6}(6k) \equiv 1 \pmod{6}$ if and only if $k(3k-1) \equiv 0 \pmod{12}$, i.e., $k \equiv 0 \pmod{3}$ and $k \equiv 0$ or $3 \pmod{4}$. Thus

$$r_{\mathbb{N},6}(6k) \equiv 1 \pmod{6} \iff k \equiv 0 \text{ or } 3 \pmod{12}. \tag{37}$$

This shows that the characterization of the elements of A, established in Lemma 2 for $m = p$ prime, fails in general if m is not prime. Here $m = 6$ and $A = \mathbb{N}$, so that all integers $k \ge 0$ are in A, but only one sixth of them satisfy the condition $r_{\mathbb{N},6}(6k) \equiv 1 \pmod{6}$.

Note however that $r_{\mathbb{N},m}$ and $r_{\mathbb{N},m+1}$ satisfy a special functional relation. Indeed, for any integer $N \geq 0$, we have

$$\sum_{n=0}^{N} r_{\mathbb{N},m}(n) = \sum_{n=0}^{N} \binom{m-1+n}{m-1} = \sum_{k=m-1}^{m-1+N} \binom{k}{m-1} =$$
$$= \binom{m+N}{m} = r_{\mathbb{N},m+1}(N).$$

Remark 2 Another m-ary representation function encountered in the literature is defined by $r'_{A,m}(n) = |R'_{A,m}(n)|$, where

$$R'_{A,m}(n) = \{(a_1, a_2, \ldots, a_m) \in A^m : a_1 \leq a_2 \leq \cdots \leq a_m,\ a_1 + a_2 + \cdots + a_m = n\}.$$

Then $r'_{A,m+1}(n)$ is a sum of terms $r'_{A,m}(n-a)$, with $a \in A$, avoiding repetitions. Thus if, for some $a \in A$, we have $n - a = a_1 + a_2 + \cdots + a_m$, with $a_1, a_2, \ldots, a_m \in A$, i.e., $n = a + a_1 + a_2 + \cdots + a_m$, which gives an element of $R'_{A,m+1}(n)$, then to get $r'_{A,m+1}(n)$, we add the $r'_{A,m}(n-a)$ for which the representations $n = a + a_1 + a_2 + \cdots + a_m$ are distinct. This allows to determine the sets $R'_{A,m}(n)$ and their cardinalities $r'_{A,m}(n)$ recursively. Then $r_{A,m}(n)$ can be obtained by applying Lemma 1 to the elements of $R'_{A,m}(n)$ (instead of the much larger set $R_{A,m}(n)$).

Alternatively, we have the simple, clear, recursion for $R'_{A,m}(n)$

$$R'_{A,m+1}(n) = \bigcup_{\substack{a\in A,\\ a\leq n-2m}} \{a\} \times R'_{A,m}(n-a). \tag{38}$$

The condition $a \leq n - 2m$ is equivalent to $n - a \geq 2m$, since $R'_{A,m}(n) = \emptyset$ if $n < 2m$.

An example follows.

Example 1 Let

$$A = \mathbb{P} = \{2, 3, 5, 7, 11, 13, 17, 19, 23, 29, 31, 37, 41, 43, 47, \ldots\}$$

be the set of prime numbers. Using the method described in Remark 2, we obtain, e.g.,

$$R'_{\mathbb{P},2}(20) = \{(3, 17), (7, 13)\},\ r'_{\mathbb{P},2}(20) = 2,\ r_{\mathbb{P},2}(20) = 4,$$
$$R'_{\mathbb{P},3}(20) = \{(2, 5, 13), (2, 7, 11)\},\ r'_{\mathbb{P},3}(20) = 2,\ r_{\mathbb{P},3}(20) = 12,$$
$$R'_{\mathbb{P},4}(20) = \{(2, 2, 3, 13), (2, 2, 5, 11), (3, 3, 3, 11), (3, 3, 7, 7), (3, 5, 5, 7), (5, 5, 5, 5)\},$$
$$r'_{\mathbb{P},4}(20) = 6,\ r_{\mathbb{P},4}(20) = 47,$$
$$R'_{\mathbb{P},5}(20) = \{(2, 2, 2, 3, 11), (2, 2, 2, 7, 7), (2, 3, 3, 5, 7), (2, 3, 5, 5, 5)\},$$

$r'_{\mathbb{P},5}(20) = 4, \ r_{\mathbb{P},5}(20) = 110,$

$R'_{\mathbb{P},6}(20) = \{(2, 2, 2, 2, 5, 7), (2, 2, 3, 3, 5, 5), (2, 2, 3, 3, 3, 7), (3, 3, 3, 3, 3, 5)\},$

$r'_{\mathbb{P},6}(20) = 4, \ r_{\mathbb{P},6}(20) = 186.$

References

1. H. Halberstam and K. F. Roth, *Sequences*, Clarendon Press, Oxford, 1966.
2. C. Helou, *Characterization of representation functions*, Integers 17 (2017), Paper No. A44.
3. M. B. Nathanson, *The inverse problem for representation functions of additive bases*, Number theory (New York, 2003), 253–262, Springer, New York, 2004.
4. M. B. Nathanson, *Inverse problems for representation functions in additive number theory*, Surveys in number theory, 89–117, Dev. Math., 17, Springer, New York, 2008.

Combinatorics of Multicompositions

Brian Hopkins and Stéphane Ouvry

Abstract Integer compositions with certain colored parts were introduced by Andrews in 2007 to address a number-theoretic problem. Integer compositions allowing zero as some parts were introduced by Ouvry and Polychronakos in 2019. We give a bijection between these two varieties of compositions and determine various combinatorial properties of these multicompositions. In particular, we determine the count of multicompositions by number of all parts, number of positive parts, and number of zeros. Then, working from three types of compositions with restricted parts that are counted by the Fibonacci sequence, we find the sequences counting multicompositions with analogous restrictions. With these tools, we give combinatorial proofs of summation formulas for generalizations of the Jacobsthal and Pell sequences.

Keywords Integer compositions · Multinomial coefficients · Integer sequences · Generating functions · Combinatorial proofs

1 Introduction

A composition of a positive integer n is an ordered collection of positive integers whose sum is n. For instance, there are four compositions of 3: $1+1+1$, $1+2$, $2+1$, and 3. We refer to the summands as parts.

The next section begins with the formal definition of multicompositions that we will use. Here are two closely related generalizations of compositions that inform the current work.

In 2007, George Andrews introduced k-compositions where k copies of each positive integer are available as parts, denoted by subscripts, sometimes considered

B. Hopkins (✉)
Saint Peter's University, Jersey City, NJ 07306, USA
e-mail: bhopkins@saintpeters.edu

S. Ouvry
LPTMS, CNRS, Université Paris-Sud, 91405 Orsay Cedex, France

M. B. Nathanson (ed.), *Combinatorial and Additive Number Theory IV*,
Springer Proceedings in Mathematics & Statistics 347,
https://doi.org/10.1007/978-3-030-67996-5_16

as colors [1]. There is a restriction that the last part must have subscript 1. There are nine 2-compositions of 3:

$$1_1 + 1_1 + 1_1,\ 1_1 + 1_2 + 1_1,\ 1_2 + 1_1 + 1_1,\ 1_2 + 1_2 + 1_1,$$
$$1_1 + 2_1,\ 1_2 + 2_1,\ 2_1 + 1_1,\ 2_2 + 1_1,\ 3_1.$$

Andrews used these multicompositions to generalize and solve a problem of Emeric Deutsch about divisors of the number of compositions with relatively prime summands, a variety of compositions introduced by Henry Gould.

In 2019, the second named author and Alexios Polychronakos introduced g-compositions where up to $g-2$ zeros can occur between positive parts [10, p. 11]. Thus standard compositions are the $g=2$ case. For $g=3$, there are nine such compositions of 3:

$$1+1+1,\ 1+1+0+1,\ 1+0+1+1,\ 1+0+1+0+1,$$
$$1+2,\ 1+0+2,\ 2+1,\ 2+0+1,\ 3.$$

Here the motivation comes from quantum mechanics and the notion of exclusion statistics. See the last section for a description along with an open problem.

In Sect. 2, we give our definition of multicompositions and determine their count and generating function, introducing a combinatorial interpretation that is the basis for subsequent results. Section 3 gives counts of multicompositions by various kinds of parts: number of all parts, number of positive parts, and number of zeros. Section 4 is informed by three types of restricted standard compositions counted by the Fibonacci sequence; the analogous multicomposition counts diverge into different families of sequences. In Sect. 5, we connect the triangular arrays of numbers from Sect. 3 with sequences from Sect. 4 and a different kind of of composition introduced in 2020 [3]—these results allow us to give combinatorial proofs of summation formulas for generalizations of Jacobsthal and Pell sequences. Section 6 gives more background for the physics motivation [10, 11] and some open questions.

2 Multicompositions and Generating Functions

Now we give our two-part definition of multicompositions, explain how these connect to the earlier definitions, and show that the two manifestations are equivalent.

Definition 1 A k-composition of n is a standard composition of n with either of the following equivalent modifications.

(a) Each part is assigned a color $1, \ldots, k$ (denoted by a subscript) except that the first part must have color 1.
(b) Each part except the first can be immediately preceded by up to $k-1$ zeros.

Write $C^k(n)$ for the set of k-compositions of n and $c^k(n)$ for the number of them. Collectively, or when k is not specified, these compositions are called multicompositions.

The connections to the earlier definitions are clear: The colored parts definition (a) just switches the restriction of Andrews's k-composition from requiring the last part to have color 1 to the first part. The internal zeros definition (b) is precisely the Ouvry–Polychronakos definition with $k = g - 1$. It remains for us to connect our (a) and (b) definitions.

We show that the (a) and (b) definitions of multicompositions are equivalent by demonstrating a bijection between the compositions they describe. The correspondence is determined part by part, explicitly

$$c_\ell \longleftrightarrow \overbrace{0 + \cdots + 0}^{\ell-1 \text{ zeros}} + c.$$

The correspondence for the two versions of 2-compositions of 3 is shown in Table 1. With the equivalence of the two definitions established, we will switch between them as needed.

All subsequent references to k-compositions indicate the parametrization of Definition 1. In Sect. 6, while discussing the motivation for g-compositions, we will return to the Ouvry–Polychronakos indexing, where the g parameter is clearly indicated.

Next, we determine $c^k(n)$, the number of k-compositions of n, and its generating function.

Proposition 2 *The generating function for $c^k(n)$ is*

$$\sum_{n\geq 1} c^k(n)x^n = \frac{\sum_{i\geq 1} x^i}{1 - k\sum_{i\geq 1} x^i} \tag{1}$$

$$= \frac{x}{1-(k+1)x} \tag{2}$$

Table 1 Correspondence between the two versions of 2-compositions of 3

Colored parts	Internal zeros
$1_1 + 1_1 + 1_1$	$1+1+1$
$1_1 + 1_1 + 1_2$	$1+1+0+1$
$1_1 + 1_2 + 1_1$	$1+0+1+1$
$1_1 + 1_2 + 1_2$	$1+0+1+0+1$
$1_1 + 2_1$	$1+2$
$1_1 + 2_2$	$1+0+2$
$2_1 + 1_1$	$2+1$
$2_1 + 1_2$	$2+0+1$
3_1	3

and $c^k(n) = (k+1)^{n-1}$.

Proof The expression (1) follows from the colored parts definition of multicompositions: The numerator accounts for the first part which must have color 1 while the denominator indicates a geometric series where each subsequent part can have color $1, \ldots,$ or k. The simplification to (2) can be seen with the intermediate steps

$$\frac{x\left(\frac{1}{1-x}\right)}{1-kx\left(\frac{1}{1-x}\right)} = \frac{x}{1-x-kx}.$$

The $c^k(n)$ formula follows directly from the generating function. However, we give a combinatorial proof instead, in order to introduce ideas used in the sequel.

To construct a k-composition of length n, consider starting with a length n board. Working from left to right, at each of the $n-1$ internal integer positions, make a choice about the relation of the two adjacent squares: they can either be joined into a longer part or separated so that a new part starts to the right. Further, a new part is assigned one of k colors. Thus there are $k+1$ choices at each internal position, denoted by markers J for join, S_1 for separate and start a part with color $1, \ldots, S_k$ for separate and start a part with color k. Figure 1 shows several examples of these tilings and the corresponding 2-compositions of 3. Note that there is no separation choice for the first part; its color is always 1. It is clear that this is a bijection between the choices and k-compositions of n. The number of possibilities for $n-1$ choices each with $k+1$ options is $(k+1)^{n-1}$. □

The simplicity of $c^k(n) = (k+1)^{n-1}$ suggests that these multicompositions are a compelling generalization of standard compositions.

The $k = 1$ case gives the 2^{n-1} standard compositions of n and here the bijection reduces to a combinatorial argument given by MacMahon; see [7] for additional background. Andrews establishes the $c^k(n)$ value also starting from MacMahon's bijection, but he proceeds in a different way [1, Lemma 3].

Note that our generating function starts from $n = 1$, rather than $n = 0$, as this leads to simpler expressions later and avoids requiring us to ponder the empty composition.

The bijection given in the proof is compatible with the internal zeros definition as well: S_m with $1 \leq m \leq k$ corresponds to inserting $m-1$ zeros in the sum before starting the next positive part. Figure 1 shows the multicompositions with both definitions.

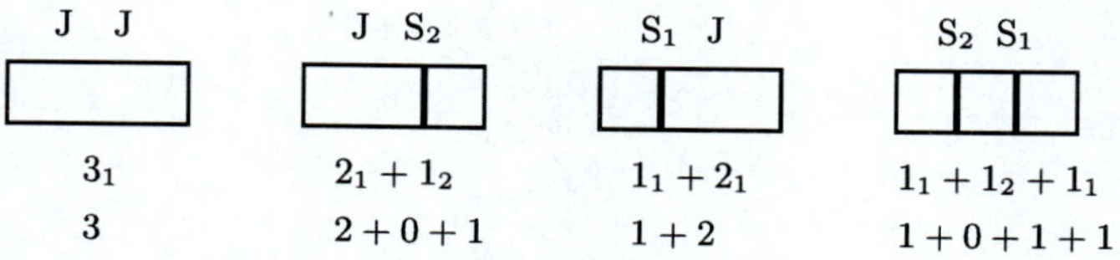

Fig. 1 Four sequences of two ordered choices from the options J, S_1, S_2, and the corresponding 2-compositions of 3 (with both definitions)

3 Counting Multicompositions by Various Parts

Counting the number of multicompositions with a given number of parts is a more nuanced undertaking than for standard compositions. Using the internal zeros definition of multicompositions, one could be interested in the number of all parts, the number of positive parts, or the number of zeros. In this section, we address each of these variants in turn. For each, we determine both a recurrence relation and a direct formula. These results provide new combinatorial interpretations for several known integer sequences.

Also, each subsection includes triangles of numbers for the $k = 2, 3$ cases, where row n partitions the k-compositions of n and the columns reference the number of various parts. Therefore the row sums are $(k+1)^{n-1}$.

First, we show that these counts have an interpretation in the colored parts definition of multicompositions, although it is less a natural statistic in that setting.

Lemma 3 *Comparing a multicomposition under the two definitions, the number of zeros equals the sum of* (*color* $-$ 1) *over the colored parts.*

Proof In the bijection of the proof of Proposition 2, each part beyond the first starts with a marker S_m. Depending on the definition in use, this contributes $m-1$ zeros before the next part or colors the next part m. The first part contributes no zeros and must have color 1. □

3.1 Number of All Parts

A standard composition of n has up to n parts. Since a k-composition of n can have up to $k-1$ zeros in up to $n-1$ positions between successive positive parts, it can have up to $nk-k+1$ parts. Write $C^k(n,\ell)$ for the set of k-compositions of n with ℓ parts and $c^k(n,\ell)$ for the number of them. Triangles of these values for $k = 2, 3$ through $n = 4$ are given in Table 2; the leftmost column in each triangle corresponds to $\ell = 1$ since every k-composition has at least one part.

Let $\binom{n}{\ell}_k = [x^\ell](1+x+\cdots+x^k)^n$ where $[x^\ell]$ denotes the coefficient of x^ℓ in the subsequent polynomial. These are known as multinomial (or polynomial) coefficients and generalize binomial coefficients, the $k = 1$ case.

Proposition 4 *The number of k-compositions of n with ℓ parts satisfies the recurrence*

$$c^k(n,\ell) = c^k(n-1,\ell-k) + \cdots + c^k(n-1,\ell-1) + c^k(n-1,\ell)$$

and $c^k(n,\ell) = \binom{n-1}{\ell-1}_k$.

Table 2 Triangles of $c^k(n,\ell)$ values where the row matches the sum n and the column indicates the number of parts ℓ, for (a) $k=2$ and (b) $k=3$. These are the trinomial [13, A027907] and quadrinomial [13, A008287] coefficients, respectively

(a)

1						
1	1	1				
1	2	3	2	1		
1	3	6	7	6	3	1

(b)

1									
1	1	1	1						
1	2	3	4	3	2	1			
1	3	6	10	12	12	10	6	3	1

Proof For the recurrence, we establish a bijection between the multicompositions counted on each side of the equation. Given a k-composition in the set $C^k(n-1,\ell-j)$ with $1\le j\le k$, append j terms, namely $j-1$ zeros and a 1. Given a k-composition in $C^k(n-1,\ell)$, increase the last part by 1. Both of these constructions give k-compositions with sum n and ℓ parts. For the inverse, if the last part of a k-composition in $C^k(n,\ell)$ is greater than 1, decrease it by 1; otherwise remove the final 1 and any zeros between it and the previous positive part.

For the direct formula, consider $(1+x+\cdots+x^k)^{n-1}$: The choice of summand for each factor corresponds to putting a J (from the choice 1) or S_m (from the choice x^m) in each of the $n-1$ internal positions of a length n board. Each x^m then contributes m parts to the resulting multicomposition: $m-1$ zeros and a new positive part. Therefore, remembering the initial part that has no J or S_m marker, $\binom{n-1}{\ell-1}_k=[x^{\ell-1}](1+x+\cdots+x^k)^{n-1}$ gives the count for k-compositions of n with ℓ parts. □

For a more combinatorial connection to multinomial coefficients, one can use Comtet's description that $\binom{n}{\ell}_k$ counts length ℓ multisets of n letters each appearing at most k times [5, p. 77]. It is not difficult to describe a bijection between these multisets and the multicomposition tilings with J and S_m markers described in the proof of Proposition 2.

3.2 Number of Positive Parts

Like standard compositions, a multicomposition of n can have up to n positive parts. But with various arrangements of uncounted zeros, there are more possibilities. Write $C_+^k(n,\ell)$ for the set of k-compositions of n with ℓ positive parts, $c_+^k(n,\ell)$ for the count. Triangles of these values for $k=2,3$ through $n=5$ are given in Table 3 ; the leftmost column in each triangle corresponds to $\ell=1$ since every k-composition has at least one positive part.

Proposition 5 *The number of k-compositions of n with ℓ positive parts satisfies the recurrence*

Table 3 Triangles of $c_+^k(n, \ell)$ values where the row matches the sum n and the column indicates the number of positive parts ℓ, for (a) $k = 2$ and (b) $k = 3$. These match [13, A013609, A013610], respectively

	1						1			
	1	2					1	3		
(a)	1	4	4			(b)	1	6	9	
	1	6	12	8			1	9	27	27
	1	8	24	32	16		1	12	54	108	81

$$c_+^k(n, \ell) = kc_+^k(n-1, \ell-1) + c_+^k(n-1, \ell)$$

and $c_+^k(n, \ell) = k^{\ell-1}\binom{n-1}{\ell-1}$.

Proof Here we use the colored parts definition of multicompositions, where each part is positive.

For the recurrence, given a k-composition in $C_+^k(n-1, \ell)$, increase the last part by 1. Given a multicomposition in $C_+^k(n-1, \ell-1)$, there are k possibilities for adding a 1, from 1_1 to 1_k. These constructions give a k-composition with sum n and ℓ (positive) parts. The inverse is clear, conditioned on whether the last part is greater than 1.

For the direct formula, we count the ways to construct a k-composition of n with ℓ parts from a length n board. To have ℓ parts, we need to place an S_m marker at $\ell - 1$ of the $n - 1$ internal positions; the positions can be chosen in $\binom{n-1}{\ell-1}$ ways. There are k choices of m for each S_m, giving the $k^{\ell-1}$ factor.

We can also modify $c^k(n, \ell) = [x^{\ell-1}](1 + x + \cdots + x^k)^{n-1}$ to determine the direct formula. Since we are not counting zeros now, each x^i contributes one positive part regardless of i. Therefore $c_+^k(n, \ell) = [x^{\ell-1}](1 + kx)^{n-1}$ and the result follows from the binomial theorem. □

3.3 Number of Zeros

A multicomposition of n can have up to $(n-1)(k-1)$ zeros. Write $C_0^k(n, \ell)$ for the set of k-compositions of n with ℓ zeros, $c_0^k(n, \ell)$ for the count. Triangles of these values for $k = 2, 3$ through $n = 5$ are given in Table 4 ; the leftmost column in each triangle corresponds to $\ell = 0$ since not all k-compositions includes zeros. Since multicompositions with no zeros are the standard compositions, $c_0^k(n, 0) = 2^{n-1}$.

Proposition 6 *The number of k-compositions of n with ℓ zeros satisfies the recurrence*

Table 4 Triangles of $c_0^k(n,\ell)$ values where the row indicates the sum n and the column indicates the number of zeros ℓ, for (a) $k = 2$ and (b) $k = 3$. These match [13, A038207, A336996], respectively

(a)

1				
2	1			
4	4	1		
8	12	6	1	
16	32	24	8	1

(b)

1								
2	1	1						
4	4	5	2	1				
8	12	18	13	9	3	1		
16	32	56	56	49	28	14	4	1

$$c_0^k(n,\ell) = c_0^k(n-1,\ell-k+1) + \cdots + c_0^k(n-1,\ell-1) + 2c_0^k(n-1,\ell)$$

and

$$c_0^k(n,\ell) = \sum_{m=0}^{n-1} \binom{n-1}{m}\binom{m}{\ell}_{k-1}.$$

Proof For the recurrence, given a k-composition in the set $C_0^k(n-1,\ell-j)$ with $1 \le j \le k-1$, append j zeros and then a 1. Given a k-composition in $C_0^k(n-1,\ell)$, construct two multicompositions: increase the last part by 1 for one and append a 1 (with no zeros) for the other. These constructions give k-compositions with sum n and ℓ zeros. For the inverse, if the last part of a k-composition in $C_0^k(n,\ell)$ is greater than 1, decrease it by 1; otherwise remove the final 1 and any zeros between it and the previous positive part.

For the direct formula, we modify $c^k(n,\ell) = [x^{\ell-1}](1+x+\cdots+x^k)^{n-1}$. Recall that the marker S_m placed on a length n board leads to $m-1$ zeros; neither J nor S_1 contribute any zeros. We recast the polynomial in z to highlight counting zeros, where x^m becomes z^{m-1}, and $1+x$ (for J and S_1) becomes $2z^0$. Thus $c_0^k(n,\ell) = [z^\ell](2+z+\cdots+z^{k-1})^{n-1}$. To determine that coefficient, write $(2+z+\cdots+z^{k-1})^{n-1}$ as $(1+(1+z+\cdots+z^{k-1}))^{n-1}$, apply the binomial theorem, and use the definition of multinomial coefficients. □

4 Restricted Part Multicompositions

There are many interesting classes of standard compositions that arise by restricting which parts are allowed. We review three such restrictions that lead to Fibonacci sequence counts and then explore their analogues for multicompositions.

Recall the Fibonacci sequence $F(0) = 0$, $F(1) = 1$, and $F(n) = F(n-1) + F(n-2)$ for $n \ge 2$. We focus on three types of restricted compositions of n: those made with just parts 1 and 2, denoted $C_{12}(n)$; those with only odd parts, $C_{\text{odd}}(n)$; and those where 1 is not allowed as a part, $C_{\hat{1}}(n)$. There will be a superscript k added for k-compositions, where zeros satisfying the definition are also allowed. Again, lower case letters refer to the sizes of the corresponding sets.

Proposition 7 *For $n \geq 1$, the numbers of restricted compositions of n as defined above are $c_{12}(n) = F(n+1)$, $c_{odd}(n) = F(n)$, and $c_{\hat{1}}(n) = F(n-1)$.*

Proofs of these counts will follow from the $k = 1$ cases of the following results on restricted multicompositions. The history here is interesting. The connection between Fibonacci numbers and sums using just 1 and 2 dates from ancient India in the study of poetry meters in languages with long and short vowels, the long vowels twice the length of short vowels [12]. De Morgan discussed compositions with only odd parts in 1846 [6, Appendix 10] and Cayley documented compositions with 1 prohibited in 1876 [4].

Proposition 8 *Let a positive integer k be given.*

(a) The number of k-compositions of n with only positive parts 1 and 2 satisfies

$$c_{12}^k(1) = 1,\ c_{12}^k(2) = k+1,\ \textit{and}\ c_{12}^k(n) = kc_{12}^k(n-1) + kc_{12}^k(n-2)\ \textit{for}\ n \geq 3.$$

(b) The number of k-compositions of n with only positive parts odd satisfies

$$c_{odd}^k(1) = 1,\ c_{odd}^k(2) = k,\ \textit{and}\ c_{odd}^k(n) = kc_{odd}^k(n-1) + c_{odd}^k(n-2)\ \textit{for}\ n \geq 3.$$

(c) The number of k-compositions of n with no parts 1 satisfies

$$c_{\hat{1}}^k(2) = 1,\ c_{\hat{1}}^k(3) = 1,\ \textit{and}\ c_{\hat{1}}^k(n) = c_{\hat{1}}^k(n-1) + kc_{\hat{1}}^k(n-2)\ \textit{for}\ n \geq 4.$$

Proof We use here the colored parts definition of multicompositions. Recall that the first part must have color 1.

(a) For multicompositions with only parts 1 and 2, the only possibility for sum 1 is 1_1. For sum 2, there are the $k+1$ possibilities, $2_1, 1_1+1_1, \ldots, 1_1+1_k$. The recurrence follows from the generating function, a modification of (1):

$$\sum_{n\geq 1} c_{12}^k(n)x^n = \frac{x+x^2}{1-k(x+x^2)}.$$

Or one can give a combinatorial argument: $C_{12}^k(n)$ consists of each element of $C_{12}^k(n-2)$ with 2_1, ..., or 2_k appended and each element of $C_{12}^k(n-1)$ with 1_1, ..., or 1_k appended. Details of verifying the bijection are left to the reader.

(b) For multicompositions with only odd parts, the only possibility for sum 1 is 1_1. For sum 2, there are the k possibilities $1_1+1_1, \ldots, 1_1+1_k$. The recurrence follows from the generating function

$$\sum_{n\geq 1} c_{\text{odd}}^k(n)x^n = \frac{x+x^3+x^5+\cdots}{1-k(x-x^3-x^5-\cdots)} = \frac{x\left(\frac{1}{1-x^2}\right)}{1-kx\left(\frac{1}{1-x^2}\right)} = \frac{x}{1-kx-x^2}$$

or a combinatorial argument: $C^k_{\text{odd}}(n)$ consists of each element of $C^k_{\text{odd}}(n-2)$ with the last part extended by 2 (i.e., converted to the next larger odd number while maintaining the same color) and each element of $C^k_{\text{odd}}(n-1)$ with 1_1, ..., or 1_k appended.

(c) For multicompositions with 1 prohibited, the only possibility for sum 2 is 2_1 and the only possibility for sum 3 is 3_1. The recurrence follows from the generating function

$$\sum_{n\geq 1} c^k_{\hat{1}}(n)x^n = \frac{x^2+x^3+x^4+\cdots}{1-k(x^2-x^3-x^4-\cdots)} = \frac{x^2\left(\frac{1}{1-x}\right)}{1-kx^2\left(\frac{1}{1-x}\right)} = \frac{x^2}{1-x-kx^2}$$

or a combinatorial argument: $C^k_{\hat{1}}(n)$ consists of each element of $C^k_{\hat{1}}(n-2)$ with $2_1, \ldots,$ or 2_k appended and each element of $C^k_{\hat{1}}(n-1)$ with the last part extended by 1 (i.e., converted to the next larger integer, at least 3, while maintaining the same color). □

Thus these three restrictions, which are all counted by the Fibonacci numbers (with different indices) for standard compositions, diverge into three families of integer sequences when applied to multicompositions. Initial terms of each sequence for $k = 2, 3, 4$ are given in Table 5.

5 Connections to Diagonal Sums of Triangles

A popular identity demonstrates how the Fibonacci numbers are "inside" Pascal's triangle: the finite sum $\sum_{i\geq 0}\binom{n-i}{i} = F(n-1)$; see [2, Identity 4] for a combinatorial proof that makes use of the restricted compositions $C_{12}(n)$. A direct generalization shows that the diagonal sums of the $\binom{n}{\ell}_2$ triangle shown in Table 2 are are the tribonacci numbers [8], etc. It is not hard to give a bijection between the 2-compositions counted by $\sum_{i\geq 0}\binom{n-i}{i}_2$ and the standard compositions with only parts 1, 2, and 3, etc.

Table 5 Restricted multicomposition counts $c^k_{\hat{1}2}(n)$, $c^k_{\text{odd}}(n)$, and $c^k_{\hat{1}}(n)$ for $2 \leq k \leq 4$ and $1 \leq n \leq 8$ with sequence identifiers

Sequence \n	1	2	3	4	5	6	7	8	Name
$c^2_{\hat{1}2}(n)$	1	3	8	22	60	164	448	1224	[13, A028859]
$c^3_{\hat{1}2}(n)$	1	4	15	57	216	819	3105	11772	[13, A125145]
$c^4_{\hat{1}2}(n)$	1	5	24	116	560	2704	13056	63040	[13, A086347]
$c^2_{\text{odd}}(n)$	1	2	5	12	29	70	169	408	Pell
$c^3_{\text{odd}}(n)$	1	3	10	33	109	360	1189	3927	[13, A006190]
$c^4_{\text{odd}}(n)$	1	4	17	72	305	1292	5473	23184	[13, A001076]
$c^2_{\hat{1}}(n)$	0	1	1	3	5	11	21	43	Jacobsthal
$c^3_{\hat{1}}(n)$	0	1	1	4	7	19	40	97	[13, A006130]
$c^4_{\hat{1}}(n)$	0	1	1	5	9	29	65	181	[13, A006131]

In this section we give two results on the sum of diagonal entries of the other triangles in Sect. 3.

One can check that the diagonal sums of the $c_+^2(n, \ell)$ and $c_+^3(n, \ell)$ triangles in Table 3 match the first few terms of the sequences $c_{\hat{1}}^2(n)$ and $c_{\hat{1}}^3(n)$ of Table 5 , respectively. We show that this holds in general. But first we give a name to that family of sequences.

Definition 9 For a given positive integer k, the k-Jacobsthal sequence has initial conditions $J^k(0) = 0$, $J^k(1) = 1$ and, for $n \geq 2$, the recurrence

$$J^k(n) = J^k(n-1) + kJ^k(n-2).$$

Note that $J^1(n) = F(n)$ and $J^2(n)$ is the standard Jacobsthal sequence [13, A001045]. Proposition 8(c) could now be written $c_{\hat{1}}^k(n) = J^k(n-1)$; Table 5 shows initial terms of $J^3(n-1)$ and $J^4(n-1)$. Realize that many different sequences in the literature are called generalized or k-Jacobsthal numbers.

Theorem 10 *Given a positive integer k, there is a bijection between the multicompositions $C_+^k(n, 1) \cup C_+^k(n-1, 2) \cup \cdots$ and $C_{\hat{1}}^k(n+1)$. Therefore,*

$$J^k(n) = \sum_{i \geq 1} k^{i-1} \binom{n-1-i}{i-1}.$$

Proof The bijection simply consists of increasing by 1 each positive part in all the multicompositions of $\bigcup_i C_+^k(n+1-i, i)$. This takes a multicomposition with sum $n+1-i$ having i positive parts to a multicomposition of $n+1$ with no parts 1 (as every part arose from adding 1 to a positive part). For the inverse map, given a multicomposition in $C_{\hat{1}}^k(n+1)$ with i positive parts, decrease each positive part by 1 to produce a multicomposition in $C_+^k(n+1-i, i)$.

The formula for $J^k(n)$ then follows from its association with $C_{\hat{1}}^k(n)$ in Proposition 8(c) and the formula for $c_+^k(n, \ell)$ in Proposition 5. □

For our last result, one can check that the diagonal sums of $c_0^2(n, \ell)$ match the first few terms of the sequence $c_{\text{odd}}^2(n)$ of Table 5 , the Pell numbers. However, the diagonal sums of $c_0^3(n, \ell)$ do not match $c_{\text{odd}}^3(n)$. Instead, we show that the diagonal sums of $c_0^k(n, \ell)$ count a different variety of colored compositions associated with the Pell numbers that were introduced recently by Bravo, Herrera, and Ramírez [3].

Definition 11 For a given positive integer k, the k-Pell sequence has initial conditions $P^k(-k+2) = \cdots = P^k(0) = 0$, $P^k(1) = 1$ and, for $n \geq 2$, the recurrence

$$P^k(n) = 2P^k(n-1) + P^k(n-2) + \cdots + P^k(n-k).$$

Note that $P^2(n)$ is the standard Pell sequence [13, A000129]. The sequences for $k = 3, 4, 5$ are [13, A077939, A103142, A141448], respectively. Bravo, Herrera, and

Ramírez establish a natural combinatorial interpretation: $P^k(n+1)$ counts compositions of n with parts 1, 1′, 2, 3, ..., k, i.e., standard compositions except that there is a second type of 1 [3, Theorem 11]. Let $B^k(n)$ be the set of these compositions of n.

Theorem 12 *Given an integer $k \geq 2$, there is a bijection between the multicompositions $C_0^k(n,0) \cup C_0^k(n-1,1) \cup \cdots$ and $B^k(n-1)$. Therefore*

$$P^k(n) = \sum_{i \geq 0} \sum_{m=0}^{n-i} \binom{n-i}{m} \binom{m}{i}_{k-1}.$$

Proof Write each multicompositions of $\bigcup_i C_0^k(n-i,i)$ in terms of the J and S_m markers introduced in the proof of Proposition 2 and then convert each J to 1′ and each S_m to m. This clearly gives a composition with parts from 1, 1′, 2, 3, ..., k; we need to show that its sum is $n-1$.

A multicomposition with sum $n-i$ and i zeros has $n-i-1$ markers. Suppose j of those markers are J. The remaining $n-i-j-1$ markers have the form S_m for $1 \leq m \leq k$. By Lemma 3, the sum of the $(m-1)$ from the various S_m must be i. Therefore, the sum of the m from the various S_m is $i + (n-i-j-1) = n-j-1$. In the image composition, there are j parts 1′ which, with the total $n-j-1$ from the m terms, give the sum $n-1$.

For the reverse map, suppose the composition of $n-1$ with parts from 1, 1′, 2, 3, ..., k has h parts, including j that are 1′. Replace each 1′ with J and each of the $h-j$ parts m with a marker S_m. Converting this into a multicomposition with internal zeros, the sum of the resulting multicomposition is $h+1$. Now the sum of the m from the various S_m is $n-1-j$, so by Lemma 3 the number of zeros in the resulting multicomposition is the sum of the $(m-1)$ which is $n-1-j-(h-j) = n-1-h$. That is, the multicomposition is in the set $C_0^k(h+1, n-1-h) = C_0^k(n-i,i)$ with the substitution $i = n-1-h$.

The formula for $P^k(n)$ comes from their association with the $B^k(n-1)$ compositions and the formula for $c_0^k(n,\ell)$ in Proposition 6. □

6 Exclusion Statistics and Further Work

As a further combinatorial investigation, one could combine the ideas of Sects. 3 and 4 by counting the restricted multicompositions $C_{12}^k(n)$, $C_{\mathrm{odd}}^k(n)$, and $C_{\bar{1}}^k(n)$ by number of all parts, number of positive parts, and number of zeros, then look for patterns in their diagonal sums as in Sect. 5 to find combinatorial derivations for additional formulas.

In the remainder of this final section, we explain how g-compositions arose and present some questions we still have about them. From here until the end, we use the g-composition definition of [10]; recall the connection to Definition 1 that $g = k+1$.

In statistical mechanics (see, for example, [9]), n-body partition functions encode the statistical equilibrium of a system of n particles at temperature T. For q quantum states, the Boltzmann factors $s(k) = \exp(-\beta\epsilon_k)$ for $1 \le k \le q$ are building blocks of the n-body partition functions where ϵ_k is a 1-body energy state and $\beta = 1/(k_B T)$ incorporates the Boltzmann constant k_B. For our purposes, though, β is irrelevant and can be set to 1. Define a n-body partition function $Z(n)$ as the nested multiple sum

$$Z(n) = \sum_{k_1=1}^{q-2n+2} \sum_{k_2=1}^{k_1} \cdots \sum_{k_n=1}^{k_{n-1}} s(k_1 + 2n - 2)s(k_2 + 2n - 4) \cdots s(k_{n-1} + 2)s(k_n).$$

Due to the +2 shifts, the arguments of the Boltzman factors $s(k)$ in $Z(n)$ differ by at least 2. Terms with particles in adjacent energy states ϵ_k and ϵ_{k+1} are excluded: $Z(n)$ is the partition function for n particles on the line obeying exclusion statistics of order $g = 2$. For g-exclusion, the shifts are $+g$.

Cluster coefficients $b(n)$ are determined by

$$\log\left(\sum_{n=0}^{\infty} Z(n)z^n\right) = \sum_{n=1}^{\infty} b(n)z^n.$$

These were introduced to give a combinatorial enumeration of closed lattice walks with a given algebraic area [10, 11]. The g-compositions naturally arise when we solve for the cluster coefficients, e.g.,

$$\begin{aligned}
b(1) &= \sum_{k=1}^{q} s^1(k), \\
-b(2) &= \frac{1}{2}\sum_{k=1}^{q} s^2(k) + \sum_{k=1}^{q} s^1(k+1)s^1(k), \\
b(3) &= \frac{1}{3}\sum_{k}^{q} s^3(k) + \sum_{k=1}^{q} s^2(k+1)s^1(k) + \sum_{k=1}^{q} s^1(k+1)s^2(k) \\
&\quad + \sum_{k=1}^{q} s^1(k+2)s^1(k+1)s^1(k),
\end{aligned}$$

with the understanding $s(k) = 0$ when $k > q$. We have written the usually-implicit exponents 1 to highlight the connection to compositions. Looking for example at the four sums which contribute to $b(3)$, notice that the exponents are precisely the four standard compositions of 3, which correspond to $g = 2$. In the g-exclusion case, the exponents in $b(n)$ correspond to g-compositions of n, with g-exclusion manifesting in the insertion of up to $g - 2$ zeros between positive parts, showing an even stronger exclusion between energy states.

What about the coefficients that appear in the $b(n)$ expressions? For instance, the 1/3, 1, 1, 1 in $b(3)$ for 2-exclusion above. Suppose a g-composition ℓ of n has parts $\ell_1, \ldots, \ell_j$ (subscripts here are indices, not colors) which may be positive integers or zeros as allowed by the definition. The coefficient of the sum corresponding to ℓ is

$$c_g(\ell) = \frac{(\ell_1 + \cdots + \ell_{g-1} - 1)!}{\ell_1! \cdots \ell_{g-1}!} \prod_{i=1}^{j-g+1} \binom{\ell_i + \cdots + \ell_{i+g-1} - 1}{\ell_{i+g-1}}$$

$$= \frac{\prod_{i=1}^{j-g+1} (\ell_i + \cdots + \ell_{i+g-1} - 1)!}{\prod_{i=1}^{j-g} (\ell_{i+1} + \cdots + \ell_{i+g-1} - 1)!} \prod_{i=1}^{j} \frac{1}{\ell_i!},$$

see [10, p. 11]. Further, summing the $c_g(\ell)$ over all g-compositions of n gives $\binom{gn}{n}/(gn)$.

What combinatorics, if any, is at work here? For $g = 2$, the binomial $\binom{2n}{n}$ counts the number of lattice walks on a line starting from a given site and making $2n$ steps, n to the right and n to the left. As in [11], consider the half with the first step to the right. For $n = 3$, writing R for right and L for left, the ten walks are

- 1 with three right-left steps on top of each other (RLRLRL),
- 3 with two right-left steps on top of each other followed by one right-left step (RRLLRL, RLRRLL, RLLRLR),
- 3 with one right-left step followed by two right-left steps on top of each other (RLLRRL, RLRLLR, RRLRLL), and
- 3 with three right-left steps following each other (RRRLLL, RRLLLR, RLLLRR);

see [11, Fig. 1]. In general, the $2nc_2(\ell)$ over all $g = 2$ compositions ℓ count the number of lattice walks on a line making $2n$ steps and having a given number of right-left steps in succession.

A combinatorial interpretation of $c_g(\ell)$ for g-compositions with $g \geq 3$ remains an open question.

Acknowledgements We would like to thank Olivier Giraud for assistance with the generating function arguments in Sect. 4. Also, we appreciate a tip from Abdelmalek Abdesselam given through the site MathOverflow. Finally, we commend the work of two conference organizers: Oleg Evnin of the Fourth Bangkok Workshop on Discrete Geometry, Dynamics, and Statistics, where this collaboration began, and Mel Nathanson of the Eighteenth Annual Workshop on Combinatorial and Additive Number Theory, where these results were (virtually) presented.

References

1. Andrews, G.E.: The theory of compositions, IV: Multicompositions. Math. Student, Special Centenary Volume, 25–31 (2007)
2. Benjamin, A.T., Quinn, J.J.: Proofs that Really Count. MAA Press, Washington, DC (2003)
3. Bravo, J.J., Herrera, J.L., Ramírez, J.L.: Combinatorial interpretation of generalized Pell numbers. J. Integer Seq. 23, 20.2.1 (2020)

4. Cayley, A.: Theorem on partitions. Messenger Math. 5, 188 (1876)
5. Comtet, L.: Advanced Combinatorics. Riedel, Dordrecht, Holland (1974)
6. De Morgan, A.: Elements of Arithmetic, fifth ed. Walton and Maberly, London (1846)
7. Heubach, S., Mansour, T.: Combinatorics of Compositions and Words. CRC Press, Boca Raton, FL (2010)
8. Hogatt, V.E., Bicknell, M.: Diagonal sums of generalized Pascal triangles. Fibonacci Quart. 7, 341–358 (1969)
9. Ma, S.-K.: Statistical Mechanics. World Scientific, Singapore (1985)
10. Ouvry, S., Polychronakos, A.P.: Exclusion statistics and lattice random walks. Nuclear Phys. B 948, 114731 (2019)
11. Ouvry, S., Wu, S.: The algebraic area of closed lattice random walks. J. Phys. A 52, 255201 (2019)
12. Singh, P.: The so-called Fibonacci numbers in ancient and medieval India. Historia Math. 12, 229–244 (1985)
13. Sloane, N.J.A., ed.: The Online Encyclopedia of Integer Sequences, oeis.org (2020)

On the Connection Between the Goldbach Conjecture and the Elliott-Halberstam Conjecture

Jing-Jing Huang and Huixi Li

Dedicated to Professor Melvyn Nathanson on the occasion of his 75th birthday.

Abstract In this paper we prove that the binary Goldbach conjecture for sufficiently large even integers would follow under the assumptions of both the Elliott-Halberstam conjecture and a variant of the Elliott-Halberstam conjecture twisted by the Möbius function, provided that the sum of their level of distributions exceeds 1. This continues the work of Pan [10]. An analogous result for the twin prime conjecture is obtained by Ram Murty and Vatwani [13].

Keywords Goldbach conjecture · Elliott-Halberstam conjecture · Level of distribution · Möbius disjointness

1 Introduction

The well-known Chen's theorem on the Goldbach conjecture states that every sufficiently large even integer N can be written as the sum of two positive integers p and q, where p is a prime and q is an almost prime with at most two prime divisors [1]. The barrier from Chen's theorem to the Goldbach conjecture has been well known as

J.-J. Huang · H. Li (✉)
Department of Mathematics and Statistics, University of Nevada, 1664 N Virginia St, 89557 Reno, NV, USA
e-mail: huixil@unr.edu

J.-J. Huang
e-mail: jingjingh@unr.edu

M. B. Nathanson (ed.), *Combinatorial and Additive Number Theory IV*,
Springer Proceedings in Mathematics & Statistics 347,
https://doi.org/10.1007/978-3-030-67996-5_17

the parity problem in sieve theory: one can not tell whether $q = N - p$ has exactly one or two prime divisors.

Let $r(N)$ be the number of representations of N as the sum of two primes, and let $\widetilde{r}(N)$ denote the number of weighted representations

$$\widetilde{r}(N) := \sum_{n<N} \Lambda(n)\Lambda(N-n),$$

where the von Mangoldt function $\Lambda(n)$ is defined as

$$\Lambda(n) = \begin{cases} \log p, & \text{if } n = p^\ell \text{ for some prime } p \text{ and integer } \ell \geq 1, \\ 0, & \text{otherwise.} \end{cases}$$

It is easy to see by partial summation that

$$r(N) = \frac{\widetilde{r}(N)}{(\log N)^2}\left(1 + O\left(\frac{\log\log N}{\log N}\right)\right) + O\left(\frac{N}{(\log N)^3}\right).$$

Since it is somewhat more convenient to deal with the smoothed sum $\widetilde{r}(N)$ rather than $r(N)$ itself, we will only focus on the study of the former in this paper.

By a heuristic argument based on the circle method, Hardy and Littlewood [6] have conjectured that

$$\widetilde{r}(N) \sim \mathfrak{S}(N)N, \tag{1}$$

where

$$\mathfrak{S}(N) = \begin{cases} 2\prod_{p>2}\left(1 - \frac{1}{(p-1)^2}\right)\prod_{\substack{p|N \\ p>2}}\left(1 + \frac{1}{p-2}\right), & \text{if } N \text{ is even,} \\ 0, & \text{if } N \text{ is odd.} \end{cases} \tag{2}$$

Pan [10] has made a new attempt on the Goldbach conjecture, which gives an alternative way to suggest the expected asymptotic formula (1). In some sense, Pan's approach is more direct and elementary than the Hardy-Littlewood circle method. Pan's main result [10, Theorem I, Theorem II] states that for sufficiently large even integers N and for $Q = N^{\frac{1}{2}}(\log N)^{-20}$, we have

$$\widetilde{r}(N) = R_0(N) + R_1(N) + R_2(N) + R_3(N) + O\left(\frac{N}{\log N}\right),$$

where

$$R_0(N) = \sum_{n<N} \left(\sum_{\substack{d_1|n \\ d_1 \le Q}} \mu(d_1) \log d_1 \right) \left(\sum_{\substack{d_2|(N-n) \\ (d_2,N)=1 \\ d_2 \le Q}} \mu(d_2) \log d_2 \right) = \mathfrak{S}(N)N + O\left(\frac{N}{\log N}\right), \quad (3)$$

$$R_1(N) = \sum_{n<N} \left(\sum_{\substack{d_1|n \\ d_1 \le Q}} \mu(d_1) \log d_1 \right) \left(\sum_{\substack{d_2|(N-n) \\ (d_2,N)=1 \\ d_2 > Q}} \mu(d_2) \log d_2 \right), \quad (4)$$

$$R_2(N) = \sum_{n<N} \left(\sum_{\substack{d_1|n \\ d_1 > Q}} \mu(d_1) \log d_1 \right) \left(\sum_{\substack{d_2|(N-n) \\ (d_2,N)=1 \\ d_2 \le Q}} \mu(d_2) \log d_2 \right), \quad (5)$$

and

$$R_3(N) = \sum_{n<N} \left(\sum_{\substack{d_1|n \\ d_1 > Q}} \mu(d_1) \log d_1 \right) \left(\sum_{\substack{d_2|(N-n) \\ (d_2,N)=1 \\ d_2 > Q}} \mu(d_2) \log d_2 \right). \quad (6)$$

Furthermore, by the Bombieri-Vinogradov theorem, we have

$$R_i(N) = O\left(\frac{N}{\log N}\right), \quad i = 1, 2.$$

As Pan points out, the difficulty in the Goldbach conjecture lies in the study of $R_3(N)$. Moreover, the Hardy-Littlewood conjecture reduces to showing that $R_3(N)$ is inferior to the expected main term $\mathfrak{S}(N)N$. Our main purpose of this paper is to show that the Goldbach conjecture for sufficiently large even integers follows under the assumptions that the Elliott-Halberstam conjecture holds with level of distribution θ and that a variant of the Elliott-Halberstam conjecture twisted by the Möbius function holds with level of distribution θ', where $\theta + \theta' > 1$.

An analogous result in the twin prime conjecture setting has been obtained by Ram Murty and Vatwani [13]. This perhaps is not surprising since we know the Goldbach conjecture and the twin prime conjecture are closely connected. For example, Chen has also proved in [1] that for every even integer $h \neq 0$, there are infinitely many pairs of integers (p, q), where p is a prime and $q = p + h$ is an almost prime with at most two prime divisors. Again, the barrier from this theorem to the twin prime conjecture is caused by the parity problem. There have been many attempts to break the parity barrier in different settings [2–4, 11, 13].

Ram Murty and Vatwani formulate in [13] a conjecture regarding the equidistribution of the Möbius function over shifted primes in arithmetic progressions (cf. [13, (1.4)]), and they prove that such a conjecture for a fixed even integer $h \neq 0$, in conjunction with the Elliott-Halberstam conjecture, can remove the parity barrier and produce infinitely many pairs of primes $(p, p+h)$. Note that if we assume the Elliott-Halberstam conjecture holds, i.e., the conjecture $EH(N^\theta)$ introduced below holds for all $\theta < 1$, then it is proved in [8] that there are infinitely many pairs of consecutive primes that differ by at most 12. Moreover, the bound has been improved in [12] to 6 assuming a generalized Elliott-Halberstam conjecture (cf. [12, Claim 2.6]). Unconditionally, after the sensational breakthrough of Yitang Zhang [17] and subsequent work of Maynard [8], the best known bound on small gaps between primes has been reduced to 246 by the Polymath group [12].

To put the main result in perspective, we state the Elliott-Halberstam conjecture and a variant of it twisted by the Möbius function.

The Elliott-Halberstam conjecture $EH(N^\theta(\log N)^C)$. For any $A > 0$, we have

$$\sum_{q \leq N^\theta(\log N)^C} \max_{y \leq N} \max_{(a,q)=1} \left| \sum_{\substack{n \leq y \\ n \equiv a \pmod q}} \Lambda(n) - \frac{y}{\phi(q)} \right| \ll \frac{N}{(\log N)^A}.$$

A variant of the Elliott-Halberstam conjecture twisted by the Möbius function $EH_\mu(N^\theta)$. For any $A > 0$, we have

$$\sum_{q \leq N^\theta} \max_{y < N} \max_{(a,q)=1} \left| \sum_{\substack{n \leq y \\ n \equiv a \pmod q}} \Lambda(n)\mu(N-n) - \frac{1}{\phi(q)} \sum_{n \leq y} \Lambda(n)\mu(N-n) \right| \ll \frac{N}{(\log N)^A}.$$

Now we are poised to state our main theorem.

Theorem 1 *For a given $A > 0$, suppose that $EH(N^\theta(\log N)^{2A+8})$ and $EH_\mu(N^{1-\theta})$ are true for some constant $0 < \theta < 1$. Then for all positive even integers N, we have*

$$\tilde{r}(N) \geq \mathfrak{S}(N)(1 - \mathcal{A}(N))N + O\left(\frac{N}{(\log N)^A}\right),$$

where

$$\mathcal{A}(N) = \prod_{\substack{p \nmid N \\ p > 2}} \left(1 - \frac{1}{p(p-1)}\right) \tag{7}$$

and the implicit constant depends on A and θ. Moreover, the assertions $\tilde{r}(N) \sim \mathfrak{S}(N)N$ and $\sum_{n<N} \Lambda(N)\mu(N-n) = o(N)$ are equivalent.

The following corollary is an immediate consequence of our main theorem.

Corollary 1 *Suppose that $EH(N^{\theta}(\log N)^{12+\epsilon})$ and $EH_{\mu}(N^{1-\theta})$ are true for some constants $0 < \theta < 1$ and $\epsilon > 0$. Then the Goldbach conjecture holds for all sufficiently large even integers, i.e., every sufficiently large even integer can be written as a sum of two prime numbers. In particular, in view of the Bombieri-Vinogradov theorem, the above conclusion holds if the conjecture $EH_{\mu}(N^{\theta'})$ is true for some $\theta' > \frac{1}{2}$.*

Our work can be naturally regarded as a continuation of Pan [10] for the following reasons. While Pan truncates the sum over d_1 and d_2 at $N^{\frac{1}{2}}(\log N)^{-20}$ due to the limitation of the Bombieri-Vinogradov theorem, it is however not necessary to do this as long as one can handle all the truncated sums properly. Moreover, Pan leaves the sum $R_3(N)$ untouched and in particular does not give any hint how to approach it. The main thrust of the current memoir is to show that we can estimate $R_3(N)$ if we have some knowledge about the equidistribution of $\Lambda(n)\mu(N-n)$ in arithmetic progressions.

As in [13], we remark that the proof goes through if we only assume equidistributions in $EH(N^{\theta}(\log N)^{2A+8})$ and $EH_{\mu}(N^{1-\theta})$ for the fixed residue class $n \equiv N \pmod{q}$ instead of taking the maximum of all residue classes coprime to q. Also by assuming other variants of the Elliott-Halberstam conjecture, our argument should give a lower bound on the number of representations of a large integer N as a linear combination $ap + bq$ of primes p and q, where a and b are fixed positive integers, provided that there is no local obstruction.

The interested readers are referred to [16] for other types of Elliott-Halberstam conjectures and their relations to the twin prime conjecture. It remains to be seen whether the conjectures mentioned in [16] are related to the Goldbach conjecture. Finally, it is worth noting that Hua has proposed another elementary approach to the Goldbach conjecture [7].

In Sect. 2 we introduce some technical lemmas that will be applied later. In Sect. 3 we prove Theorem 1 and Corollary 1 after we first evaluate two important sums assuming the conjectures $EH(N^{\theta}(\log N)^{2A+8})$ and $EH_{\mu}(N^{1-\theta})$ hold. Throughout the paper, we will use Vinogradov's symbol $f(x) \ll g(x)$ and Landau's symbol $f(x) = O(g(x))$ to mean there exists a constant C such that $|f(x)| \leq Cg(x)$. We use ϵ to denote any sufficiently small positive number, which may not necessarily be the same in each occurrence.

2 Preliminary Lemmata

Our first lemma is a result of Goldston and Yıldırım involving the singular series $\mathfrak{S}(N)$ defined in (2).

Lemma 1 ([5, Lemma 2.1]) *For any positive integer N and any real number R satisfying $\log N \ll \log R$, there exists some constant $c_1 > 0$ such that*

$$\sum_{\substack{d\le R\\(d,N)=1}} \frac{\mu(d)}{\phi(d)}\log\left(\frac{R}{d}\right) = \mathfrak{S}(N) + O\left(e^{-c_1\sqrt{\log R}}\right)$$

and

$$\sum_{\substack{d\le R\\(d,N)=1}} \frac{\mu(d)}{\phi(d)} = O\left(e^{-c_1\sqrt{\log R}}\right).$$

Our next lemma has similar flavor as Lemma 1. It can be viewed as a quantitative refinement of [13, Proposition 3.2, Proposition 3.3] by Ram Murty and Vatwani.

Lemma 2 *For any positive even integer N and any real number R satisfying $\log N \ll \log R$, there exists some constant $c_2 > 0$ such that*

$$\mathcal{A}(N) \sum_{\substack{d\le R\\(d,N)=1}} \frac{\mu(d)\log(1/d)g_N(d)}{\phi(d)} = \mathfrak{S}(N) + O\left(e^{-c_2\sqrt{\log R}}\right), \tag{8}$$

where

$$g_N(d) = \prod_{\substack{p\mid d\\p\nmid N}} \left(\frac{(p-1)^2}{p^2-p-1}\right). \tag{9}$$

Proof Let

$$f_N(s) = \sum_{\substack{d=1\\(d,N)=1}}^{\infty} \frac{\mu(d)g_N(d)}{\phi(d)d^s}.$$

Noting that the series on the right hand side admits the Euler product

$$f_N(s) = \prod_{p\nmid N}\left(1-\frac{g_N(p)}{(p-1)p^s}\right) = \prod_{p\nmid N}\left(1-\frac{p-1}{(p^2-p-1)p^s}\right),$$

we immediately see that $f_N(s)$ converges absolutely for $\Re(s) > 0$ and therefore defines an analytic function there. Actually we may write

$$f_N(s) = \zeta(s+1)^{-1}h_N(s),$$

where

$$h_N(s) = \prod_{p\mid N}\left(1-\frac{1}{p^{s+1}}\right)^{-1} \prod_{p\nmid N}\left(\left(1-\frac{1}{p^{s+1}}\right)^{-1}\left(1-\frac{p-1}{(p^2-p-1)p^s}\right)\right).$$

Then, it is readily verified that $h_N(s)$ is analytic for $\Re(s) > -1$. So $f_N(s)$ can be meromorphically continued to $\Re(s) > -1$ with only poles at zeros of $\zeta(s+1)$.

Hence,

$$f'_N(s) = \sum_{\substack{d=1 \\ (d,N)=1}}^{\infty} \frac{\mu(d)\log(1/d)g_N(d)}{\phi(d)d^s} = \frac{h'_N(s)}{\zeta(s+1)} - h_N(s)\frac{\zeta'(s+1)}{\zeta(s+1)^2}$$

is also a meromorphic function for $\Re(s) > -1$ with only poles at zeros of $\zeta(s+1)$. Let

$$a_N(d) = \begin{cases} \frac{\mu(d)\log(1/d)g_N(d)}{\phi(d)}, & \text{if } (d,N) = 1, \\ 0, & \text{otherwise.} \end{cases}$$

Thus noting that $g_N(d) \ll 1$ and $\phi(d) \gg d/\log\log d$ by [15, Theorem 5.6], we have

$$|a_N(d)| \ll \frac{(\log d)^2}{d}.$$

Here we adopt the conventional notation $s = \sigma + it$. By Perron's formula ([9, Corollary 5.3]), we have

$$\sum_{d \le R} a_N(d) = \frac{1}{2\pi i} \int_{\sigma_0 - iT}^{\sigma_0 + iT} f'_N(s)\frac{R^s}{s} ds + E_N,$$

where $\sigma_0 > 0$ and

$$E_N \ll \sum_{\frac{R}{2}<d<2R} |a_N(d)| \min\left(1, \frac{R}{T|d-R|}\right) + \frac{4^{\sigma_0} + R^{\sigma_0}}{T} \sum_{d=1}^{\infty} \frac{|a_N(d)|}{d^{\sigma_0}}.$$

Suppose that $2 \le T \le R$ and that $\sigma_0 = \frac{1}{\log R}$. Then we obtain via some elementary calculation that

$$E_N \ll \frac{(\log R)^3}{T}.$$

Let $\sigma_1 = -c_2/\log T$, where c_2 a small positive constant to be chosen later. The integrand $f'_N(s)\frac{R^s}{s}$ is a meromorphic function for $\Re(s) > -1$ with a simple pole at $s = 0$. It follows from the residue theorem that

$$\frac{1}{2\pi i}\int_{\sigma_0-iT}^{\sigma_0+iT} f_N'(s)\frac{R^s}{s}ds$$

$$=\mathrm{Res}_{s=0}\left(f_N'(s)\frac{R^s}{s}\right)+\frac{1}{2\pi i}\left(\int_{\sigma_0-iT}^{\sigma_1-iT}+\int_{\sigma_1-iT}^{\sigma_1+iT}+\int_{\sigma_1+iT}^{\sigma_0+iT}\right)f_N'(s)\frac{R^s}{s}ds.$$

First of all, it is easy to check that for positive even integers N we have

$$\mathrm{Res}_{s=0}\left(f_N'(s)\frac{R^s}{s}\right)=f_N'(0)=h_N(0)=\frac{\mathfrak{S}(N)}{\mathcal{A}(N)}.$$

Recalling the classical zero-free region of the zeta function, we will choose c_2 small enough such that $\zeta(s)$ is nonvanishing when

$$\sigma\ge 1-\frac{c_2}{\log|t|},\quad |t|\ge 2,$$

and moreover, in this region we have

$$\frac{1}{\zeta(s)}\ll \log|t|,\quad \frac{\zeta'(s)}{\zeta(s)}\ll\log|t|.$$

Also when $\sigma\ge -\frac{1}{4}$, we have

$$\log|h_N(s)|\ll -\sum_{p|N}\log\left(1-\frac{1}{p^{\sigma+1}}\right)\ll\sum_{p|N}\frac{1}{p^{\sigma+1}}\ll\sum_{p<2\log(2N)}\frac{1}{p^{\sigma+1}}\ll(\log N)^{\frac{1}{4}},$$

therefore

$$|h_N(s)|\ll e^{\sqrt[4]{\log N}}$$

and

$$|h_N'(s)|\ll e^{\sqrt[4]{\log N}}\sum_{p|N}\frac{\log p}{p^{\sigma+1}}\ll e^{\sqrt[4]{\log N}}\log N.$$

So when $\sigma\ge\sigma_1$ and $|t|\le T$, we have

$$|f_N'(s)|\ll e^{\sqrt[4]{\log N}}(\log N+\log T)^2\ll e^{c_4\sqrt[4]{\log R}}$$

in view of the assumption that $\log N\ll\log R$. Then

$$\int_{\sigma_1+iT}^{\sigma_0+iT} f_N'(s)\frac{R^s}{s}ds \ll e^{c_4\sqrt[4]{\log R}}\frac{R^{\sigma_0}}{T}(\sigma_0-\sigma_1) \ll \frac{e^{c_4\sqrt[4]{\log R}}}{T}.$$

Similarly, we may treat the integral from $\sigma_0 - iT$ to $\sigma_1 - iT$. Lastly, we observe that

$$\begin{aligned}&\int_{\sigma_1-iT}^{\sigma_1+iT} f_N'(s)\frac{R^s}{s}ds\\ \ll & e^{c_4\sqrt[4]{\log R}}R^{\sigma_1}\left(\int_{-T}^{T}\frac{1}{|t|+1}dt+\int_{-1}^{1}\frac{1}{|\sigma_1-it|}dt\right)\\ \ll & e^{c_4\sqrt[4]{\log R}}R^{\sigma_1}\left(\log T+\sigma_1^{-1}\right)\\ \ll & e^{c_4\sqrt[4]{\log R}}R^{\sigma_1}\log T.\end{aligned}$$

Combining the above estimates, we hitherto obtain

$$\sum_{d\le R} a_N(d) = \frac{\mathfrak{S}(N)}{\mathcal{A}(N)} + O\left((\log R)^3\left(\frac{1}{T}+R^{-\frac{c_2}{\log T}}e^{c_4\sqrt[4]{\log R}}\right)\right).$$

Now we may optimize the error on choosing $T = e^{\sqrt{c_2\log R}}$ and obtain that the error term satisfies

$$(\log R)^3\left(\frac{1}{T}+R^{-\frac{c_2}{\log T}}e^{c_4\sqrt[4]{\log R}}\right) \ll (\log R)^3 e^{-\sqrt{c_2\log R}+c_4\sqrt[4]{\log R}} \ll e^{-c_2\sqrt{\log R}},$$

as we may take $c_2 < 1$. Since $\mathcal{A}(N) \le 1$, this completes the proof of the lemma. □

The following two lemmas are variations of some results in [13]. They will be applied in Sects. 3.1 and 3.2, respectively.

Lemma 3 *Let $A > 0$ and $0 < \theta < 1$ be given. Then*

$$\sum_{d\le N^\theta}\mu(d)\log(1/d)\sum_{\substack{b\le B\\(bd,N)=1}}\frac{\mu(b)}{\phi([b^2,d])} = \mathfrak{S}(N) + O\left(\frac{1}{(\log N)^A}\right)$$

holds for all positive even integers N, where $B = (\log N)^{A+4}$.

Proof The proof is similar to the proof of the asymptotic formula for

$$S_{11}(\beta,x,h) = \sum_{d\le\beta}\mu(d)\log(1/d)\sum_{\substack{b\le B\\(bd,h)=1}}\frac{\mu(b)}{\phi([b^2,d])}$$

defined on [13, P. 653], where x is a large real number, $\beta = x^{\theta''}$ for some $0 < \theta'' < 1$, and $h \neq 0$ is a fixed even integer. Our case here is trickier, since as shown later in the proof, we need to worry about the rate of convergence of the series (8), which is taken care of by Lemma 2.

For square-free integers $b \le B$ and $d \le N^\theta$, we have $\left(b^2, \frac{d}{(b,d)}\right) = 1$ and $\left((b,d), \frac{d}{(b,d)}\right) = 1$, and thus

$$\phi([b^2, d]) = \phi\left(b^2 \frac{d}{(b,d)}\right) = \frac{b\phi(b)\phi(d)}{\phi((b,d))}.$$

So

$$\sum_{d \le N^\theta} \mu(d) \log(1/d) \sum_{\substack{b \le B \\ (bd,N)=1}} \frac{\mu(b)}{\phi([b^2,d])} = \sum_{d \le N^\theta} \frac{\mu(d)\log(1/d)}{\phi(d)} \sum_{\substack{b \le B \\ (bd,N)=1}} \frac{\mu(b)\phi((b,d))}{b\phi(b)}.$$

We extend the sum over b to infinity with an error

$$\sum_{d \le N^\theta} \frac{\mu(d)\log(1/d)}{\phi(d)} \sum_{\substack{b > B \\ (bd,N)=1}} \frac{\mu(b)\phi((b,d))}{b\phi(b)} \ll \log N \sum_{d \le N^\theta} \frac{\mu^2(d)}{\phi(d)} \sum_{b > B} \frac{\mu^2(b)(b,d)}{b\phi(b)}.$$

We apply $(b,d) = \sum_{r|(b,d)} \phi(r)$, and then make the substitution $b = b_1 r$ and $d = d_1 r$ to obtain that

$$\sum_{d \le N^\theta} \frac{\mu^2(d)}{\phi(d)} \sum_{b > B} \frac{\mu^2(b)(b,d)}{b\phi(b)} = \sum_{d \le N^\theta} \frac{\mu^2(d)}{\phi(d)} \sum_{b > B} \frac{\mu^2(b)}{b\phi(b)} \sum_{r|(b,d)} \phi(r)$$
$$\ll \sum_{r \le N^\theta} \frac{1}{r\phi(r)} \sum_{d_1 \le \frac{N^\theta}{r}} \frac{1}{\phi(d_1)} \sum_{b_1 > \frac{B}{r}} \frac{1}{b_1\phi(b_1)} \ll \frac{(\log N)^2 \log\log N}{B},$$

where in the last line some summation results of [14] are applied. Since $B = (\log N)^{A+4}$, we have

$$\begin{aligned}
&\sum_{d\le N^\theta}\mu(d)\log(1/d)\sum_{\substack{b\le B\\(bd,N)=1}}\frac{\mu(b)}{\phi([b^2,d])}\\
=&\sum_{d\le N^\theta}\frac{\mu(d)\log(1/d)}{\phi(d)}\sum_{\substack{b\ge 1\\(bd,N)=1}}\frac{\mu(b)\phi((b,d))}{b\phi(b)}+O\left(\frac{1}{(\log N)^A}\right)\\
=&\sum_{\substack{d\le N^\theta\\(d,N)=1}}\frac{\mu(d)\log(1/d)}{\phi(d)}\prod_{p\nmid N}\left(1-\frac{\phi((p,d))}{p(p-1)}\right)+O\left(\frac{1}{(\log N)^A}\right)\\
=&\mathcal{A}(N)\sum_{\substack{d\le N^\theta\\(d,N)=1}}\frac{\mu(d)\log(1/d)g_N(d)}{\phi(d)}+O\left(\frac{1}{(\log N)^A}\right),
\end{aligned}$$

where $\mathcal{A}(N)$ is defined in (7) and $g_N(d)$ is defined in (9). Therefore, it follows by Lemma 2 that

$$\begin{aligned}
&\sum_{d\le N^\theta}\mu(d)\log(1/d)\sum_{\substack{b\le B\\(bd,N)=1}}\frac{\mu(b)}{\phi([b^2,d])}\\
=&\mathfrak{S}(N)+O\left(e^{-c_2\sqrt{\log(N^\theta)}}\right)+O\left(\frac{1}{(\log N)^A}\right)\\
=&\mathfrak{S}(N)+O\left(\frac{1}{(\log N)^A}\right).
\end{aligned}$$

Lemma 4 *Let $A>0$ and $0<\theta<1$ be given. Suppose the conjecture $EH_\mu(N^{1-\theta})$ holds, then we have*

$$\sum_{q\le N^{1-\theta}}\max_{y<N}\max_{(a,q)=1}\left|\sum_{\substack{n\le y\\ n\equiv a\pmod q}}\Lambda(n)\mu(N-n)\log(N-n)-\frac{1}{\phi(q)}\sum_{n\le y}\Lambda(n)\mu(N-n)\log(N-n)\right|\ll\frac{N}{(\log N)^A}.$$

Proof The proof is similar to the proof of [13, Proposition 4.2]. For any $(a,q)=1$ and $y<N$, denote

$$\sum_{\substack{n\le y\\ n\equiv a\pmod q}}\Lambda(n)\mu(N-n)=M(y)+E_\mu(y,q,a),$$

where

$$M(y)=\frac{1}{\phi(q)}\sum_{n\le y}\Lambda(n)\mu(N-n).$$

By partial summation, we have

$$\sum_{\substack{n\le y\\ n\equiv a \pmod q}} \Lambda(n)\mu(N-n)\log(N-n)$$

$$= \left(\sum_{\substack{n\le y\\ n\equiv a \pmod q}} \Lambda(n)\mu(N-n)\right)\log(N-y) + \int_2^y \frac{1}{N-t}\left(\sum_{\substack{n\le t\\ n\equiv a \pmod q}} \Lambda(n)\mu(N-n)\right)dt$$

$$= M(y)\log(N-y) + \int_2^y \frac{M(t)}{N-t}dt + O\left(\max_{t\le y}|E_\mu(t,q,a)|\log N\right).$$

Using partial summation again, the sum of the first two terms above is

$$\frac{1}{\phi(q)}\sum_{n\le y}\Lambda(n)\mu(N-n)\log(N-n).$$

Now the lemma immediately follows from our assumption $EH_\mu(N^{1-\theta})$.

3 Proof of Theorem 1 and Corollary 1

By the Möbius inversion formula, we have

$$\Lambda(n) = \sum_{d|n}\mu(d)\log(1/d). \tag{10}$$

As in (3)–(6), we split the sum on the right hand side of (10) into $\Lambda_\alpha(n)$ and $\widetilde{\Lambda}_\alpha(n)$, where

$$\Lambda_\alpha(n) = \sum_{\substack{d|n\\ d\le\alpha}}\mu(d)\log(1/d),$$

$$\widetilde{\Lambda}_\alpha(n) = \sum_{\substack{d|n\\ d>\alpha}}\mu(d)\log(1/d),$$

and α is a real number that will be determined later. Therefore, we have

$$\sum_{n<N}\Lambda(n)\Lambda(N-n) = \sum_{n<N}\Lambda(n)\Lambda_\alpha(N-n) + \sum_{n<N}\Lambda(n)\widetilde{\Lambda}_\alpha(N-n). \tag{11}$$

Note that if we take $\alpha = Q = N^{\frac{1}{2}}(\log N)^{-20}$, then Pan's result implies the first sum on the right hand side of (11) is asymptotic to $R_0(N) + R_2(N) + O\left(\frac{N}{\log N}\right) =$

$\mathfrak{S}(N)N + O\left(\frac{N}{\log N}\right)$ and the second sum on the right hand side of (11) is asymptotic to $R_1(N) + R_3(N) = R_3(N) + O\left(\frac{N}{\log N}\right)$. This indicates that the sum $\sum_{n<N} \Lambda(n)\widetilde{\Lambda}_\alpha(N-n)$ is harder to deal with, which represents the main obstacle we are facing.

To make the estimate of $\sum_{n<N} \Lambda(n)\widetilde{\Lambda}_\alpha(N-n)$ more accessible, we only sum up those $n < N$ such that $N-n$ is square-free, as has been done on [13, P. 647]. Since $\Lambda(N-n) = \mu^2(N-n)\Lambda(N-n)$ except when $n = N - p^{\ell_1} > 0$ for some prime p and integer $\ell_1 \geq 2$, and there are only $O(N^{\frac{1}{2}} \log N)$ such $n < N$, we can write (11) as

$$\begin{aligned}
&\sum_{n<N} \Lambda(n)\Lambda(N-n) \\
=&\sum_{n<N} \Lambda(n)\mu^2(N-n)\Lambda(N-n) + O(N^{\frac{1}{2}}(\log N)^3) \\
=&\sum_{n<N} \Lambda(n)\mu^2(N-n)\Lambda_\alpha(N-n) + \sum_{n<N} \Lambda(n)\mu^2(N-n)\widetilde{\Lambda}_\alpha(N-n) + O(N^{\frac{1}{2}}(\log N)^3) \\
=&S_1(\alpha) + S_2(\alpha) + O(N^{\frac{1}{2}}(\log N)^3),
\end{aligned} \tag{12}$$

where

$$S_1(\alpha) = \sum_{n<N} \Lambda(n)\mu^2(N-n) \sum_{\substack{d|(N-n) \\ d\leq\alpha}} \mu(d)\log(1/d)$$

and

$$\begin{aligned}
S_2(\alpha) &= \sum_{n<N} \Lambda(n)\mu^2(N-n) \sum_{\substack{d|(N-n) \\ d>\alpha}} \mu(d)\log(1/d) \\
&= \sum_{n<N} \Lambda(n)\mu^2(N-n) \sum_{\substack{k|(N-n) \\ k<\frac{N-n}{\alpha}}} \mu\left(\frac{N-n}{k}\right)\log\left(\frac{k}{N-n}\right).
\end{aligned}$$

The $\mu^2(N-n)$ term, i.e., the assumption that we only deal with square-free $N-n$, will make it easier when we evaluate $S_2(\alpha)$, since we can apply the equation $\mu^2(N-n)\mu\left(\frac{N-n}{k}\right) = \mu(N-n)\mu(k)$ later in (18) when $N-n$ is square-free and $k \mid (N-n)$. Meanwhile, the term $\mu^2(N-n)$ does not change the asymptotic behavior of $S_1(\alpha)$ as shown in the next subsection.

3.1 Evaluation of $S_1(\alpha)$ Using $EH(N^\theta(\log N)^{2A+8})$

In this subsection we prove that

$$S_1(\alpha) = \mathfrak{S}(N)N + O\left(\frac{N}{(\log N)^A}\right) \tag{13}$$

under the assumption $EH(N^{\theta}(\log N)^{2A+8})$.

Since

$$\mu^2(N-n) = \sum_{b^2|(N-n)} \mu(b) = \begin{cases} 1, & \text{if } N-n \text{ is square-free,} \\ 0, & \text{otherwise,} \end{cases}$$

we have

$$\begin{aligned} S_1(\alpha) &= \sum_{n<N} \Lambda(n)\mu^2(N-n) \sum_{\substack{d|(N-n) \\ d\le\alpha}} \mu(d)\log(1/d) \\ &= \sum_{n<N} \Lambda(n) \sum_{b^2|(N-n)} \mu(b) \sum_{\substack{d|(N-n) \\ d\le\alpha}} \mu(d)\log(1/d) \\ &= \sum_{d\le\alpha} \mu(d)\log(1/d) \sum_{\substack{b<\sqrt{N} \\ [b^2,d]<N}} \mu(b) \sum_{\substack{n<N \\ n\equiv N \pmod{[d,b^2]}}} \Lambda(n). \end{aligned} \tag{14}$$

We truncate the sum over $b < \sqrt{N}$ into two parts $b \le B$ and $B < b < \sqrt{N}$ with $B = (\log N)^{A+4}$. Let $(b, d) = \delta_1$, $b = b_2\delta_1$, and $d = d_2\delta_1$. Then $[b^2, d] = b_2^2 d_2 \delta_1^2$. For the sum over $B < b < \sqrt{N}$, we have

$$\begin{aligned} &\sum_{d\le\alpha} \mu(d)\log(1/d) \sum_{\substack{B<b<\sqrt{N} \\ [b^2,d]<N}} \mu(b) \sum_{\substack{n<N \\ n\equiv N \pmod{[b^2,d]}}} \Lambda(n) \\ &\ll (\log N)^2 \sum_{d<N} \sum_{\substack{b>B \\ [b^2,d]<N}} \left(\frac{N}{[b^2,d]} + 1\right) \\ &\ll (\log N)^2 \sum_{\delta_1<N} \sum_{d_2<\frac{N}{\delta_1}} \sum_{\substack{b_2>\frac{B}{\delta_1} \\ (b_2)^2 d_2 (\delta_1)^2 \le N}} \frac{N}{(b_2)^2 d_2 (\delta_1)^2} \\ &\ll N(\log N)^2 \sum_{d_2\le N} \frac{1}{d_2} \sum_{b_2\le\sqrt{N}} \frac{1}{(b_2)^2} \sum_{\delta_1>\frac{B}{b_2}} \frac{1}{(\delta_1)^2} \\ &\ll \frac{N(\log N)^4}{B} \ll \frac{N}{(\log N)^A}. \end{aligned}$$

Thus we can write (14) as

$$S_1(\alpha) = \sum_{d \le \alpha} \mu(d) \log(1/d) \sum_{\substack{b \le B \\ [b^2,d] < N}} \mu(b) \sum_{\substack{n < N \\ n \equiv N \pmod{[b^2,d]}}} \Lambda(n) + O\left(\frac{N}{(\log N)^A}\right). \tag{15}$$

For b or d that are not relatively prime to N on the right hand side of (15), suppose $(bd, N) = \delta_2 > 1$. If an integer $n < N$ satisfies $\Lambda(n) \neq 0$ and $n \equiv N \pmod{[b^2, d]}$, then n is a prime power $n = p^{\ell_2}$ for some integer $\ell_2 \geq 1$, and moreover, we have $n \equiv p^{\ell_2} \equiv \delta_2 \frac{N}{\delta_2} \pmod{\delta_2 \frac{[b^2,d]}{\delta_2}}$. This implies $p \mid \delta_2$ and thus $p \mid N$. Therefore, taking $\alpha = N^\theta$ for a fixed constant $0 < \theta < 1$, we know

$$\begin{aligned}
&\sum_{d \le \alpha} \mu(d) \log(1/d) \sum_{\substack{b \le B \\ [b^2,d] < N \\ (bd,N) > 1}} \mu(b) \sum_{\substack{n < N \\ n \equiv N \pmod{[b^2,d]}}} \Lambda(n) \\
&\ll \log N \sum_{d \le \alpha} \sum_{\substack{b \le B \\ [b^2,d] < N \\ (bd,N) > 1}} \sum_{p \mid N} \sum_{\ell_2 \le \frac{\log N}{\log p}} \Lambda(p^{\ell_2}) \\
&\ll \alpha B (\log N)^2 \ll \frac{N}{(\log N)^A}.
\end{aligned}$$

Note that we can drop the restriction $[b^2, d] < N$ when $b \leq B = (\log N)^{A+4}$ and $d \leq \alpha = N^\theta$. Therefore, we have

$$\begin{aligned}
S_1(\alpha) &= \sum_{d \le \alpha} \mu(d) \log(1/d) \sum_{\substack{b \le B \\ (bd,N) = 1}} \mu(b) \sum_{\substack{n < N \\ n \equiv N \pmod{[b^2,d]}}} \Lambda(n) + O\left(\frac{N}{(\log N)^A}\right) \\
&= N \sum_{d \le \alpha} \mu(d) \log(1/d) \sum_{\substack{b \le B \\ (bd,N) = 1}} \frac{\mu(b)}{\phi([b^2, d])} + E_1(\alpha) + O\left(\frac{N}{(\log N)^A}\right),
\end{aligned} \tag{16}$$

where

$$\begin{aligned}
E_1(\alpha) &= \sum_{d \le \alpha} \mu(d) \log(1/d) \sum_{\substack{b \le B \\ (bd,N) = 1}} \mu(b) \left(\sum_{\substack{n < N \\ n \equiv N \pmod{[b^2,d]}}} \Lambda(n) - \frac{N}{\phi([b^2, d])} \right) \\
&\ll \log N \sum_{d \le \alpha} \mu^2(d) \sum_{\substack{b \le B \\ (bd,N) = 1}} \mu^2(b) \left| \sum_{\substack{n < N \\ n \equiv N \pmod{[b^2,d]}}} \Lambda(n) - \frac{N}{\phi([b^2, d])} \right|.
\end{aligned}$$

For $b \leq B$ and $d \leq \alpha$ that are square-free, let $z = [b^2, d]$. Since the number of pairs of square-free natural numbers b and d such that $[b^2, d] = z$ is bounded by $\tau_3(z)$, the number of ways of writing z as a product of three positive integers, we can bound the error term $E_1(\alpha)$ as

$$E_1(\alpha) \ll \log N \sum_{z \le \alpha B^2} \tau_3(z) \left| \sum_{\substack{n<N \\ n \equiv N \pmod z}} \Lambda(n) - \frac{N}{\phi(z)} \right|.$$

For all $z \le \alpha B^2 = N^\theta (\log N)^{2A+8}$, since $\sum_{\substack{n<N \\ n \equiv N \pmod z}} \Lambda(n) \ll \frac{N \log N}{z}$ and $\frac{N}{\phi(z)} \ll \frac{N \log\log z}{z} \ll \frac{N \log N}{z}$, we have a trivial bound

$$\left| \sum_{\substack{n<N \\ n \equiv N \pmod z}} \Lambda(n) - \frac{N}{\phi(z)} \right| \ll \frac{N \log N}{z}.$$

By the Cauchy-Schwarz inequality, we know

$$E_1(\alpha) \ll N^{\frac{1}{2}} (\log N)^{\frac{3}{2}} \left(\sum_{z \le \alpha B^2} \frac{\tau_3^2(z)}{z} \right)^{\frac{1}{2}} \left(\sum_{z \le \alpha B^2} \left| \sum_{\substack{n<N \\ n \equiv N \pmod z}} \Lambda(n) - \frac{N}{\phi(z)} \right| \right)^{\frac{1}{2}}.$$

By [15, (1.80)] and partial summation, we can bound the first sum by

$$\sum_{z \le \alpha B^2} \frac{\tau_3^2(z)}{z} \ll (\log N)^9.$$

By our assumption $EH(N^\theta (\log N)^{2A+8})$, we bound the second sum by

$$\begin{aligned} \sum_{z \le \alpha B^2} \left| \sum_{\substack{n<N \\ n \equiv N \pmod z}} \Lambda(n) - \frac{N}{\phi(z)} \right| &\ll \sum_{q \le N^\theta (\log N)^{2A+8}} \max_{y \le N} \max_{\substack{a \\ (a,q)=1}} \left| \sum_{\substack{n \le y \\ n \equiv a \pmod q}} \Lambda(n) - \frac{y}{\phi(q)} \right| \\ &\ll \frac{N}{(\log N)^{2A+12}}. \end{aligned}$$

Therefore, we obtain

$$E_1(\alpha) \ll N^{\frac{1}{2}} (\log N)^{\frac{3}{2}} \left((\log N)^9 \right)^{\frac{1}{2}} \left(\frac{N}{(\log N)^{2A+12}} \right)^{\frac{1}{2}} \ll \frac{N}{(\log N)^A}.$$

Finally, inserting the above bound into (16), we have

$$S_1(\alpha) = N \sum_{d \le \alpha} \mu(d) \log(1/d) \sum_{\substack{b \le B \\ (bd,N)=1}} \frac{\mu(b)}{\phi([b^2, d])} + O\left(\frac{N}{(\log N)^A}\right),$$

which implies (13) after applying Lemma 3.

3.2 Evaluation of $S_2(\alpha)$ Using $EH_\mu(N^{1-\theta})$

In this subsection we show that

$$S_2(\alpha) = -\mathfrak{S}(N) \sum_{n<N} \Lambda(n)\mu(N-n) + O\left(\frac{N}{(\log N)^A}\right) \tag{17}$$

under the assumption $EH_\mu(N^{1-\theta})$.

Suppose $N-n$ is square-free and let k be a divisor of $N-n$ such that $k < \frac{N-n}{\alpha}$. Since $\mu^2(N-n)\mu\left(\frac{N-n}{k}\right) = \mu(N-n)\mu(k)$, we have

$$\begin{aligned} S_2(\alpha) &= \sum_{n<N} \Lambda(n)\mu^2(N-n) \sum_{\substack{d|(N-n) \\ d>\alpha}} \mu(d)\log(1/d) \\ &= \sum_{n<N} \Lambda(n)\mu^2(N-n) \sum_{\substack{k|(N-n) \\ k<\frac{N-n}{\alpha}}} \mu\left(\frac{N-n}{k}\right) \log\left(\frac{k}{N-n}\right) \\ &= \sum_{n<N} \Lambda(n) \sum_{\substack{k|(N-n) \\ k<\frac{N-n}{\alpha}}} \mu(N-n)\mu(k) \log\left(\frac{k}{N-n}\right) \\ &= \sum_{k<\frac{N-1}{\alpha}} \mu(k) \sum_{\substack{n<N \\ n\equiv N \pmod k}} \Lambda(n)\mu(N-n)\log\left(\frac{k}{N-n}\right). \end{aligned} \tag{18}$$

For those integers $k < \frac{N-1}{\alpha}$ such that $(k, N) = \delta_3 > 1$, suppose an integer $n < N$ satisfies $\Lambda(n) \neq 0$ and $n \equiv N \pmod k$, then n is a prime power $n = p^{\ell_3}$ for some integer $\ell_3 \ge 1$, and the condition $n \equiv p^{\ell_3} \equiv N \pmod k$ would imply $p \mid \delta_3$ and hence $p \mid N$. Therefore, we have

$$\begin{aligned} &\sum_{\substack{k<\frac{N-1}{\alpha} \\ (k,N)>1}} \mu(k) \sum_{\substack{n<N \\ n\equiv N \pmod k}} \Lambda(n)\mu(N-n)\log\left(\frac{k}{N-n}\right) \\ &\ll \log N \sum_{k<\frac{N}{\alpha}} \sum_{p|N} \sum_{\ell_3 \le \frac{\log N}{\log p}} \Lambda(p^{\ell_3}) \ll \frac{N}{\alpha}(\log N)^2, \end{aligned}$$

which implies

$$S_2(\alpha) = \sum_{\substack{k<\frac{N-1}{\alpha} \\ (k,N)=1}} \mu(k) \sum_{\substack{n<N \\ n\equiv N \pmod k}} \Lambda(n)\mu(N-n)\log\left(\frac{k}{N-n}\right) + O\left(\frac{N}{(\log N)^A}\right)$$

$$= S_3(\alpha) - S_4(\alpha) + O\left(\frac{N}{(\log N)^A}\right), \tag{19}$$

where

$$S_3(\alpha) = \sum_{\substack{k<\frac{N-1}{\alpha} \\ (k,N)=1}} \mu(k)\log k \sum_{\substack{n<N \\ n\equiv N \pmod k}} \Lambda(n)\mu(N-n)$$

and

$$S_4(\alpha) = \sum_{\substack{k<\frac{N-1}{\alpha} \\ (k,N)=1}} \mu(k) \sum_{\substack{n<N \\ n\equiv N \pmod k}} \Lambda(n)\mu(N-n)\log(N-n).$$

To estimate $S_3(\alpha)$, we first write

$$S_3(\alpha) = \sum_{\substack{k<\frac{N-1}{\alpha} \\ (k,N)=1}} \frac{\mu(k)\log k}{\phi(k)} \sum_{n<N} \Lambda(n)\mu(N-n) + E_3(\alpha),$$

where

$$E_3(\alpha) = \sum_{\substack{k<\frac{N-1}{\alpha} \\ (k,N)=1}} \mu(k)\log k \left(\sum_{\substack{n<N \\ n\equiv N \pmod k}} \Lambda(n)\mu(N-n) - \frac{1}{\phi(k)} \sum_{n<N} \Lambda(n)\mu(N-n) \right).$$

Our assumption $EH_\mu(N^{1-\theta})$ implies that

$$E_3(\alpha) \ll \log N \sum_{\substack{k<\frac{N-1}{\alpha} \\ (k,N)=1}} \left| \sum_{\substack{n<N \\ n\equiv N \pmod k}} \Lambda(n)\mu(N-n) - \frac{1}{\phi(k)} \sum_{n<N} \Lambda(n)\mu(N-n) \right|$$

$$\ll \log N \sum_{q\le N^{1-\theta}} \max_{y<N} \max_{\substack{a \\ (a,q)=1}} \left| \sum_{\substack{n\le y \\ n\equiv a \pmod q}} \Lambda(n)\mu(N-n) - \frac{1}{\phi(q)} \sum_{n\le y} \Lambda(n)\mu(N-n) \right|$$

$$\ll \frac{N}{(\log N)^A}.$$

Therefore, we have

$$S_3(\alpha)=\sum_{\substack{k<\frac{N-1}{\alpha}\\(k,N)=1}}\frac{\mu(k)\log k}{\phi(k)}\sum_{n<N}\Lambda(n)\mu(N-n)+O\left(\frac{N}{(\log N)^A}\right).$$

By Lemma 1, we may write the sum on the right hand side as

$$\begin{aligned}&\left(-\mathfrak{S}(N)+O\left(e^{-c_4\sqrt{\log(N/\alpha)}}\right)\right)\sum_{n<N}\Lambda(n)\mu(N-n)\\&=-\mathfrak{S}(N)\sum_{n<N}\Lambda(n)\mu(N-n)+O\left(\frac{N}{e^{c_4\sqrt{\log(N/\alpha)}}}\right)\\&=-\mathfrak{S}(N)\sum_{n<N}\Lambda(n)\mu(N-n)+O\left(\frac{N}{(\log N)^A}\right),\end{aligned}$$

where in the second line the trivial bound $\sum_{n<N}|\Lambda(n)\mu(N-n)|=O(N)$ is applied in the error term and c_4 is some positive constant. So we arrive at the estimate

$$S_3(\alpha)=-\mathfrak{S}(N)\sum_{n<N}\Lambda(n)\mu(N-n)+O\left(\frac{N}{(\log N)^A}\right). \tag{20}$$

Next we treat $S_4(\alpha)$ in a similar fashion. Again, we write

$$\begin{aligned}S_4(\alpha)&=\sum_{\substack{k<\frac{N-1}{\alpha}\\(k,N)=1}}\mu(k)\sum_{\substack{n<N\\n\equiv N\pmod k}}\Lambda(n)\mu(N-n)\log(N-n)\\&=\sum_{\substack{k<\frac{N-1}{\alpha}\\(k,N)=1}}\frac{\mu(k)}{\phi(k)}\sum_{n<N}\Lambda(n)\mu(N-n)\log(N-n)+E_4(\alpha),\end{aligned}$$

where

$$E_4(\alpha)=\sum_{\substack{k<\frac{N-1}{\alpha}\\(k,N)=1}}\mu(k)\left(\sum_{\substack{n<N\\n\equiv N\pmod k}}\Lambda(n)\mu(N-n)\log(N-n)-\frac{1}{\phi(k)}\sum_{n<N}\Lambda(n)\mu(N-n)\log(N-n)\right).$$

By Lemma 4, we have

$$
\begin{aligned}
E_4(\alpha) &\ll \sum_{\substack{k<\frac{N-1}{\alpha}\\(k,N)=1}} \left| \sum_{\substack{n<N\\n\equiv N \pmod k}} \Lambda(n)\mu(N-n)\log(N-n) - \frac{1}{\phi(k)}\sum_{n<N}\Lambda(n)\mu(N-n)\log(N-n) \right| \\
&\ll \sum_{q\le N^{1-\theta}} \max_{y<N}\max_{(a,q)=1} \left| \sum_{\substack{n\le y\\n\equiv a \pmod q}} \Lambda(n)\mu(N-n)\log(N-n) - \frac{1}{\phi(q)}\sum_{n\le y}\Lambda(n)\mu(N-n)\log(N-n) \right| \\
&\ll \frac{N}{(\log N)^A}.
\end{aligned}
$$

Moreover, it follows by Lemma 1 that

$$
\left(\sum_{\substack{k<\frac{N-1}{\alpha}\\(k,N)=1}} \frac{\mu(k)}{\phi(k)} \right) \left(\sum_{n<N} \Lambda(n)\mu(N-n)\log(N-n) \right) \ll \frac{N(\log N)^2}{e^{c_1\sqrt{\log(N/\alpha)}}} \ll \frac{N}{(\log N)^A},
$$

where c_1 is a positive constant.

Now, merging the above two estimates yields

$$
S_4(\alpha) = O\left(\frac{N}{(\log N)^A} \right). \tag{21}
$$

Therefore, we deduce the desired estimate (17) from (19)–(21).

3.3 *Proof of Theorem 1*

Recall from (12) that

$$
\sum_{n<N} \Lambda(n)\Lambda(N-n) = S_1(\alpha) + S_2(\alpha) + O(N^{\frac{1}{2}}(\log N)^3).
$$

Moreover, for any $A > 0$, if we suppose that $EH(N^{\theta}(\log N)^{2A+8})$ and $EH_{\mu}(N^{1-\theta})$ hold for some constant $0 < \theta < 1$, then we recall from (13) and (17) that

$$
S_1(\alpha) = \mathfrak{S}(N)N + O\left(\frac{N}{(\log N)^A} \right)
$$

and

$$
S_2(\alpha) = -\mathfrak{S}(N)\sum_{n<N} \Lambda(n)\mu(N-n) + O\left(\frac{N}{(\log N)^A} \right).
$$

Therefore, we have

$$\sum_{n<N} \Lambda(n)\Lambda(N-n) = \mathfrak{S}(N)\left(N - \sum_{n<N} \Lambda(n)\mu(N-n)\right) + O\left(\frac{N}{(\log N)^A}\right). \tag{22}$$

Thus to find a lower bound on $\sum_{n<N} \Lambda(n)\Lambda(N-n)$, we only need an upper bound on $\left|\sum_{n<N} \Lambda(n)\mu(N-n)\right|$. We first apply the triangle inequality and obtain that

$$\left|\sum_{n<N} \Lambda(n)\mu(N-n)\right| \le \sum_{n<N} \Lambda(n)\mu^2(N-n).$$

Since $\mu^2(N-n) = \sum_{b^2|(N-n)} \mu(b)$, we have

$$\begin{aligned}
&\sum_{n<N} \Lambda(n)\mu^2(N-n) \\
&= \sum_{n<N} \Lambda(n) \sum_{b^2|(N-n)} \mu(b) \\
&= \sum_{b<(N-1)^{\frac{1}{2}}} \mu(b) \sum_{\substack{n<N \\ n\equiv N \pmod{b^2}}} \Lambda(n) \\
&= \sum_{b\le N^{\frac{1}{4}-\epsilon}} \mu(b) \sum_{\substack{n<N \\ n\equiv N \pmod{b^2}}} \Lambda(n) + \sum_{N^{\frac{1}{4}-\epsilon}<b<N^{\frac{1}{2}}} \mu(b) \sum_{\substack{n<N \\ n\equiv N \pmod{b^2}}} \Lambda(n). \qquad (23)
\end{aligned}$$

The sum over $N^{\frac{1}{4}-\epsilon} < b < N^{\frac{1}{2}}$ can be bounded by

$$\sum_{N^{\frac{1}{4}-\epsilon}<b<N^{\frac{1}{2}}} \mu(b) \sum_{\substack{n<N \\ n\equiv N \pmod{b^2}}} \Lambda(n) \ll \log N \sum_{N^{\frac{1}{4}-\epsilon}<b\le N^{\frac{1}{2}}} \frac{N}{b^2} \ll N^{\frac{3}{4}+\epsilon}. \tag{24}$$

The contribution of those $b \le N^{\frac{1}{4}-\epsilon}$ such that $(b, N) > 1$ is

$$\sum_{\substack{b\le N^{\frac{1}{4}-\epsilon} \\ (b,N)>1}} \mu(b) \sum_{\substack{n<N \\ n\equiv N \pmod{b^2}}} \Lambda(n) \ll N^{\frac{1}{4}-\epsilon}. \tag{25}$$

On the other hand, for those $b \le N^{\frac{1}{4}-\epsilon}$ with $(b, N) = 1$, we apply the Bombieri-Vinogradov theorem and obtain

$$\begin{aligned}
&\sum_{\substack{b\le N^{\frac14-\epsilon}\\(b,N)=1}} \mu(b) \sum_{\substack{n<N\\ n\equiv N \pmod{b^2}}} \Lambda(n)\\
&= N \sum_{\substack{b\le N^{\frac14-\epsilon}\\(b,N)=1}} \frac{\mu(b)}{\phi(b^2)} + \sum_{\substack{b\le N^{\frac14-\epsilon}\\(b,N)=1}} \mu(b) \left(\sum_{\substack{n<N\\ n\equiv N \pmod{b^2}}} \Lambda(n) - \frac{N}{\phi(b^2)} \right)\\
&= N \sum_{\substack{b\le N^{\frac14-\epsilon}\\(b,N)=1}} \frac{\mu(b)}{\phi(b^2)} + O\left(\sum_{\substack{b\le N^{\frac14-\epsilon}\\(b,N)=1}} \left| \sum_{\substack{n<N\\ n\equiv N \pmod{b^2}}} \Lambda(n) - \frac{N}{\phi(b^2)} \right| \right)\\
&= N \sum_{\substack{b\le N^{\frac14-\epsilon}\\(b,N)=1}} \frac{\mu(b)}{\phi(b^2)} + O\left(\frac{N}{(\log N)^{A+1}}\right). \qquad (26)
\end{aligned}$$

It is easy to see that the infinite series

$$\sum_{(d,N)=1} \frac{\mu(d)}{\phi(d^2)} = \prod_{\substack{p\nmid N\\ p>2}} \left(1 - \frac{1}{p(p-1)}\right) = \mathcal{A}(N) \qquad (27)$$

converges and that its tail is bounded by

$$\sum_{\substack{b>N^{\frac14-\epsilon}\\(d,N)=1}} \frac{\mu(d)}{\phi(d^2)} \ll \sum_{b>N^{\frac14-\epsilon}} \frac{\log\log d}{d^2} \ll N^{-\frac14+\epsilon}. \qquad (28)$$

Therefore, by (23)–(28), we have

$$\sum_{n<N} \Lambda(n)\mu^2(N-n) = N\mathcal{A}(N) + O\left(\frac{N}{(\log N)^{A+1}}\right).$$

Thus, we deduce from (22) that

$$\begin{aligned}
\sum_{n<N} \Lambda(n)\Lambda(N-n) &\ge \mathfrak{S}(N) \left(N - N\mathcal{A}(N) + O\left(\frac{N}{(\log N)^{A+1}}\right)\right) + O\left(\frac{N}{(\log N)^{A}}\right)\\
&\ge \mathfrak{S}(N)\,(1 - \mathcal{A}(N))\,N + O\left(\frac{N}{(\log N)^{A}}\right),
\end{aligned}$$

where in the line above we use the bound

$$\begin{aligned}\mathfrak{S}(N) =& 2\prod_{p>2}\left(1-\frac{1}{(p-1)^2}\right)\prod_{\substack{p\mid N\\ p>2}}\left(1+\frac{1}{p-2}\right)\\ &\ll \prod_{2<m\le \log N}\frac{m-1}{m-2}\\ &\ll \log N.\end{aligned}$$

Finally, it follows directly from (22) that the assertions

$$\sum_{n<N}\Lambda(n)\Lambda(N-n)\sim \mathfrak{S}(N)N$$

and

$$\sum_{n<N}\Lambda(n)\mu(N-n)=o(N)$$

are equivalent. This completes the proof of Theorem 1.

3.4 *Proof of Corollary 1*

To prove the corollary, we just take $A = 2 + \frac{\epsilon}{2}$ in our theorem and obtain

$$\widetilde{r}(N)\ge \mathfrak{S}(N)(1-\mathcal{A}(N))N+O\left(\frac{N}{(\log N)^{2+\frac{\epsilon}{2}}}\right).$$

Since $\mathfrak{S}(N)\gg 1$, it suffices to show that

$$1-\mathcal{A}(N)\gg \frac{1}{(\log N)^2}$$

in order to prove $\widetilde{r}(N)>0$ for sufficiently large even N.

Since for sufficiently large N, the least prime not dividing N is at most $2\log N$, we have

$$1-\mathcal{A}(N)=1-\prod_{\substack{p\nmid N\\ p>2}}\left(1-\frac{1}{p(p-1)}\right)>\frac{1}{2\log N(2\log N-1)},$$

which confirms the above claim and therefore completes the proof of the corollary.

References

1. Jing-Run Chen, *On the representation of a large even integer as the sum of a prime and the product of at most two primes*, Scientia Sinica. Zhongguo Kexue, 21(4), pp 421–430, 1978.
2. Etienne Fouvry and Henryk Iwaniec, *Gaussian primes*, Acta Arithmetica, 79(3), pp 249–287, 1997.
3. John Friedlander and Henryk Iwaniec, *Asymptotic sieve for primes*, Annals of Mathematics. Second Series, 148(3), pp 1041–1065, 1998.
4. John Friedlander and Henryk Iwaniec, *The polynomial $X^2 + Y^4$ captures its primes*, Annals of Mathematics. Second Series, 148(3), pp 945–1040, 1998.
5. Daniel Goldston and Cem Yıldırım, *Higher correlations of divisor sums related to primes. I. Triple correlations*, Integers. Electronic Journal of Combinatorial Number Theory, 3, pp A5, 66, 2003.
6. Godfrey Hardy and John Littlewood, *Some problems of 'Partitio numerorum'; III: On the expression of a number as a sum of primes*, Acta Mathematica, 44(1), pp 1–70, 1923.
7. Loo-Keng Hua, *A direct attempt to Goldbach problem*, Acta Mathematica Sinica. New Series, 5(1), pp 1–8, 1989.
8. James Maynard, *Small gaps between primes*, Annals of Mathematics. Second Series, 181(1), pp 383–413, 2015.
9. Hugh Montgomery and Robert Vaughan, *Multiplicative number theory. I. Classical theory*, Cambridge Studies in Advanced Mathematics, 97, 2007.
10. Cheng-Dong Pan, *A new attempt on Goldbach conjecture*, Chinese Annals of Mathematics, 3(4), pp 555–560, 1982.
11. János Pintz, *An approximation to the twin prime conjecture and the parity phenomenon*, Indagationes Mathematicae. New Series, 26(5), pp 883–896, 2015.
12. D. H. J. Polymath, *Variants of the Selberg sieve, and bounded intervals containing many primes*, Research in the Mathematical Sciences, 1, pp Art. 12, 83, 2014.
13. Ram Murty and Akshaa Vatwani, *Twin primes and the parity problem*, Journal of Number Theory, 180, pp 643–659, 2017.
14. Sitaramachandra Rao, *On an error term of Landau. II*, The Rocky Mountain Journal of Mathematics, 15(2), pp 579–588, 1985.
15. Gérald Tenenbaum, *Introduction to analytic and probabilistic number theory*, Graduate Studies in Mathematics, 163, 2015.
16. Akshaa Vatwani, *Variants of equidistribution in arithmetic progression and the twin prime conjecture*, Mathematische Zeitschrift, 293, pp 285–317, 2019.
17. Yi-Tang Zhang, *Bounded gaps between primes*, Annals of Mathematics. Second Series, 179(3), pp 1121–1174, 2014.

Part-Frequency Matrices, II: Recent Work

William J. Keith

Abstract We illustrate the use of part-frequency matrices as a tool for combinatorial proofs of partition theorems. Several new theorems are rapidly proved, and progress is made toward replacing the proof of a known theorem proved by modular forms with a more combinatorial approach. A few lines of investigation are suggested. All currently-published papers known to the author that employ the idea are cited here.

Keywords Partitions · Partition rank · Glaisher's bijection

1 Introduction

The talk on which this article is based was, largely, expository; although new theorems are herein proved, the main purpose was to make known to more professionals a technique which has been "in the air" for a few years in widely scattered papers, which seems promising for future work in partition theory.

A partition of n, denoted $\lambda \vdash n$, is a weakly decreasing sequence of positive integers $(\lambda = (\lambda_1, \lambda_2, \ldots, \lambda_i)$ such that $\sum \lambda_j = n$. We say n is the *weight* of the partition, and write $|\lambda| = n$. Denote the number of partitions of n by $p(n)$. Their generating function is

$$P(n) := \sum_{n=0}^{\infty} p(n)q^n = \prod_{i=1}^{\infty} \frac{1}{1-q^i}.$$

Herein we assume for convergence concerns that $q \in \mathbb{C}$ with $0 < |q| < 1$ and never need to concern ourself further.

One of the first and most useful theorems learned by a student of partition theory is that the number of partitions of n into odd parts (e.g., $3+1$ or $1+1+1+1$ for $n=4$) is equal to the number of partitions of n into distinct parts (4 or $3+1$). This

W. J. Keith (✉)
Michigan Technological University, Houghton, MI, USA
e-mail: wjkeith@mtu.edu

M. B. Nathanson (ed.), *Combinatorial and Additive Number Theory IV*,
Springer Proceedings in Mathematics & Statistics 347,
https://doi.org/10.1007/978-3-030-67996-5_18

theorem, interpreted in generating functions, gives the q-series identity

$$\prod_{k=1}^{\infty} \frac{1}{1-q^{2k-1}} = \prod_{k=1}^{\infty}(1+q^k),$$

and it was proved in this form by Euler [6].

This theorem was given a combinatorial proof by James Glaisher: if in a partition into odd parts the part i appears $a_{i,0}2^0 + a_{i,1}2^1 + \cdots$ times, $a_{i,\ell} \in \{0, 1\}$, then construct the distinct parts $a_{i,\ell}\, i \cdot 2^\ell$. The reverse map collects parts with common odd factor i. The identity and Glaisher's map both generalize in the obvious way to partitions into parts not divisible by m and partitions in which parts appear fewer than m times.

The core notion in this paper is the fact that Glaisher's maps can be extended from these specialized sets to a collection of involutions on all partitions in a natural way: construct a list of matrices M_j indexed by the numbers j not divisible by m. Enumerate columns and rows starting with 0. If the part $N = jm^k$ appears $a_{N,0}m^0 + a_{N,1}m^1 + a_{N,2}m^2 + \cdots$ times as written in the ascending m-ary notation, then assign row k of matrix M_j to be

$$a_{k,1}\ a_{k,2}\ a_{k,3}\ \ldots \quad := \quad a_{N,0}\ a_{N,1}\ a_{N,2}\ \ldots.$$

For example, if $m = 2$, the partition $(20, 5, 5, 4, 2, 2, 1, 1, 1, 1, 1)$ of 43 would be depicted (with all other entries 0):

$$\begin{array}{c|ccc} 1 & & & \\ \hline & 1 & 0 & 1 \\ & 0 & 1 & 0 \\ & 1 & 0 & 0 \end{array} \quad \begin{array}{c|ccc} 3 & & & \\ \hline & 0 & 0 & 0 \\ & 0 & 0 & 0 \\ & 0 & 0 & 0 \end{array} \quad \begin{array}{c|ccc} 5 & & & \\ \hline & 0 & 1 & 0 \\ & 0 & 0 & 0 \\ & 1 & 0 & 0 \end{array}.$$

Now it is easy to see that Glaisher's map is simply transposition of this sequence of matrices, restricted to partitions into odd or distinct parts, where either the first row or the first column are the only nonzero entries.

The observation with which this paper is mainly concerned is thus:

Theorem 1 *Any weight-preserving transformation of the matrices M_j corresponds to an involution of the set of partitions of n.*

This much was illustrated in the first paper which elucidated this idea, (K) [8], and also used in similar ways in Flowers, Neville and Sellers [7], Chern and Qiu [4], and Pathak [10]. Pathak in [11] cites Yang [15], who makes a partial step in this direction with a modification of Glaisher's map.

In this paper, in addition to further examples of the same idea, we will consider new uses, including:

1. exchanging roles between frequencies of appearance, and index, under conditions when this is allowable;

2. parity theorems involving involutions which preserve or map subsets of interest;
3. combining these maps with more common maps such as conjugation to prove stronger identities.

In Sect. 2 we apply these ideas to m-ary partitions, for which we suggest they are particularly useful. In Sect. 3 we apply the idea to theorems previously proved by modular forms techniques; although we cannot completely reprove these theorems combinatorially yet, early progress strongly indicates that such a proof should be possible, and would certainly be interesting. In Sect. 4 we discuss theorems involving partitions with parts separated by parity. Finally, in Sect. 5 we offer a few potentially useful lines of investigation with this tool.

2 On m-Ary Partitions

An *m-ary partition* is a partition in which parts are only powers of m. They have been studied for some time and research is ongoing: see for instance [3, 12]. Let the number of m-ary partitions of n be $t_m(n)$. They have generating function

$$\sum_{n=0}^{\infty} t_m(n)q^n = \prod_{k=0}^{\infty} \frac{1}{1-q^{m^k}}.$$

The m-ary partitions have a very specific feature that makes them interact well with part-frequency ideas: only the matrix M_1 has occupied entries in an m-ary partition. This has the interesting consequence that transformations of the matrix are transformations preserving the property of being an m-ary partition. For example, transposition is an involution on this set. Hence, for instance, the set of m-ary partitions of n has the same parity as their subset which has matrix representation that is symmetrical about the main diagonal of the matrix.

This gives the theorem

Theorem 2

$$\sum_{n=0}^{\infty} t_m(n)q^n \equiv \left(\prod_{k=0}^{\infty} \frac{1-q^{m^{2k+1}}}{1-q^{m^{2k}}}\right)\left(\prod_{k=0}^{\infty} \frac{1}{1-q^{2m^{2k+1}}}\right) \pmod 2.$$

Proof The first term on the right-hand side is easily seen to be the generating function for the partitions whose parts lie only on the main diagonal of M_1. For the second part, consider the diagonals immediately above and below the main diagonal; call them D_1 and D_{-1} for convenience. If they are filled with equal amounts in corresponding positions, the amount contributed to the size of the partition is $2 \cdot (0, 1, \ldots, m-1) \cdot m^{2k+1}, k \geq 0$. Entries in diagonals D_2 and D_{-2} contribute amounts $2 \cdot (0, 1, \ldots, m-1) \cdot m^{2k+2}, k \geq 0$.

A little algebraic cancellation gives that the only surviving factor from these diagonals is $\frac{1}{1-q^{2m}}$. The next two diagonals yield $\frac{1}{1-q^{2m^3}}$, and the pattern continues. □

Remark: Of course transposition is not the only transformation that can be employed. Consider the *orbit* action: rotate all diagonals one place south-west, looping entries on the leftmost column to the top. Every m-ary partition has some finite orbital period. If we group and discard those with even orbit, we are left with the subset that have odd orbits.

We could also reduce the subset of m-ary partitions symmetric across the main diagonal with Kathy Ji and Herb Wilf's idea of *recursively palindromic* words—pair up partitions that are symmetric across the main diagonal but have left and right halves that do not reverse within themselves symmetrically, etc.

In particular, we note that the binary partition function satisfies $t_m(n) \equiv 0 \pmod 2$ for all $n \geq 2$, so it is more interesting to ask about its residue mod 4. For this, we might classify partitions by whether they are asymmetric at the first transposition, symmetric at the first but asymmetric at the second, half-diagonal transposition, or symmetric at both.

Question: Can we identify a "core" set of conditions, analogous to partitions into distinct odd parts, giving a subset of m-ary partitions of n specifying the parity (or residue mod higher powers of 2) of $t_m(n)$?

The following theorem does not strictly require the framework of part-frequency matrices to be proved, but that framework gives an easy visualization. What we note here is that it is allowable to exchange the role of *frequencies of appearance*, and the modulus-*m index of part sizes*, since our original part sizes are powers of m that can be handled in a convenient way.

Theorem 3 *The number of m-ary partitions of n is equal to the number of partitions of n into parts of any type, whose frequencies of appearance, written in base-m notation, contain only 1s, no two parts having a 1 in the same digit place.*

Corollary 1 *The number of binary partitions of n is equal to the number of partitions of n into parts of any type, wherein at most one frequency of appearance f_j of any part size may satisfy any given congruence $f \equiv x \pmod{2^k}$ for any x, k.*

Proof Suppose that power m^j appears f_j times. Replace this part by part size f_j appearing $1 \cdot m^j$ times. This gives the partition of the second type. With $m = 2$, the second clause of the description degenerates and any frequency is allowable for any one part size, but no collection of sizes may overlap in any digit, giving the corollary. □

3 Partitions into Two or Three Part Sizes

In this section the Ferrers diagrams will be important. The Ferrers diagram of a partition $(\lambda_1, \ldots, \lambda_j)$ is a stack of unit-size squares justified to the origin in the fourth quadrant, having λ_i squares in the i-th row from the top. Thus for instance the Ferrers diagram of (4, 4, 2, 1, 1) is

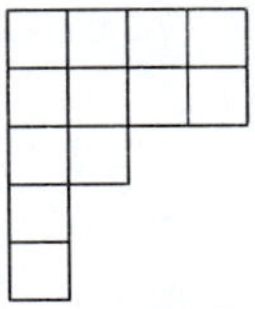

The *conjugate* of a partition λ is the partition with Ferrers diagram that of λ reflected across the diagonal. The conjugate of (4, 4, 2, 1, 1) is (5, 3, 2, 2).

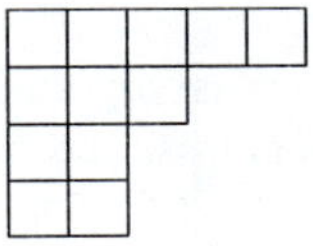

A partition fixed under conjugation is *self-conjugate*.

Let $\nu_k(n)$ be the number of partitions of n into exactly k different part sizes. For example, $\nu_2(6) = 6$, counting (5, 1), (4, 2), (4, 1, 1), (3, 1, 1, 1), (2, 2, 1, 1), (2, 1, 1, 1, 1). Note that conjugation preserves the number of part sizes in a partition. It was established in [9] that

$$\nu_2(16n + 14) \equiv 0 \pmod 4.$$

However, this proof required a modular-forms argument, and the question was raised of a combinatorial proof.

We begin by noting that there are no self-conjugate partitions of any number $4n + 2$ into exactly two part sizes, by the fact that a self-conjugate partition into exactly two part sizes has weight which is a difference of squares, and numbers $4n + 2$ cannot be written as a difference of squares.

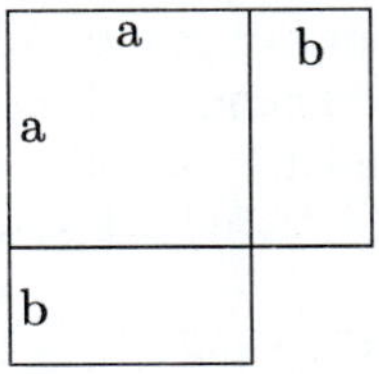

This means that $\nu_2(4n + 2) \equiv 0 \pmod 2$, and we could in theory show $\nu_2(16n + 14) \equiv 0 \pmod 4$ more combinatorially by pairing up different conjugate pairs. In this section we will not fully answer this problem but will make progress on doing so, by a simple map on the part-frequency matrices of the partitions involved. The reader is invited to complete the attack.

Begin by defining some classes. Describe them by the parity of the larger part, the parity of its number of appearances, the parity of the smaller part, and the parity of its number of appearances: for instance, class $O^E E^E$ consists of partitions in which the larger part is odd, it appears an even number of times, the smaller part is even, and it too appears an even number of times.

For $n \equiv 14 \pmod{16}$, any classes not listed below are trivially empty, e.g. $O^O E^O$ would partition an odd number, $E^E E^E$ one divisible by 4. The populated classes are related thus by conjugation:

$O^O O^O$	conjugate to	$E^O O^E$
$E^O E^O$		$E^E O^E$
$E^O E^E$		$O^E O^E$
$O^E E^E$		$E^E E^O$
$O^E E^O$		self

To be clear, in the last line we are saying that any element of the class $O^E E^O$ conjugates to another element in that class, not to itself.

The easiest class to deal with is $O^O O^O$. Its conjugates are all in $E^O O^E$. The 2-modular matrices for a partition in $E^O O^E$ and $O^E E^O$ have the following features (with k_1 possibly equal to k_2):

$$\begin{array}{c|lll} k_1 & & & \\ \hline & 0 & 1 & \dots \\ & 0 & 0 & 0 \\ & 0 & 0 & 0 \end{array} \qquad \begin{array}{c|lll} k_2 & & & \\ \hline & 0 & \dots & 0 \\ & \dots & 0 & 0 \\ & 1 & \dots\dots & \end{array}$$

or

$$\begin{array}{c|llll} k_1 & & & & \\ \hline & 0 & \dots & 1 & \dots \\ & 0 & 0 & 0 & 0 \end{array} \qquad \begin{array}{c|lll} k_2 & & & \\ \hline & 0 & 0 & 0 \\ & 1 & \dots\dots & \end{array}$$

Here any finite number of entries in the same row following a 1 (and only they) may also be 1s.

It will be noted that, since $n \equiv 2 \pmod 4$, it must be the case that exactly one of the two matrices has a single entry 1 in either the first row, second element (the odd number appearing an even number of times) or the second row, first element (the even number appearing an odd number of times). This contributes an amount 2 mod 4 to n, and so the other initial 1 cannot be in either of these two positions. However, one of the initial 1s is always in the first row and the other is always in the first column.

The involution here is obvious: we interchange the two roles of the rows with some nonzero entries. The initial 1 in the first row will move to the first column, and vice versa, keeping the initial 1 in the same diagonal and all other entries of the row in the same relative position. Call this map ϕ.

If $\lambda \in E^O O^E$, then $\phi(\lambda)$ will be *either* still in $E^O O^E$, or possibly in $O^E E^O$, depending on which is now the larger part.

For example, $\phi(10, 1^4) = (5^2, 4)$.

$$\begin{array}{c|ccc} 1 & & & \\ \hline & 0 & 0 & 1 \\ & 0 & 0 & 0 \\ & 0 & 0 & 0 \end{array} \quad \begin{array}{c|ccc} 5 & & & \\ \hline & 0 & 0 & 0 \\ & 1 & 0 & 0 \\ & 0 & 0 & 0 \end{array} \quad \rightarrow \quad \begin{array}{c|ccc} 1 & & & \\ \hline & 0 & 0 & 0 \\ & 0 & 0 & 0 \\ & 1 & 0 & 0 \end{array} \quad \begin{array}{c|ccc} 5 & & & \\ \hline & 0 & 1 & 0 \\ & 0 & 0 & 0 \\ & 0 & 0 & 0 \end{array}$$

It is clear that this is a reversible map. In fact, this map has no fixed points, for the occupied entry in the second diagonal (i.e. the part contributing 2 mod 4) always switches position. Second, observe that $O^E E^O$ always maps to $E^O O^E$ by ϕ, since if the odd part is the larger it will only increase moving down its diagonal, while the even part is reduced. So the map sends $O^E E^O$ to a subset of $E^O O^E$.

As a side note, this gives a simple incidental corollary:

Corollary 2 $|OEEO(4n+2)| < |EOOE(4n+2)|$.

We now pair up pairs involving $O^O O^O$ in the following fashion: if their conjugate in $E^O O^E$ mapped by *phi* to another in $E^O O^E$ which conjugates back to $O^O O^O$, pair up those conjugate pairs. If the image under ϕ lands in $O^E E^O$, conjugate the result and pair up the matching pair.

Here are examples of the two types of pairings:

$$\left((13,1) \underset{conj.}{\rightarrow} (2,1^{12})\right) \underset{\phi}{\rightarrow} \left((4^3,1^2) \underset{conj.}{\rightarrow} (5,3^3)\right)$$

$$\left((5,1^9) \underset{conj.}{\rightarrow} (10,1^4)\right) \underset{\phi}{\rightarrow} (5^2,4) \underset{conj.}{\rightarrow} (3^4,2) \underset{\phi}{\rightarrow} \left((12,1^2) \underset{conj.}{\rightarrow} (3,1^9)\right)$$

So we have established:

Lemma 1 $|O^O O^O| = |E^O O^E| \equiv 0 \pmod 2$ *and thus* $|O^O O^O| + |E^O O^E| \equiv 0 \pmod 4$.

The case $E^O E^E$ is similarly easy to handle. It conjugates to $O^E O^E$, in which two different matrices must have entries in the first rows. Send these to the first column and you get that subset of $E^O E^O$ which has its entries in two different matrices.

Remaining cases are then:

1. $O^E E^O$ (still to be counted)
2. $O^E E^E$ and its conjugate $E^E E^O$
3. The subset of $E^O E^O$ with both occupied rows in the same matrix, and its conjugates in $E^E O^E$.

Until now we have only invoked so far the fact that n is 2 mod 4. But observe the remaining subset of $E^O E^O$. Suppose that the parts of λ are of size $\lambda_1 = 2^j m$ and $\lambda_2 = 2m$, m odd, $j > 1$, and say that they have frequencies o_1 and o_2 respectively. If $o_1 = o_2$, we claim that $o_i \neq m$, for in this case

$$|\lambda| = m^2(2+2^j), \quad j > 1.$$

But $m^2 \equiv 1, 9 \pmod{16}$, and with $j > 1$, $2+2^j \equiv 2, 6, 10 \pmod{16}$, and hence $m^2(2+2^j) \not\equiv 14 \pmod{16}$.

Thus elements of $E^O E^O$ whose parts have equal index and equal odd frequencies may exchange the role of the index and frequency without fixed point, and this is an involution on $E^O E^O$.

Now we are left with:

1. $O^E E^O$
2. $O^E E^E$ and its conjugate $E^E E^O$
3. The subset of $E^O E^O$ with both occupied rows in the same matrix but having different odd frequencies, and its conjugates in $E^E O^E$.

Partitions of the latter type must be of the form $(2^j m)^{2a+1}(2m)^{2b+1}$, $j > 1, a \neq b$, and given the types of partitions remaining to relate, this seems suggestive of potential proof strategies.

At this point we stop and pose two problems for the reader.

Problem 1: Complete the proof strategy above.
Problem 2: While $\nu_3(16n+14) \equiv 0 \pmod 2$ (note: but is not always divisible by 4), the numbers $16n+14$ may have some self-conjugate partitions into three part sizes. This too was proven in [9] by modular forms. Give a combinatorial proof.

4 Partitions with Parts Separated by Parity

In a recent series of papers starting with [2], George Andrews and pursuing authors have been considering partitions in which all even parts are larger than all odd parts, or vice versa, with other conditions. For example, $P_{ou}^{ed}(n)$ is the set of partitions of n with any even parts larger than any odd parts, with the further restriction that even parts are distinct. These sets have connections with third-order mock theta functions.

We observe that the 2-modular part-frequency matrices have a more complicated but still visualizable interaction with the property of parts being separated by parity. (In general, part-frequency matrices do suffer from the weakness that "differences among parts" conditions are difficult to concisely capture.)

For example, in P_{ou}^{ed}, in the 2-modular matrices, only entries in the first row or first column may be nonzero. Any odd parts may appear any number of times, and almost any even part distinctly, but there is a "forbidden zone" of rows above the first that may not be occupied, namely those for which $m2^j < k$, where k is the largest odd part appearing.

We illustrate with the partition $(20, 16, 10, 5, 5, 1, 1, 1, 1, 1) \in P_{ou}^{ed}$. Since the largest odd part appearing is 5, the even parts 2 and 4 may not appear at all.

$$
\begin{array}{c|ccc}
1 & & & \\ \hline
 & 1 & 0 & 1 \\
 & 0 & 0 & 0 \\
 & 0 & 0 & 0 \\
 & 1 & 0 & 0
\end{array}
\quad
\begin{array}{c|ccc}
3 & & & \\ \hline
 & 0 & 0 & 0 \\
 & 0 & 0 & 0 \\
 & 0 & 0 & 0
\end{array}
\quad
\begin{array}{c|ccc}
5 & & & \\ \hline
 & 0 & 1 & 0 \\
 & 1 & 0 & 0 \\
 & 1 & 0 & 0
\end{array}
.
$$

Transposition of all occupied matrices gives us an equinumerous set: almost the same, but now the forbidden zone is in the odds. We have the following theorem:

Theorem 4 *The set $P_{ou}^{ed}(n)$ is equinumerous with the set of partitions of n with the following condition: write all parts in λ as $k2^j$, m odd, and denote by $k(\lambda)$ the largest odd factor thus appearing. Then odd parts m in λ may only appear a number of times congruent to 0 or 1 modulo $2^{\lceil \log_2 (k/m) \rceil}$.*

However, we do not have to transpose the entire matrix. If we only transpose the portion above the first row, we obtain the following:

Theorem 5 *The set $P_{ou}^{ed}(n)$ is equinumerous with the set of partitions into parts not divisible by 4, in which the even parts $2m$ must appear a number of times divisible by 2^{j-1} with $m2^j > k$, k the largest odd part.*

One can play this game essentially infinitely with similar conditions. The following are theorems related to partitions with odd parts less than twice the smallest part [1]:

Theorem 6 *Partitions into odd parts $< 2N$ are equinumerous with partitions into parts $k \cdot 2^i$ where $k + i \leq 2N - 1$, and part $k \cdot 2^i$ is distinct if $k + i < 2N - 1$.*

Theorem 7 *Partitions into parts not divisible by m, of size $\leq Nm$, are equinumerous with partitions into parts $\leq Nm$ among which parts of size $\leq N$ appear less than m times.*

5 Investigations

Hopefully this paper has sparked some interest in employing this tool. We conclude with a few open questions or suggestions of research direction.

1. Herb Wilf [14] asked for *any* interesting theorems about the set of partitions with distinct multiplicities. Having the feature that all occupied rows are distinguishable, might some part-frequency operation be able to say something about this notoriously unruly set?
2. We reiterate here a question from the earlier paper [8] which still seems to be of considerable interest: a very special subset of partitions with respect to part-frequency matrix for any modulus are those with nonzero entries only in the upper-left corner. These are partitions in which parts appearing are not divisible

by m, and appear less than m times. Denote the number of these by $p_{m,m}(n)$. These share features with all partitions: for instance, they possess the same Ramanujan-type congruence $p_{5,5}(5n+4) \equiv 0 \pmod 5$. Perhaps a combinatorial proof of this fact could be given, thereby establishing Ramanujan's congruence in another way. A general study of such partitions could be fruitful.

Acknowledgements The author would like to thank Melvyn Nathanson and all other staff and organizers for the opportunity to speak at *Combinatorial and Additive Number Theory 2020*, and for their handling of these *Proceedings*.

References

1. G. E. Andrews, A. Dixit, and A. J. Yee, *Partitions associated with the Ramanujan/Watson mock theta functions* $\omega(q)$, $\nu(q)$, and $\phi(q)$. Research in Number Theory (2015) 1:19. https://doi.org/10.1007/s40993-015-0020-8
2. G. E. Andrews, *Partitions with parts separated by parity*. Annals of Combinatorics **23**, 241–248 (2019)
3. G. E. Andrews, A. S. Fraenkel, and J. A. Sellers, *Characterizing the number of m -ary partitions modulo m*. Amer. Math. Monthly **122** 9 (2015) 880–885
4. S. Chern and S. Qiu, *Partitions, geometric progressions, and a Putnam problem*. Math. Gazette, 103 (557) (2019): 337–343. https://doi.org/10.1017/mag.2019.71
5. G. Andrews, F. Garvan, *Dyson's crank of a partition*, Bull. Amer. Math. Soc. (N.S.) 18 (1988) 2:167–171
6. L. Euler, Introductio in analysin infinitorum, volume 2. Marcum Michaelem Bousquet, 1748.
7. T. B. Flowers, S. Neville, and J. A. Sellers, *An m -ary generalization of a past Putnam problem*. Australasian J. Comb. **72** (2) (2018) 369–375
8. W. J. Keith, *The part-frequency matrices of a partition*, Proceedings of the 2015 Kliakhandler Conference, J. Algebra Comb. Discrete Appl. 3(3) (2016)
9. W. J. Keith, *Restricted k -colored partitions*. Ramanujan J., 2016, Volume 40, Issue 1, pp 71–92. https://doi.org/10.1007/s11139-015-9704-x
10. A. Patak, *Glaisher's partition problem*. To appear, Math. Student. arXiv:2003.08220
11. A. Pathak, *On certain partition bijections related to Euler's partition problem*. arXiv:2004.03596
12. Ø. Rødseth, *Some arithmetical properties of m -ary partitions*. Math. Proc. of the Cambridge Phil. Soc., **68** 2 (1970), 447–453
13. Kathy Q. Ji & Herbert S. Wilf (2008) *Extreme Palindromes*. The American Mathematical Monthly, 115:5, 447–451, https://doi.org/10.1080/00029890.2008.11920548
14. H. S. Wilf, *Some unsolved problems*. https://www.math.upenn.edu/~wilf/website/UnsolvedProblems.pdf
15. J. X. Y. Yang, *Combinatorial proofs and generalizations of conjectures related to Euler's partition theorem*. European Journal of Combinatorics 76 (2019) 62–72.

A Conjectural Inequality for Visible Points in Lattice Parallelograms

Gabriel Khan, Mizan R. Khan, Joydip Saha, and Peng Zhao

Abstract Let $a, n \in \mathbb{Z}^+$, with $a < n$ and $\gcd(a, n) = 1$. Let $P_{a,n}$ denote the lattice parallelogram spanned by $(1, 0)$ and (a, n), that is,

$$P_{a,n} = \{t_1(1,0) + t_2(a,n) \,:\, 0 \le t_1, t_2 \le 1\},$$

and let

$$V(a,n) = \#\text{ of visible lattice points in the interior of } P_{a,n}.$$

In this paper we prove some elementary (and straightforward) results for $V(a, n)$. The most interesting aspects of the paper are in Section 5 where we discuss some numerics and display some graphs of $V(a, n)/n$. (These graphs resemble an integral sign that has been rotated counter-clockwise by 90°.) The numerics and graphs suggest the conjecture that for $a \neq 1, n-1$, $V(a, n)/n$ satisfies the inequality

$$0.5 < V(a,n)/n < 0.75.$$

G. Khan (✉)
Department of Mathematics, Iowa State University, Ames, IA 50011, USA
e-mail: gkhan@iastate.edu

M. R. Khan
Department of Mathematical Sciences, Eastern Connecticut State University, Willimantic, CT 06226, USA
e-mail: khanm@easternct.edu

J. Saha
Narayanganj, Bangladesh
e-mail: joydip.t.saha@gmail.com

P. Zhao
Department of Mathematics and Computer Science, Indiana State University, 47809 Terre Haute, IN, USA
e-mail: peng.zhao@indstate.edu

M. B. Nathanson (ed.), *Combinatorial and Additive Number Theory IV*, Springer Proceedings in Mathematics & Statistics 347,
https://doi.org/10.1007/978-3-030-67996-5_19

Keywords Lattice parallelograms · Visible points · Unimodular transformations · Floor and ceiling functions · Zeta values

1 Introduction and Notation

In this short paper we describe an ongoing project on visible points in lattice parallelograms. We start with some definitions and notation.

A lattice point $(u_1, u_2) \in \mathbb{Z}^2$ is said to be *visible* (from the origin) if $\gcd(u_1, u_2) = 1$. Following Apostol [1, Theorem 3.9, p. 63], for $r > 0$ let $N(r)$ denote the number of lattice points in the square $[-r, r]^2$, and let $N'(r)$ denote the number of the lattice points in this square that are visible, then

$$\lim_{r\to\infty} \frac{N'(r)}{N(r)} = \frac{6}{\pi^2}; \tag{1}$$

that is, if you pick two integers at random, then there is just over a 60% chance that they are relatively prime.

In this paper we count the number of visible lattice points lying inside lattice parallelograms of a certain form. Specifically, lattice parallelograms that satisfy the following two conditions : (i) other than the four vertices, there are no other lattice points on the boundary of the parallelogram; (ii) one of the vertices is the origin. Such a parallelogram is of the form

$$P = \{t_1\vec{u} + t_2\vec{v} \ : \ 0 \le t_1, t_2 \le 1\},$$

where $\vec{u}, \vec{v} \in \mathbb{Z}^2$ are two visible lattice points that are linearly independent. (Following [3] we call a lattice polygon a *clean* lattice polygon if the only lattice points on its boundary are the vertices.) Our initial goal was to determine the asymptotic estimates that arose in counting the number of visible points lying inside P. We elaborate on this goal in Sect. 2.1.

We will use $P_{a,n}$ to denote the clean parallelogram with vertices $(0, 0)$, $(1, 0)$, (a, n) and $(a + 1, n)$, where $1 \le a < n$ and $\gcd(a, n) = 1$. For a lattice parallelogram, P (with one vertex at the origin) let

$$V(P) = \#\text{ of visible lattice points in the interior of } P.$$

In the special case when $P = P_{a,n}$, we simply write $V(a, n)$, instead of $V\left(P_{a,n}\right)$.

2 A Reduction Result

We begin by showing that questions about visible lattice points inside a clean lattice parallelogram (with one vertex at the origin) can be reduced to studying the visible lattice points inside the appropriate $P_{a,n}$. We also describe the minimum and maximum value of $V(P)$. We prove our result via unimodular transformations. These are linear transformations $T : \mathbb{R}^n \to \mathbb{R}^n$ with the key property that they preserve lattice points, that is, $T(\mathbb{Z}^n) = \mathbb{Z}^n$. Furthermore, unimodular transformations preserve visibility.

Definition 1 A $n \times n$ matrix M is said to be unimodular if $M \in \mathbb{M}_n(\mathbb{Z})$ with $\det(M) = \pm 1$. A linear transformation $T : \mathbb{R}^n \to \mathbb{R}^n$ is unimodular if it can be represented by a unimodular matrix.

Theorem 1 *Let P be the clean lattice parallelogram spanned by the lattice points $\vec{u}, \vec{v}$ with area$(P) = n$. Then there is an unimodular transformation $T : \mathbb{R}^2 \to \mathbb{R}^2$ such that*

$$T(P) = P_{a,n} \text{ with } 1 \le a < n \text{ and } \gcd(a, n) = 1.$$

Furthermore,

$$V(P) = 1 \Longleftrightarrow a = n - 1$$

and

$$V(P) = n - 1 \Longleftrightarrow a = 1.$$

Geometrically, this means that

(i) All but one of the lattice points in the interior of P are not visible if and only if all of these $n - 1$ lattice points lie on the diagonal $\vec{u} + \vec{v}$.
(ii) All the lattice points in the interior of P are visible if and only if all of these lattice points lie on the diagonal

$$t\vec{u} + (1 - t)\vec{v}, \ 0 \le t \le 1.$$

Proof We begin by observing that there are precisely $(n - 1)$ lattice points in the interior of P. This follows by combining Pick's theorem with the hypotheses that area$(P) = n$ and that the only lattice points on the boundary of P are the vertices.

Without loss of generality we may assume that the pair of lattice points

$$\vec{u} = (u_1, u_2), \ \vec{v} = (v_1, v_2)$$

are positively oriented, that is $\det(\vec{u}, \vec{v}) > 0$. Since $\gcd(u_1, u_2) = 1$, there exist $m_1, m_2 \in \mathbb{Z}$ such that

$$m_1 u_1 + m_2 u_2 = 1.$$

We now consider the unimodular matrix

$$\begin{pmatrix} m_1 & m_2 \\ -u_2 & u_1 \end{pmatrix}.$$

This unimodular matrix maps P to the parallelogram

$$P' = \{t_1(1,0) + t_2(m_1 v_1 + m_2 v_2, n) \ : \ 0 \le t_1, t_2 \le 1\}.$$

Since (v_1, v_2) is visible, the same holds for $(m_1 v_1 + m_2 v_2, n)$. If

$$0 < (m_1 v_1 + m_2 v_2) < n, \ \text{ then } a = m_1 v_1 + m_2 v_2.$$

If $(m_1 v_1 + m_2 v_2)$ does not satisfy this inequality, then we find $k \in \mathbb{Z}$ such that

$$0 \le (m_1 v_1 + m_2 v_2 + kn) < n$$

and act on the parallelogram P' by the unimodular matrix

$$\begin{pmatrix} 1 & k \\ 0 & 1 \end{pmatrix}$$

to obtain $P_{a,n}$.

We now consider the special cases $a = 1$ and $a = n - 1$. All of the lattice points inside $P_{1,n}$ lie on the diagonal $y + x = n$. All of these lattice points are visible. Turning to $a = n - 1$ we note that all of the lattice points inside $P_{n-1,n}$ lie on the diagonal $y = x$. Consequently the only visible point is $(1, 1)$.

To complete the proof of the theorem we need to show the following two things. For $2 \le a \le n - 2$, $P_{a,n}$ contains: (a) at least one non-visible point; (b) at least two visible points.

The proof of item (a) is as follows. Since $2 \le a < n$ and $\gcd(a, n) = 1$, there is a positive integer l such that $n/2 < la < n$. The lattice point

$$\frac{2(n - la)}{n}(1,0) + \frac{2l}{n}(a, n) = (2, 2l)$$

is a non-visible lattice point in the interior of $P_{a,n}$.

We turn to the proof of item (b). Since $\gcd(a + 1, n) \le n/2$, there is a lattice point $\vec{s}$ inside $P_{a,n}$ which does not lie on the diagonal $y = x$. We now consider the group $G = (\mathbb{Z} \oplus \mathbb{Z})/(\mathbb{Z}e_1 \oplus \mathbb{Z}(a, n))$. Let S denote the coset of G that is represented by $\vec{s}$. The lattice point representatives for the cosets S and $-S$ lie on half-rays that are distinct from each other. Each of these half-rays contains a visible lattice point inside $P_{a,n}$. □

2.1 *Motivation for Studying This Problem*

Having proved Theorem 1, we are in a position to expand on the motivation underlying this paper. Unlike the square $[-r, r]^2$, the parallelograms $P_{a,n}$ are long and thin. Thus, it seems plausible to infer that most values of $V(a, n)/n$ should be quite different from $6/\pi^2$. What we will see is that this inference is false much of the time. That is, for most values of a, $V(a, n)/n$ is close to $6/\pi^2$. However, when a is close to 1 or close to n the numerics for $V(a, n)/n$ differ significantly from $6/\pi^2$. The reader may want to jump ahead to Sect. 5 titled *Some Numerics* to see some graphs of $V(a, n)/n$.

3 $V(a, n) = V(a^{-1} \mod n, n)$

Proposition 1 *Let $b = a^{-1} \mod n$. Then the parallelograms $P_{a,n}$ and $P_{b,n}$ are unimodularly equivalent. Consequently, $V(a, n) = V(b, n)$.*

Proof There exists $k \in \mathbb{Z}$ such that $ab + kn = 1$. The unimodular transformation $L : \mathbb{Z}^2 \to \mathbb{Z}^2$, via,

$$L((1, 0)) = (b, n) \text{ and } L((0, 1)) = (k, -a)$$

maps $P_{a,n}$ to $P_{b,n}$. □

The above proposition shows that different values of a that are far apart can take on the same V-value. Let us elaborate. Consider the following set in $[0, 1]^2$:

$$S = \left\{ \left(\frac{a}{n}, \frac{b}{n} \right) \; : \; a, b, n \in \mathbb{Z}^+, 1 \le a, b, < n, ab \equiv 1 \pmod{n} \right\}.$$

A standard argument via Kloosterman sums shows that the points in S are uniformly distributed in the unit square (see [2]). A consequence of this is that an element and its multiplicative inverse modulo n can be very far apart. These two elements of $\mathbb{Z}_n^*$ take on the same V-value.

4 Counting Visible Points in $P_{a,n}$

Theorem 2 *The lattice points in the interior of $P_{a,n}$ are*

$$\left\langle \frac{k(n-a)}{n} \right\rangle (1, 0) + \frac{k}{n}(a, n) = \left(\left\lceil \frac{ka}{n} \right\rceil, k \right), \tag{2}$$

with $k = 1, \ldots, n-1$. We note that

$$\left(\left\lceil \frac{ka}{n} \right\rceil, k\right) = \left(\frac{ka}{n} - \left\langle \frac{ka}{n} \right\rangle + 1, k\right).$$

From Theorem 2 we get that the lattice points inside the parallelogram $P_{n-a,n}$ are of the form $(\lceil k(n-a)/n \rceil, k)$ for $k = 1, \ldots, n-1$. Consequently $(\lceil k(n-a)/n \rceil, k)$ is visible if and only if

$$\gcd(\lceil k(n-a)/n \rceil, k) = 1.$$

At this juncture we interject a pretty result that allows us to view the quantity $\gcd(\lceil k(n-a)/n \rceil, k)$ in a couple of different ways.

Proposition 2 *Let a, n be positive integers with $a < n$ and $\gcd(a, n) = 1$. For $k = 1, \ldots, n-1$,*

$$\gcd\left(\left\lceil \frac{k(n-a)}{n} \right\rceil, k\right) = \gcd\left(\left\lceil \frac{k(n-a)}{n} \right\rceil, \left\lfloor \frac{ka}{n} \right\rfloor\right) = \gcd\left(\left\lfloor \frac{ka}{n} \right\rfloor, k\right). \quad (3)$$

Furthermore if n is a prime p, then

$$\gcd\left(\left\lceil \frac{k(p-a)}{p} \right\rceil, k\right) = \gcd(ka \bmod p, k). \quad (4)$$

Proof We recall the identity $1 - \langle x \rangle = \langle -x \rangle$ when $x \notin \mathbb{Z}$. On setting

$$x = \frac{k(n-a)}{n}$$

we get

$$\left\lceil \frac{k(n-a)}{n} \right\rceil + \left\lfloor \frac{ka}{n} \right\rfloor = \frac{k(n-a)}{n} - \left\langle \frac{k(n-a)}{n} \right\rangle + 1 + \frac{ka}{n} - \left\langle \frac{ka}{n} \right\rangle = k.$$

Consequently,

$$\gcd\left(\left\lceil \frac{k(n-a)}{n} \right\rceil, \left\lfloor \frac{ka}{n} \right\rfloor\right) = \gcd\left(\left\lceil \frac{k(n-a)}{n} \right\rceil, k\right) = \gcd\left(\left\lfloor \frac{ka}{n} \right\rfloor, k\right).$$

We now consider when $n = p$ is prime. Let $r = ka \mod p$. Then

$$ka = p\left\lfloor \frac{ka}{p} \right\rfloor + r,$$

and we infer that

$$\gcd\left(p\left\lfloor ka/p \right\rfloor, k\right) \mid \gcd(r, k) \text{ and } \gcd(r, k) \mid \gcd(p\left\lfloor ka/p \right\rfloor, k).$$

Consequently, $\gcd(r, k) = \gcd(p\lfloor ka/p\rfloor, k)$. We conclude by observing that since $\gcd(p, k) = 1$, $\gcd(p\lfloor ka/p\rfloor, k) = \gcd(\lfloor ka/p\rfloor, k)$. □

We now pose a general question.

Question 1 We have proved that $V(1, n) = n - 1$ and $V(n - 1, n) = 1$. We pose the *vague* question of what can we say about $V(a, n)$ for $a \neq 1, n - 1$?

The next result gives specific examples of a where $V(a, n)/n$ is "far" away from 0.6.

Theorem 3 *Let n be odd. Then we have the following:*

$$V(2, n) = V((n + 1)/2, n) = \begin{cases} (3n - 3)/4, \ n \equiv 1 \pmod 4 \\ (3n - 5)/4, \ n \equiv 3 \pmod 4. \end{cases} \tag{5}$$

$$V((n - 1)/2, n) = V(n - 2, n) = (n + 1)/2. \tag{6}$$

Proof Since $(n + 1)/2 = 2^{-1} \mod n$, we have $V((n + 1)/2, n) = V(2, n)$ by Proposition 1. We now invoke (2). The lattice points in the interior of $P_{2,n}$ are

$$\{(\lceil 2k/n\rceil, k) \ : \ k = 1, \ldots, n - 1\}$$

$$= \{(1, 1), (1, 2), \ldots, (1, (n - 1)/2), (2, (n + 1)/2), \ldots, (2, n - 1)\}.$$

Thus the only non-visible lattice points inside $P_{2,n}$ are of the form $(2, k)$ with k even and $n/2 < k \leq (n - 1)$. A simple counting argument then gives us (5).

Since $n - 2 = ((n - 1)/2)^{-1} \mod n$, we have $V((n - 1)/2, n) = V(n - 2, n)$ by Proposition 1. By (2) the lattice points in the interior of $P_{(n-1)/2,n}$ are of the form

$$\left\langle \frac{k(n + 1)}{2n} \right\rangle (1, 0) + \frac{k}{n}\left(\frac{n - 1}{2}, n\right) = \begin{cases} (k/2, k), & k \text{ even} \\ ((k + 1)/2, k), & k \text{ odd}, \end{cases} \tag{7}$$

for $k = 1, 2, \ldots, n - 1$. Now, if k is odd, then $(\gcd((k + 1)/2, k) = 1$; and if k is even, then $\gcd(k/2, k) = 1$ if and only if $k = 2$. By combining these observations we conclude that $V((n - 1)/2, n) = (n + 1)/2$. □

We remark that we can extend the above result for any fixed value of a. Our next result gives an expression for $V(a, n)/n$ in terms of the Euler phi function and the Mobius function.

Theorem 4 *Let* $\gcd(a, n) = 1$. *Then*

$$\frac{V(a, n)}{n} = \frac{1}{a}\sum_{s=1}^{a} \frac{\varphi(s)}{s} - \frac{1}{n}\sum_{s=1}^{a}\sum_{d\mid s}\left(\left\langle\frac{(s - 1)n}{ad}\right\rangle - \left\langle\frac{sn}{ad}\right\rangle\right)\mu(d) - \frac{2}{n}, \tag{8}$$

where φ is the Euler phi function and μ is the Mobius function.

Proof We begin by writing $V(a, n)$ as a sum of a terms and then apply an asymptotic formula for each of these terms. For $s = 1, 2, 3, \ldots, a$, let l_s denote the vertical line (in $\mathbb{R}^2$) $x = s$. We have that

$$P_{a,n} \cap \mathbb{Z}^2 = \bigcup_{s=1}^{a} \left(P_{a,n} \cap l_s \cap \mathbb{Z}^2 \right),$$

and consequently

$$V(a, n) + 2 = \sum_{s=1}^{a} V(s, a, n), \tag{9}$$

where

$$V(s, a, n) = \#\text{ of visible lattice points in the intersection } (P_{a,n} \cap l_s).$$

The inclusion of 2 in the LHS of (9) arises for the following reason. The term $V(1, a, n)$ counts the lattice point $(1, 0)$ and $V(a, a, n)$ counts the lattice point (a, n). Both these lattice points are boundary lattice points of $P_{a,n}$. Since $V(a, n)$ only counts the visible points in the interior of $P_{a,n}$, we need to make an adjustment.

From (2) we get

$$V(s, a, n) = \#\left\{ k \ : \ \frac{(s-1)n}{a} < k \leq \frac{sn}{a},\ \gcd(s, k) = 1 \right\}.$$

Let $\varphi(x, N)$ denote the Legendre totient function, that is, the number of positive integers $\leq x$ that are relatively prime to N. Then

$$V(s, a, n) = \varphi\left(\frac{sn}{a}, s\right) - \varphi\left(\frac{(s-1)n}{a}, s\right), \tag{10}$$

and consequently

$$V(a, n) + 2 = \sum_{s=1}^{a} V(s, a, n) = \sum_{s=1}^{a} \left(\varphi\left(\frac{sn}{a}, s\right) - \varphi\left(\frac{(s-1)n}{a}, s\right) \right). \tag{11}$$

A standard identity for the Euler phi function is

$$\varphi(m) = \sum_{d|m} \frac{m}{d} \mu(d).$$

(See [1, Theorem 2.3].) It is a simple matter to extend this identity to the following:

$$\varphi(x, N) = \sum_{d|N} \left[\frac{x}{d}\right] \mu(d) = \sum_{d|N} \left(\frac{x}{d} - \left\langle\frac{x}{d}\right\rangle\right) \mu(d).$$

Consequently

$$\varphi\left(\frac{sn}{a}, s\right) - \varphi\left(\frac{(s-1)n}{a}, s\right) = \frac{n}{a} \sum_{d|s} \frac{\mu(d)}{d} + \sum_{d|s} \left(\left\langle\frac{(s-1)n}{ad}\right\rangle - \left\langle\frac{sn}{ad}\right\rangle\right) \mu(d).$$

Thus,

$$V(s, a, n) = \frac{n}{a} \cdot \frac{\varphi(s)}{s} - \sum_{d|s} \left(\left\langle\frac{(s-1)n}{ad}\right\rangle - \left\langle\frac{sn}{ad}\right\rangle\right) \mu(d). \tag{12}$$

We obtain (8) by substituting (12) into (11). □

Corollary 1 *If* $a = O(n^{1-\epsilon})$ *for some* $0 < \epsilon < 1$, *then*

$$\frac{V(a, n)}{n} = \frac{1}{a} \sum_{s=1}^{a} \frac{\varphi(s)}{s} + o(1). \tag{13}$$

Proof The secondary term in (8) is

$$\frac{1}{n} \sum_{s=1}^{a} \sum_{d|s} \left(\left\langle\frac{(s-1)n}{ad}\right\rangle - \left\langle\frac{sn}{ad}\right\rangle\right) \mu(d) - \frac{2}{n}.$$

We obtain our asymptotic by showing that the double sum is bounded above by $a \log(a)$. Now,

$$\left|\sum_{s=1}^{a} \sum_{d|s} \left(\left\langle\frac{(s-1)n}{ad}\right\rangle - \left\langle\frac{sn}{ad}\right\rangle\right) \mu(d)\right| \le \sum_{s=1}^{a} \sum_{d|s} |\mu(d)| = \sum_{s=1}^{a} \sum_{d|s, \mu(d)\neq 0} 1.$$

Now,

$$\sum_{s=1}^{a} \sum_{d|s, \mu(d)\neq 0} 1 = \sum_{d \le a, \mu(d)\neq 0} \sum_{t \le a/d} 1 = \sum_{d \le a, \mu(d)\neq 0} \left\lfloor \frac{a}{d} \right\rfloor = O(a \log(a)) = o(n).$$

□

We now raise the question of the behavior of $V(a, n)/n$ when n and a are roughly of the same size. We have made very little progress on this. We only have the following simple result.

Proposition 3 *Let* $a, n \in \mathbb{Z}^+$ *with* $1 < a < n$ *and* $\gcd(a, n) = 1$. *We make the further assumptions that both* a *and* $\lceil n/a \rceil$ *are even. Then*

$$V(a,n) \leq 0.75a \cdot \left\lceil \frac{n}{a} \right\rceil.$$

Proof For $s = 1, \ldots a$ the length of the vertical line segment $P_{a,n} \cap l_s$ equals n/a. Since $n/a \notin \mathbb{Z}$ we can infer that for $s = 1, \ldots, a$,

$$\#\left(P_{a,n} \cap l_s \cap \mathbb{Z}^2\right) = \left\lceil \frac{n}{a} \right\rceil.$$

For $s = 1, \ldots, a$, we have the trivial inequalities

$$V(s,a,n) \leq \left\lceil \frac{n}{a} \right\rceil, \ s \text{ odd}; \ V(s,a,n) \leq \frac{1}{2}\left\lceil \frac{n}{a} \right\rceil, \ s \text{ even}.$$

For the second inequality we are simply using the fact that the y-coordinate of every other lattice point on a vertical line $x = s$ is even. Using these inequalities we obtain

$$V(a,n) \leq \sum_{s=1}^{a} V(s,a,n) \leq \frac{a}{2} \cdot \frac{3}{2}\left\lceil \frac{n}{a} \right\rceil = 0.75a \cdot \left\lceil \frac{n}{a} \right\rceil.$$

□

5 Some Numerics

We have some numerical results for $V(a,n)/n$. For $n \in \mathbb{Z}^+$, let

$$f_n : \{a \ : \ a = 1, \ldots, n-1, \gcd(a,n) = 1\} \to [0,1] \text{ via } f_n(a) = V(a,n)/n.$$

The following graphs are the graphs of f_{499}, f_{500}, f_{1000} and f_{9973}. (499 and 9973 are primes.)

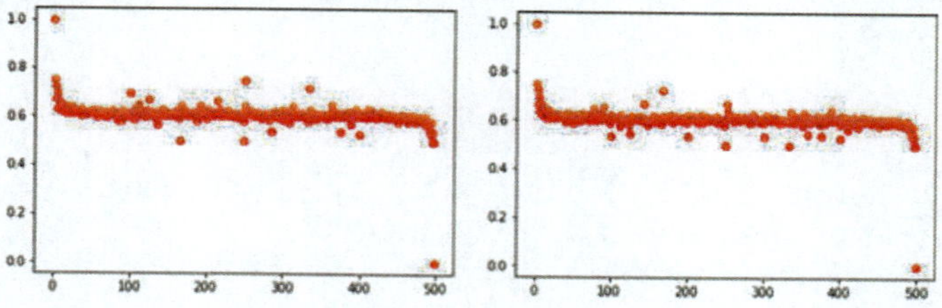

The graphs of f_{499} and f_{500}.

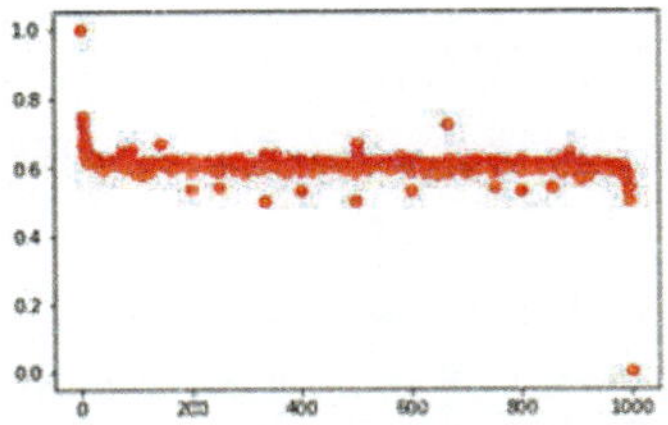

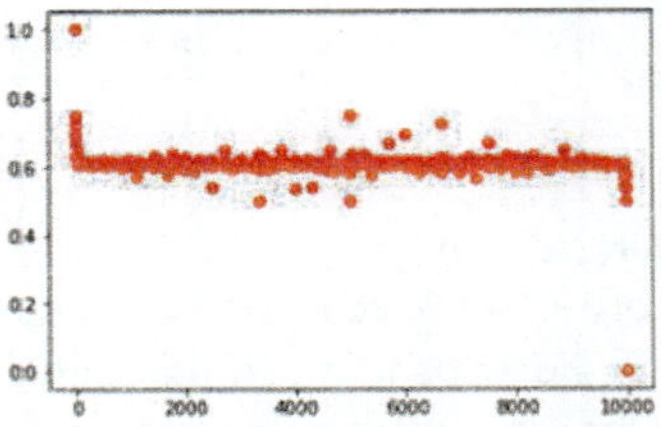

The graphs of f_{1000} and f_{9973}.

Proposition 3 supports the heuristic that one-fourth of the lattice points in the interior of $P_{a,n}$ have both even x and y co-ordinates. Thus the upper bound for $V(a,n)$ should be at most $0.75n$. Unfortunately, we do not have a heuristic for a lower bound. Our computational work suggests the following conjecture.

Conjecture 1 For $a \neq 1, n-1$,

$$\frac{1}{2} < \frac{V(a,n)}{n} < \frac{3}{4}.$$

6 Further Speculative Remarks

The values of f_n oscillate, but only slightly. However, if one decides to view the graph of f_n from a distance, then it can be described as consisting of 3 parts. Initially it resembles a sharply decreasing function for $a \lessapprox \sqrt{n}$. The $\sqrt{n}$ follows by an application of the well-known *eyeball* metric. It then levels off around $6/\pi^2$ and is "constant" for most a-values. Then when $a \gtrapprox n - \sqrt{n}$ it again decreases sharply. (Here we once again invoke the eyeball metric.) There is some ambiguity among the authors about what the eyeball metric tells us. At least, one of us believes that the $\sqrt{n}$ term should be replaced by $\log(n)$.

In the middle of the graph we see a few outliers where f_n is noticeably bigger or smaller than $6/\pi^2$. At present, we have identified two factors influencing these effects. However, our attempts to make this intuition rigorous have not matured and will be continued in future work.

Firstly, Proposition 1 shows that

$$V(a,n) = V(a^{-1} \mod n, n).$$

As such, a-values near the endpoints of $[1,n]$ whose modular inverses are in the middle of $[1,n]$ partially explain this outlying phenomena.

Alternatively, we can try to explain this phenomena by trying to get more precise estimates on the terms in Theorem 4. The graphs of f_n indicate that we have not obtained sharp estimates for the error analysis in Corollary 1. Using straightforward estimates, we have the asymptotic

$$\sum_{n\leq x}\frac{\varphi(n)}{n}\approx\frac{x}{\zeta(2)}+O(\log(x)).$$

(See [1, exercise 5, p. 70].)

With more careful analysis we believe that it is possible to get more refined estimates on the error term. In particular, whenever a is sufficiently large and a/n is not too close to a rational number p/q with $p\ll n$, some preliminary heuristics suggest that the error term

$$\frac{1}{n}\sum_{s=1}^{a}\sum_{d|s}\left(\left\langle\frac{(s-1)n}{ad}\right\rangle-\left\langle\frac{sn}{ad}\right\rangle\right)\mu(d)\ll 1.$$

More precisely, we assume that the set

$$\left\{\left\langle\frac{kn}{a}\right\rangle \mid 1\leq k\leq Q\right\}$$

is sufficiently "equidistributed" on [0, 1] for $1\ll Q\ll a$. Under this assumption, it is possible to use probabilistic estimates to show that the above expression is $O(n^{-1/2})$. Since our results currently rely on this equidistributional assumption, we postpone a more complete discussion for future work. In any case, we are led to the following (possibly overly optimistic) conjecture.

Conjecture 2 Let $0<\rho<1$ be an irrational number with irrationality measure 2. Then,

$$\lim_{n\to\infty}\frac{V(\lceil\rho n\rceil,n)}{n}=\frac{6}{\pi^2}.$$

In this conjecture, the irrationality measure serves as an attempt to capture the sufficient genericity condition between a and n, which quantitatively reflects that a/n is not too close to a sequence of rational numbers with small denominator. Intuitively speaking, this conjecture seeks to formalize the observation that for "most" pairs of a and n, we expect that

$$\frac{V(a,n)}{n}\approx\frac{6}{\pi^2}.$$

Another possible way to make this observation rigorous is to consider the function $f_n(a)$ and try to show that it converges (after renormalization) to $6/\pi^2$ as n goes to infinity. We know that it does not converge pointwise, but it may converge in L^p or some other norm.

Acknowledgements We received some interesting feedback from Kevin Ford and Igor Shparlinski. In particular Igor told us how to obtain a stronger version of Corollary 1.

The first author was partially supported by DARPA/ARO Grant W911NF-16-1-0383 (PI: Jun Zhang).

References

1. T. Apostol, *Introduction to Analytic Number Theory*, Springer-Verlag, 1976.
2. J. Beck and M. R. Khan, On the uniform distribution of inverses modulo *n*, *Periodica Mathematica Hungarica* **44** (2002), no. 2, 147–155.
3. B. Reznick, Clean lattice tetrahedra, arxiv.org/abs/math/0606227.

On a Two-Dimensional Exponential Sum

Angel V. Kumchev

Abstract In this paper we study the two-dimensional analog of the classical quadratic Weyl sum. We establish a major arc approximation with a strong error term akin to celebrated results of R.C. Vaughan from the early 1980s and the late 2000s. Our main result has applications to the study of discrete maximal operators related to triangular configurations.

Keywords Exponential sums · Weyl sums · Major arc approximations · Minor arc bounds · Applications of the circle method

1 Introduction

The classical Weyl sum

$$f_k(\boldsymbol{\alpha}) = \sum_{1 \le x \le N} e(\alpha_k x^k + \cdots + \alpha_1 x), \tag{1}$$

has been the subject of intense investigations for more than a century, going back to Weyl's pioneering work [8] on uniform distribution. Here, N is a large parameter, $\boldsymbol{\alpha} \in \mathbb{R}^k$, $k \ge 2$, and $e(x) = e^{2\pi i x}$. In general, there are two regimes to the behavior of such sums. When all the coefficients $\alpha_2, \dots, \alpha_k$ are close to rationals with small denominators, one can establish an asymptotic formula for $f_k(\boldsymbol{\alpha})$; otherwise, one has a non-trivial upper bound on $f_k(\boldsymbol{\alpha})$. Borrowing terminology from the Hardy–Littlewood circle method, we will refer to these two regimes as $\boldsymbol{\alpha}$ being on a major arc or on a minor arc, respectively.

When $\alpha_1 = \cdots = \alpha_{k-1} = 0$, Vaughan [5] obtained a major arc approximation that was strong enough to give a unified treatment of the quadratic and cubic Weyl sums across both regimes. More recently, Vaughan [7] extended that work to the

A. V. Kumchev (✉)
Department of Mathematics, Towson University, 8000, York Road, Towson, MD 21252, USA
e-mail: akumchev@towson.edu

M. B. Nathanson (ed.), *Combinatorial and Additive Number Theory IV*,
Springer Proceedings in Mathematics & Statistics 347,
https://doi.org/10.1007/978-3-030-67996-5_20

full quadratic Weyl sum $f_2(\boldsymbol{\alpha})$, while Brüdern and Robert [4] and Brandes, Parsell, Poulias, Shakan, and Vaughan [3] obtained variants in the case $\alpha_2 = \cdots = \alpha_{k-1} = 0$, $k \geq 3$.

In this paper, we are interested in the two-dimensional quadratic variant of (1),

$$F(\boldsymbol{\alpha}; \boldsymbol{\theta}) = \sum_{1\leq x\leq N} \sum_{1\leq y\leq N} e(\alpha_1 x^2 + 2\alpha_2 xy + \alpha_3 y^2 + \theta_1 x + \theta_2 y).$$

For this sum, $\boldsymbol{\alpha}$ is on a major arc if there exist integers a_1, a_2, a_3 and q_1, q_2, q_3 such that

$$|q_i\alpha_i - a_i| \leq QN^{-2}, \quad 1 \leq q_i \leq Q, \quad (a_i, q_i) = 1, \tag{2}$$

for some parameter $Q \leq N$. When $\boldsymbol{\alpha}$ is on a minor arc, a simple argument (see Proposition 1 below) yields the bound

$$F(\boldsymbol{\alpha}; \boldsymbol{\theta}) \ll_\varepsilon N^{2+\varepsilon} Q^{-1/2} \tag{3}$$

for any fixed $\varepsilon > 0$. On the other hand, when $\boldsymbol{\alpha}$ is on the major arc defined by conditions (2), a different routine argument (see Proposition 2) leads to an approximation of the form

$$F(\boldsymbol{\alpha}; \boldsymbol{\theta}) = F^*(\boldsymbol{\alpha}; \boldsymbol{\theta}) + O(qN(1 + N^2|\boldsymbol{\beta}|)), \tag{4}$$

where

$$q = \operatorname{lcm}[q_1, q_2, q_3], \quad \beta_i = \alpha_i - \frac{a_i}{q_i}, \tag{5}$$

and $F^*(\boldsymbol{\alpha}; \boldsymbol{\theta})$ is a local approximant that satisfies the inequality

$$F^*(\boldsymbol{\alpha}; \boldsymbol{\theta}) \ll \frac{N^2 \log N}{(q + qN^2|\boldsymbol{\beta}|)^{1/2}}. \tag{6}$$

The exponential sum $F(\boldsymbol{\alpha}; \boldsymbol{\theta})$ appears in a recent work [1] by Anderson, Palsson and the author on discrete maximal functions. Motivated by the applications in [1], we will be concerned with the error term in (4), which can be as large as $O(NQ^3)$ under conditions (2). As one wants that error term to be smaller than the trivial bound for $F(\boldsymbol{\alpha}; \boldsymbol{\theta})$, this restricts the choice of Q to $Q \leq N^{1/3}$. Indeed, this error term is weaker than the minor arc bound (3) when $Q \geq N^{2/7}$. Our main result establishes an approximation of the form (4) with an error term of size $O(NQ^{1/2} \log N)$, provided that $Q \leq \eta N^{1/2}$ for a sufficiently small absolute constant $\eta > 0$.

Given $q \in \mathbb{N}$, $\mathbf{a} \in \mathbb{Z}^3$, $\mathbf{m} \in \mathbb{Z}^2$, $\boldsymbol{\beta} \in \mathbb{R}^3$, and $\boldsymbol{\theta} \in \mathbb{R}^2$, we define

$$S(q;\mathbf{a},\mathbf{m}) = \frac{1}{q^2}\sum_{x=1}^{q}\sum_{y=1}^{q} e_q\big(a_1x^2 + 2a_2xy + a_3y^2 + m_1x + m_2y\big),$$

$$I(N;\boldsymbol{\beta},\boldsymbol{\theta}) = \int_0^N\int_0^N e\big(\beta_1x^2 + 2\beta_2xy + \beta_3y^2 + \theta_1x + \theta_2y\big)\,dxdy,$$

where $e_q(x) = e(x/q)$. Also, for vectors $\mathbf{a},\mathbf{q} \in \mathbb{Z}^3$ with $(a_i, q_i) = 1, i = 1, 2, 3$, we write

$$b_i = \frac{a_iq}{q_i}, \quad Q_1 = \mathrm{lcm}[q_1, q_2], \quad Q_2 = \mathrm{lcm}[q_2, q_3], \tag{7}$$

where $q = \mathrm{lcm}[q_1, q_2, q_3]$, as in (5). With this notation, we can state our theorem as follows.

Theorem 1 *Let* $1 \le Q \le 0.1N^{1/2}$, $\boldsymbol{\alpha} \in \mathbb{R}^3$, *and* $\boldsymbol{\theta} \in \mathbb{R}^2$. *Suppose that* $\mathbf{a},\mathbf{q} \in \mathbb{Z}^3$ *satisfy* (2) *and that* $\mathbf{m} \in \mathbb{Z}^2$ *is such that*

$$\left|\theta_i - \frac{m_i}{Q_i}\right| \le \frac{1}{2Q_i}.$$

Then

$$F(\boldsymbol{\alpha};\boldsymbol{\theta}) = H(\mathbf{q};\mathbf{a},\mathbf{m})I(N;\boldsymbol{\beta};\boldsymbol{\delta}) + O\big(NQ^{1/2}\log N\big), \tag{8}$$

where q *and* $\boldsymbol{\beta}$ *are defined by* (5), $\delta_i = \theta_i - m_i/Q_i$, *and*

$$H(\mathbf{q};\mathbf{a},\mathbf{m}) = \frac{1}{q^2}\sum_{h=1}^{q}\sum_{k=1}^{q} e_q(2b_2q_1q_3hk + \tilde{m}_1q_1h + \tilde{m}_2q_3k)S(q;\mathbf{b},\mathbf{n}_{h,k}),$$

with $\tilde{m}_i = m_iq/Q_i$ *and* $\mathbf{n}_{h,k} = (\tilde{m}_1 + 2b_2q_3k, \tilde{m}_2 + 2b_2q_1h)$.

The main term in (8) is not exactly what one may expect—that would be the main term in (11) below (see also Proposition 2). While this is somewhat disappointing, it is not unprecedented—in [3], the authors of that paper also obtain a major arc approximation to $f_k(\alpha_1, \alpha_k)$ that attains a sharp error term at the expense of including additional main terms (albeit of a slightly different nature). The main strength of our theorem is that the main term satisfies (6), while the error term is much smaller than that in (4). This allows us to combine the above result and the basic bounds in Propositions 1 and 2 below to prove the following theorem.

Theorem 2 *Let* $1 \le Q \le N^{1/2}$, $\boldsymbol{\alpha} \in \mathbb{R}^3$, *and* $\boldsymbol{\theta} \in \mathbb{R}^2$. *Then either*

$$F(\boldsymbol{\alpha};\boldsymbol{\theta}) \ll N^2Q^{-1/2}\log N, \tag{9}$$

or there exist $q \in \mathbb{N}$ *and* $\mathbf{a} \in \mathbb{Z}^3$ *such that*

$$|q\alpha_i - a_i| \le QN^{-2}, \quad 1 \le q \le Q, \quad (q, a_1, a_2, a_3) = 1, \tag{10}$$

and

$$F(\boldsymbol{\alpha};\boldsymbol{\theta}) = S(q;\mathbf{a},\mathbf{m})I(N;\boldsymbol{\beta};\boldsymbol{\delta}) + O(NQ), \tag{11}$$

where $\boldsymbol{\beta}$ *is defined by* (5) *and* $\mathbf{m} \in \mathbb{Z}^2$ *is chosen so that* $\boldsymbol{\delta} = \boldsymbol{\theta} - q^{-1}\mathbf{m}$ *satisfies* $|\boldsymbol{\delta}| \le (2q)^{-1}$.

For $r \in \mathbb{N}$, define

$$\mathcal{V}_r = \left\{(\mathbf{m},\mathbf{n}) \in \mathbb{Z}^{2d} : |\mathbf{m}|^2 = |\mathbf{n}|^2 = |\mathbf{m}-\mathbf{n}|^2 = r\right\},$$

where $|\cdot|$ denotes the Euclidean norm on $\mathbb{R}^d$. In [1], a special case of Theorem 2 (see Lemma 7 in [1]) was used to study the properties of the maximal operator

$$T^*(f,g)(\mathbf{x}) = \sup_{r\in\mathbb{N}} r^{3-d} \sum_{(\mathbf{m},\mathbf{n})\in\mathcal{V}_r} |f(\mathbf{x}-\mathbf{m})g(\mathbf{x}-\mathbf{n})|,$$

where $f \in \ell^p(\mathbb{Z}^d)$ and $g \in \ell^q(\mathbb{Z}^d)$. In particular, the main result of that paper establishes that T^* is bounded from $\ell^p(\mathbb{Z}^d) \times \ell^\infty(\mathbb{Z}^d)$ to $\ell^p(\mathbb{Z}^d)$ whenever $d \ge 9$ and $p > \max\left(\frac{32}{d+8}, \frac{d+4}{d-2}\right)$. The lower bound for p in this theorem depends on the saving over the trivial bound in the minor arc bound (3). Theorem 2 with $Q = N^{1/2}$ yields the bound

$$F(\boldsymbol{\alpha};\boldsymbol{\theta}) \ll_\varepsilon N^{7/4+\varepsilon} \tag{12}$$

outside the set of major arcs defined by conditions (10). By comparison, if we relied only on the basic approximation (4), we would be restricted to $Q \le N^{2/7}$, leading to a weaker range for p in the above result.

In [1], we proved also an asymptotic formula for the Fourier multiplier of the underlying bilinear operator,

$$\widehat{T}_r(\boldsymbol{\xi},\boldsymbol{\eta}) = r^{3-d} \sum_{(\mathbf{m},\mathbf{n})\in\mathcal{V}_r} e(\boldsymbol{\xi}\cdot\mathbf{m} + \boldsymbol{\eta}\cdot\mathbf{n}).$$

The preliminary version of Theorem 2 that was used in [1] was not sufficiently general to apply in the proof of that result (Theorem 2 in [1]), and so the bound for the error term in the asymptotic formula is not as sharp as one may hope. Using Theorem 2 as stated above, one may improve on that result. We omit the complete statement of the improvement, since the asymptotic formula for $\widehat{T}_r(\boldsymbol{\xi},\boldsymbol{\eta})$ is too technical to include here for the mere purpose of this brief remark. However, for the benefit of the interested reader, we note that one only need to apply (12) above in place of (6.1) in [1].

Remark 1 The condition $Q \le 0.1N^{1/2}$ in the statement of Theorem 1 can be replaced by $Q \le \eta N^{1/2}$ for a slightly larger $\eta \in (0, \frac{1}{4})$. The above choice, while not optimal, is relatively close to the optimal one and also allows us to justify easily several claims of the form

$$q_i|\beta_j| \le Q^2N^{-1} \le 0.01,$$

where $1 \le q_i, q_j \le Q$ and $q_j|\beta_j| \le QN^{-2}$. Such estimates are implicit in several places in the proof of Theorem 1 (for example, see (26)).

2 Preliminaries

To begin, we state and prove the main minor arc bound for $F(\boldsymbol{\alpha}; \boldsymbol{\theta})$. Although its proof is standard, we include it for completeness. Here and through the remainder of the paper, we write $L = \log N$. We also abbreviate $\gcd(a, b, \dots)$ and $\mathrm{lcm}[a, b, \dots]$ as $(a, b, \dots)$ and $[a, b, \dots]$, respectively.

Proposition 1 *Let $1 \le Q \le N$ and suppose that $\boldsymbol{\alpha} \in \mathbb{R}^3$ has no rational approximation of the form (2). Then, for all $\boldsymbol{\theta} \in \mathbb{R}^2$, one has*

$$|F(\boldsymbol{\alpha}; \boldsymbol{\theta})| \ll N^2Q^{-1/2}L. \tag{13}$$

Proof By Dirichlet's theorem on Diophantine approximation, there exist rationals a_i/q_i, $i = 1, 2, 3$, such that

$$|q_i\alpha_i - a_i| \le QN^{-2}, \quad 1 \le q_i \le N^2Q^{-1}, \quad (a_i, q_i) = 1. \tag{14}$$

By our assumption that $\boldsymbol{\alpha}$ does not have a rational approximation satisfying (2), we must have $q_i > Q$ for at least one index i, and by symmetry, we may assume that $i = 1$ or 2.

By Cauchy's inequality,

$$\begin{aligned}
|F(\boldsymbol{\alpha}; \boldsymbol{\theta})|^2 &\le N \sum_{y \le N} \Big| \sum_{x \le N} e\big(\alpha_1x^2 + 2\alpha_2xy + \theta_1x\big)\Big|^2 \\
&\le N \sum_{y \le N} \sum_{|h| \le N} \sum_{x \in I(h)} e\big(\alpha_1h(2x+h) + 2\alpha_2hy + \theta_1h\big) \\
&\le N \sum_{|k| \le 2N} \Big| \sum_{x \in I(k/2)} e(\alpha_1kx)\Big| \cdot \Big| \sum_{y \le N} e(\alpha_2ky)\Big| \\
&\ll N^3 + N \sum_{k \le 4N} \prod_{j=1}^{2} \min\big(N, \|\alpha_jk\|^{-1}\big).
\end{aligned}$$

where $I(h)$ is a subinterval of $[1, N]$ that depends on h and $\|x\|$ denotes the distance from x to the nearest integer. Recalling that (14) holds with $q_i > Q$ for $i = 1$ or 2, we can now apply Lemma 2.2 in Vaughan [6] to deduce that

$$|F(\boldsymbol{\alpha}; \boldsymbol{\theta})|^2 \ll N^4\big(q_i^{-1} + N^{-1} + q_iN^{-2}\big)L,$$

and (13) follows. □

The next proposition establishes the basic major arc approximation of type (4). Note that unless $\boldsymbol{\alpha}$ satisfies conditions (10) for some $Q \le N$, the error term is weaker than the trivial bound for $F(\boldsymbol{\alpha}; \boldsymbol{\theta})$. However, when we pair this result with Proposition 1, we apply this approximation on the major arcs defined by (2). In that context, we have $q = [q_1, q_2, q_3]$ and $\mathbf{a} = \mathbf{b}$, where $b_i = a_i q / q_i$; in particular, the error term is $O(NQ^3)$.

Proposition 2 *Let $q \in \mathbb{N}$, $\mathbf{a} \in \mathbb{Z}^3$, $\mathbf{m} \in \mathbb{Z}^2$, $\boldsymbol{\alpha} \in \mathbb{R}^3$, and $\boldsymbol{\theta} \in \mathbb{R}^2$ satisfy*

$$(q, a_1, a_2, a_3) = 1, \quad |q\theta_i - m_i| \le \tfrac{1}{2}.$$

Then

$$F(\boldsymbol{\alpha}; \boldsymbol{\theta}) = S(q; \mathbf{a}, \mathbf{m}) I(N; \boldsymbol{\beta}; \boldsymbol{\delta}) + O\big(qN(1 + N^2|\boldsymbol{\beta}|)\big), \tag{15}$$

where $\boldsymbol{\beta} = \boldsymbol{\alpha} - q^{-1}\mathbf{a}$, $\boldsymbol{\delta} = \boldsymbol{\theta} - q^{-1}\mathbf{m}$.

Proof The result follows by partial summation from the asymptotic formula

$$\sum_{\substack{X < n \le Y \\ n \equiv a \ (\mathrm{mod}\, q)}} e(\theta n) = \frac{1}{q} \int_X^Y e(\theta x)\, \mathrm{d}x + O(1), \tag{16}$$

where $a, q \in \mathbb{N}$ and $|\theta| \le (2q)^{-1}$.

Let

$$F_{r,s}(\boldsymbol{\alpha}; \boldsymbol{\theta}) = \sum_{\substack{x \le N \\ x \equiv r \ (\mathrm{mod}\, q)}} \sum_{\substack{y \le N \\ y \equiv s \ (\mathrm{mod}\, q)}} e\big(\alpha_1 x^2 + 2\alpha_2 xy + \alpha_3 y^2 + \theta_1 x + \theta_2 y\big).$$

By splitting the terms in $F(\boldsymbol{\alpha}; \boldsymbol{\theta})$ according to their residues modulo q, we get

$$\begin{aligned} F(\boldsymbol{\alpha}; \boldsymbol{\theta}) &= \sum_{r,s=1}^{q} F_{r,s}(\boldsymbol{\alpha}; \boldsymbol{\theta}) \\ &= \sum_{r,s=1}^{q} e_q\big(a_1 r^2 + 2a_2 rs + a_3 s^2 + m_1 r + m_2 s\big) F_{r,s}(\boldsymbol{\beta}; \boldsymbol{\delta}). \end{aligned} \tag{17}$$

We apply partial summation and (16) successively to the sums over x and y to get

$$F_{r,s}(\boldsymbol{\beta}; \boldsymbol{\delta}) = q^{-2} I(N; \boldsymbol{\beta}, \boldsymbol{\delta}) + O\big(q^{-1}N(1 + N^2|\boldsymbol{\beta}|)\big),$$

and the proposition follows from (17). □

The next two lemmas provide basic bounds that we will use to bound the main term in (15). In particular, together they establish that if $(q, a_1, a_2, a_3) = 1$, one has

$$S(q; \mathbf{a}, \mathbf{m}) I(N; \boldsymbol{\beta}, \boldsymbol{\delta}) \ll \frac{N^2 L}{(q + qN^2|\boldsymbol{\beta}|)^{1/2}}, \tag{18}$$

uniformly in $\mathbf{m}$ and $\boldsymbol{\delta}$.

Lemma 1 *Suppose that* $(q, a_1, a_2, a_3) = 1$. *Then*

$$|S(q; \mathbf{a}, \mathbf{m})| \ll q^{-1}(q, a_1 a_3 - a_2^2)^{1/2}.$$

Proof This is Lemma 1 in [1], so we provide only a brief sketch. After squaring out, one can use the orthogonality of the additive characters on $\mathbb{Z}_q$ to show that

$$q^4 |S(q; \mathbf{a}, \mathbf{m})|^2 \le q^2 \nu(q; 2\mathbf{a}), \tag{19}$$

where $\nu(q; \mathbf{a})$ denotes the number of solutions $(h, k) \in \mathbb{Z}_q^2$ of the pair of congruences

$$a_1 h + a_2 k \equiv a_2 h + a_3 k \equiv 0 \pmod{q}.$$

The lemma follows from (19) and the bound

$$\nu(q; \mathbf{a}) \le (q, a_1 a_3 - a_2^2),$$

which can be proved using the basic properties of linear congruences. □

We remark that while in this paper we need only the bound $S(q; \mathbf{a}, \mathbf{m}) \ll q^{-1/2}$, the full strength of Lemma 1 plays an important role in [1]. In that work, we apply the above bound to averages of $S(q; \mathbf{a}, \mathbf{m})$ over $\mathbf{a}$, and are able to make use of the following inequality (see Lemma 2 in [1]): if $s \ge 1$, then

$$\sum_{\substack{1 \le a_1, a_2, a_3 \le q \\ (q, a_1, a_2, a_3) = 1}} (q, a_1 a_3 - a_2^2)^s \le \tau(q)^2 q^{s+2}.$$

Lemma 2 *One has*

$$|I(N; \boldsymbol{\beta}, \boldsymbol{\theta})| \ll N^2 L (1 + N^2 |\boldsymbol{\beta}| + N|\boldsymbol{\theta}|)^{-1/2}.$$

Proof This is a special case of Theorem 1.5 in [2]. □

Lemma 3 *Let* $\boldsymbol{\alpha} \in \mathbb{R}^2$, $q \in \mathbb{N}$, *and* $\mathbf{a} \in \mathbb{Z}^2$ *satisfy*

$$(q, a_2) = 1, \quad |q\alpha_1 - a_1| \le \tfrac{1}{2}.$$

Then

$$f_2(\boldsymbol{\alpha}) = S_1(q; \mathbf{a}) I_1(N; \boldsymbol{\beta}) + O\big((q + qN^2|\beta_2|)^{1/2}\big),$$

where

$$S_1(q; \mathbf{a}) = \frac{1}{q}\sum_{x=1}^{q} e_q(a_2x^2 + a_1x), \quad I_1(N; \boldsymbol{\beta}) = \int_0^N e(\beta_2x^2 + \beta_1x)\,dx,$$

and $\boldsymbol{\beta} = \boldsymbol{\alpha} - q^{-1}\mathbf{a}$.

Proof This Theorem 8 in [7]. □

Recall also the well-known bound for the Gauss sum:

$$S_1(q; \mathbf{a}) \ll q^{-1/2} \tag{20}$$

when $(a_2, q) = 1$, and the first-derivative estimate for trigonometric integrals: if $\phi : [a, b] \to \mathbb{R}$ is monotone and $\inf_{a \le x \le b} |\phi'(x)| \ge \lambda$, then

$$\int_a^b e(\phi(x))\,dx \ll \lambda^{-1}. \tag{21}$$

3 Proof of Theorem 1

The basic idea of the proof is to apply Lemma 3 successively to each of the summations in $F(\boldsymbol{\alpha}; \boldsymbol{\theta})$. We recall (7) and also define

$$q_{ij} = \frac{q_i}{(q_i, q_j)}.$$

for $1 \le i, j \le 3$.

Fix $u \in \mathbb{Z}$ and $y \in [1, N]$ with

$$-\tfrac{1}{2}q_{21} < u \le \tfrac{1}{2}q_{21}, \qquad 2a_2q_{12}y + m_1 \equiv u \pmod{q_{21}}. \tag{22}$$

We note that when $q_{21} = 1$ (i.e., when $Q_1 = q_1$), these conditions reduce to setting $u = 0$, and when $q_{21} = 2$ to selecting $u \in \{0, 1\}$ based on the parity of m_1. In all other cases, conditions (22) specify the residue class of y modulo $q_{21}/(2, q_{21})$. We also choose the unique integer k_y with

$$\frac{-1}{2q_1} \le \gamma_y := 2\alpha_2y + \theta_1 - \frac{k_y}{q_1} < \frac{1}{2q_1}.$$

Lemma 3 with $q = q_1$ then gives

$$f_2(\alpha_1, 2\alpha_2 y + \theta_1) = S_1(q_1; a_1, k_y) I_1(N; \beta_1, \gamma_y) + O(Q^{1/2}). \tag{23}$$

Suppose first that $u \neq 0$. We have

$$\begin{aligned}\inf_{x\in[0,N]} |2\beta_1 x + \gamma_y| &\geq \left| \frac{2a_2 q_{12} y + m_1}{Q_1} - \frac{k_y}{q_1} \right| - |\delta_1| - 2(|\beta_1| + |\beta_2|)N \\ &\geq \left| \frac{2a_2 q_{12} y + m_1 - u}{Q_1} - \frac{k_y}{q_1} + \frac{u}{Q_1} \right| - \frac{0.6}{Q_1}.\end{aligned}$$

By (22), the fraction

$$w = \frac{2a_2 q_{12} y + m_1 - u}{Q_1} - \frac{k_y}{q_1}$$

is an integer multiple of $1/q_1$. So, either $w = 0$ and

$$\inf_{x\in[0,N]} |2\beta_1 x + \gamma_y| \geq \frac{|u| - 0.6}{Q_1} > \frac{|u|}{3Q_1},$$

or $|w| \geq 1/q_1$ and

$$\inf_{x\in[0,N]} |2\beta_1 x + \gamma_y| \geq \frac{1}{q_1} - \frac{|u| + 0.6}{Q_1} > \frac{|u|}{3Q_1}.$$

Thus, by (21),

$$I(N; \beta_1, \gamma_y) \ll Q_1 |u|^{-1},$$

and (20) and (23) give

$$f_2(\alpha_1, 2\alpha_2 y + \theta_1) \ll q_1^{1/2} q_{21} |u|^{-1} + Q^{1/2}. \tag{24}$$

Summing this inequality over all pairs u, y, with $u \neq 0$, that satisfy (22), we find that the total contribution to $F(\boldsymbol{\alpha}; \boldsymbol{\theta})$ from such pairs is

$$\ll \sum_{1 \leq |u| \leq \frac{1}{2} q_{21}} (N/q_{21}) q_1^{1/2} q_{21} |u|^{-1} + N Q^{1/2} \ll N Q^{1/2} L. \tag{25}$$

When $u = 0$, we relate k_y, γ_y to m_1, δ_1. We have

$$\left| \frac{2a_2 y}{q_2} + \frac{m_1}{Q_1} - \frac{k_y}{q_1} \right| \leq |\gamma_y| + |\delta_1| + 2|\beta_2| N < \frac{1}{2q_1} + \frac{0.6}{Q_1}. \tag{26}$$

Since the left side of this inequality is a multiple of $1/q_1$, we find that in this case

$$\frac{k_y}{q_1} = \frac{2a_2 y}{q_2} + \frac{m_1}{Q_1} + \frac{e_y}{q_1}, \quad \gamma_y = \delta_1 + 2\beta_2 y - \frac{e_y}{q_1},$$

with $e_y \in \{0, \pm 1\}$. In particular, when $e_y = 0$, we may rewrite (23) as

$$\begin{aligned} f_2(\alpha_1, 2\alpha_2 y + \theta_1) &= S_1(Q_1; a_1 q_{21}, 2a_2 q_{12} y + m_1) I_1(N; \beta_1, \gamma_y^*) + O(Q^{1/2}), \\ &= S_1(q; b_1, 2b_2 y + \tilde{m}_1) I_1(N; \beta_1, \gamma_y^*) + O(Q^{1/2}), \end{aligned} \tag{27}$$

where $\gamma_y^* = 2\beta_2 y + \delta_1$.

On the other hand, it is clear from (26) that the case $e_y = \pm 1$ may occur only when $Q_1 = q_1$ and $0.4q_1^{-1} \le |\delta_1| \le 0.5q_1^{-1}$. Under these assumptions, we have

$$\inf_{x \in [0,N]} \left| 2\beta_1 x + \gamma_y \right| \ge \frac{1}{q_1} - |\delta_1| - 2(|\beta_1| + |\beta_2|)N > \frac{1}{3q_1}$$

for all $x \in [0, N]$. Thus, by (21),

$$I_1(N; \beta_1, \gamma_y) \ll q_1,$$

and (20) and (23) yield

$$f_2(\alpha_1, 2\alpha_2 y + \theta_1) \ll Q^{1/2}.$$

Moreover, since in this case

$$\inf_{x \in [0,N]} \left| 2\beta_1 x + \gamma_y^* \right| \ge |\delta_1| - 2(|\beta_1| + |\beta_2|)N > \frac{1}{3q_1},$$

we have also

$$I_1(N; \beta_1, \gamma_y^*) \ll q_1.$$

We conclude that (27) holds also when $e_y = \pm 1$.

Having established (27) for all y with $2a_2 q_{12} y \equiv -m_1 \pmod{q_{21}}$ and recalling the bound (25) for the contribution of the remaining $y \in [1, N]$, we obtain

$$F(\boldsymbol{\alpha}; \boldsymbol{\theta}) = \int_0^N G(\alpha_3; \lambda_x) e(\beta_1 x^2 + \delta_1 x)\, dx + O(NQ^{1/2}L), \tag{28}$$

where $\lambda_x = \lambda(0, x)$, $\lambda(r, x) = 2a_2 r/q_2 + 2\beta_2 x + \theta_2$, and

$$\begin{aligned} G(\alpha; \lambda_x) &= \frac{1}{q} \sum_{r=1}^{q} e_q(b_1 r^2 + \tilde{m}_1 r) \sum_{\substack{1 \le y \le N \\ 2a_2 q_{12} y \equiv -m_1 \ (\mathrm{mod}\ q_{21})}} e(\alpha y^2 + \lambda(r, x) y) \\ &= \frac{1}{qq_{21}} \sum_{h=1}^{q_{21}} e_{q_{21}}(m_1 h) \sum_{r=1}^{q} e_q(b_1 r^2 + \tilde{m}_1 r) f_2(\alpha, \lambda(r + q_1 h, x)). \end{aligned} \tag{29}$$

Next, we approximate the sum $G(\alpha_3, \lambda_x)$ in (28). For a fixed $r \in \mathbb{Z}$ and $x \in [0, N]$, we choose $k_{r,x}$ with

$$\frac{-1}{2q_3} \le \gamma_{r,x} := \lambda(r, x) - \frac{k_{r,x}}{q_3} < \frac{1}{2q_3}.$$

We also choose an integer $u = u_r$ with

$$-\tfrac{1}{2}q_{23} < u \le \tfrac{1}{2}q_{23}, \quad 2a_2q_{32}r + m_2 \equiv u \pmod{q_{23}}. \tag{30}$$

Similarly to (23), we have

$$f_2(\alpha_3, \lambda(r, x)) = S_1(q_3; a_3, k_{r,x})I_1(N; \beta_3, \gamma_{r,x}) + O(Q^{1/2}). \tag{31}$$

When $u \ne 0$, we have

$$\begin{aligned}\inf_{y\in[0,N]} |2\beta_3 y + \gamma_{r,x}| &\ge \left|\frac{2a_2q_{32}r + m_2}{Q_2} - \frac{k_{r,x}}{q_3}\right| - |\delta_2| - 2(|\beta_2| + |\beta_3|)N \\ &\ge \left|\frac{2a_2q_{32}r + m_2 - u}{Q_2} - \frac{k_{r,x}}{q_2} + \frac{u}{Q_2}\right| - \frac{0.6}{Q_2} > \frac{|u|}{3Q_2}.\end{aligned}$$

Hence, similarly to (24), we get

$$f_2(\alpha_3, \lambda(r, x)) \ll q_3^{1/2}q_{23}|u|^{-1} + Q^{1/2}.$$

Therefore, the total contribution to the right side of (29) from terms subject to

$$2a_2q_{32}(r + q_1h) \equiv u - m_2 \pmod{q_{23}} \tag{32}$$

with $u \ne 0$, is bounded by

$$\ll \sum_{1\le|u|\le\frac{1}{2}q_{23}} \frac{1}{qq_{21}} \sum_{h=1}^{q_{21}} \sum_{r=1}^{q}{}^{(u)} q_3^{1/2}q_{23}|u|^{-1} + Q^{1/2} \ll Q^{1/2}L,$$

where the notation $\sum^{(u)}$ indicates that r satisfies the congruence (32) for the specified value of u. Thus,

$$\begin{aligned}G(\alpha; \lambda_x) = \frac{1}{qq_{21}} \sum_{h=1}^{q_{21}} \sum_{r=1}^{q}{}^{(0)} & e_q\left(b_1r^2 + \tilde{m}_1(r + q_1h)\right) \\ &\times f_2(\alpha, \lambda(r + q_1h, x)) + O(Q^{1/2}L).\end{aligned} \tag{33}$$

On the other hand, we have

$$\left|\frac{2a_2q_{32}r+m_2}{Q_2}-\frac{k_{r,x}}{q_3}\right| \le |\gamma_{r,x}|+|\delta_2|+2|\beta_2|N < \frac{1}{2q_3}+\frac{0.6}{Q_2},$$

so when r satisfies (30) with $u=0$ we get

$$\frac{k_{r,x}}{q_3}=\frac{2a_2r}{q_2}+\frac{m_2}{Q_2}+\frac{e_r}{q_3}, \quad \gamma_{r,x}=\delta_2+2\beta_2x-\frac{e_r}{q_3},$$

with $e_r\in\{0,\pm1\}$. When $e_r=0$, (31) becomes

$$f_2(\alpha_3,\lambda(r,x))=S_1(q;b_3,2b_2r+\tilde{m}_2)I_1(N;\beta_3,2\beta_2x+\delta_2)+O(Q^{1/2}). \tag{34}$$

As before, the alternative case $e_r=\pm1$ may occur only if $Q_2=q_3$ and $0.4q_3^{-1}\le|\delta_2|\le 0.5q_3^{-1}$. Again, under these conditions, we retain (34) because both sides are $O(Q^{1/2})$. Combining (33) and (34) (with $r+q_1h$ in place of r), we deduce that

$$G(\alpha_3,\lambda_x)=T(\mathbf{q};\mathbf{a},\mathbf{m})I_1(N;\beta_3,2\beta_2x+\delta_2)+O(Q^{1/2}L), \tag{35}$$

where

$$\begin{aligned} T(\mathbf{q};\mathbf{a},\mathbf{m}) &= \frac{1}{qq_{21}}\sum_{h=1}^{q_{21}} e_{q_{21}}(m_1h)\sum_{r=1}^{q}{}^{(0)} e_q\left(b_1r^2+\tilde{m}_1r\right) \\ &\quad\times S_1(q;b_3,2b_2(r+q_1h)+\tilde{m}_2) \\ &= \frac{1}{q_{21}q_{23}}\sum_{h=1}^{q_{21}}\sum_{k=1}^{q_{23}} e_{q_{21}}(m_1h)e_{q_{23}}(m_2k)e_{q_2}(2a_2q_1q_3hk) \\ &\quad\times\frac{1}{q^2}\sum_{r=1}^{q}\sum_{s=1}^{q} e_q\left(b_1r^2+2b_2rs+b_3s^2+n_{1,k}r+n_{2,h}s\right) \\ &= H(\mathbf{q};\mathbf{a},\mathbf{m}), \end{aligned} \tag{36}$$

with $n_{1,k}=\tilde{m}_1+2b_2q_3k$ and $n_{2,h}=\tilde{m}_2+2b_2q_1h$.

The theorem is a direct consequence of (28), (35), and (36).

4 Proof of Theorem 2

Let $Q_0=0.1N^{1/2}$. By Dirichlet's theorem on Diophantine approximation, there exist rationals b_i/q_i, $i=1,2,3$, such that

$$|q_i\alpha_i-b_i|\le Q_0N^{-2}, \quad 1\le q_i\le N^2Q_0^{-1}, \quad (b_i,q_i)=1. \tag{37}$$

Proposition 1 yields (9) unless $1 \le q_1, q_2, q_3 \le Q_0$. Let $q = [q_1, q_2, q_3]$ and define $a_i = b_i q/q_i$. Under conditions (37) with $q_1, q_2, q_3 \le Q_0$, Theorem 1 and (18) give

$$F(\boldsymbol{\alpha}; \boldsymbol{\theta}) \ll \frac{N^2 L}{(q + N^2|q\boldsymbol{\alpha} - \mathbf{a}|)^{1/2}} + N^{5/4} L.$$

The right side of this inequality is $O(NQ^{-1/2}L)$ unless

$$1 \le q \le Q, \quad |q\alpha_i - a_i| \le QN^{-2}.$$

Since (37) and the definition of the a_i's ensure that $(q, a_1, a_2, a_3) = 1$, we conclude that (9) may only fail if $\boldsymbol{\alpha}$ has a rational approximation of the form (10). Under those conditions, (11) follows from Proposition 2.

References

1. T. C. Anderson, A. V. Kumchev, and E. A. Palsson, *Discrete maximal operators for surfaces of higher codimension*, preprint arXiv:2006.09968.
2. G. I. Arkhipov, V. N. Chubarikov, and A. A. Karatsuba, *Trigonometric Sums in Number Theory and Analysis*, Walter de Gruyter GmbH & Co., Berlin, 2004.
3. J. Brandes, S. T. Parsell, k. Poulias, G. Shakan, and R. C. Vaughan, *On generating functions in additive number theory, II: Lower-order terms and applications to PDEs*, Math. Ann. **379** (2021), 347–376.
4. J. Brüdern and O. Robert, *Rational points on linear slices of diagonal hypersurfaces*, Nagoya Math. J. **218** (2015), 51–100.
5. R. C. Vaughan, *Some remarks on Weyl sums*, in "Topics in Classical Number Theory," Vol. I, II (Budapest, 1981), North-Holland, Amsterdam, 1984, pp. 1585–1602.
6. R. C. Vaughan, *The Hardy–Littlewood Method*, Second ed., Cambridge University Press, Cambridge, 1997.
7. R. C. Vaughan, *On generating functions in additive number theory I*, in "Analytic Number Theory. Essays in Honour of Klaus Roth", Cambridge University Press, Cambridge, 2009, pp. 436–448.
8. H. Weyl, *Über die Gleichverteilung von Zahlen mod Eins*, Math. Ann. **77** (1916), no. 3, 313–352.

On Consecutive Perfect Powers with Elementary Methods

Paolo Leonetti

Abstract Catalan's conjecture states that the equation $x^p - y^q = 1$ admits the unique solution $3^2 - 2^3 = 1$ in integers $x, y, p, q \geq 2$. The conjecture has been proved by Mihăilescu in 2002 using the theory of cyclotomic fields and Galois modules. Here, relying only on elementary methods, we prove several instances of this result. In particular, we show it in the following cases: p even, q is even, x divides q, y divides $x - 1$, y is a power of a prime, and $y \leq p^{p/2}$.

Keywords Perfect powers · Catalan's conjecture · Diophantine equation · Mihăilescu's theorem

1 Introduction

In 1844 the Belgian mathematician Eugène Charles Catalan [4] conjectured that 8 and 9 are the only (positive) consecutive perfect powers. That is to say, the unique solution of the Diophantine equation

$$x^p - y^q = 1 \tag{1}$$

in integers $x, y, p, q \geq 2$ is given by $3^2 - 2^3 = 1$. In this regard, he only made some empirical observations, which he stated without proof: e.g., he considered the special cases $(x + 1)^x - x^x = 1$ and $x^y - y^x = 1$.

The history of the problem dates back to at least Levi ben Gerson (also known as Gersonides), who considered in 1343 the Diophantine equations

$$2^p - 3^q = 1 \quad \text{and} \quad 3^p - 2^q = 1.$$

P. Leonetti (✉)
Department of Decision Sciences, Università Luigi Bocconi, via Roentgen 1, 20136 Milano (IT), Italy
e-mail: leonetti.paolo@gmail.com

M. B. Nathanson (ed.), *Combinatorial and Additive Number Theory IV*, Springer Proceedings in Mathematics & Statistics 347,
https://doi.org/10.1007/978-3-030-67996-5_21

Some years after Catalan made his conjecture, Lebesgue [8] (not to be confused with his more famous namesake H.L. Lebesgue) solved the case $q = 2$. In the early 1900's, Euler [5] and Chao Ko [6] gave a solution of the conjecture in the case $p = 2$ (it is worth noting that a particular instance of this result, i.e., the case $p = 4$, has been previously shown by Selberg). Their solutions are surveyed in Sects. 2 and 3, respectively.

After years of study, Tijdeman [16] made an important breakthrough in 1976: by means of Baker's method in trascendence theory, he was able to prove that Catalan's equation (1) has at most finitely many solutions. In addition, he shows the existence of an effective computable constant C such that $p, q \leq C$ whenever $x^p - y^q = 1$ for some $x, y \geq 2$. Then, Langevin [7] computed the explicit upper bound

$$C \leq e^{e^{e^{e^{730}}}}.$$

More recently, Mignotte [10] improved this upper bound, showing that every solution of Equation (1) satisfies

$$p < 7.15 \cdot 10^{11} \quad \text{and} \quad q < 7.78 \cdot 10^{16}.$$

Unfortunately, it appeared that the required computation to complete the proof was too large to be feasible (at this time, the conjecture could be proved by case checking only for $p, q < 10^7$).

Finally, Mihăilescu [11] solved completely the conjecture in April 2002, with a brilliant proof which relies on cyclotomic fields and Galois modules (the result was published two years later). In this regard, Catalan's conjecture is nowadays known as *Mihăilescu's theorem*.

Here, we are not going to present Mihăilescu's proof, which can be found, e.g., in Schoof's monograph [14]. The aim of this work is to provide and collect elementary proofs for several interesting cases of Catalan's equation (1). The arguments used can be followed by diligent students who have taken a first course in number theory.

First, a remark is in order: it can be assumed without loss of generality that p and q are distinct primes. Indeed, if p' and q' are primes dividing p and q, respectively, then

$$x^p - y^q = X^{p'} - Y^{q'},$$

where $X := x^{p/p'}$ and $Y := y^{q/q'}$ are integers ≥ 2. In addition, if $p = q$ then

$$x^p - y^p \geq (y+1)^p - y^p \geq 1 + py > 1, \tag{2}$$

i.e., there are no two consecutive p-th powers.

Theorem 1 *Catalan's conjecture holds in each of the following cases:*

(i) $q = 2$;
(ii) $p = 2$;

(iii) *x is a power of* 2;
(iv) $x \equiv 3, 5, 7 \pmod 8$;
(v) *x divides q*;
(vi) $x \equiv 1 \pmod y$;
(vii) *y is a power of a prime*;
(viii) $y \le \min\left(\frac{(pq)^p}{q}, \left(\frac{q\sqrt{p}}{2}\right)^p\right)$.

In particular, case (viii) implies that Catalan's conjecture holds if $y \le p^{p/2}$.

1.1 Notation

Let $\mathbf{N}$ and $\mathbf{N}^+$ represent the set non-negative and positive integers, respectively; in particular, $0 \in \mathbf{N}$. Given a prime p and $n \in \mathbf{N}^+$, let $v_p(n)$ be the p-adic valuation of n, that is, the greatest $k \in \mathbf{N}$ such that p^k divides n.

2 Proof of Theorem 1(i): $q = 2$

Here, we are going to show that the Diophantine equation

$$y^2 + 1 = x^p \tag{3}$$

has no integral solutions ≥ 2.

The case $p = 2$ does not have any solution, as it follows from inequality (2). Hence, let us assume that $p \ge 3$ (here, we recall that there is no loss of generality to say that p is a prime). Clearly, x and y have different parity. If x is even then y is odd and $y^2 + 1$ has to be divisible by 4, which is turn impossible because

$$0 \equiv x^p = y^2 + 1 = \underbrace{(y+1)}_{\text{even}}\underbrace{(y-1)}_{\text{even}} + 2 \equiv 2 \pmod 4.$$

This implies that x has to be odd and y even.

At this point, let us rewrite Eq. (3) in $\mathbf{Z}[i]$ as

$$(y+i)(y-i) = x^p,$$

where i is the imaginary number for which, by definition, $i^2 = -1$. Then, on the one hand, the greatest common divisor $\gcd(y+i, y-i) = \gcd(y+i, 2i)$ divides $2i$. On the other hand, i is unit in $\mathbf{Z}[i]$ and x is odd. It follows that $y+i$ and $y-i$ have to be (coprime) p-th powers in $\mathbf{Z}[i]$. Therefore there exist integers a, b for which

$$y + i = (a + bi)^p = \sum_{j=0}^{p} \binom{p}{j} a^j (bi)^{p-j}.$$

In particular, the imaginary parts of the both sides are equal, i.e.,

$$1 = \sum_{j=0}^{\frac{p-1}{2}} \binom{p}{2j} a^{2j} b^{p-2j} i^{p-2j-1} = b \sum_{j=0}^{\frac{p-1}{2}} \binom{p}{2j} a^{2j} b^{p-2j-1} (-1)^{\frac{p-1}{2}-j}. \quad (4)$$

Since the right hand side is divisible by b, then $|b| = 1$. Multiplying both sides by $(-1)^{\frac{p-1}{2}} b$, it follows that we can rewrite Eq. (4) as

$$\sum_{j=0}^{\frac{p-1}{2}} \binom{p}{2j} (-a^2)^j = (-1)^{\frac{p-1}{2}} b. \quad (5)$$

Considering also that

$$y - i = \overline{y + i} = \overline{(a + bi)^p} = \left(\overline{a + bi}\right)^p = (a - bi)^p,$$

we obtain that, for some odd x, it holds

$$x^p = (a + bi)^p (a - bi)^p = (a + i)^p (a - i)^p = (a^2 + 1)^p.$$

In particular, a is even and different from 0 (because $a^2 + 1 = x \geq 2$). It follows that

$$\sum_{j=0}^{\frac{p-1}{2}} \binom{p}{2j} (-a^2)^j \equiv 1 \pmod 4,$$

which implies, together with (5), that

$$\sum_{j=0}^{\frac{p-1}{2}} \binom{p}{2j} (-a^2)^j = 1. \quad (6)$$

At this point, if $p = 3$ then the above equation implies $a = 0$, which is impossible. Otherwise $\frac{p-1}{2} \geq 2$, hence Eq. (6) can be further rewritten as

$$\sum_{j=2}^{\frac{p-1}{2}} \binom{p}{2j} (-a^2)^j = a^2 \binom{p}{2}. \quad (7)$$

Finally, it is claimed that the number of factors 2 dividing each term in the left hand side of the above equation is strictly greater than the one on the right hand side. Indeed, considering that the identity

$$\binom{p}{2j} = \frac{1}{j(2j-1)}\binom{p-2}{2j-2}\binom{p}{2}$$

holds for all $j = 2, \ldots, \frac{p-1}{2}$, we obtain that

$$\begin{aligned} \upsilon_2\left(a^{2j}\binom{p}{2j}\right) &= \upsilon_2\left(\frac{a^{2j}}{j(2j-1)}\binom{p-2}{2j-2}\binom{p}{2}\right) \\ &\geq 2j\upsilon_2(a) - \upsilon_2(j) + \upsilon_2\left(\binom{p}{2}\right) \\ &\geq (2j - \upsilon_2(j))\,\upsilon_2(a) + \upsilon_2\left(\binom{p}{2}\right), \end{aligned}$$

which in turn is strictly greater than

$$\upsilon_2\left(a^2\binom{p}{2}\right) = 2\upsilon_2(a) + \upsilon_2\left(\binom{p}{2}\right).$$

This implies that Eq. (7) has no integral solutions, concluding the proof.

3 Proof of Theorem 1(ii): $p = 2$

Since two squares of positive integers cannot be consecutive, it is enough to check the cases $q = 3$ and $q \geq 5$.

3.1 *Case* $q = 3$.

In this case, Eq. (1) can be rewritten as

$$y^3 = (x+1)(x-1). \tag{8}$$

Notice that $x + 1$ and $x - 1$ are positive integers which differ by 2, hence they cannot be both cubes. Since their greatest common divisor divides their difference, then $\gcd(x + 1, x - 1)$ is exactly 2. In particular, x is odd and y is even: let us say $x = 2m + 1$ and $y = 2n$ for some $m, n \in \mathbf{N}^+$. Then Eq. (8) can be rewritten as

$$\binom{m+1}{2} = n^3.$$

Setting $m = 1$ we obtain the solution $(x, y) = (3, 2)$. Then, it is claimed that every triangular number greater than 1 is not a cube.

If m is even, say $m = 2k$ for some $k \in \mathbf{N}^+$, then $k(2k+1) = n^3$ and $\gcd(k, 2k+1) = 1$ so that k and $2k+1$ need to be both cubes. Otherwise m is odd, say $m = 2k-1$ for some $k \in \mathbf{N}^+$, then $k(2k-1) = n^3$ and $\gcd(k, 2k-1) = 1$ so that k and $2k-1$ need to be (coprime) cubes.

Therefore, in both cases we have to find k and $\varepsilon \in \{-1, 1\}$ such that $k = a^3$ and $2k + \varepsilon = b^3$ for some coprime $a, b \in \mathbf{N}^+$. By construction, it follows that

$$b^3 + (-\varepsilon)^3 = b^3 - \varepsilon = 2k = 2a^3,$$

hence it is sufficient to show that the Diophantine equation

$$\alpha^3 + \beta^3 = 2\gamma^3 \tag{9}$$

admits no non-trivial solutions in $\mathbf{Z}$ (a step-by-step solution of a slightly more general equation than (9) can be found in [15]). If (α, β, γ) is a solution of (9) and there exists a prime r dividing both α and β then r divides also γ and $(\alpha/r, \beta/r, \gamma/r)$ still satisfies (9). That is why we can assume without loss of generality that $\gcd(\alpha, \beta) = 1$ and by symmetry $\alpha \le \beta$.

In particular, α and β have to be odd, hence we can define coprime integers $u, v \in \mathbf{N}$ such that

$$u = \frac{\alpha + \beta}{2} \quad \text{and} \quad v = \frac{\alpha - \beta}{2}.$$

Therefore, by construction, we get $u(u^2 + 3v^2) = \gamma^3$, and it is claimed that there are no solutions whenever v is positive.

- Case 3 does not divide u: If u is not divisible by 3 then $\gcd(u, u^2 + 3v^2) = 1$, so there exist coprime integers $c, d \in \mathbf{N}$ such that $u = c^3$ and $u^2 + 3v^2 = d^3$. Defining $t = d - c^2$, we obtain $t(t^2 + 3tc^2 + 3c^4) = 3v^2$.
  Checking the remainders modulo 3, we deduce that t is divisible by 3, v is divisible by 3, and t is then divisible by 9. Substituting $t = 9e$ and $v = 3f$, we can rewrite the equation as $e(27e^2 + 9ec^2 + c^4) = f^2$. At this point, it is straightforward to verify that $\gcd(e, 27e^2 + 9ec^2 + c^4) = 1$, hence they have to be both squares. It means that we end to solve in $\mathbf{Z}$ an equation in the form

$$\mathfrak{a}^4 + 9\mathfrak{a}^2\mathfrak{b}^2 + 27\mathfrak{b}^4 = \mathfrak{c}^2. \tag{10}$$

- Case 3 divides u: If u is divisible by 3, with a similar reasoning, we obtain that there exist $w, z \in \mathbf{N}$ such that $u = 9w$ and $v = 3z$, so that $w(27w^2 + v^2)$ is a cube. Moreover $\gcd(w, 27w^2 + v^2) = 1$, therefore they have to be both cubes, namely $w = \chi^3$ and $27w^2 + v^2 = \delta^3$ for some $\chi, \delta \in \mathbf{N}$. Define $\nu = \delta - 3\chi^2$;

then $\nu(\nu^2 + 9\chi^2\nu + 27\chi^4)$ is a square. Again, these factors are coprime and we obtain an equation of the type (10).

It means that it is enough to show that Eq. (10) has no solutions in non-zero integers. Without loss of generality we can assume that $\gcd(\mathfrak{a}, \mathfrak{b}) = 1$. At this point, if $\mathfrak{a}$ is even then 4 divides $27\mathfrak{b}^4 - \mathfrak{c}^2$. Looking at this divisibility modulo 4, it implies that also $\mathfrak{b}$ is even, contradicting our coprimality assumption $\gcd(\mathfrak{a}, \mathfrak{b}) = 1$. In addition, if $\mathfrak{a}$ and $\mathfrak{b}$ are both odd then, by reasoning modulo 8, it is easily seen that Eq. (10) has no integral solutions.

It follows that if $(\mathfrak{a}, \mathfrak{b}, \mathfrak{c})$ is a solution of Equation (10) then $\mathfrak{a}$ is odd and $\mathfrak{b}$ is even. In particular, we can define the integer $\mathfrak{m} = \mathfrak{b}/2$. Note that $\mathfrak{a}$ is not divisible by 3, otherwise $\mathfrak{b}$ would be divisible by 3 too, looking at the equation modulo 81. Substituting we can rewrite the equation as

$$27\mathfrak{m}^4 = \left(\frac{\mathfrak{c} + \mathfrak{a}^2}{2} + 9\mathfrak{m}^2\right)\left(\frac{\mathfrak{c} - \mathfrak{a}^2}{2} - 9\mathfrak{m}^2\right). \tag{11}$$

These factors are coprime and positive since their sum and product are both positive.

Only two cases are possible: in the first case, the factors are $27a^4$ and b^4, respectively, for some coprime integers $a, b \in \mathbf{N}$. It implies 3 divides $27a^4 - 18\mathfrak{m}^2 = b^4 + \mathfrak{a}^2$, which is impossible since -1 is not a quadratic residue modulo 3.

In the second case, the factors are a^4 and $27b^4$, respectively, for some coprime integers $a, b \in \mathbf{N}$. It implies that $a^4 - 18\mathfrak{m}^2 = 27b^4 + \mathfrak{a}^2$, with $ab = \mathfrak{m}$. The variable a cannot be even, looking at the equation modulo 8. Moreover a and b cannot be both odd since $\mathfrak{a}$ is odd, hence by force b is even. To sum up, we can rewrite Eq. (11) as

$$27b^4 = \left(\frac{a^2 + \mathfrak{a}}{2} - \frac{9}{2}b^2\right)\left(\frac{a^2 - \mathfrak{a}}{2} - \frac{9}{2}b^2\right) \tag{12}$$

with $a, b, \mathfrak{a}$ integers such that a and $\mathfrak{a}$ are odd and b is even.

Similarly, it is not difficult to check that the factors in Eq. (12) are coprime and strictly positive, implying that they are, in some order, in the form $27c^4$ and d^4 for some $c, d \in \mathbf{N}$. This means that $c^4 + 9c^2d^2 + 27c^4 = a^2$, which is again in the form of Eq. (10).

Notice that $a \le \mathfrak{m} < \mathfrak{b} < \mathfrak{c}$ whenever $\mathfrak{c} \ge 1$. It implies that, if $(\mathfrak{a}^\star, \mathfrak{b}^\star, \mathfrak{c}^\star)$ is a solution of (10) which minimizes $\mathfrak{c}$ with $\mathfrak{c} \ge 1$, then we can construct another solution $(\mathfrak{a}', \mathfrak{b}', \mathfrak{c}')$ such that $\mathfrak{c}'$ is smaller than $\mathfrak{c}^\star$. Therefore the unique solution of Equation (10) is $(0, 0, 0)$.

3.2 *Case $q \ge 5$.*

We have to solve the equation

$$x^2 - y^q = 1 \tag{13}$$

in integers ≥ 2, where q is a prime ≥ 5.

If x is even then $\gcd(x+1, x-1) = 1$ and $(x+1)(x-1) = y^q$, implying that $x+1$ and $x-1$ are two coprime q-powers that differ by 2, which is impossible. Hence x has to be odd and y even.

The following elementary result, which we are not going to prove here, belongs to the folkore and it is commonly known as *Lifting the Exponent*. Typically it is attributed to Lucas [9] and Carmichael [2], the latter having fixed an error in Lucas' original work in the 2-adic case.

Lemma 1 *For all integers a, b, n and primes p such that n is positive, p does not divide ab and p divides $a - b$, the following ones hold:*

- *If $p \geq 3$, then $v_p(a^n - b^n) = v_p(a-b) + v_p(n)$;*
- *If $v_2(a-b) \geq 2$, then $v_2(a^n - b^n) = v_2(a-b) + v_2(n)$;*
- *If $p = 2$ and n is even, then $v_2(a^n - b^n) = v_2(a-b) + v_2(a+b) + v_2(n) - 1$.*

Based on this result, which will be useful also for other cases, we prove the following:

Lemma 2 *Let p, q be primes which are not both equal to 2 and a, b be distinct coprime integers such that $a^p - b^p$ is a q-th power not divisible by p. Then $a - b$ is a q-th power.*

Proof By hypothesis, there exists an integer c not divisible by p such that

$$(a-b)\left(\frac{a^p - b^p}{a-b}\right) = c^q. \tag{14}$$

Then, let us suppose that there exists a prime r such that

$$r \text{ divides } \gcd\left(a-b, \frac{a^p - b^p}{a-b}\right).$$

According to Euler-Fermat's theorem, it holds $a^p - b^p \equiv a - b \pmod{p}$. Since p does not divide $a^p - b^p$, then $r \neq p$. Hence, we obtain by Lemma 1 that, if r divides $a - b$, then

$$v_r\left(\frac{a^p - b^p}{a-b}\right) = v_r(p) = 0.$$

This implies that the factors in Eq. (14) are coprime.

If $q \geq 3$, then we are done because each of them has to be a q-th power. Otherwise, we are just left to check the case $q = 2$ and $p \geq 3$, together with $a - b = -d^2$, for some non-zero integer d. But this is impossible since it would imply that

$$\frac{a^p - b^p}{a-b} = \frac{c^2}{-d^2} < 0,$$

and in particular $a^p - b^p$ and $a - b$ should have different signs.

At this point, since $\gcd(x+1, x-1) = 2$, we can define integers $\varepsilon \in \{-1, 1\}$ and coprime $a, b \in \mathbf{N}^-$ such that $x \equiv \varepsilon \pmod 4$, $y = 2ab$, $x + \varepsilon = 2a^q$, and $x - \varepsilon = 2^{q-1}b^q$.

Since $q \geq 5$ and $x \geq 2$, we deduce

$$\left(\frac{a}{b}\right)^q = 2^{q-2}\frac{x+\varepsilon}{x-\varepsilon} \geq 8\,\frac{x-1}{x+1} \geq 2,$$

implying, in particular, that a is greater than b. Moreover, we have by construction

$$a^{2q} - (2\varepsilon b)^q = \left(\frac{x+\varepsilon}{2}\right)^2 - 2\varepsilon(x-\varepsilon) = \left(\frac{x-3\varepsilon}{2}\right)^2.$$

Then, according to Lemma 2, if q does not divide $\frac{1}{2}(x-3\varepsilon)$ then $a^2 - 2\varepsilon b$ has to be a square: this is not possible since $a^2 \neq a^2 - 2\varepsilon b$ and $|2\varepsilon b| = 2b \leq 2(a-1)$, with the consequence that

$$(a-1)^2 < a^2 - 2\varepsilon b < (a+1)^2.$$

It follows that q divides $\frac{1}{2}(x-3\varepsilon)$ and in particular q does not divide x since $q \geq 5$. Again by Lemma 2, rewriting Eq. (13) as $x^2 = y^q - (-1)^q$, there exists an integer $\zeta \geq 1$ such that $y - (-1) = \zeta^2$. In particular ζ is a odd integer and y is not a square, since by assumption $y \geq 2$. It means that $(\zeta, 1)$ and $(x, y^{\frac{1}{2}(q-1)})$ are two solutions of the Pell-equation $\alpha^2 - y\beta^2 = 1$. Since $(\zeta, 1)$ is its fundamental solution (see for example [1] for the theory underlying Pell-equations), there exists a positive integer m such that

$$x + y^{\frac{q-1}{2}}\sqrt{y} = (\zeta + \sqrt{y})^m. \tag{15}$$

Then, in $\mathbf{Z}/y\mathbf{Z}[\sqrt{y}]$ we obtain $x = \zeta^m + m\zeta^{m-1}\sqrt{y}$, which implies that y divides $m\zeta^{m-1}$. In particular, y is even, ζ is odd, and m is even.

Looking finally at Eq. (15) in $\mathbf{Z}/\zeta\mathbf{Z}[\sqrt{y}]$, we obtain $x + y^{(q-1)/2}\sqrt{y} = y^{m/2}$, so that ζ divides $y^{(q-1)/2}$. Supposing that $\zeta \geq 2$, there exists a prime r such that r divides ζ, which in turn divides y, and by construction r divides $\zeta^2 - y = -1$, which is impossibile. Therefore we have shown that, given a prime $q \geq 5$, if x, y, ζ, m verify Eq. (15) then $\zeta = 1$, with the consequence that Eq. (13) has no solutions.

4 Proof of Theorem 1(iii): x is a power of 2

Assuming that q is an odd prime by case (i), it is sufficient to rewrite Eq. (1) as

$$x^p = (1+y)\underbrace{(1 - y + y^2 - \cdots + y^{q-1})}_{\text{odd} \geq 3},$$

which does not have clearly any solution.

5 Proof of Theorem 1(iv): $x \equiv 3, 5$ or $7 \pmod 8$

Thanks to cases (i) and (ii), we can assume that p and q are distinct odd primes. Since x has remainder 3, 5, or 7 modulo 8, then x is odd. Hence, y has to be even. It follows that 8 divides $x^p - 1$, which is impossible since

$$x^p - 1 = x \underbrace{\left(x^{\frac{p-1}{2}} + 1\right)\left(x^{\frac{p-1}{2}} - 1\right)}_{\text{two consecutive even numbers}} + (x-1) \equiv x - 1 \pmod 8$$

and 8 does not divide $x - 1$ by hypothesis.

6 Proof of Theorem 1(v): x divides q

Notice that if $x^p - y^q = 1$ for some positive integers x, y, p, q greater than 1 and x divides q then

$$x^p - Y^x = 1,$$

where $Y := y^{q/x}$. Accordingly, we can assume without loss of generality that $x = q$ and solve the Diophantine equation $x^p - y^x = 1$. By the case (ii) we know that x has to be odd, hence y is even.

Since $y + 1$ divides $y^x + 1 = x^p$ then there exists an odd prime r dividing both x and $y + 1$. It follows by Lemma 1 that

$$\upsilon_r(x^p) = \upsilon_r\left((y+1) \cdot \frac{y^x + 1}{y + 1}\right) = \upsilon_r(y+1) + \upsilon_r(x),$$

so that $(p-1)\upsilon_r(x) = \upsilon_r(y+1)$. In particular, we obtain

$$y + 1 \geq r^{\upsilon_r(y+1)} = r^{(p-1)\upsilon_r(x)} \geq 3^{p-1} \geq 2^p + 1,$$

where the last inequality holds for all $p \geq 3$. Therefore y is greater than or equal to 2^p, with the consequence that

$$x^p = y^x + 1 \geq 2^{px} + 1 > 2^{px}.$$

We conclude that $x > 2^x$ for some integer $x \geq 2$, which is impossible.

7 Proof of Theorem 1(vi): $x \equiv 1 \pmod{y}$

Let us assume that there exists $z \in \mathbf{N}^+$ such that $x = yz + 1$. Again, we can assume that p and q are distinct odd primes in Eq. (1), which in turn can be rewritten as $y^q = (yz+1)^p - 1$, i.e.,

$$y^{q-1} = z\left(1 + (yz+1) + \cdots + (yz+1)^{p-1}\right).$$

Since the second factor has remainder p modulo z, then

$$\gcd\left(z, 1 + (yz+1) + \cdots + (yz+1)^{p-1}\right) \quad \text{divides} \quad p. \tag{16}$$

Let us first assume that the greatest common divisor in (16) is 1. Then there exists coprime $\alpha, \beta \in \mathbf{N}$ such that

$$z = \alpha^{q-1} \quad \text{and} \quad 1 + (yz+1) + \cdots + (yz+1)^{p-1} = \beta^{q-1}.$$

Since $\alpha\beta = y$ by construction, then $yz = \alpha^q \beta$, which implies in turn that

$$\sum_{j=0}^{p-1} (1 + \alpha^q \beta)^j = \beta^{q-1}.$$

At this point, β is an integer greater than 1 and the right hand side has remainder p modulo β. Therefore β has to be exactly p, so that

$$p^{q-1} = \sum_{j=0}^{p-1} (1 + \alpha^q p)^j = \sum_{j=0}^{p-1} \sum_{k=0}^{j} \binom{j}{k} \alpha^{qk} p^k = p + \sum_{j=1}^{p-1} \sum_{k=1}^{j} \binom{j}{k} \alpha^{qk} p^k.$$

One the one hand, it is easily seen that

$$\begin{aligned}\sum_{j=1}^{p-1} \sum_{k=1}^{j} \binom{j}{k} \alpha^{qk} p^k &= (1 + 2 + \cdots + (p-1))\, \alpha^q p + \sum_{j=2}^{p-1} \sum_{k=2}^{j} \binom{j}{k} \alpha^{qk} p^k \\ &= p^2 \frac{p-1}{2} \alpha^q + p^2 \sum_{j=2}^{p-1} \sum_{k=2}^{j} \binom{j}{k} \alpha^{qk} p^{k-2} \equiv 0 \pmod{p^2}.\end{aligned}$$

On the other hand, since q is an odd prime then p^2 divides p^{q-1}, the above sum cannot be a multiple of p^2.

Let us assume now that the greatest common divisor in (16) is p. Then there exist $u, v \in \mathbf{N}$ such that

$$z = pu, \qquad 1 + (yz+1) + \cdots + (yz+1)^{p-1} = pv, \quad \text{and} \quad y^{q-1} = p^2 uv.$$

Hence, y is a divisible by p, and there exist positive integers h, k such that $y = p^h k$ such that p does not divide k. As in the previous case, we can see that p does not divide v, indeed

$$\sum_{j=0}^{p-1} (yz+1)^j = \sum_{j=0}^{p-1} \left(\underbrace{p^{h+1}ku}_{\text{divisible by } p^2} + 1 \right)^j \equiv p \pmod{p^2}.$$

It implies that there exist coprime positive integers γ, δ, not divisible by p, such that

$$z = p^{h(q-1)-1}\gamma^{q-1}, \quad 1 + (yz+1) + \cdots + (yz+1)^{p-1} = p\delta^{q-1}, \quad \text{and} \quad y = p^h\gamma\delta.$$

Accordingly, we deduce that

$$p\delta^{q-1} = \sum_{j=0}^{p-1} (yz+1)^j = \sum_{j=0}^{p-1} \left(p^{hq-1}\gamma^q\delta + 1\right)^j.$$

Multiplying both sides by $p^{hq-1}\gamma^q\delta$ we get

$$y^q = \left(p^{hq-1}\gamma^q\delta + 1\right)^p - 1 = \sum_{j=1}^{p} \binom{p}{j} p^{j(hq-1)}\gamma^{qj}\delta^j. \tag{17}$$

If $\delta > 1$, then it has to be divisible by some prime r, so that

$$\begin{aligned} qv_r(\delta) &= v_r\left(\sum_{j=1}^{p} \binom{p}{j} p^{j(hq-1)}\gamma^{qj}\delta^j\right) \\ &= v_r\left(p^{hq-1}\gamma^q\delta + \sum_{j=2}^{p} \binom{p}{j} p^{j(hq-1)}\gamma^{qj}\delta^j\right). \end{aligned}$$

Recalling that δ, γ, and p are (pairwise) coprime, then the right hand side is exactly $v_r(\delta)$, which is impossible since $q > 1$. Therefore $\delta = 1$, and Eq. (17) simplifies to

$$\sum_{j=2}^{p} \binom{p}{j} p^{j(hq-1)}\gamma^{jq} = 0,$$

which cannot hold because it is a (non-empty) sum of positive integers.

8 Proof of Theorem 1(vii): y is Power of a Prime

Thanks to cases (i) and (ii), if (x, y, p, q) is a solution of Eq. (1) different from $(3, 2, 2, 3)$ then p and q are [distinct] odd primes. Let us rewrite (1) as

$$y^q = (x-1) \cdot \left(\frac{x^p - 1}{x - 1}\right). \tag{18}$$

The first factor has to be greater than 1 by case (iii). In addition, the second factor (which is greater than $x - 1$) has to be odd, indeed

$$\frac{x^p - 1}{x - 1} = 1 + x + \cdots + x^{p-1} \equiv 1 + x\underbrace{(p-1)}_{\text{even}} \equiv 1 \pmod 2.$$

In particular, also $x - 1$ has to be odd: indeed, supposing that $y = \ell^c$ for some prime ℓ and positive integer c, then $x - 1 = \ell^a$ and $\frac{x^p-1}{x-1} = \ell^b$ for some integers $1 \le a < b$ such that $a + b = cq$.

Then, according to Lemma 1, we obtain

$$v_\ell\left(\frac{x^p - 1}{x - 1}\right) = v_\ell(p).$$

It implies that the greatest common divisor between $x - 1$ and $\frac{x^p-1}{x-1}$ is exactly p, which in turn is equal to ℓ. Therefore, we can say that

$$x - 1 = p \quad \text{and} \quad \frac{x^p - 1}{x - 1} = p^b$$

for some integer $b \ge 2$. We can conclude that

$$\begin{aligned} p^{b+1} &= (x-1)p^b = x^p - 1 = (p+1)^p - 1 \\ &= \sum_{j=1}^{p} \binom{p}{j} p^j = p^2 + p^3 \cdot \frac{p-1}{2} + \sum_{j=3}^{p} \binom{p}{j} p^j, \end{aligned}$$

which is impossible since the left hand side is divisible by p^3 while the right hand side is not.

9 Proof of Theorem 1(viii): $y \le \min\left(\frac{(pq)^p}{q}, \left(\frac{q\sqrt{p}}{2}\right)^p\right)$

As before, assume that p and q are distinct odd primes. Reasoning as in the case (vii), we rewrite Eq. (1) in the form (18), and notice that for each odd prime r dividing

$x-1$ we get by Lemma 1

$$v_r\left(\frac{x^p-1}{x-1}\right)=v_r(p),$$

which implies in turn that

$$\gcd\left(x-1,\frac{x^p-1}{x-1}\right) \text{ divides } p. \tag{19}$$

Let us suppose, at first, that the greatest common divisor in (19) is 1, i.e., there exist two coprime integers $a, b \geq 2$ with product y such that

$$x-1=a^q \quad \text{and} \quad \frac{x^p-1}{x-1}=b^q.$$

In addition, let us assume the following Cassell's result [3], which can be proved by means of elementary methods.

Lemma 3 *Fix integers $x, y \geq 2$ and primes $p, q \geq 3$ such that $|x^p-y^q|=1$. Then p divides y and q divides x.*

Lemma 3 immediately implies that p divides $y^q=(x-1)\left(\frac{x^p-1}{x-1}\right)$, hence p divides at least one between $x-1$ and $\frac{x^p-1}{x-1}$. If p divides $x-1$ then, thanks to Lemma 1, p divides also $\frac{x^p-1}{x-1}$. Conversely, if p divides also $\frac{x^p-1}{x-1}$, then $x^p \equiv 1 \pmod{p}$ and, by Euler-Fermat's theorem, $x \equiv 1 \pmod{p}$ as well. This contradicts the coprimality assumption between $x-1$ and $\frac{x^p-1}{x-1}$.

On the other hand, let us suppose that the greatest common divisor in (19) is p, i.e., there exist two coprime integers $c, d \geq 2$ with product y such that

$$x-1=p^{q-1}c^q \quad \text{and} \quad \frac{x^p-1}{x-1}=pd^q.$$

Recalling that q divides x by Lemma 3, we obtain

$$0 \equiv x = 1+p^{q-1}c^q \equiv 1+1\cdot c^q \equiv 1+c \pmod{q},$$

so that there exists $X \in \mathbf{N}^+$ such that

$$x=1+p^{q-1}(Xq-1)^q.$$

Rewriting Eq. (1) as $(y+1)\left(\frac{y^q+1}{y+1}\right)=x^p$, it can be seen with a similar argument that there exists $Y \in \mathbf{N}$ such that

$$y=-1+q^{p-1}(pY+1)^p.$$

Let us suppose that $Y \geq 1$. Then the hypothesis $y \leq q^{p-1} p^p$ implies

$$q^{p-1} p^p \geq y \geq -1 + q^{p-1}(p+1)^p > q^{p-1}((p+1)^p - 1) > q^{p-1} p^p,$$

which is false. This forces $Y = 0$, so that $y = q^{p-1} - 1$.

Finally, the hypothesis $y \leq p^{p/2}(q/2)^p$ implies

$$\begin{aligned} 1 + p^{q-1}(q-1)^q \leq x = (y^q + 1)^{1/p} &\leq \left(\left(p^{\frac{p}{2}}(q/2)^p\right)^q + 1\right)^{1/p} \\ &< \left(\left(p^{p\frac{q-1}{q}}(q/2)^p\right)^q + 1\right)^{1/p} \\ &< \left(p^{p\frac{q-1}{q}}(q-1)^p\right)^{q/p} = p^{q-1}(q-1)^q, \end{aligned}$$

which is a contradiction, completing the proof.

10 Closing Remarks

There are many open questions related to Catalan's equation. For instance, Pillai [12, 13] conjectured in 1936 that, for each $k \in \mathbf{N}^+$, the Diophantine equation

$$x^p - y^q = k$$

has only finitely many positive integer solutions (x, y, p, q), with $p \geq 2$ and $q \geq 2$. It is worth noting that Erdös formulated an even stronger conjecture, i.e., there exist $\varepsilon, C > 0$ such that

$$a_{n+1} - a_n \geq Cn^{\varepsilon}$$

for all n, where a_n stands for the n-th element of the increasing sequence of perfect powers, so that $a_1 = 1, a_2 = 4, a_3 = 8, a_4 = 9, \ldots$.

On a similar note about perfect powers, the banker A. Beal offered in 1993 a one million prize for a proof of the following conjecture: Given $x, y, z, p, q, r \in \mathbf{N}^+$ with $p, q, r > 2$ such that

$$x^p + y^q = z^r$$

then $\gcd(x, y, z) > 1$. Actually, the prize remains unclaimed (however, the abc conjecture would imply that there are at most finitely many counterexamples to Beal's conjecture). A beautiful survey of open problems and history of classical questions related to perfect powers can be found in [17].

References

1. Barbeau, E.J.: Pell's Equation, Problem Books in Mathematics. Springer-Verlag (2003)
2. Carmichael, R.D.: On the Numerical Factors of Certain Arithmetic Forms. Amer. Math. Monthly, **16**, 153–159 (1909)
3. Cassell, J.W.S.: On the equation $a^x - b^y = 1$, II. Proc. Cambridge Soc., **56**, 97–113 (1960)
4. Catalan, E.C.: Note extraite d'une lettre adressée à l'éditeur. J. reine angew. Mathematik, **27**, 192 (1844)
5. Euler, L.: Theorematum quorundam arithmeticorum demonstrationes. Commentationes Arithmeticae, Opera Omnia, Series I, Vol.II, Teubner, 56–58 (1915)
6. Ko, Chao: On the diophantine equation $x^2 = y^n + 1, xy \neq 0$. Sci. Sinica, **14**, 457–460 (1965)
7. Langevin, M.: Quelques applications de nouveaux résultats de Van der Poorten. In: Sém. Delange-Pisot-Poitou (1975/76). Théorie des nombres: Fasc. 2, Exp. No. G12 (1977)
8. Lebesgue, V.A.: Sur l'impossibilité, en nombres entiers, de l'équation $x^m = y^2 + 1$. Nouv. Ann. Math., **9**, 178–181 (1850)
9. Lucas, É.: Théorie des Fonctions Numériques Simplement Periodiques. Amer. J. Math., **1**, 184–196, 197–240, 289–321 (1878)
10. Mignotte, M.: Une remarque sur l'équation de Catalan. In: Number Theory in Progress (K. Györy et al., eds.), Proc. Conf. Zakopane, Poland, June 30–July 9, 1997, vol. 1, Walter de Gruyter, Berlin–New York (1999)
11. Mihăilescu, P.: Primary Cyclotomic Units and a proof of Catalan's Conjecture. J. reine angew. Mathematik, **572**, 167–195 (2004)
12. Pillai, S.S.: On $a^x - b^y = c$. J. Indian Math. Soc. (N.S.), **2**, 119–122 (1936)
13. Pillai, S.S.: On the equation $2^x - 3^y = 2^X + 3^Y$. Bull. Calcutta Math. Soc., **37**, 15–20 (1945)
14. Schoof, R.: Catalan's Conjecture. Springer (2007)
15. Sierpinki, W.: Elementary theory of numbers. Monografie Matematyczne (1964)
16. Tijdeman, R.: On the equation of Catalan. Acta Arith., **29**, 197–209 (1976)
17. Waldschmidt, M.: Perfect powers: Pillai's works and their developments. Preprint, http://arxiv.org/abs/0908.4031 (2009)

Sidon Sets and Perturbations

Melvyn B. Nathanson

Abstract A subset A of an additive abelian group is an h-Sidon set if every element in the h-fold sumset hA has a unique representation as the sum of h not necessarily distinct elements of A. Let $\mathbf{F}$ be a field of characteristic 0 with a nontrivial absolute value, and let $A = \{a_i : i \in \mathbf{N}\}$ and $B = \{b_i : i \in \mathbf{N}\}$ be subsets of $\mathbf{F}$. Let $\varepsilon = \{\varepsilon_i : i \in \mathbf{N}\}$, where $\varepsilon_i > 0$ for all $i \in \mathbf{N}$. The set B is an ε-perturbation of A if $|b_i - a_i| < \varepsilon_i$ for all $i \in \mathbf{N}$. It is proved that, for every $\varepsilon = \{\varepsilon_i : i \in \mathbf{N}\}$ with $\varepsilon_i > 0$, every set $A = \{a_i : i \in \mathbf{N}\}$ has an ε-perturbation B that is an h-Sidon set. This result extends to sets of vectors in $\mathbf{F}^n$.

Keywords Sidon set · Sumset · Representation function · Additive number theory

2020 Mathematics Subject Classification 11B13 11B24 11B75 11P99

1 Sidon Sets

Let $\mathbf{N} = \{1, 2, 3, \ldots\}$ be the set of positive integers and $\mathbf{N}_0 = \mathbf{N} \cup \{0\} = \{0, 1, 2, 3, \ldots\}$ the set of nonnegative integers. Denote the cardinality of the set S by $|S|$.

Let G be an additive abelian group, and let $A = \{a_i : i \in I\}$ be a nonempty subset of G with $a_i \neq a_j$ for $i \neq j$. For every positive integer h, define the *h-fold sumset*

$$hA = \{a_{i_1} + \cdots + a_{i_h} : a_{i_1}, \ldots, a_{i_h} \in A\}.$$

For every group element $b \in G$, we have $b \in hA$ if and only if there is a set of nonnegative integers $\{u_i : i \in I\}$ such that

Supported in part by a grant from the PSC-CUNY Research Award Program.

M. B. Nathanson (✉)
Lehman College (CUNY), Bronx, New York 10468, USA
e-mail: melvyn.nathanson@lehman.cuny.edu

M. B. Nathanson (ed.), *Combinatorial and Additive Number Theory IV*,
Springer Proceedings in Mathematics & Statistics 347,
https://doi.org/10.1007/978-3-030-67996-5_22

$$h = \sum_{i \in I} u_i$$

and

$$b = \sum_{i \in I} u_i a_i.$$

Define $0A = \{0\}$.

An *h-Sidon set* (also called a *B_h-set*) is a subset $A = \{a_i : i \in I\}$ of G such that every element in the sumset hA has a unique representation as the sum of h elements of A. Equivalently, an h-Sidon set is a set A that satisfies the condition: For all $b \in hA$, if $\{u_i : i \in I\}$ and $\{v_i : i \in I\}$ are sets of nonnegative integers such that

$$h = \sum_{i \in I} u_i = \sum_{i \in I} v_i \tag{1}$$

and

$$b = \sum_{i \in I} u_i a_i = \sum_{i \in I} v_i a_i \tag{2}$$

then

$$u_i = v_i \quad \text{for all } i \in I. \tag{3}$$

A 2-Sidon set is usually called a *Sidon set*.

A subset of an h-Sidon set is also an h-Sidon set, and an h-Sidon set is an r-Sidon set for every positive integer $r < h$. However, an h-Sidon set is not necessarily an $(h+1)$-Sidon set. For example, let $G = \mathbf{Z}$ be the additive group of integers. For every integer $h \geq 2$, the set $A = \{h^i : i \in \mathbf{N}_0\}$ is an h-Sidon set because $0 \notin A$ and every positive integer has a unique h-adic representation. However, for all nonnegative integers $a < b$, the identity

$$h^{a+1} + h^{b+1} = \underbrace{h^a + \cdots + h^a}_{h \text{ summands}} + h^{b+1} = h^{a+1} + \underbrace{h^b + \cdots + h^b}_{h \text{ summands}}$$

proves that A is not an $(h+1)$-Sidon set.

Most work on Sidon sets has been restricted to subsets of the integers or other discrete groups. Cilleruelo and Ruzsa [1] have studied Sidon sets of real, complex, and p-adic numbers. O'Bryant [2] is a survey of Sidon sets.

2 Perturbations of Countably Infinite Sets

Let G be an additive abelian group and let A and B be subsets of G. Define the *difference set*

$$A - B = \{a - b : a \in A \text{ and } b \in B\}$$

and, for all nonnegative integers r and s, the (r, s)*-sum-difference set*

$$rA - sA = \left\{ \sum_{i=1}^{r} a_i - \sum_{i=r+1}^{r+s} a_i : a_i \in A \text{ for } i = 1, \ldots, r+s \right\}.$$

For $c \in G$, define the *translate*

$$A + c = \{a + c : a \in A\}.$$

For $b \in G$ and $r \in \{0, 1, \ldots, h\}$, let

$$A_{r,h}(b) = rA + (h - r)b.$$

We have

$$h(A \cup \{b\}) = \bigcup_{r=0}^{h} (rA + (h - r)b) = \bigcup_{r=0}^{h} A_{r,h}(b).$$

Lemma 1 *Let G be an additive abelian group and let A be a subset of G that is an h-Sidon set. Let $b \in G$. The set $A \cup \{b\}$ is an h-Sidon set if and only if the sets*

$$A_{r,h}(b) = rA + (h - r)b$$

are pairwise disjoint for all $r \in \{0, 1, \ldots, h\}$.

Proof Let $r, s \in \{0, 1, \ldots, h\}$, and let $a_1, a_2, \ldots, a_{r+s}$ be a sequence of elements of A. We have

$$a_1 + \cdots + a_r + (h - r)b \in A_{r,h}(b)$$

and

$$a_{r+1} + \cdots + a_{r+s} + (h - s)b \in A_{s,h}(b).$$

Suppose that

$$c = a_1 + \cdots + a_r + (h - r)b = a_{r+1} + \cdots + a_{r+s} + (h - s)b. \tag{4}$$

Note that

$$c \in A_{r,h}(b) \cap A_{s,h}(b).$$

Because A is an h-Sidon set, if $r = s$, then $a_1 + \cdots + a_r = a_{r+1} + \cdots + a_{2r}$ and there is a permutation σ of $\{1, \ldots, r\}$ such that $a_{r+i} = a_{\sigma(i)}$ for all $i \in \{1, \ldots, r\}$.

It follows that $A \cup \{b\}$ is not an h-Sidon set if and only if there exist integers $r, s \in \{0, 1, \ldots, h\}$ with $r \neq s$, and elements $a_1, a_2, \ldots, a_{r+s}$ in A that satisfy (4).

This is equivalent to the condition that the sets $A_{r,h}(b)$ are not pairwise disjoint for $r \in \{0, 1, \ldots, h\}$. This completes the proof. □

Let $\mathbf{F}$ be a field. For $A \subseteq \mathbf{F}$ and $c \in \mathbf{F}$, define the *dilate*

$$c * A = \{ca : a \in A\}.$$

An *absolute value* $|\ |$ on the field $\mathbf{F}$ is trivial if $|0| = 0$ and $|x| = 1$ for all $x \in \mathbf{F} \setminus \{0\}$. A field $\mathbf{F}$ with a nontrivial absolute value is infinite, and

$$\inf\{|x| : x \in \mathbf{F} \setminus \{0\}\} = 0. \tag{5}$$

The usual absolute values on $\mathbf{Q}$, $\mathbf{R}$, and $\mathbf{C}$ and the p-adic absolute values on $\mathbf{Q}$ and $\mathbf{Q}_p$ are nontrivial.

Let $\mathbf{F}$ be a field with a nontrivial absolute value $|\ |$. Let I be a nonempty set, and let $\{a_i : i \in I\}$ and $\{b_i : i \in I\}$ be sets of elements of the field $\mathbf{F}$. Let $\{\varepsilon_i : i \in I\}$ be a set of positive real numbers. The set $\{b_i : i \in I\}$ is an *ε-perturbation* of the set $\{a_i : i \in I\}$ if

$$|b_i - a_i| < \varepsilon_i$$

for all $i \in I$.

Lemma 2 *Let $\mathbf{F}$ be a field of characteristic 0 with a nontrivial absolute value $|\ |$, and let A be a finite subset of $\mathbf{F}$. For every $a \in \mathbf{F}$ and $\delta > 0$, there exists an element $b \in \mathbf{F}$ such that*

$$|b - a| < \delta$$

and the $h + 1$ sets

$$A_{r,h}(b) = rA + (h - r)b$$

for all $r \in \{0, 1, 2, \ldots, h\}$ are pairwise disjoint.

Proof Let $a \in \mathbf{F}$ and $x \in \mathbf{F}$. For all $r, s \in \{0, 1, \ldots, h\}$ with $s < r$, we have

$$\begin{aligned}\emptyset \neq\ & A_{r,h}(a + x) \bigcap A_{s,h}(a + x) \\ &= (rA + (h - r)(a + x)) \bigcap (sA + (h - s)(a + x))\end{aligned}$$

if and only if there is a sequence $a_1, a_2, \ldots, a_{r+s}$ of elements of A such that

$$a_1 + \cdots + a_r + (h - r)(a + x) = a_{r+1} + \cdots + a_{r+s} + (h - s)(a + x).$$

Equivalently,

$$(r - s)(a + x) = (a_1 + \cdots + a_r) - (a_{r+1} + \cdots + a_{r+s}) \in rA - sA. \tag{6}$$

Because $\mathbf{F}$ has characteristic 0, the positive integer $r - s$ is a unit in $\mathbf{F}$ and

$$a + x \in (r - s)^{-1} * (rA - sA). \tag{7}$$

Therefore,

$$x \in \left((r - s)^{-1} * (rA - sA)\right) - a.$$

The set A is finite, and so the sets

$$C_{r,s} = \left((r - s)^{-1} * (rA - sA)\right) - a$$

and

$$C = \bigcup_{0 \leq s < r \leq h} C_{r,s}$$

are also finite. The set $\mathbf{F} \setminus C$ is infinite.

For all $x \in \mathbf{F} \setminus C$, the sets $A_{r,h}(a + x)$ for $r \in \{0, 1, \ldots, h\}$ are pairwise disjoint. Because the set C is finite, we have

$$\delta_1 = \min\left(|c| : c \in C \text{ and } c \neq 0\right) > 0.$$

The absolute value on $\mathbf{F}$ is nontrivial, and so, by (5), there exists $x \in \mathbf{F}$ with $0 < |x| < \min(\delta_1, \delta)$. The inequality $0 < |x| < \delta_1$ implies that $x \notin C$, and so the $h + 1$ sets $A_{r,h}(a + x)$ are pairwise disjoint. Let $b = a + x$. We have $|b - a| = |x| < \delta$ and the sets $A_{r,h}(b)$ are pairwise disjoint for $r \in \{0, 1, \ldots, h\}$. This completes the proof. □

Theorem 1 *Let* $\mathbf{F}$ *be a field of characteristic 0 with a nontrivial absolute value* $|\ |$*. Let* A *be a finite subset of* $\mathbf{F}$ *that is an* h*-Sidon set. For every* $a \in \mathbf{F}$ *and* $\delta > 0$*, there exists an element* $b \in \mathbf{F}$ *such that*

$$|b - a| < \delta$$

and $A \cup \{b\}$ *is an* h*-Sidon set.*

Proof This follows immediately from Lemmas 1 and 2. □

Theorem 2 *Let* $\mathbf{F}$ *be a field of characteristic 0 with a nontrivial absolute value. Let* $\{\varepsilon_i : i \in \mathbf{N}\}$ *be a set of positive real numbers. For every subset* $\{a_i : i \in \mathbf{N}\}$ *of* $\mathbf{F}$*, there is an* h*-Sidon set* $\{b_i : i \in \mathbf{N}\}$ *in* $\mathbf{F}$ *that is an* ϵ*-perturbation of* A*.*

Proof We construct the set B by induction. Let $b_1 = a_1$. Let $k \geq 1$, and let $\{b_1, \ldots, b_k\}$ be an h-Sidon set in $\mathbf{F}$ such that $|b_i - a_i| < \varepsilon_i$ for all $i \in \{1, \ldots, k\}$. Applying Theorem 1 with $A = \{b_1, \ldots, b_k\}$ and $a = a_{k+1}$, we obtain an element $b_{k+1} \in \mathbf{F}$ such that $|b_{k+1} - a_{k+1}| < \varepsilon_{k+1}$ and $\{b_1, \ldots, b_k, b_{k+1}\}$ an h-Sidon set. This completes the proof. □

Theorem 3 *Let* $\mathbf{F}$ *be a field with a nontrivial absolute value. For every set* $A = \{a_i : i \in \mathbf{N}\}$ *in* $\mathbf{F}$ *there exists an* h*-Sidon set* $B = \{b_i : i \in \mathbf{N}\}$ *in* $\mathbf{F}$ *such that* $\lim_{i\to\infty} |a_i - b_i| = 0$.

Proof Apply Theorem 2 with a set $\{\varepsilon_i : i \in \mathbf{N}\}$ such that $\lim_{i\to\infty} \varepsilon_i = 0$. □

3 h-Sidon Sets of Vectors in $\mathbf{F}^n$

Let $\mathbf{F}$ be a field with a nontrivial absolute value $|\ |$. For vectors $\mathbf{a} = \begin{pmatrix} a_1 \\ \vdots \\ a_n \end{pmatrix} \in \mathbf{F}^n$ and $\mathbf{b} = \begin{pmatrix} b_1 \\ \vdots \\ b_n \end{pmatrix} \in \mathbf{F}^n$, define

$$\|\mathbf{a} - \mathbf{b}\| = \sum_{j=1}^{n} |a_j - b_j|.$$

Let $\mathbf{0} = \begin{pmatrix} 0 \\ \vdots \\ 0 \end{pmatrix} \in \mathbf{F}^n$ denote the zero vector. It follows from (5) that

$$\inf\{\|\mathbf{x}\| : \mathbf{x} \in \mathbf{F}^n \setminus \{\mathbf{0}\}\} = 0. \tag{8}$$

Let $c\mathbf{x}$ denote scalar multiplication of the vector $\mathbf{x}$ by the scalar c. Let $\mathbf{X}$ be a set of vectors. The *dilation of* $\mathbf{X}$ *by* c is the set $c * \mathbf{X} = \{c\mathbf{x} : \mathbf{x} \in \mathbf{X}\}$. Note that if h is a positive integer, then the dilated set $h * \mathbf{X}$ is a (usually proper) subset of the sumset $h\mathbf{X}$.

Lemma 3 *Let* $\mathbf{F}$ *be a field of characteristic 0 with a nontrivial absolute value* $|\ |$*, and let* $\mathbf{A}$ *be a finite set of vectors in* $\mathbf{F}^n$*. For every vector* $\mathbf{a} \in \mathbf{F}^n$ *and* $\delta > 0$*, there exists a vector* $\mathbf{b} \in \mathbf{F}^n$ *such that*

$$\|\mathbf{b} - \mathbf{a}\| < \delta$$

and the $h + 1$ *sets*

$$\mathbf{A}_{r,h}(\mathbf{b}) = r\mathbf{A} + (h - r)\mathbf{b}$$

for all $r \in \{0, 1, 2, \ldots, h\}$ *are pairwise disjoint sets of vectors in* $\mathbf{F}^n$.

Proof This is similar to the proof of Lemma 2. Let $\mathbf{a} \in \mathbf{F}^n$ and $\mathbf{x} \in \mathbf{F}^n$. For all $r, s \in \{0, 1, \ldots, h\}$ with $s < r$, we have

$$\begin{aligned} \emptyset &\neq \mathbf{A}_{r,h}(\mathbf{a} + \mathbf{x}) \bigcap \mathbf{A}_{s,h}(\mathbf{a} + \mathbf{x}) \\ &= (r\mathbf{A} + (h - r)(\mathbf{a} + \mathbf{x})) \bigcap (s\mathbf{A} + (h - s)(\mathbf{a} + \mathbf{x})) \end{aligned}$$

if and only if there is a sequence $\mathbf{a}_1, \mathbf{a}_2, \ldots, \mathbf{a}_{r+s}$ of vectors in $\mathbf{A}$ such that

$$\mathbf{a}_1 + \cdots + \mathbf{a}_r + (h-r)(\mathbf{a}+\mathbf{x}) = \mathbf{a}_{r+1} + \cdots + \mathbf{a}_{r+s} + (h-s)(\mathbf{a}+\mathbf{x}).$$

Equivalently,

$$(r-s)(\mathbf{a}+\mathbf{x}) = (\mathbf{a}_1 + \cdots + \mathbf{a}_r) - (\mathbf{a}_{r+1} + \cdots + \mathbf{a}_{r+s}) \in r\mathbf{A} - s\mathbf{A}.$$

Because $\mathbf{F}$ has characteristic 0, the positive integer $r-s$ is a unit in $\mathbf{F}$ and

$$\mathbf{a}+\mathbf{x} \in (r-s)^{-1} * (r\mathbf{A} - s\mathbf{A}).$$

Therefore,

$$\mathbf{x} \in \left((r-s)^{-1} * (r\mathbf{A} - s\mathbf{A})\right) - \mathbf{a}.$$

The set $\mathbf{A}$ is finite, and so the sets

$$\mathbf{C}_{r,s} = \left((r-s)^{-1} * (r\mathbf{A} - s\mathbf{A})\right) - \mathbf{a}$$

and

$$\mathbf{C} = \bigcup_{0 \leq s < r \leq h} \mathbf{C}_{r,s}$$

are also finite. The set $\mathbf{F}^n \setminus \mathbf{C}$ is infinite.

For all vectors $\mathbf{x} \in \mathbf{F}^n \setminus \mathbf{C}$, the sets $\mathbf{A}_{r,h}(\mathbf{a}+\mathbf{x})$ for $r \in \{0, 1, \ldots, h\}$ are pairwise disjoint. Because the set $\mathbf{C}$ is finite, we have

$$\delta_1 = \min\left(|\mathbf{c}| : \mathbf{c} \in \mathbf{C} \text{ and } \mathbf{c} \neq \mathbf{0}\right) > 0.$$

The absolute value on $\mathbf{F}^n$ is nontrivial, and so, by (8), there exists $\mathbf{x} \in \mathbf{F}^n$ with $0 < \|\mathbf{x}\| < \min(\delta_1, \delta)$. The inequality $0 < \|\mathbf{x}\| < \delta_1$ implies that $\mathbf{x} \notin \mathbf{C}$, and so the $h+1$ sets $\mathbf{A}_{r,h}(\mathbf{a}+\mathbf{x})$ are pairwise disjoint. Let $\mathbf{b} = \mathbf{a}+\mathbf{x}$. We have $|\mathbf{b}-\mathbf{a}| = |\mathbf{x}| < \delta$ and the sets $\mathbf{A}_{r,h}(\mathbf{b})$ are pairwise disjoint for $r \in \{0, 1, \ldots, h\}$. This completes the proof. □

Theorem 4 *Let* $\mathbf{F}$ *be a field of characteristic 0 with a nontrivial absolute value* $|\ |$. *Let* $\mathbf{A}$ *be a finite set of vectors in* $\mathbf{F}^n$ *that is an h-Sidon set. For every vector* $\mathbf{a} \in \mathbf{F}^n$ *and* $\delta > 0$, *there is a vector* $\mathbf{b} \in \mathbf{F}^n$ *such that*

$$|\mathbf{b} - \mathbf{a}| < \delta$$

and $\mathbf{A} \cup \{\mathbf{b}\}$ *is an h-Sidon set.*

Proof This follows immediately from Lemmas 1 and 3. □

Theorem 5 *Let* $\mathbf{F}$ *be a field of characteristic 0 with a nontrivial absolute value. Let* $\varepsilon = \{\varepsilon_i : i \in \mathbf{N}\}$ *be a set of positive real numbers. For every set* $\mathbf{A} = \{\mathbf{a}_i : i \in \mathbf{N}\}$ *of vectors in* $\mathbf{F}^n$*, there is an h-Sidon set* $\mathbf{B} = \{\mathbf{b}_i : i \in \mathbf{N}\}$ *in* $\mathbf{F}^n$ *that is an* ε*-perturbation of* $\mathbf{A}$.

Proof We construct the set $\mathbf{B}$ by induction. Let $\mathbf{b}_1 = \mathbf{a}_1$. Let $k \geq 1$, and let $\{\mathbf{b}_1, \ldots, \mathbf{b}_k\}$ be an h-Sidon set in $\mathbf{F}$ such that $|\mathbf{b}_i - \mathbf{a}_i| < \varepsilon_i$ for all $i \in \{1, \ldots, k\}$. Applying Theorem 4 with $\mathbf{A} = \{\mathbf{b}_1, \ldots, \mathbf{b}_k\}$ and $\mathbf{a} = \mathbf{a}_{k+1}$, we obtain an element $\mathbf{b}_{k+1} \in \mathbf{F}$ such that $|\mathbf{b}_{k+1} - \mathbf{a}_{k+1}| < \varepsilon_{k+1}$ and $\{\mathbf{b}_1, \ldots, \mathbf{b}_k, \mathbf{b}_{k+1}\}$ an h-Sidon set. This completes the proof. □

Theorem 6 *Let* $\mathbf{F}$ *be a field with a nontrivial absolute value. For every set of vectors* $\mathbf{A} = \{\mathbf{a}_i : i \in \mathbf{N}\}$ *in* $\mathbf{F}^n$ *there is an h-Sidon set* $\mathbf{B} = \{\mathbf{b}_i : i \in \mathbf{N}\}$ *in* $\mathbf{F}^n$ *such that* $\lim_{i\to\infty} |\mathbf{a}_i - \mathbf{b}_i| = 0$.

Proof Apply Theorem 2 with a set $\{\varepsilon_i : i \in \mathbf{N}\}$ such that $\lim_{i\to\infty} \varepsilon_i = 0$. □

4 Open Problems

1. The condition that the field $\mathbf{F}$ has characteristic 0 is used only to deduce (7) from (6). Let A be a countably infinite set in a field of characteristic $p > 0$ with a nontrivial absolute value. Is there an ε-perturbation of A that is an h-Sidon set?
2. Let A be an uncountably infinite set. Is there an ε-perturbation of A that is an h-Sidon set?

References

1. J. Cilleruelo and I. Z. Ruzsa, *Real and p-adic Sidon sequences*, Acta Sci. Math. (Szeged) **70** (2004), no. 3-4, 505–510.
2. K. O'Bryant, *A complete annotated bibliography of work related to Sidon sequences*, Electronic J. Combinatorics (2004), Dynamic Surveys DS 11.

Multiplicative Representations of Integers and Ramsey's Theorem

Melvyn B. Nathanson

Abstract Let $\mathcal{B} = (B_1, \ldots, B_h)$ be an h-tuple of sets of positive integers. Let $g_{\mathcal{B}}(n)$ count the number of representations of n in the form $n = b_1 \cdots b_h$, where $b_i \in B_i$ for all $i \in \{1, \ldots, h\}$. It is proved that $\liminf_{n\to\infty} g_{\mathcal{B}}(n) \geq 2$ implies $\limsup_{n\to\infty} g_{\mathcal{B}}(n) = \infty$.

Keywords Multiplicative representations · Representation functions · Erdős-Turán conjecture · Ramsey theory

2020 Mathematics Subject Classification 11B34 · 11B75 · 11N25 · 11B13 · 05C55 · 05D10

1 Does One Solution Imply Many Solutions?

A general problem: Suppose that an infinite set of equations has the property that every equation, or all but finitely many of the equations, or, perhaps, infinitely many of the equations have at least one solution. Does some equation have at least two solutions? Three solutions? Arbitrarily many solutions? Infinitely many solutions?

If the answer is 'no,' then modify the question. Suppose that an infinite set of equations has the property that every equation, or all but finitely many of the equations, or, perhaps, infinitely many of the equations have at least two solutions. Does some equation have at least three solutions? Four solutions? Arbitrarily many solutions? Infinitely many solutions?

And so on.

This paper considers a problem of this kind in multiplicative number theory.

Supported in part by a grant from the PSC-CUNY Research Award Program.

M. B. Nathanson (✉)
Lehman College (CUNY), Bronx, New York 10468, USA
e-mail: melvyn.nathanson@lehman.cuny.edu

M. B. Nathanson (ed.), *Combinatorial and Additive Number Theory IV*,
Springer Proceedings in Mathematics & Statistics 347,
https://doi.org/10.1007/978-3-030-67996-5_23

2 Additive Bases

Let $\mathbf{N} = \{1, 2, 3, \ldots\}$ be the set of positive integers and $\mathbf{N}_0 = \mathbf{N} \cup \{0\} = \{0, 1, 2, 3, \ldots, \}$ the set of nonnegative integers. Let $\mathbf{P} = \{2, 3, 5, 7, \ldots\}$ be the set of prime numbers.

The set A of nonnegative integers is an *additive basis of order h* if every nonnegative integer can be represented as the sum of h elements in the set. The set A is an *asymptotic additive basis of order h* if every sufficiently large integer is the sum of h elements in the set.

For every positive integer h, the *additive representation function* $r_{A,h}(n)$ counts the number of ordered representations of n as a sum of h elements of A:

$$r_{A,h}(n) = \text{card}\left\{(a_1, \ldots, a_h) \in A^h : a_1 + \cdots + a_h = n\right\}.$$

Thus, A is an additive basis of order h if $r_{A,h}(n) \geq 1$ for all $n \geq 0$, and an asymptotic additive basis of order h if $r_{A,h}(n) \geq 1$ for all $n \geq n_0$. In 1941, Erdős and Turán [1] asked if the representation function of every asymptotic additive basis of order 2 is unbounded. Equivalently, is it true that

$$\liminf_{n\to\infty} r_{A,2}(n) \geq 1 \quad \text{implies} \quad \limsup_{n\to\infty} r_{A,2}(n) = \infty?$$

This problem is still unsolved. More generally, for $h \geq 2$, is it true that

$$\liminf_{n\to\infty} r_{A,h}(n) \geq 1 \quad \text{implies} \quad \limsup_{n\to\infty} r_{A,h}(n) = \infty?$$

3 Multiplicative Bases and Multiplicative Systems

Let B be a set of positive integers. For every positive integer h, the *multiplicative representation function* $g_{B,h}(n)$ counts the number of ordered representations of n as a product of h elements of B:

$$g_{B,h}(n) = \text{card}\left\{(b_1, \ldots, b_h) \in B^h : b_1 \cdots b_h = n\right\}.$$

The set B is a *multiplicative basis of order h* if $g_{B,h}(n) \geq 1$ for all $n \geq 1$, and an *asymptotic multiplicative basis of order h* if $g_{B,h}(n) \geq 1$ for all sufficiently large integers n.

In 1964, Erdős [2] proved the following multiplicative analogue of the Erdős-Turán conjecture.

Theorem 1 (Erdős) *The multiplicative representation function of an asymptotic multiplicative basis of B of order 2 is unbounded.*

Equivalently,

$$\liminf_{n\to\infty} g_{B,2}(n) \geq 1 \qquad \text{implies} \qquad \limsup_{n\to\infty} g_{B,2}(n) = \infty.$$

Erdős' proof was graph theoretic. In 1985, Nešetřil and Rödl [5] gave a different proof using Ramsey's theorem. In Sect. 4, the Nešetřil-Rödl method is used to prove Theorem 1 for asymptotic multiplicative bases of order h for all $h \geq 2$.

There is a natural generalization of a multiplicative basis. Let $h \geq 2$ and let $\mathcal{B} = (B_1, \ldots, B_h)$ be an h-tuple of sets of positive integers. For every positive integer h, consider the *multiplicative representation function*

$$g_{\mathcal{B}}(n) = \operatorname{card}\{(b_1, \ldots, b_h) \in B_1 \times \cdots \times B_h : b_1 \cdots b_h = n\}.$$

The h-tuple $\mathcal{B}$ is a *multiplicative system of order h* if $g_{\mathcal{B}}(n) \geq 1$ for all $n \geq 1$, and an *asymptotic multiplicative system of order h* if $g_{\mathcal{B}}(n) \geq 1$ for all sufficiently large integers n.

For example, let $\mathbf{P} = \{2, 3, 5, \ldots\}$ be the set of prime numbers, and let

$$\mathbf{P} = \mathbf{P}_1 \cup \cdots \cup \mathbf{P}_h$$

be a partition of the prime numbers into h pairwise disjoint nonempty sets. For all $i = \{1, \ldots, h\}$, let B_i be the set of positive integers all of whose prime factors are in $\mathbf{P}_i$, and let $\mathcal{B} = (B_1, \ldots, B_h)$. Note that $1 \in B_i$ for all i. The fundamental theorem of arithmetic is equivalent to the statement that $g_{\mathcal{B}}(n) = 1$ for all $n = 1, 2, 3, \ldots$. Thus,

$$\liminf_{n\to\infty} g_{\mathcal{B}}(n) = \limsup_{n\to\infty} g_{\mathcal{B}}(n) = 1$$

and the analogue of Erdős' theorem on asymptotic multiplicative bases is false for multiplicative systems.

This example, however, is misleading. For $h \geq 2$, let $M(h)$ consist of all pairs (s, t) such that

$$s = \liminf_{n\to\infty} g_{\mathcal{B}}(n) \qquad \text{and} \qquad t = \limsup_{n\to\infty} g_{\mathcal{B}}(n)$$

for some asymptotic multiplicative system $\mathcal{B} = (B_1, \ldots, B_h)$ of order h. In Sect. 6 we prove the following result of Nathanson [4]:

$$M(h) = \{(1, t) : t \in \mathbf{N} \cup \{\infty\}\} \cup \{(s, \infty) : s = 2, \ldots, h\}.$$

This implies that

$$\liminf_{n\to\infty} g_{\mathcal{B}}(n) \geq 2 \qquad \text{implies} \qquad \limsup_{n\to\infty} g_{\mathcal{B}}(n) = \infty.$$

4 An Iterated Ramsey's Theorem and Multiplicative Bases

Let X be a set and let $[X]^k$ be the set of all subsets of X of cardinality k. We have $[X]^0 = \{\emptyset\}$ and $[X]^1 = \{\{x\} : x \in X\}$. Let $[X]^{<\omega}$ be the set of all finite subsets of X.

The infinite pigeonhole principle states that if an infinite set X is the union of a finite number of subsets, then at least one of the subsets is infinite. This is equivalent to Ramsey's theorem in the case $k = 1$.

Theorem 2 (Ramsey's theorem) *Let X be a countably infinite set. If k and r_k are nonnegative integers and*

$$[X]^k = C_0^k \cup C_1^k \cup \cdots \cup C_{r_k}^k$$

then there exists an infinite subset X_k of X and an integer $\varepsilon_k \in \{0, 1, \ldots, r_k\}$ such that

$$[X_k]^k \subseteq C_{\varepsilon_k}^k.$$

A standard reference is *Ramsey Theory* by Graham, Rothschild, and Spencer [3]. We use an iterated form of Ramsey's theorem to prove that the representation function of a multiplicative basis is unbounded (Theorem 5).

Theorem 3 (Iterated Ramsey's theorem) *Let X be a countably infinite set. For every nonnegative integer k, let $C_0^k, C_1^k, \ldots, C_{r_k}^k$ be (not necessarily pairwise disjoint) subsets of $[X]^k$ such that*

$$[X]^k = C_0^k \cup C_1^k \cup \cdots \cup C_{r_k}^k. \tag{1}$$

There exist an infinite decreasing sequence of infinite subsets of X

$$X \supseteq X_0 \supseteq X_1 \supseteq \cdots \supseteq X_k \supseteq \cdots \supseteq X_n \supseteq \cdots \tag{2}$$

and a sequence $(\varepsilon_k)_{k \in \mathbf{N}_0}$ with $\varepsilon_k \in \{0, 1, \ldots, r_k\}$ such that

$$[X_k]^k \subseteq C_{\varepsilon_k}^k \tag{3}$$

for all $k \in \mathbf{N}_0$.

It follows from (2) and (3) that $[X_n]^k \subseteq [X_k]^k \subseteq C_{\varepsilon_k}^k$ for all $k \in \{0, 1, \ldots, n\}$.

Proof The proof is by induction on k.

Let $X_0 = X$. We have

$$[X_0]^0 = [X]^0 = \{\emptyset\} = C_{\varepsilon_0}^0$$

for some $\varepsilon_0 \in \{0, 1, \ldots, r_0\}$. This proves the theorem for $k = 0$. The case $k = 1$ follows from the infinite pigeonhole principle.

Let $k \geq 1$ and let X_k be an infinite subset of X that satisfies (3). For $i \in \{0, 1, \ldots, r_{k+1}\}$, let

$$D_i^{k+1} = [X_k]^{k+1} \cap C_i^{k+1}.$$

Because $[X_k]^{k+1} \subseteq [X]^{k+1}$, we have

$$[X_k]^{k+1} = D_0^{k+1} \cup D_1^{k+1} \cup \cdots \cup D_{r_{k+1}}^{k-1}.$$

By Ramsey's theorem, there is an infinite subset X_{k+1} of X_k such that

$$[X_{k+1}]^{k+1} \subseteq D_{\varepsilon_{k+1}}^{k+1} \subseteq C_{\varepsilon_{k+1}}^{k+1}$$

for some $\varepsilon_{k+1} \in \{0, 1, \ldots, r_{k+1}\}$. This completes the proof. □

Theorem 4 *Let $h \geq 2$. Let X be a countably infinite set, and let $\mathcal{A}$ be a set of finite subsets of X. For every finite subset S of X, let $g_{\mathcal{A},h}(S)$ count the number of h-tuples $(A_1, \ldots, A_h)$ in $\mathcal{A} \times \cdots \times \mathcal{A}$ such that*

$$A_i \cap A_j = \emptyset \quad \text{for } 1 \leq i < j \leq h \tag{4}$$

and

$$S = \bigcup_{i=1}^{h} A_i. \tag{5}$$

Let W be an infinite set of positive integers such that $g_{\mathcal{A},h}(S) \geq 1$ for all $n \in W$ and for all but finitely many sets $S \in [X]^n$. For every integer n, there is a finite subset S of X with $g_{\mathcal{A},h}(S) \geq n$.

Proof For all $k \in \mathbf{N}_0$, let

$$C_1^k = [X]^k \cap \mathcal{A} \qquad \text{and} \qquad C_0^k = [X]^k \setminus C_1^k.$$

Thus,

$$[X]^k = C_0^k \cup C_1^k \qquad \text{and} \qquad C_0^k \cap C_1^k = \emptyset.$$

We apply Theorem 3 with $r_k = 1$ for all $k \in \mathbf{N}_0$. For every positive integer n, there is an infinite subset X_n of X such that, for all $k \in \{0, 1, \ldots, n\}$, there exists $\varepsilon_k \in \{0, 1\}$ with

$$[X_n]^k \subseteq C_{\varepsilon_k}^k.$$

Let $n \in W$. The infinite set X_n contains infinitely many subsets of cardinality n. Choose a subset S of X_n of cardinality n for which there exists an h-tuple $(A_1, \ldots, A_h)$ in $\mathcal{A} \times \cdots \times \mathcal{A}$ satisfying conditions (4) and (5). We have $k_i = |A_i| \in \{0, 1, \ldots, n\}$ for $i \in \{1, \ldots, h\}$. The sets A_i are pairwise disjoint, and so

$$\sum_{i=1}^{h} k_i = \sum_{i=1}^{h} |A_i| = |S| = n.$$

Because $A_i \subseteq S \subseteq X_n \subseteq X$ and $A_i \in \mathcal{A}$, we have

$$A_i \in [X_n]^{k_i} \cap \mathcal{A} \subseteq [X]^{k_i} \cap \mathcal{A} = C_1^{k_i}.$$

It follows that $\varepsilon_{k_i} = 1$ and

$$[X_n]^{k_i} \subseteq C_1^{k_i} \subseteq \mathcal{A} \tag{6}$$

for all $i \in \{1, \ldots, h\}$. Thus, every subset of X_n of cardinality k_i belongs to $\mathcal{A}$. The multinomial coefficient

$$\binom{n}{k_1, \ldots, k_h} = \frac{n!}{k_1! \cdots k_h!}$$

counts the number of partitions of S into pairwise disjoint subsets $A'_1, \ldots, A'_h$ of cardinalities $k_1, \ldots, k_h$, respectively. It follows from (6) that $A'_i \in \mathcal{A}$ for all $i \in \{1, \ldots, h\}$, and so

$$g_{\mathcal{A}}(S) \geq \binom{n}{k_1, \ldots, k_h}.$$

If $k_i < n$ for all i, then

$$g_{\mathcal{A}}(S) \geq \binom{n}{k_1, \ldots, k_h} \geq n.$$

If $k_j = n$ for some j, then $k_i = 0$ for all $i \neq j$ and $\binom{n}{k_1,\ldots,k_h} = 1$. In this case, $S \in \mathcal{A}$ and $[X_n]^n \subseteq C_1^n$. Thus, every n-element subset of X_n belongs to $\mathcal{A}$. Also, $\emptyset \in \mathcal{A}$. Let S' be a subset of X_n of cardinality $2n$. There are $\binom{2n}{n}$ partitions of S of the form $T = A_1 \cup A_2 \cup \cdots \cup A_h$, where $|A_1| = |A_2| = n$ and $|A_i| = 0$ for $i \in \{3, \ldots, h\}$. It follows that

$$g_{\mathcal{A}}(S') \geq \binom{2n}{n} \geq n.$$

This completes the proof. □

Theorem 5 *The multiplicative representation function of an asymptotic multiplicative basis B of order h is unbounded.*

Equivalently,

$$\liminf_{n\to\infty} g_{B,h}(n) \geq 1 \quad \text{implies} \quad \limsup_{n\to\infty} g_{B,h}(n) = \infty.$$

Proof A *square-free (or quadratfrei) integer* is a positive integer that is the product of distinct primes. Equivalently, a positive integer n is square-free if it is not divisible by the square of a prime. Let $Q = \{1, 2, 3, 5, 6, 7, 10, 11, \ldots\}$ be the set of square-free integers. Define the function $\Phi : Q \to [\mathbf{P}]^{<\omega}$ as follows: For $q \in Q$,

$$\Phi(q) = \{p \in \mathbf{P} : p \text{ divides } q\}.$$

The function Φ is a bijection, with inverse function $\Phi^{-1} : [\mathbf{P}]^{<\omega} \to Q$ defined as follows: For $S \in [\mathbf{P}]^{<\omega}$,

$$\Phi^{-1}(S) = \prod_{p \in S} p.$$

If n is square-free and if $b_1, \ldots, b_h$ are positive integers such that $n = b_1 \cdots b_h$, then the integers $b_1, \ldots, b_h$ are square-free and pairwise relatively prime. Let $S = \Phi(n)$ and let $S_i = \Phi(b_i)$ for $i \in \{1, \ldots, h\}$. It follows that $S = \bigcup_{i=1}^{h} S_i$ and that $S_1, \ldots, S_h$ are pairwise disjoint sets in $[\mathbf{P}]^{<\omega}$.

Conversely, let $S_1, \ldots, S_h$ be pairwise disjoint sets in $[\mathbf{P}]^{<\omega}$ and $S = \bigcup_{i=1}^{h} S_i$. Let $n = \Phi^{-1}(S)$ and let $b_i = \Phi^{-1}(S_i)$ for $i \in \{1, \ldots, h\}$. We have

$$n = \Phi^{-1}(S) = \prod_{i=1}^{h} \Phi^{-1}(S_i) = \prod_{i=1}^{h} b_i.$$

Thus, there is a one to one correspondence between partitions of a finite set of primes into h pairwise disjoint subsets and representations of a square-free integer as a product of h factors.

Let B be an asymptotic multiplicative basis of order h. Let

$$\mathcal{A} = \{\Phi^{-1}(b) : b \in B \cap Q\} \subseteq [\mathbf{P}]^{<\omega}.$$

Because B is an asymptotic multiplicative basis, every square-free integer, with a finite number of exceptions, has a representation as a product of h pairwise relatively prime square-free integers in B. Thus, with a finite number of exceptions, every set $S \in [\mathbf{P}]^{<\omega}$ has a representation as the union of h pairwise disjoint sets in $\mathcal{A}$. Theorem 4 implies that for every integer n there is a set $S \in [\mathbf{P}]^{<\omega}$ with at least n representations as the union of h pairwise disjoint sets in $\mathcal{A}$. It follows that the integer $q = \Phi(S) \in Q$ has at least n representations as a product of h pairwise relatively prime square-free integers in B, and so $g_{B,h}(q) \geq n$. This completes the proof. □

5 A Doubly Iterated Ramsey's Theorem

We use an iterated form of the iterated Ramsey's theorem to study the representation function of a multiplicative system (Theorem 8). We need the following elementary Boolean identity.

Lemma 1 *Let Y and I be sets. For all $i \in I$, let J_i be a set, and let $\{C_{i,j_i} : j_i \in J_i\}$ be a set of subsets of Y such that*

$$Y = \bigcup_{j_i \in J_i} C_{i,j_i}.$$

Then

$$Y = \bigcup_{(j_i) \in \prod_{i \in I} J_i} \bigcap_{i \in I} C_{i,j_i}$$

where $\prod_{i \in I} J_i$ is the Cartesian product of the sets J_i.

Proof Because the sets C_{i,j_i} are subsets of Y, we have

$$\bigcup_{(j_i) \in \prod_{i \in I} J_i} \bigcap_{i \in I} C_{i,j_i} \subseteq Y.$$

Let $y \in Y$. For all $i \in I$, there exists $j_i \in J_i$ such that $y \in C_{i,j_i}$ and so

$$y \in \bigcap_{i \in I} C_{i,j_i}.$$

Therefore,

$$y \in \bigcup_{(j_i) \in \prod_{i \in I} J_i} \bigcap_{i \in I} C_{i,j_i}$$

and

$$Y \subseteq \bigcup_{(j_i) \in \prod_{i \in I} J_i} \bigcap_{i \in I} C_{i,j_i}.$$

This completes the proof. □

Theorem 6 (Doubly iterated Ramsey's theorem) *Let X be a countably infinite set. Let I_k be a finite set for every nonnegative integer k. For all $i \in I_k$, let J_i^k be a finite set and let $\{C_{i,j_i}^k : j_i \in J_i^k\}$ be a set of (not necessarily pairwise disjoint) subsets of $[X]^k$ such that*

$$[X]^k = \bigcup_{j_i \in J_i} C_{i,j_i}^k. \tag{7}$$

There is an infinite decreasing sequence of infinite subsets of X

$$X \supseteq X_0 \supseteq X_1 \supseteq \cdots \supseteq X_k \supseteq X_{k+1} \supseteq \cdots \tag{8}$$

and, for all $k \in \mathbf{N}_0$ and for all $i \in I_k$, there is an integer $\varepsilon_{i,k} \in J_i^k$ such that

$$[X_k]^k \subseteq C^k_{i,\varepsilon_{i,k}}. \tag{9}$$

It follows from (8) and (9) that $[X_n]^k \subseteq [X_k]^k \subseteq C^k_{i,\varepsilon_{i,k}}$ for all $k \in \{0, 1, \ldots, n\}$.

Proof It suffices to prove that there exists $(\varepsilon_{i,k}) = (\varepsilon_{i,k})_{i \in I_k} \in \prod_{i \in I_k} J_i^k$ such that

$$[X_k]^k \subseteq \bigcap_{i \in I_k} C^k_{i,\varepsilon_{i,k}}.$$

Applying Lemma 1 with $Y = [X]^k$, we obtain

$$[X]^k = \bigcup_{(\varepsilon_{i,k}) \in \prod_{i \in I_k} J_i^k} \bigcap_{i \in I} C_{i,\varepsilon_{i,k}}$$

The set $\prod_{i \in I_k} J_i^k$ is finite because the sets I_k and J_i^k are finite. The doubly iterated Ramsey's theorem follows immediately from the iterated Ramsey's theorem (Theorem 3). This completes the proof. □

Theorem 7 *Let $h \geq 2$. Let X be a countably infinite set, and let $\mathcal{A}^* = (\mathcal{A}_1, \ldots, \mathcal{A}_h)$ be an h-tuple of sets of finite subsets of X. For every subset S of X, let $g_{\mathcal{A}^*}(S)$ count the number of h-tuples $(A_1, \ldots, A_h)$ in $\mathcal{A}_1 \times \cdots \times \mathcal{A}_h$ such that*

$$A_i \cap A_j = \emptyset \quad \text{for } 1 \leq i < j \leq h \tag{10}$$

and

$$S = \bigcup_{i=1}^{h} A_i. \tag{11}$$

Let W be an infinite set of positive integers such that $g_{\mathcal{A}^*}(S) \geq 2$ for all $n \in W$ and for all but finitely many sets $S \in [X]^n$. For every integer n, there is a finite subset S of X with $g_{\mathcal{A}^*}(S) \geq n$.

Proof For all $k \in \mathbf{N}_0$, let $I_k = \{1, \ldots, h\}$ and, for all $i \in I_k$, let $J_i^k = \{0, 1\}$. The sets

$$C^k_{i,1} = [X]^k \cap \mathcal{A}_i$$

and

$$C^k_{i,0} = [X]^k \setminus C^k_{1,i}$$

partition $[X]^k$. Thus,

$$[X]^k = C^k_{i,0} \cup C^k_{i,1} \quad \text{and} \quad C^k_{i,0} \cap C^k_{i,1} = \emptyset.$$

By Theorem 6, there is an infinite decreasing sequence

$$X \supseteq X_0 \supseteq X_1 \supseteq \cdots \supseteq X_k \supseteq \cdots \supseteq X_n \supseteq \cdots$$

of infinite subsets of X such that, for all $n \in \mathbf{N}_0$, the set X_n has the following property: For all $k \in \{0, 1, \ldots, n\}$ and $i \in I_k$, there exists $\varepsilon_{i,k} \in \{0, 1\}$ with

$$[X_n]^k \subseteq [X_k]^k \subseteq C^k_{i,\varepsilon_{i,k}}.$$

If there is a subset A_i of X_n of cardinality $k_i \leq n$ with $A_i \in \mathcal{A}_i$, then $A_i \in C^{k_i}_{i,1}$ and so $[X_n]^{k_i} \subseteq C^{k_i}_{i,1}$, that is, every subset of X_n of cardinality k_i is in $\mathcal{A}_i$.

Let $n \in W$. Because X_n is an infinite subset of X, there is a set $S \in [X_n]^n$ with $g_{\mathcal{A}^*}(S) \geq 2$. Thus, there are distinct h-tuples $(A_1, \ldots, A_h)$ and $(A'_1, \ldots, A'_h)$ in $\mathcal{A}_1 \times \cdots \times \mathcal{A}_h$ that satisfy conditions (10) and (11). Let

$$|A_i| = k_i \qquad \text{and} \qquad |A'_i| = k'_i$$

for $i \in \{1, \ldots, h\}$. The sets $A_1, \ldots, A_h$ are pairwise disjoint and the sets $A'_1, \ldots, A'_h$ are pairwise disjoint, and so

$$\sum_{i=1}^{h} k_i = \sum_{i=1}^{h} k'_i = n.$$

Because $A_i \in [X_n]^{k_i} \cap \mathcal{A}_i$ and $A'_i \in [X_n]^{k'_i} \cap \mathcal{A}_i$, every subset of X_n of cardinality k_i or k'_i is in $\mathcal{A}_i$ It follows that if $S = \bigcup_{i=1}^{h} B_i$ is a partition of S with $|B_i| = k_i$ for all $i \in \{1, \ldots, h\}$, then $B_i \in \mathcal{A}_i$ for all $i \in \{1, \ldots, h\}$. The number of such partitions is the multinomial coefficient

$$\binom{n}{k_1, \ldots, k_h}.$$

If $k_i < n$ for all $i \in \{1, \ldots, h\}$, then

$$\binom{n}{k_1, \ldots, k_h} \geq n.$$

Similarly, if $S = \bigcup_{i=1}^{h} B'_i$ is a partition of S with $|B'_i| = k'_i$ for all $i \in \{1, \ldots, h\}$, then $B'_i \in \mathcal{A}_i$ for all $i \in \{1, \ldots, h\}$. The number of such partitions is the multinomial coefficient

$$\binom{n}{k'_1, \ldots, k'_h}.$$

If $k'_i < n$ for all $i \in \{1, \dots, h\}$, then

$$\binom{n}{k'_1, \dots, k'_h} \geq n.$$

If $k_i = k'_j = n$ for some $i, j \in \{1, \dots, h\}$, then $B_i = B'_j = S$ and $B_\ell = \emptyset$ for all $\ell \neq i$ and $B'_\ell = \emptyset$ for all $\ell \neq j$. Because $(B_1, \dots, B_h) \neq (B'_1, \dots, B'_h)$, we have $i \neq j$. Every subset of X_n of cardinality n is in both $\mathcal{A}_i$ and $\mathcal{A}_j$. Also, $\emptyset \in \mathcal{A}_i$ for all $i \in \{1, \dots, h\}$.

Let S_1 and S_2 be disjoint subsets of $[X_n]^n$, and let $S = S_1 \cup S_2$. We have $|S| = 2n$. There are $\binom{2n}{n}$ subsets U of S of cardinality n. For each of these sets U, let $B_i = U$, $B_j = S \setminus U$, and $B_\ell = \emptyset$ for $\ell \neq i, j$. This gives $\binom{2n}{n} > n$ ordered h-tuples $(B_1, \dots, B_h) \in \mathcal{A}_1 \times \cdots \times \mathcal{A}_h$ that satisfy conditions (10) and (11). This completes the proof. □

6 Representation Functions of Multiplicative Systems

Theorem 8 *For $h \geq 2$, let $M(h)$ consist of all pairs (s, t) such that*

$$s = \liminf_{n\to\infty} g_{\mathcal{B}}(n) \quad \text{and} \quad t = \limsup_{n\to\infty} g_{\mathcal{B}}(n)$$

for some asymptotic multiplicative system $\mathcal{B}$ of order h. Then

$$M(h) = \{(1, t) : t \in \mathbf{N} \cup \{\infty\}\} \cup \{(s, \infty) : s \in \{2, \dots, h\}\}.$$

Proof For every asymptotic multiplicative system $\mathcal{B} = (B_1, \dots, B_h)$ of order h, we have

$$s = \liminf_{n\to\infty} g_{\mathcal{B}}(n) \geq 1.$$

Let p be a prime number. For all $(b_1, \dots, b_h) \in B_1 \times \cdots \times B_h$, we have $p = b_1 \cdots b_h$ if and only if $b_j = p$ for some $j \in \{1, \dots, h\}$ and $b_i = 1$ for all $i \neq j$. It follows that $g_{\mathcal{B}}(p) \leq h$, and so

$$s = \liminf_{n\to\infty} g_{\mathcal{B}}(n) \leq h.$$

For every positive integer t, define the sets

$$B_1 = \mathbf{N}, \quad B_2 = \{2^k : 0 \leq k \leq t - 1\}, \quad B_i = \{1\} \ \text{ for } i = 3, \dots, h$$

and let $\mathcal{B} = (B_1, \ldots, B_h)$. If n is an odd positive integer, then

$$g_{\mathcal{B}}(n) = 1$$

and so

$$\liminf_{n\to\infty} g_{\mathcal{B}}(n) = 1.$$

Let $n = 2^{\ell-1}m$, where m is odd. We have $(b_1, b_2, \ldots, b_h) \in B_1 \times B_2 \times \cdots \times B_h$ and $n = b_1 b_2 \cdots b_h$ if and only if

$$(b_1, b_2, \ldots, b_h) = \left(2^{\ell-1-i}m, 2^i, 1, \ldots, 1\right)$$

for some $i \in \{0, 1, \ldots, \min(\ell-1, t-1)\}$. It follows that $g_{\mathcal{B}}(n) = \min(\ell, t)$ and so

$$\limsup_{n\to\infty} g_{\mathcal{B}}(n) = t.$$

Therefore, $(1, t) \in M(h)$ for all $t \in \mathbf{N}$.

Let $\mathcal{B} = (B_1, \ldots, B_h)$, where

$$B_1 = \mathbf{N}, \quad B_2 = \{2^k : k \in \mathbf{N}_0\}, \quad B_i = \{1\} \quad \text{for } i = 3, \ldots, h.$$

We have

$$\liminf_{n\to\infty} g_{\mathcal{B}}(n) = 1 \qquad \text{and} \qquad \limsup_{n\to\infty} g_{\mathcal{B}}(n) = \infty$$

and so $(1, \infty) \in M(h)$.

For $s \in \{2, 3, \ldots, h\}$, let $\mathcal{B} = (B_1, \ldots, B_h)$ be the multiplicative system defined by

$$B_1 = \mathbf{N}, \qquad B_i = \mathbf{P} \cup \{1\} \quad \text{for } i = 2, \ldots, s,$$

and

$$B_i = \{1\} \quad \text{for } i = s+1, \ldots, h.$$

We have

$$g_{\mathcal{B}}(p) = s \qquad \text{for all } p \in \mathbf{P}$$

and

$$g_{\mathcal{B}}(n) \geq s \qquad \text{for all } n \geq 2.$$

Therefore,

$$\liminf_{n\to\infty} g_{\mathcal{B}}(n) = s.$$

If $n \in \mathbf{N}$ has k distinct prime factors, then $g_{\mathcal{B}}(n) \geq k$ and so

$$\limsup_{n\to\infty} g_{\mathcal{B}}(n) = \infty.$$

It follows that

$$\{(s, \infty) : s \in \{2, \ldots, h\}\} \subseteq M(h).$$

We must prove that if $(s, t) \in M(h)$ and $s \geq 2$, then $t = \infty$. As in the proof of Theorem 5, we consider the set Q of square-free integers, and the functions

$$\Phi : Q \to [\mathbf{P}]^{<\omega} \qquad \text{and} \qquad \Phi^{-1} : [\mathbf{P}]^{<\omega} \to Q$$

defined by

$$\Phi(q) = \{p \in \mathbf{P} : p \text{ divides } q\} \qquad \text{for all } q \in Q$$

and

$$\Phi^{-1}(S) = \prod_{p\in S} p \qquad \text{for all } S \in [\mathbf{P}]^{<\omega}.$$

There is a one to one correspondence between ordered partitions of a finite set of primes into h pairwise disjoint subsets and ordered representations of a square-free integer as a product of h factors. Let $S_1, \ldots, S_h$ be pairwise disjoint sets in $[\mathbf{P}]^{<\omega}$, and let $S = \bigcup_{i=1}^h S_i$. Let $n = \Phi^{-1}(S)$ and $b_i = \Phi^{-1}(S_i)$ for all $i \in \{1, \ldots, h\}$. We have

$$n = \Phi^{-1}(S) = \prod_{i=1}^h \Phi^{-1}(S_i) = \prod_{i=1}^h b_i.$$

Conversely, if $b_1, \ldots, b_h$ are pairwise relatively prime square-free integers and $n = b_1 \cdots b_h$ with $b_i \in B_i$, then the sets $\Phi(b_1), \ldots, \Phi(b_h)$ are pairwise disjoint and $\Phi(n) = \bigcup_{i=1}^h \Phi(b_i)$.

For all $i \in \{1, \ldots, h\}$, let

$$\mathcal{A}_i = \{\Phi(b_i) : b_i \in Q \cap B_i\} \subseteq [\mathbf{P}]^{<\omega}$$

where $Q \cap B_i$ is the set of square-free integers in the set B_i. Let $\mathcal{A}^* = (\mathcal{A}_1, \ldots, \mathcal{A}_h)$. We have $b_i \in Q \cap B_i$ if and only if $\Phi(b_i) \in \mathcal{A}_i$, and so

$$g_{\mathcal{B}}(q) = g_{\mathcal{A}^*}(\Phi(q))$$

for all square-free integers q. The inequality $\liminf_{q\in Q} g_{\mathcal{B}}(q) \geq 2$ is equivalent to $\liminf_{S\in[\mathbf{P}]^{<\omega}} g_{\mathcal{A}^*}(S) \geq 2$. Theorem 7 implies $\limsup_{S\in[\mathbf{P}]^{<\omega}} g_{\mathcal{A}^*}(S) = \infty$, and so

$$\begin{aligned}\limsup_{n\to\infty} g_{\mathcal{B}}(n) &= \limsup_{q\in Q} g_{\mathcal{B}}(q) \\ &= \limsup_{q\in Q} g_{\mathcal{A}^*}(\Phi(q)) \\ &= \limsup_{S\in[\mathbf{P}]^{<\omega}} g_{\mathcal{A}^*}(S) \\ &= \infty.\end{aligned}$$

This completes the proof. □

Corollary 1 *For every asymptotic multiplicative system $\mathcal{B}$ of order h,*

$$\liminf_{n\to\infty} g_{\mathcal{B}}(n) \geq 2 \quad \textit{implies} \quad \limsup_{n\to\infty} g_{\mathcal{B}}(n) = \infty.$$

7 Open Problems

Theorems 5 and 8 are answers to questions of the following kind. Suppose that a representation function has at least one, or two, or three, or ...solutions. Does this imply that there are occasionally even more solutions? Does this imply that the representation function can be arbitrarily large? Here are related open problems.

1. Let $\mathcal{A}$ be a set of nonnegative integers that is an asymptotic additive basis of order 2. Suppose that every sufficently large integer has at least 1000 representations as the sum of two elements of $\mathcal{A}$? Does some integer have 1010 representations?
2. Let $\mathcal{A}$ be a set of vectors in $\mathbf{Z}^n$. The set $\mathcal{A}$ is an *inner product basis* for a set W of integers if every integer in W is the inner product of two vectors in $\mathcal{A}$. For what infinite sets W with inner product basis $\mathcal{A}$ is the representation function bounded? For what infinite sets W with inner product basis $\mathcal{A}$ is the representation function unbounded?
3. Associated to every $n \times n$ matrix M is the quadratic form Q_M, defined by $Q_M(\mathbf{x}) = \mathbf{x}^t M\mathbf{x}$ for $\mathbf{x} \in \mathbf{R}^n$. Let M be an $n \times n$ matrix with nonnegative integral coordinates. Let $\mathcal{A}$ be a set vectors with nonnegative integral coordinates, and let W be an infinite set of nonnegative integers such that $W \subseteq Q_M(\mathcal{A})$. When is the representation function bounded? When is the representation function unbounded?

References

1. P. Erdős and P. Turán, *On a problem of Sidon in additive number theory, and on some related problems*, J. London Math. Soc. **16** (1941), 212–215.
2. P. Erdős, *On the multiplicative representation of integers*, Israel J. Math. **2** (1964), 251–261.

3. R. L. Graham, B. L. Rothschild, and J. H. Spencer, *Ramsey Theory*, second ed., Wiley-Interscience Series in Discrete Mathematics and Optimization, John Wiley & Sons, Inc., New York, 1990, A Wiley-Interscience Publication.
4. M. B. Nathanson, *Multiplicative representations of integers*, Israel J. Math. **57** (1987), no. 2, 129–136.
5. J. Nešetřil and V. Rödl, *Two proofs in combinatorial number theory*, Proc. Amer. Math. Soc. **93** (1985), 185–188.

On Distinct Consecutive Differences

Imre Ruzsa, George Shakan, József Solymosi, and Endre Szemerédi

Abstract We show that if $A = \{a_1 < a_2 < \ldots < a_k\}$ is a set of real numbers such that the differences of the consecutive elements are distinct, then for and finite $B \subset \mathbb{R}$,

$$|A + B| \gg |A|^{1/2}|B|.$$

The bound is tight up to the constant.

Keywords Sumset inequalities · Convex sets · Additive combinatorics

1 Introduction

Given two sets $A, B \subset \mathbb{R}$, the *sumset* of A and B is

$$A + B = \{a + b : a \in A \text{ and } b \in B\}.$$

We say a finite set $A = \{a_1 < a_2 < \cdots < a_k\}$ of real numbers with the property that

$$a_i - a_{i-1} < a_{i+1} - a_i, \tag{1}$$

I. Ruzsa · E. Szemerédi
Alfréd Rényi Institute of Mathematics, Budapest, Hungary
e-mail: ruzsa@renyi.hu

E. Szemerédi
e-mail: szemered@renyi.hu

G. Shakan
Department of Mathematics at University of Oxford, Oxford, England
e-mail: shakan@maths.ox.ac.uk

J. Solymosi (✉)
Department of Mathematics, University of British Columbia, Vancouver, Canada
e-mail: solymosi@math.ubc.ca

M. B. Nathanson (ed.), *Combinatorial and Additive Number Theory IV*,
Springer Proceedings in Mathematics & Statistics 347,
https://doi.org/10.1007/978-3-030-67996-5_24

for any $1 < i < k$ is *convex*. There is the following conjecture of Erdős, which motivates the current work. We use Vinogradov's notation so that $a \ll b$ means $a = O(b)$.

Conjecture 1 Let $A \subset \mathbb{R}$ be convex. Then for any $\epsilon > 0$,

$$|A + A| \gg_\epsilon |A|^{2-\epsilon}.$$

Conjecture 1 asserts that the local hypothesis of being convex implies the global consequence of having a large sumset. The following example which shows that some form of the ϵ is necessary.

Example 1 Let k be a positive integer and $A = \{i^2 : 1 \leq i \leq k\}$. Then $A + A$ is contained in the set of integers of size $\leq 2k^2$ that can be represented as the sum of two squares. Fermat showed that such integers must have a prime factorization where all the primes equivalent to 3 modulo 4 appear to an even power. The sieve implies

$$|A + A| \ll k^2 \prod_{\substack{p \equiv 1 \pmod 4 \\ p \leq k}} (1 - 1/p) \ll k^2/\log(k)^{1/2}.$$

One trivial obstruction to a sumset being small is that it is a large subset of an arithmetic progression. It is easy to see that any convex subset of an arithmetic progression has size $\ll \sqrt{n}$, which supports Conjecture 1. On the other hand, no such argument can establish the growth demanded by Conjecture 1.

The first progress towards Conjecture 1 is due to Hegyvári [5], who proved that if A is convex then

$$|A + A| \gg k \frac{\log k}{\log \log k}.$$

Hegyvári's result was later improved by Elekes et al. [3], who showed if A is convex then

$$|A + B| \gg k^{3/2}, \tag{2}$$

for any set B with $|B| = k$. Garaev [4] later provided a different proof in the case $B = A$. Solymosi and Szemerédi[1] proved that there is a constant $c > 0$ such that if A is a large enough convex set of numbers then

$$|A + A| \gg |A|^{3/2+c}.$$

Schoen and Shkredov improved the result in [9] by showing that the constant c in the above inequality can be arbitrarily close to $1/18$ (and $1/10$ is sumset is replaced by difference set). The current best result towards Conjecture 1 is that c can be taken arbitrarily close to $5/74$, which follows from the Schoen-Shkredov argument and a later paper of Shkredov [10, Theorem 2].

[1] Unpublished result.

We extend this result of Elekes, Nathanson and Ruzsa, (2), to sets with distinct consecutive differences. We say a set A has distinct consecutive differences if for any $1 \leq i, j \leq k$, $a_{i+1} - a_i = a_{j+1} - a_j$ implies $i = j$.

Theorem 1 *Let A and B be finite sets of real numbers. If A has distinct consecutive differences, then*

$$|A + B| \gg |A||B|^{1/2}$$

In particular, if $|A| = |B|$ then

$$|A + B| \gg |A|^{3/2}$$

The basic idea behind the proof is the following. The sumset $A + B$ consists of $|B|$ translates of A. The translates of two consecutive elements of A are typically not "far" from each other in the sumset $A + B$. Also, from a translate of two consecutive elements, $b + a_i$,$b + a_{i+1}$ we can recover the value of b, since all of the consecutive differences are distinct. Then the number of "close" pairs in $A + B$ should be large, around $|A||B|$, therefore $A + B$ is also large.

In the second part of the paper we extend Theorem 1 for two sets. As an application we show that for any convex function F, and finite sets of real numbers, A, B, and C, if $|A| = |B| = |C| = n$, then $\max\{|A + B|, |F(A) + C \geq cn^{5/4}$.

Along the same lines of the proof, one can prove a statistical analog of Theorem 1 which we state without working out the details of the proof.

Theorem 2 *Let $A = \{a_1 < a_2, \cdots < a_k\} \subset \mathbb{R}$. Suppose the set of consecutive differences*

$$D = \{a_{i+1} - a_i : 1 \leq i \leq k - 1\},$$

is large, that is $|D \geq \delta|A|$. Then for any finite $B \subset \mathbb{R}$,

$$|A + B| \gg_\delta |A||B|^{1/2}.$$

Parts of this work was available earlier in unpublished manuscripts, so it received some references in further works, such as [7–9].

2 Distinct Consecutive Differences

2.1 *Proof of Theorem 1*

Proof Let

$$A = \{a_1 < \cdots < a_k\},$$

and

$$S = A + B = \{c_1 < \cdots < c_r\}.$$

Thus $r = |A + B|$. Fix $b \in B$ and $1 \leq i \leq k - 1$ and set

$$J_b(i) := (a_i + b, a_{i+1} + b] \cap S.$$

Thus the $J_b(i)$ are disjoint as i varies and

$$\sum_{i=1}^{k-1} |J_b(i)| \leq |A + B| \tag{3}$$

We now say $J_b(i)$ is *good* if $|J_b(i)| \leq 100|A + B|k^{-1}$. By (3) and pigeon-hole, we see that for a fixed b, the number of good $J_b(i)$ is $\gg k$. Thus the total number of good $J_b(i)$ is $\gg k|B|$. Furthermore the sets $J_b(i)$ are distinct. Indeed, given the interval $(a_i + b, a_{i+1} + b]$, we can recover $a_{i+1} - a_i$ by subtracting the two end points. Since A has distinct consecutive differences, this allows as to recover a_i, a_{i+1} and then b.

On the other hand, the number of choices of $\leq 100|A + B|k^{-1}$ consecutive elements in S is

$$\ll |A + B|^2 k^{-1},$$

which is also an upper bound for the number of good $J_b(i)$. Combining our upper and lower bounds for the number of good $J_b(i)$, we find

$$k|B| \ll |A + B|^2 k^{-1},$$

as desired.

2.2 *Distinct Pairs of Consecutive Differences*

For an application, it is useful to extend Theorem 1 to a more general setting. Let

$$A = \{a_1 < a_2 < \cdots < a_k\},$$

and

$$A' = \{a'_1 < a'_2 < \cdots < a'_k\},$$

be nonempty sets of real numbers. For $1 \leq i \leq k - 1$, let

$$d_i = a_{i+1} - a_i, \quad d'_i = a'_{i+1} - a'_i.$$

The sets A and A' have *distinct pairs of consecutive differences* if the ordered pairs (d_i, d_i') are distinct.

Theorem 3 *Let A and A' be finite sets of real numbers with k elements and distinct pairs of consecutive differences. Let B, and B' be arbitrary finite sets of real numbers. Then*

$$|A+B|\cdot|A'+B'| \gg \left(k^3|B||B'|\right)^{1/2}.$$

If $k = |A| = |A'| = |B| = |B'|$, then

$$|A+B|\cdot|A'+B'| \gg k^{5/2}.$$

The proof is a two dimensional version of the proof of Theorem 1, which we now give.

Proof Let

$$S = A+B = \{c_1 < \cdots < c_r\},$$

and

$$S' = A'+B' = \{c_1' < \cdots < c_{r'}'\}.$$

Fix $b \in B$ and $b' \in B'$ and $1 \leq i \leq k-1$ and set

$$J_{b,b'}(i) = (a_i + b, a_{i+1} + b] \cap S \times (a_i' + b', a_{i+1}' + b'] \cap S' \subset S \times S'.$$

We let $\pi_1 : \mathbb{R}^2 \to \mathbb{R}$ be projection onto the first coordinate and π_2 be projection to the second. We say $J_{b,b'}(i)$ is *good* if

$$|\pi_1(J_{b,b'}(i))| \leq 100|A+B|k^{-1}, \quad |\pi_2(J_{b,b'}(i))| \leq 100|A'+B'|k^{-1}.$$

For a fixed $b \in B$ and $b' \in B'$,

$$\sum_{i=1}^{k-1} |\pi_1(J_{b,b'}(i))| \leq |A+B|, \quad \sum_{i=1}^{k-1} |\pi_2(J_{b,b'}(i))| \leq |A'+B'|.$$

Thus by pigeon-hole, the number of good $J_{b,b'}(i)$ is $\gg k$. Allowing b and b' to vary, we conclude the total number of good $J_{b,b'}(i)$ is $\gg k|B||B'|$. Furthermore, the sets $J_{b,b'}(i)$ are distinct. Indeed given intervals $(a_i + b, a_{i+1} + b]$ and $(a_i' + b', a_{i+1}' + b']$, we can recover $a_{i+1} - a_i$ and $a_{i+1}' - a_i'$. Since A and A' have distinct pairs of consecutive differences, we may recover $a_i, a_{i+1}, a_i', a_{i+1}'$ and then b, b'.

On the other hand, the number boxes $I \times I'$ in $S \times S'$ satisfying

$$|I \leq 100|S|k^{-1}, \quad |I'| \leq 100|S'|k^{-1},$$

is $\ll |S|^2|S'|^2k^{-2}$. Thus the number of good $J_{b,b'}(i)$ is $\ll |S|^2|S'|^2k^{-2}$. Combining our lower and upper bounds for the number of good $J_{b,b'}(i)$ we find

$$k|B||B'| \ll |A+B|^2|A'+B'|^2k^{-2},$$

which completes the proof.

A simple consequence of Theorem 3 is the following result, which was first proved by Elekes et al. [3].

Theorem 4 *For any strictly convex real function F, and finite sets of real numbers, A, B, and C, if $|A| = |B| = |C| = k$, then*

$$\max\{|A+B|, |F(A)+C| \gg k^{5/4}.$$

In particular,

$$|A+F(A)| \gg k^{5/4}.$$

Proof The two sets, A and $F(A)$, have distinct pairs of consecutive differences. For the second inequality set $B = F(A)$ and $C = A$.

Based on the work of Schoen and Shkredov [9], Li and Roche-Newton [8] improved the bound in Theorem 4.

3 A Construction for the Lower Bound

In this section we show that the bound in Theorem 1 is tight up to a constant multiplier.

Let S be a Sidon set, that is a set for which all the nonzero differences are distinct. Suppose further that $|S|$ is odd and let us choose the elements of S to be positive and also satisfying

$$\max_{s\in S}(s) < 1/2,$$

for all $s \in S$. Then there is a list L of the elements of S with repetitions consisting of $k = 2\binom{|S|}{2}$ elements, such that the consecutive elements have distinct differences. ($L = (s_1, s_2, \ldots, s_k)$ where $s_{i+1} - s_i = s_{j+1} - s_j$ implies that $i = j$.) Indeed, we may follow a directed eulerian circuit in the complete graph $K_{|S|}$ where the vertices are labeled by the elements of S.

Now we are ready to define A which is the sumset of S and $[k]$:

$$A = \{i + s_i : 1 \le i \le k\}$$

The set A has the property that the consecutive differences are distinct, as they are of the form $1 + (s_{i+1} - s_i)$. Let us set $B = [k]$ so that $|A| = |B|$. Then

$$A + B \subset [2k] + S,$$

and so

$$|A + B| \leq 2k|S| \ll |A|^{3/2}.$$

Note in the above example B has a much different structure than A. This motivates the following question.

Question How small can $|A + A|$ be for sets A of size k with distinct consecutive differences? □

4 Convex Sets and $|A + A - A|$

In this section we provide a simple argument that shows a convex set cannot have additive structure.

Proposition 1 *Suppose A is convex and $A' \subset A$. Then*

$$|A' + A - A| \gg |A'||A|. \tag{4}$$

In particular

$$|A - A||A + A| \gg |A|^3.$$

Note that (4) is best possible, as is seen from Example 1.

Proof We let

$$A = \{a_1 < \cdots < a_k\}.$$

We prove the first statement in the case $A' = A$ and the general case follows similarly. Let $1 \leq j \leq k - 1$. Then the j elements

$$a_j + a_2 - a_1 < \cdots < a_j + a_{j+1} - a_j,$$

all lie in the interval $(a_j, a_{j+1}]$. Thus

$$|A + A - A| \geq \sum_{j=1}^{k-1} j \gg k^2.$$

For the second statement, by [6, Corollary 1.5], there is a set $A' \subset A$ such that $|A'| \geq |A|/2$ and

$$|A' + A - A| \ll \left(\frac{|A + A||A - A|}{|A|} \right),$$

and the result follows now from (4).

While the argument is simple, it is not robust. For instance, we cannot prove a statistical analog of (4). The proof of Proposition 1 can be modified to handle the case where A is only assumed to have distinct consecutive differences, already hinting at Theorem 1.

5 Difference Sets of Convex Sets

In this section we prove Schoen and Shkredov's [9] bound for difference sets of convex sets, slightly modifying some details. We choose to work with difference sets, as there are additional technicalities for sumsets. We say $b \gtrsim a$ if $a = O(b \log^c |A|)$ for some $c > 0$.

Theorem 5 ([9]) *Let A be a convex set. Then*

$$|A - A| \gtrsim |A|^{8/5}$$

Before beginning the proof, we recall the kth order energy of sets A and B is defined as

$$E_k(A, B) := \sum_x r_{A-B}(x)^k,$$

where

$$r_{A-B}(x) = \#\{(a, b) \in A \times B : x = a - b\}.$$

We set $E_k(A) := E_k(A, A)$. Using Szemerédi-Trotter, it was shown in [9], building upon the main idea of [3], that if A is convex then

$$E_3(A, B) \lesssim |A||B|^2. \tag{5}$$

Note that (5) is not true if we merely assume that A has distinct consecutive differences as the following example demonstrates.

Example 2 Let k be a positive integer (divisible by 10) and

$$A = d \cdot [k/10] \cup d' \cdot [k/10].$$

For appropriately chosen d and d' (i.e. $d = k$ and $d' = k + 1$), we have that A has distinct consecutive differences. On the other hand

$$E_3(A) \geq E_3([k]) \gg |A|^4.$$

Proof (Theorem 5) We set $K = |A - A||A|^{-1}$. By (5), we have

$$E_3(A) \lesssim |A|^3. \tag{6}$$

On the other hand, $E_3(A)$ is the number of solutions to

$$a - b = c - d = e - f, \quad a, b, c, d, e, f \in A. \tag{7}$$

We let

$$\Delta(A) = \{(a, a) : a \in A\} \subset A \times A.$$

Then (7) implies

$$E_2(\Delta(A), A \times A) = E_3(A). \tag{8}$$

By a dyadic decomposition there is a $\Delta \geq 1$ such that

$$|P|\Delta \gtrsim |A|^2, \quad P := \{x : \Delta \leq r_{A-A}(x) \leq 2\Delta\}.$$

We define a graph $G = G(\Delta)$ on $A \times A$ such that the edges are

$$G = \{(a, b) \in A^2 : a - b \in P\}.$$

Then it follows that

$$|G| \gtrsim |A|^2 \quad \Delta \gtrsim \frac{|A|}{K}. \tag{9}$$

By Cauchy-Schwarz,

$$|A|^6 \lesssim \left(\sum_{x,y} r_{\Delta(A) - A \times_G A}(x, y) \right)^2 \leq E_2(\Delta(A), A \times A)|\Delta(A) - (A \times_G A)|.$$

Thus by (8) and (6), we find that

$$|\Delta(A) - A \times_G A| \gtrsim |A|^3. \tag{10}$$

Now given $(u, v) \in \Delta(A) - A \times_G A$, we write

$$(u, v) = (c, c) - (a, b), \quad a, b, c \in A, \quad (a, b) \in G$$

and it follows that

$$u - v = a - b.$$

But by the definition of G,

$$r_{A-A}(u - v) = r_{A-A}(a - b) \geq \Delta.$$

Combining with (9), we conclude for each $(u, v) \in \Delta(A) - A \times_G A$,

$$r_{A-A}(u-v) \gtrsim |A|K^{-1},$$

and so by (10),

$$|A|^4K^{-1} \lesssim \sum_{u,v\in\Delta(A)-A\times_G A} r_{A-A}(u-v). \tag{11}$$

We set $D = A - A$ and since $u, v \in D$, we find

$$\frac{|A|^4}{K} \lesssim E_2(A, D) \leq E_3(A, D)^{1/2}|A|^{1/2}|D|^{1/2}.$$

Applying (5) to the right hand size, we conclude

$$\frac{|A|^4}{K} \lesssim |A|^{5/2}K^{3/2}. \tag{12}$$

Theorem 5 now follows from simplification.

It is only in (12) of Theorem 5 that we utilize (5) for a set $B \neq \pm A$.

Acknowledgements Research was supported by the OTKA K 119528 grant. The work of the third author has received funding from the European Research Council (ERC) under the European Unions Horizon 2020 research and innovation programme (grant agreement No 741420, 617747, 648017). His research is also supported an NSERC grant. The second author is supported by Ben Green's Simons Investigator Grant 376201.The authors thank Mel Nathanson for his help on writing this paper.

References

1. Gy. Elekes, On the number of sums and products. *Acta Arith.* 81 (1997), no. 4, 365–367.
2. Gy. Elekes, Sums versus products in number theory, algebra and Erdős geometry. in: Paul Erdős and his Mathematics. II, Budapest, Bolyai Society Mathematical Studies, 11. (2002), .
3. Gy. Elekes, M. Nathanson, and I, Ruzsa, Convexity and sumsets. *J. Number Theory* 83 (2000), no. 2, 194–201.
4. M. Garaev, On lower bounds for the L_1-norm of exponential sums. *Mathematical Notes* 68 (2000), 713–720.
5. N. Hegyvári, On consecutive sums in sequences. *Acta Math. Hungar.* 48 (1986), no. 1-2, 193–200.
6. N. Katz and C. Shen. A slight improvement to Garaev's sum product estimate. Proc. Amer. Math. Soc., 136 (2008) 2499–2504.
7. J. Li, G. Shakan, On distinct consecutive r-differences, *Journal of Number Theory*, Volume 199, (2019), 363–376.
8. L. Li, O. Roche-Newton Convexity and a sum-product type estimate, *Acta Arithmetica* 156 (2012), 247–255.
9. T. Schoen, I. Shkredov, On Sumsets of Convex Sets. *Combinatorics, Probability and Computing*, (2011), 20(5), 793–798.
10. I. Shkredov, "On sums of Szemerédi–Trotter sets" in Proc. Steklov Inst. Math. 289:1 (2015), 300–309.

Limit Points of Nathanson's Lambda Sequences

Satyanand Singh

Abstract We consider the set $A_n = \cup_{j=0}^{\infty}\{\varepsilon_j(n) \cdot n^j : \varepsilon_j(n) \in \{0, \pm 1, \pm 2, \ldots, \pm \lfloor n/2 \rfloor\}\}$. Let $\mathscr{S}_{\mathscr{A}} = \bigcup_{a \in \mathscr{A}} A_a$ where $\mathscr{A} \subseteq \mathbb{N}$. We denote by $\lambda_{\mathscr{A}}(h)$ the smallest positive integer that can be represented as a sum of h, and no less than h, elements in $\mathscr{S}_{\mathscr{A}}$. Nathanson studied the properties of the $\lambda_{\mathscr{A}}(h)$-sequence and posed the problem of finding the values of $\lambda_{\mathscr{A}}(h)$. When $\mathscr{A} = \{2, i\}$, we represent $\lambda_{\mathscr{A}}(h)$ by $\lambda_{2,i}(h)$. Only the values $\lambda_{2,3}(1) = 1$, $\lambda_{2,3}(2) = 5$, $\lambda_{2,3}(3) = 21$ and $\lambda_{2,3}(4) = 150$ are known. In this paper, we extend this result. For odd $i > 1$ and $h \in \{1, 2, 3\}$, we find an extensive set of values for $\lambda_{2,i}(h)$. Furthermore, for fixed $h \in \{1, 2, 3\}$, we find certain values of $\lambda_{2,i}(h)$ that occur infinitely many times as i runs over the odd integers bigger than 1. We call these numbers the *limit points of Nathanson's lambda sequences*.

Keywords Diophantine equations · Lambda sequences · Limit points · Special g-adic representations

1 Introduction

We begin with the generating set

$$A_n = \cup_{j=0}^{\infty}\{\varepsilon_j(n) \cdot n^j : \varepsilon_j(n) \in \{0, \pm 1, \pm 2, \ldots, \pm \lfloor n/2 \rfloor\}\}.$$

For any finite set $\mathscr{A}$ of positive integers bigger than 1 and $\mathscr{S}_{\mathscr{A}} = \bigcup_{a \in \mathscr{A}} A_a$, we can represent any positive integer k as a sum of elements of the set $\mathscr{S}_{\mathscr{A}}$ in the form

S. Singh (✉)
Department of Mathematics, New York City College of Technology, CUNY, 300 Jay Street, Brooklyn, NY 11201, USA
e-mail: ssingh@citytech.cuny.edu

M. B. Nathanson (ed.), *Combinatorial and Additive Number Theory IV*,
Springer Proceedings in Mathematics & Statistics 347,
https://doi.org/10.1007/978-3-030-67996-5_25

$$k = \sum_{\alpha \in \mathscr{A}} \left(\sum_{j=0}^{\infty} \varepsilon_j(\alpha)\alpha^j \right).$$

Associated with the above representation of k is the length function which is defined as

$$l_{\mathscr{A}}(k) = \sum_{j \in \{0,1,2,\ldots\},\, \alpha \in \mathscr{A}} |\varepsilon_j(\alpha)|.$$

This representation of integers follows naturally from the special g-adic representations of positive integers outlined in Sects. 1.1 and 1.2 below. We will now proceed to define the $\lambda_{\mathscr{A}}(h)$ function.

Definition 1 For every positive integer h, let $\lambda_{\mathscr{A}}(h)$ denote the smallest positive integer that can be represented as the sum of elements of $\mathscr{S}_{\mathscr{A}}$ with length h, but that cannot be represented as a sum with length less than h.

For ease of notation, we will write $\lambda_{\{2,n\}}(h) = \lambda_{2,n}(h)$ and $l_{\{2,n\}}(k) = l_{2,n}(k)$.

The lambda values are very important in geometric group theory and additive number theory. Nathanson in [8–10] did extensive work in solving some of these problems and derived important properties of the $\lambda_{\mathscr{A}}(h)$ sequences in the process. These $\lambda_{\mathscr{A}}(h)$ sequences originate from the study of groups, generators and metric spaces and their existence can be traced back to Jarden and Narkiewicz in [7] and Hajdu in [4]. This work was further extended by Hajdu and Tijdeman in [5] where they found non-trivial bounds for the terms of the $\lambda_A(h)$ sequences and in [6] where they considered sums and differences of power products. Additional extensions were done by Bertók in [1] by considering representations as power products with non-prime bases. Very few terms are known for Nathanson's lambda sequences even in the simplest of cases such as $\lambda_{2,3}(h)$, where one needs to consider representations of integers only as sums and differences of powers of two's and three's. In fact, only the first four terms, $\lambda_{2,3}(1) = 1$, $\lambda_{2,3}(2) = 5$, $\lambda_{2,3}(3) = 21$ and $\lambda_{2,3}(4) = 150$ are known; see [9, 11]. A similar function is considered by Dimitrov and Howe in [3], where representations are generated from sums and differences of the product of powers of 2's and 3's. In this paper, we will extend work on Nathanson's lambda sequences by finding the values of $\lambda_{2,n}(h)$ for $h \in \{1, 2, 3\}$ and various odd $n > 1$. We will also find some of the limit points of these sequences which are those values of $\lambda_{2,n}(h)$ which occur for infinitely many values of n with $h \in \{1, 2, 3\}$ fixed.

We will now discuss the case when the cardinality of $\mathscr{A}$, denoted by $|\mathscr{A}|$ is 1, i.e. $\mathscr{A} = \{m\}$ and $\mathscr{S}_{\mathscr{A}} = A_m$. This will provide us with a useful tool to compute lambda values. In this case, Nathanson's algorithms in [9] allow us to represent positive integers in special m-adic form uniquely and with shortest length in a finite number of steps. We will denote these lengths by $\widehat{l_m}(k)$. To find terms of the lambda sequences, we only require shortest length representations, uniqueness is not necessary. This is prevalent for $|\mathscr{A}| > 1$ and in particular for representations of integers from $A_2 \cup A_n$ for odd $n > 1$. We now outline the algorithms.

1.1 Generating Special 2-Adic Representations

(a) We begin by writing the positive integer n as a standard binary expansion in the unique binary form $m = \sum_{i=0}^{\infty} \varepsilon_i 2^i$, where $\varepsilon_i \in \{0, 1\}$.
(b) If there are t consecutive sums of powers of 2, i.e. $\varepsilon_i = \varepsilon_{i+1} = \cdots = \varepsilon_{i+t-1} = 1$ for $t \geq 2$, then we rewrite $\sum_{j=i}^{i+t-1} 2^j = 2^{i+t} - 2^i$. Notice that this shortens the number of terms from t to 2 for $t > 2$. After performing the above step on all consecutive blocks of powers of 2, if we end up with s consecutive negative terms, of powers of 2 then by a similar argument we see that $\varepsilon_i = \varepsilon_{i+1} = \cdots = \varepsilon_{i+s-1} = -1$ and $-\sum_{j=i}^{i+s-1} 2^j = -2^{i-s} + 2^i$.
(c) If we have two consecutive terms with opposite signs, say the k-th and $(k+1)$-th terms, then we combine them in the following manner: $-2^k + 2^{k+1} = 2^k$ or $2^k - 2^{k+1} = -2^k$ to convert them into a single term.
(d) If identical powers of 2's occur with opposite signs after implementing the above steps, cancel them. If a term of the form $m2^j$ occurs, where $m = 2^r$, then replace the product by 2^{j+r}. If $m \neq 2^j$ then write the binary expansion of m and multiply each power of two in this expansion by 2^j.

After repeating these shortening steps, there will be no consecutive terms and the process will terminate after a finite number of steps to give us a minimum special 2-adic expansion of our integer, see [9]. For example, using the above algorithm we can write $473 = 2^9 - 2^5 - 2^3 + 2^0$ and it follows that $\widehat{l_2}(473) = 4$.

1.2 Generating Special g-Adic Representations for Odd $g > 1$

(a) Take any positive integer n and write n it in the standard base g representation, $n = \sum_{i=0}^{\infty} \varepsilon_i g^i$ with $\varepsilon_i \in \{0, 1, 2, \ldots, g-1\}$ for all i.
(b) If $(g+1)/2 \leq \varepsilon_i \leq g-1$ for some i, then choose the smallest such i and apply the identity
$$\varepsilon_i g^i = -(g - \varepsilon_i)g^i + g^{i+1}.$$
(c) If $-(g-1) \leq \varepsilon_i \leq -(g+1)/2$ for some i, then apply the identity
$$\varepsilon_i g^i = -(g + \varepsilon_i)g^i - g^{i+1}.$$
(d) Delete all occurrences of 0.
(e) If g^i and $-g^i$ occur, delete them.
(f) If g^i (resp $-g^i$) occurs g times, replace these g summands with the one summand g^{i+1} (resp $-g^{i+1}$).

By iterating the above shortening process, we get a special g-adic representation of n which must terminate after a finite number of steps, see for example [9]. For example 39797 written as a special 9-adic representation is $39797 = 1 \cdot 9^5 - 3 \cdot 9^4 + 1 \cdot 9^3 - 4 \cdot 9^2 + 3 \cdot 9^1 - 1 \cdot 9^0$ and it follows that $\widehat{l_9}(39797) = 13$.

In the next section we will examine representations of positive integers with elements from $\mathscr{S}_{\{2,n\}}$ for various odd $n > 1$. In particular, we will find values of $\lambda_{2,n}(h)$ for $h = 1, 2$ and 3.

2 Computing $\lambda_{2,n}(h)$ for $h \in \{1, 2, 3\}$ and Various Odd $n > 1$.

It is clear that representations of numbers from A_2 and A_n are also representations from $A_{2,n}$. In addition $\widehat{l_{2,n}}(k) \leq \widehat{l_2}(n)$, where $\widehat{l_{2,n}}(k)$ denotes the minimal length of k in $\mathscr{S}_{\{2,n\}} = A_2 \cup A_n$. We will now proceed to compute the values of $\lambda_{2,n}(h)$ for $h \in \{1, 2, 3\}$ and certain odd $n > 1$. We begin with an example to compute some lambda values.

Example 1 $\lambda_{2,5}(1) = 1, \lambda_{2,5}(2) = 3$ and $\lambda_{2,5}(3) = 19$. We represent the first twenty positive integers using elements from $A_2 \cup A_5 = \{0\} \cup \{\pm 2^j, j = 0, 1, 2, 3, ...\} \cup \{\pm 1 \cdot 5^m : m = 0, 1, 2, ...\} \cup \{\pm 2 \cdot 5^k : k = 0, 1, 2, ...\}$ and list their corresponding lengths in Table 1. Our task is to identify the first number with a representation of shortest length one to get $\lambda_{2,5}(1)$, which is trivially 1. Similarly we identify the first number with shortest length 2, to get $\lambda_{2,5}(2)$ and so on.

We observe that we can represent 1 and 2 as numbers with lengths 1; namely $1 = 2^0$ and $2 = 2^1$. Clearly 3 cannot have length 1, but it has several representations with length 2, which follows since the equations $2^{\gamma_1} = 3,\ 2 \cdot 5^{\gamma_2} = 3$ and $5^{\gamma_3} = 3$ are insoluble in **N**, the set of positive integers and one representation of 3 with length 2 is $-1 \cdot 2^1 + 1 \cdot 5^1$. The first number with length three is 19. To see this we need to establish that $l_{2,5}(19) \geq 3$, and as such that $\pm 2^{\alpha_1} = 19, \pm 2^{\alpha_2} \pm 2^{\alpha_3} = 19, \pm 5^{\alpha_4} = 19, \pm 2 \cdot 5^{\alpha_5} = 19, \pm 5^{\alpha_6} \pm 5^{\alpha_7} = 19, \pm 2^{\alpha_8} \pm 5^{\alpha_9} = 19$ are all insoluble. By Nathanson's algorithms, we see that $\widehat{l_2}(19) = \widehat{l_2}(1 \cdot 2^4 + 1 \cdot 2^1 - 1 \cdot 2^0) = 3$ and $\widehat{l_5}(19) = \widehat{l_2}(1 \cdot 5^2 - 1 \cdot 5^1 - 1 \cdot 5^0) = 3$, which establish the insolubility of all the equations except for $\pm 2^{\alpha_8} \pm 5^{\alpha_9} = 19$. This latter system's insolubility is established in the proof of Theorem 2 and it follows that $\widehat{l_{2,5}}(19) = 3$.

For the rest of the paper, we will write representations of the form $\pm 1 \cdot a^n$ as $\pm a^n$. We will now prove some results that are essential for the calculation of $\lambda_{2,n}(h)$ for $h \in \{1, 2, 3\}$ and odd $n > 1$.

Remark 1 Let P be a finite or infinite set of integers > 1, for all $k \in P, \widehat{l_P}(k) = 1$.

This is trivial, since $k \in P$, we can represent k as $1 \cdot k^1$ which has length 1.

Lemma 1 *For any $n \geq 3$, $\widehat{l_{2,n}}(t) \leq 2$ for $t \in Q$ where,*

$$Q = \{1, 2, 3, 4, 5, 6, 7, 8, 9, 10, 12, 14, 15, 16, 17, 18, 20, 24\}.$$

Table 1 Representations in $A_2 \cup A_5$

n	$l_{2,5}(n)$	n	$l_{2,5}(n)$
$1=1\cdot 2^0$	$\boxed{1}$	$11=1\cdot 2^4-1\cdot 5^1$	2
$2=1\cdot 2^1$	1	$12=1\cdot 2^4-1\cdot 2^2$	2
$3=-1\cdot 2^0+1\cdot 2^2$	$\boxed{2}$	$13=1\cdot 2^3+1\cdot 5^1$	2
$4=1\cdot 2^2$	1	$14=1\cdot 2^4-1\cdot 2^1$	2
$5=1\cdot 5^1$	1	$15=1\cdot 2^4-1\cdot 2^0$	2
$6=-1\cdot 2^1+1\cdot 2^3$	2	$16=1\cdot 2^4$	1
$7=1\cdot 2^1+1\cdot 5^1$	2	$17=1\cdot 2^4+1\cdot 2^0$	2
$8=1\cdot 2^3$	1	$18=1\cdot 2^4+1\cdot 2^1$	2
$9=1\cdot 2^0+1\cdot 2^3$	2	$19=-1\cdot 2^0-1\cdot 5^1+1\cdot 5^2$	$\boxed{3}$
$10=2\cdot 5^1$	2	$20=1\cdot 2^4+1\cdot 2^2$	2

Proof For $t \in Q$, $\widehat{l_2}(t) \leq 2$ which is seen from the special 2-adic algorithm in Sect. 1.1. We list the shortest length representations of the elements of Q from A_2. $1 = 1, 2 = 2, 3 = -1 + 2^2, 4 = 2^2, 5 = 1 + 2^2, 6 = -2 + 2^3, 7 = -1 + 2^3, 8 = 2^3, 9 = 1 + 2^3, 10 = 2^1 + 2^3, 12 = -2^2 + 2^4, 14 = -2 + 2^4, 15 = -1 + 2^4, 16 = 2^4, 17 = 1 + 2^4, 18 = 2 + 2^4, 20 = 2^2 + 2^4, 24 = -2^3 + 2^5$. It readily follows that we can find representations of t such that $\widehat{l_{2,n}}(t) \leq 2$ since $\widehat{l_{2,n}}(t) \leq \widehat{l_2}(t)$ □

Experimental results suggest that the numbers 11 and 13 are the only limit points of $\lambda_{2,n}(3)$, we consider their representations and lengths in the special n-adic form for odd n. We will now classify all odd n's with $\widehat{l_{2,n}}(11) \leq 2$.

Lemma 2 *For odd $n > 1$, $\widehat{l_{2,n}}(11) = 1$ if and only if $n = 11$. Further, $\widehat{l_{2,n}}(11) = 2$ for odd n's if and only if $n \in S$, where $S_1 = \{w : w = -2^\gamma + 11, \gamma = 1, 2, 3\}$, $S_2 = \{t : t = 2^\alpha + 11, \alpha \geq 1, \alpha \in \mathbf{N}\}$, $S_3 = \{s : s = 2^\beta - 11, \beta \geq 4, \beta \in \mathbf{N}\}$ and $S = S_1 \cup S_2 \cup S_3$.*

Proof The first statement is trivial. For the second part, when $u \in S$ it readily follows that $\widehat{l_{2,u}}(11) = 2$, since 11 can be written as $|2^c - u|$ and from the first part it follows that $\widehat{l_{2,u}}(11) \neq 1$. □

Lemma 3 *For odd $n > 1$, $\widehat{l_{2,n}}(13) = 1$ if and only if $n = 13$. Further, $\widehat{l_{2,n}}(13) = 2$ for odd n's if and only if $n \in T$, where $T_1 = \{l : l = -2^{m_1} + 13, m_1 = 1, 2, 3\}$, $T_2 = \{m : m = 2^{\gamma_1} + 13, \gamma_1 \geq 1, \gamma_1 \in \mathbf{N}\}$, $T_3 = \{k : k = 2^{\phi+2} - 13, \phi \geq 4, \phi \in \mathbf{N}\}$ and $T = T_1 \cup T_2 \cup T_3$.*

Proof The first statement is trivial. For the second part, when $v \in S$ it readily follows that $\widehat{l_{2,v}}(13) = 2$, since 13 can be written as $|2^c - v|$ and from the first part it follows that $\widehat{l_{2,v}}(13) \neq 1$.

Lemma 4 $T \cap S = \{3, 5, 9, 15, 19, 21\}$.

Proof It is an easy exercise to show that this is equivalent to solving equations of the shape $2^n \pm 2^m = 2$ or 24 and from these equations the result follows. □

Proposition 1 $\widehat{l_{2,n}}(11) = 3$ *for any* $n \notin S \wedge n \equiv 1, 7(mod\ 8)$.

Proof For any $n \notin S \wedge n \equiv 1, 7 \pmod 8$, we need to show that $n^\alpha \notin S \wedge n \equiv 1, 7 \pmod 8$. This requires that we establish that the equations of the shape $\pm n^\alpha \pm 2^\beta = 11$ are insoluble for $\beta \geq 4$. The case $-n^\alpha - 2^\beta = 11$ is trivial and $n^\alpha + 2^\beta = 11$ is insoluble since $n^\alpha < 0$. For the equations $|n^\alpha - 2^\beta| = 11$, when $\beta \in \{0, 1, 2\}$, $n^\alpha \in \{-10, -9, -7, 12, 13, 15\}$ which is ruled out, and for $\beta \geq 3$, we get obstructions in the ring $\mathbb{Z}/8\mathbb{Z}$ when $n \equiv 1, 7 \pmod 8$.

By Lemmas 1 and 2 and the fact that $\widehat{l_{2,n}}(11) = \widehat{l_{2,n}}(-1 - 2^2 + 2^4) = 3$ the claim follows.

Remark 2 In a similar manner to the proof outlined in Proposition 1, we can reduce the study of the insolubility of the equations $\pm n^\alpha \pm 2^\beta = 13$ to $|n^\alpha - 2^\beta| = 13$, where $\beta \geq 3$.

We will now proceed to show that for infinitely many $n \in S$, $n^\alpha \notin T$. This is equivalent to establishing the insolubility of certain diophantine equations and will lead to the important result that 13 is a limit point of $\lambda_{2,n}(3)$.

Lemma 5 *Let* $n = 2^\gamma + 11, \gamma \geq 3$ *and* $n \in S \setminus \{3, 5, 9, 11, 13, 15, 19, 21\}$ *then for* $\alpha \geq 3$, $|n^\alpha - 2^\beta| = 13$ *is insoluble for* $\gamma \equiv 1, 2(mod\ 3)$ *or* $\gamma \equiv 6(mod\ 12)$.

Proof Case1. $n^\alpha - 2^\beta = 13$ is insoluble for $\alpha, \gamma \geq 3$. $n^\alpha = (2^\gamma + 11)^\alpha \equiv \{1, 3\}$ mod 8, $2^\beta + 13 \equiv \{5\}$ mod 8 and we have an obstruction in $\mathbb{Z}/8\mathbb{Z}$.
Case 2. $n^\alpha - 2^\beta = -13$ is insoluble for $\alpha \geq 3$. For $\gamma = 3k + 1$, $n = (2^\gamma + 11) = (2^{3k+1} + 11) \equiv \{6\}$ mod 7, $n^\alpha = (2^{3t+1} + 11)^\alpha + 13 \equiv \{0, 5\}$ mod 7, $2^\beta \equiv \{1, 2, 4\}$ mod 7 and we have an obstruction in $\mathbb{Z}/7\mathbb{Z}$.
Case 3. $n^\alpha - 2^\beta = -13$ is insoluble for $\alpha \geq 3$. For $\gamma = 3k + 2$, $n = (2^\gamma + 11) = (2^{3k+2} + 11) \equiv \{1\}$ mod 7, $n^\alpha = (2^{3t+2} + 11)^\alpha + 13 \equiv \{0\}$ mod 7, $2^\beta \equiv \{1, 2, 4\}$ mod 7 and we have an obstruction in $\mathbb{Z}/7\mathbb{Z}$.
Case 4. $n^\alpha - 2^\beta = -13$ is insoluble for $\alpha \geq 3$. For $\gamma = 12k + 6$, $k \geq 1$, $n = (2^\gamma + 11) = (2^{12k+6} + 11) \equiv \{0\}$ mod 15, $n^\alpha = (2^{12k+6} + 11)^\alpha + 13 \equiv \{13\}$ mod 15, $2^\beta \equiv \{1, 2, 4, 8\}$ mod 15 and we have an obstruction in $\mathbb{Z}/15\mathbb{Z}$. □

Lemma 6 *Let* $n = 2^\gamma - 11, \gamma \geq 4$ *and* $n \in S \setminus \{3, 5, 9, 11, 13, 15, 19, 21\}$ *then for* $\alpha \geq 4$, $|n^\alpha - 2^\beta| = 13$, *is insoluble for* $\gamma \equiv 0, 1, 3(mod\ 6)$.

Proof Case1. $n^\alpha - 2^\beta = -13$ is insoluble. For $\alpha, \gamma \geq 4$, $n^\alpha = (2^\gamma - 11)^\alpha \equiv \{1, 5\}$ mod 8, $2^\beta - 13 \equiv \{3\}$ mod 8 and we have an obstruction in $\mathbb{Z}/8\mathbb{Z}$.
Case 2. $n^\alpha - 2^\beta = 13$ is insoluble for $\gamma = 6t$, $t \geq 1$. For $\alpha \geq 4$, $n = (2^\gamma - 11) = (2^{6t} - 11) \equiv \{53\}$ mod 63, $n^\alpha = (2^{6t} - 11)^\alpha \equiv \{1, 8, 37, 44, 46, 53\}$ mod 63, $2^\beta + 13 \equiv \{14, 15, 17, 21, 29, 45\}$ mod 63 and we have an obstruction in $\mathbb{Z}/63\mathbb{Z}$.
Case 3. $n^\alpha - 2^\beta = 13$ is insoluble $\gamma = 6t + 1$, $t \geq 1$. For $\alpha \geq 4$, $n = (2^\gamma - 11) =$

$(2^{6t+1} - 11) \equiv \{0, 30\} \mod 39$, $n^\alpha = (2^{6t+1} - 11)^\alpha \equiv \{0, 1, 3, 9, 12, 27, 30, 36\} \mod 39$, $2^\beta + 13 \equiv \{6, 14, 15, 17, 18, 21, 23, 24, 29, 33, 35, 38\} \mod 39$ and we have an obstruction in $\mathbb{Z}/39\mathbb{Z}$.
Case 4. $n^\alpha - 2^\beta = 13$ is insoluble for $\gamma = 6t + 3$, $t \geq 1$. For $\alpha \geq 4$, $n = (2^\gamma - 11) = (2^{6t+3} - 11) \equiv \{60\} \mod 63$, $n^\alpha = (2^{6t+3} - 11)^\alpha \equiv \{1, 9, 18, 36, 60\} \mod 63$, $2^\beta + 13 \equiv \{14, 15, 17, 21, 29, 45\} \mod 63$ and we have an obstruction in $\mathbb{Z}/63\mathbb{Z}$. □

We now define $\widehat{S_2} = \{t : t = 2^\alpha + 11, \alpha \equiv 1, 2 (\mathrm{mod}\ 3) \text{ or } \alpha \equiv 6 (\mathrm{mod}\ 12), \alpha \in \mathbf{N}\}$ and $\widehat{S_3} = \{s : s = 2^\beta - 11, \beta \geq 4, \beta \equiv 0, 1, 3 (\mathrm{mod}\ 6), \beta \in \mathbf{N}\}$.

Proposition 2 *$\widehat{l_{2,n}}(13) = 3$ for any $n \in \widehat{S_2} \cup \widehat{S_3} \setminus \{3, 5, 9, 11, 13, 15, 19, 21\}$.*

Proof For any $n \geq 3$, $n \in \widehat{S_2} \cup \widehat{S_3} \setminus \{3, 5, 9, 11, 13, 15, 19, 21$, we claim that $n^\alpha \notin T$. To show this, we must establish that the equations of the shape $\pm n^\alpha \pm 2^\beta = 13$ are insoluble for $\beta \geq 4$. By Remark 2, it is enough to consider only the equations $|n^\alpha - 2^\beta| = 13$, where $\beta \geq 3$. These equations are insoluble by Lemmas 5 and 6.

By Lemmas 1, 2, 3 and the fact that $\widehat{l_{2,n}}(11) = \widehat{l_{2,n}}(1 - 2^2 + 2^4) = 3$ the claim follows.

To find all the limit points of $\lambda_{2,n}(2) = 3$, we would need to solve the problem below.

Problem 1 For all odd $n \geq 3$, find all solutions to $|n^\alpha - 2^\beta| = 11$ and $|n^\alpha - 2^\beta| = 13$. These are generalized Ramanujan-Nagel equations, see for example [2]. There are partial solutions to some of these equations and in particular by Theorem 3.1 in [2], it readily follows that for even β, $n^\alpha - 2^\beta = 11$ has the only solution $(n, \alpha, \beta) = (3, 3, 4)$.

We will now proceed to establish $\lambda_{2,n}(h)$ for $h \in \{1, 2, 3\}$ and various odd $n > 1$. It is a routine calculation to show that $\lambda_{2,3}(2) = 5$, see for example [9].

Proposition 3 *$\lambda_{2,n}(1) = 1$, and $\lambda_{2,n}(2) = 3$ for any odd integer $n \geq 5$, $n \in \mathbf{N}$.*

Proof $\lambda_{2,n}(1) = 1$ is trivial and $\lambda_{2,n}(2) = 3$ follows from the fact that $\widehat{l_2}(1) = \widehat{l_2}(2) = 1$, $\widehat{l_2}(3) = 2$ and the Diophantine equations: $\pm 2^a = 3$, and $\pm n^b = 3$ are insoluble. □

Corollary 1 *The only limit point of $\lambda_{2,n}(1)$ is 1 and that of $\lambda_{2,n}(2)$ is 3.*

Proof The corollary follows immediately from Proposition 3. □

We have seen in Lemma 4 that $T \cap S = \{3, 5, 9, 15, 19, 21\}$ and that $\{11, 13\}$ are also points of interest since $\widehat{l_{2,11}}(11) = 1$, $\widehat{l_{2,11}}(13) = \widehat{l_{2,11}}(2 + 11) = 3$, $\widehat{l_{2,13}}(13) = 1$ and $\widehat{l_{2,13}}(11) = \widehat{l_{2,13}}(-2 + 13) = 2$ which readily follows from Remark 1 and the insolubility of $\pm 13^t = 11$ and $\pm 11^s = 13$. It follows that in the computation of $\lambda_{2,n}(3)$ we need to consider the set of values for $n \in \{3, 5, 9, 11, 13, 15, 19, 21\}$ separately.

We will now state our main result which gives two limit points of $\lambda_{2,n}(3)$.

Theorem 1 *For odd $n > 1$, where S is the set defined in Lemma 2 we have that*

$$\lambda_{2,n}(3) = \begin{cases} 11 \text{ for } n \notin S \cup \{11\} \wedge n \equiv 1, 7 (mod\ 8) \\ 13 \text{ for } n \in \widehat{S}_2 \cup \widehat{S}_3 \setminus \{3, 5, 9, 11, 13, 15, 19, 21\} \end{cases}$$

Proof From Lemma 1, we have that $\widehat{l}_{2,n}(t) \leq 2$ for $t \in \{1, 2, 3, 4, 5, 6, 7, 8, 9, 10, 12\}$. Our Theorem follows by using this fact, Lemmas 2 and 3, and Propositions 1 and 2. □

Corollary 2 *Two limit points of $\lambda_{2,n}(3)$ are* 11 *and* 13.

Proof The corollary follows immediately from Theorem 1. □

Theorem 2 $\lambda_{2,3}(3) = 21,\ \lambda_{2,5}(3) = 19,\ \lambda_{2,9}(3) = 19,\ \lambda_{2,11}(3) = 23,\ \lambda_{2,13}(3) = 22,\ \lambda_{2,15}(3) = 21,\ \lambda_{2,19}(3) = 22,\ \lambda_{2,21}(3) = 26.$

Proof $\lambda_{2,3}(3) = 21$, see [9]. Now consider the pairs $(g, k_{2,g})$, where $(g, k_{2,g}) \in \{(5, 19), (9, 19), (11, 23), (13, 22), (15, 21), (19, 22), (21, 26)\}$.

For integer valued x, where $1 \leq x < k_{2,g}$ we will first establish that $l_{2,g}(x) \leq 2$. An application of Lemmas 1 and 4, shows that $\widehat{l}_{2,g}(x) \leq 2$ for $x \in \widehat{Q} = Q \cup \{11, 13\}$. For each $g \in \{5, 9, 15, 19, 21\}$ we will now consider the values of $x \notin \widehat{Q}$ but in the interval $1 \leq x < k_{2,g}$ and their appropriate representations to show that $\widehat{l}_{2,g}(x) \leq 2$.

- For $g \in \{5, 9\}$, $x \in \widehat{Q}$ includes all values in the interval $1 \leq x < 19$.
- For $g = 11$, $x \in \{11, 19 = 2^3 + 11, 21 = 2^5 - 11, 22 = 2 \cdot 11\} \cup \widehat{Q}$ shows that $\widehat{l}_{2,11}(x) \leq 2$ for values in the interval $1 \leq x < 23$.
- For $g = 13$, $x \in \{13, 19 = 2^5 - 13, 21 = 2^3 + 13\} \cup \widehat{Q}$ shows that $\widehat{l}_{2,13}(x) \leq 2$ for values in the interval $1 \leq x < 22$.
- For $g = 15$, $\{x = 19 = 2^2 + 15\} \cup \widehat{Q}$ shows that $\widehat{l}_{2,15}(x) \leq 2$ for values in the interval $1 \leq x < 21$.
- For $g=19$, $x \in \{19, 21 = 2+19\} \cup \widehat{Q}$ shows that $\widehat{l}_{2,19}(x) \leq 2$ for values in the interval $1 \leq x < 22$.
- For $g = 21$, $x \in \{19 = -2 + 21, 21, 22 = 1 + 21, 23 = 2 + 21, 25 = 2^2 + 21\} \cup \widehat{Q}$ shows that $\widehat{l}_{2,21}(x) \leq 2$ for values in the interval $1 \leq x < 26$.

We will now establish that the Diophantine equations listed below which are representations of $k_{2,g}$ with $l_{2,g}(k_{2,g}) = 2$ are insoluble in non-negative integers for each pair $(g, k_{2,g})$ to rule out the possibility that $l_{2,g}(k_{2,g}) \leq 2$. In addition, we will produce a representation of $k_{2,g}$ with $l_{2,g}(k_{2,g}) = 3$ to show that $\widehat{l}_{2,g}(k_{2,g}) = 3$ and complete the proof of our Theorem.

$$\pm 2^{a_1} = k_{2,g} \tag{1}$$

$$\pm g^{a_2} = k_{2,g} \tag{2}$$

$$\pm 2 \cdot g^{a_3} = k_{2,g} \tag{3}$$

Table 2 $k_{2,g}$ and its corresponding $\widehat{l_2}(k_{2,g})$ and $\widehat{l_g}(k_{2,g})$

g	$k_{2,g}$	Representations in A_2	Representations in A_g	$\widehat{l_2}(k_{2,g})$	$\widehat{l_g}(k_{2,g})$
5	19	$-1+2^2+2^4$	$-1-5+5^2$	3	3
9	19	$-1+2^2+2^4$	$1+2\cdot 9$	3	3
11	23	$-1-2^3+2^5$	$1+2\cdot 11$	3	3
13	22	2^5-2^3-2	$-2^2+2\cdot 13$	3	3
15	21	$1+2^2+2^4$	$6+15$	3	7
19	22	$-2-2^3+2^5$	$3+19$	3	4
21	26	$2-2^3+2^5$	$5+21$	3	6

$$\pm 2^{a_4} \pm 2^{a_5} = k_{2,g} \tag{4}$$

$$\pm g^{a_6} \pm g^{a_7} = k_{2,g} \tag{5}$$

$$\pm 2^{a_8} \pm g^{a_9} = k_{2,g} \tag{6}$$

We now compute $\widehat{l_2}(k_{2,g})$ and $l_g(k_{2,g})$ and display in Table 2 below.

Observe that $\widehat{l_2}(k_{2,g}) = 3$ and $\widehat{l_g}(k_{2,g}) \geq 3$, and as a consequence of Nathanson's minimal representations discussed in Sects. 1.1 and 1.2, equations represented by (10) through (14) above are insoluble. We will now consider the eight remaining cases for $g \in \{5, 9, 11, 13, 15, 19, 21\}$ individually to establish insolubility of equations represented by (6) above by finding obstructions in certain rings.

1. (a) For $2^{c_1} + 5^{c_2} = 19$, we get insolubility by substituting for $c_1 = 0, 1, 2, 3, 4$.
   (b) For $2^{c_3} - 5^{c_4} = 19$, we get obstructions in the ring $\mathbb{Z}/15\mathbb{Z}$.
   (c) For $-2^{c_5} + 5^{c_6} = 19$. Clearly for $c_5 = 0, 1$ the equation is insoluble. For $c_5 > 1$ we get obstructions in the ring $\mathbb{Z}/4\mathbb{Z}$.
2. (a) For $2^{e_1} + 9^{e_2} = 19$, we get insolubility by substituting for $e_1 = 0, 1, 2, 3, 4$.
   (b) For $2^{e_3} - 9^{e_4} = 19$, when $e_3 = 0, 1$ or for $e_4 = 0$ there are no solutions and for $e_3 > 2$ and $e_4 > 0$, we get obstructions in the ring $\mathbb{Z}/8\mathbb{Z}$.
   (c) For $-2^{e_5} + 9^{e_6} = 19$, clearly for $e_6 = 0, 1$ the equation is insoluble and for $d_6 > 0$ we get obstructions in the ring $\mathbb{Z}/4\mathbb{Z}$.
3. (a) For $2^{f_1} + 11^{f_2} = 23$, we get insolubility by substituting for $f_1 = 0, 1, 2, 3, 4$. We see that there are no corresponding values for f_2.
   (b) For $-2^{f_3} + 11^{f_4} = 23$, we get obstructions in $\mathbb{Z}/91\mathbb{Z}$.
   (c) For $2^{f_3} - 11^{f_4} = 23$, we get obstructions in $\mathbb{Z}/133\mathbb{Z}$.
4. (a) For $2^{g_1} + 13^{g_2} = 22$, we get insolubility by substituting for $g_1 = 0, 1, 2, 3, 4$.
   (b) For $2^{g_3} - 13^{g_4} = 22$, we see that there is no solution for $g_3 = 0$ or $g_4 = 0$ and we get obstructions in the ring $\mathbb{Z}/2\mathbb{Z}$ for $g_3 > 0$ or $g_4 > 0$.
   (c) For $-2^{g_5} + 13^{g_6} = 22$, clearly for $g_5 = 0$ or for $g_6 = 0$, the equation is insoluble and for $g_5 > 0$ and $g_6 > 0$ we get obstructions in the ring $\mathbb{Z}/2\mathbb{Z}$.

5. (a) For $2^{m_1} + 15^{m_2} = 21$, we get insolubility by substituting for $m_1 = 0, 1, 2, 3, 4$.
(b) For $2^{m_3} - 15^{m_4} = 21$, we see that there is no solution for $k_4 = 0$. We get obstructions in the ring $\mathbb{Z}/15\mathbb{Z}$ for $k_4 > 0$.
(c) For $-2^{k_5} + 15^{k_6} = 21$, clearly for $k_6 = 0$, the equation is insoluble. For $k_6 > 0$ we get obstructions in the ring $\mathbb{Z}/15\mathbb{Z}$.
6. (a) For $2^{u_1} + 19^{u_2} = 22$, we get insolubility by substituting for $u_1 = 0, 1, 2, 3, 4$.
(b) For $2^{u_3} - 19^{u_4} = 22$, we see that there is no solution for $u_3 = 0$ or $u_4 = 0$. We get obstructions in the ring $\mathbb{Z}/2\mathbb{Z}$ for $u_3 > 0$ or $u_4 > 0$.
(c) For $-2^{u_5} + 19^{u_6} = 22$, clearly for $u_5 = 0$ or for $u_6 = 0$, the equation is insoluble. For $u_5 > 0$ and $u_6 > 0$ we get obstructions in the ring $\mathbb{Z}/2\mathbb{Z}$.
7. (a) For $2^{z_1} + 21^{z_2} = 26$, we get insolubility by substituting for $z_1 = 0, 1, 2, 3, 4$.
(b) For $2^{z_3} - 21^{z_4} = 26$, we see that there is no solution for $z_3 = 0$ or $z_4 = 0$. We get obstructions in the ring $\mathbb{Z}/2\mathbb{Z}$ for $z_3 > 0$ or $z_4 > 0$.
(c) For $-2^{z_5} + 21^{z_6} = 26$, clearly for $z_5 = 0$ or for $z_6 = 0$, the equation is insoluble. For $z_5 > 0$ and $z_6 > 0$ we get obstructions in the ring $\mathbb{Z}/2\mathbb{Z}$.

We have shown that $l_{2,g}(k_{2,g}) \geq 3$, but since $l_2(k_{2,g}) = 3$, it follows that $l_{2,g}(k_{2,g}) = 3$. This is the first instance (i.e. $x = k_{2,g}$) where the length $l_{2,g}(x)$ is 3 for each g. Therefore we have that $\lambda_{2,3}(3) = 21$, $\lambda_{2,5}(3) = 19$, $\lambda_{2,9}(3) = 19$, $\lambda_{2,11}(3) = 23$, $\lambda_{2,13}(3) = 22$, $\lambda_{2,15}(3) = 21$, $\lambda_{2,19}(3) = 22$, $\lambda_{2,21}(3) = 26$. □

3 Additional Results and Open Problems

In [11] it is shown that for any fixed odd $g \geq 3$, $h \in \mathbf{N}$ and $j = 1, 2, 3, ..., (g-1)/2$, $\lambda_g\left(\frac{(h-1)(g-1)+2j}{2}\right) = \frac{(2j-1)g^{h-1}+1}{2}$. An explicit expression is also established for $\lambda_2(h)$ and explicit upper bounds are given for $\lambda_g(h)$, where g is an even integer greater than 2. It is also shown that $\lambda_{2,3}(4) = 150$ and $\lambda_{2,5}(4) = 83$. It is an open problem to find additional terms of $\lambda_{2,n}(h)$ for odd $n > 1$ and $h \geq 3$.

Acknowledgements I would like to thank Melvyn B. Nathanson for introducing me to this problem.

References

1. Cs. Bertók, *Representing integers as sums or differences of general power products*, Acta Math. Hungar. **141** (2013), 291–300
2. A. Bérczes and I. Pink, *On generalized Lebesgue-Ramanujan-Nagell equations*, An. Şt. Univ. Ovid. Cons. **22(1)**, (2014), 51–71
3. V. S. Dimitrov and E. W. Howe, *Lower bounds on the lengths of double-base representations*, Proc. Am. Math. Soc. **139**(10)(2011), 3423–3430
4. L. Hajdu, *Arithmetic progressions in linear combinations of S-units*, Period. Math. Hungar **54**, (2007), 175–181
5. L. Hajdu and R. Tijdeman, *Representing integers as linear combinations of powers*, Publ. Math. Debrecen **79**, (2011), 461–468

6. L. Hajdu and R. Tijdeman, *Representing integers as linear combinations of power products*, Arch. Math. **98** (2012), 527–533
7. M. Jarden and W. Narkiewicz, *On sums of units*, Monatschefte für Mathematik **150**, (2007), 327–332
8. M. B. Nathanson, *Phase transitions in infinitely generated groups, and related problems in additive number theory*, Integers **11A**, (2011)
9. M. B. Nathanson, *Problems in additive number theory, iv: Nets in groups and shortest length g-adic representations, and minimal additive complements*, Int. J. Number Theory **7**, (2011), 1999–2017
10. M. B. Nathanson, *Geometric group theory and arithmetic diameter*, Publ. Math. Debrecen **79**, (2011), 563–572
11. S. Singh, *Special representations, Nathanson's Lambda Sequences and Explicit Bounds*, Ph.D Thesis, CUNY Graduate Center, (2014)